POLYMERS FOR PACKAGING APPLICATIONS

POLYMERS FOR PACKAGING APPLICATIONS

Edited by

Sajid Alavi, PhD, Sabu Thomas, PhD, K. P. Sandeep, PhD, Nandakumar Kalarikkal, PhD, Jini Varghese, and Srinivasarao Yaragalla

Apple Academic Press Inc.
3333 Mistwell Crescent
Oakville, ON L6L 0A2
Canada

Apple Academic Press Inc.
9 Spinnaker Way
Waretown, NJ 08758
USA

First issued in paperback 2021

Exclusive worldwide distribution by CRC Press, a member of Taylor & Francis Group

ISBN 13: 978-1-77463-304-5 (pbk)
ISBN 13: 978-1-926895-77-2 (hbk)

Library of Congress Control Number: 2014940534

Library and Archives Canada Cataloguing in Publication

Polymers for packaging applications/editors: Sajid Alavi, PhD, Sabu Thomas, PhD, K.P. Sandeep, PhD, Nandakumar Kalarikkal, PhD, Jini Varghese, and Srinivasarao Yaragalla.

Includes bibliographical references and index.
ISBN 978-1-926895-77-2 (bound)
1. Food--Packaging. 2. Polymers. 3. Plastics. 4. Polymers--Biodegradation.
5. Plastics--Biodegradation. 6. Nanocomposites (Materials)--Biodegradation.
7. Food--Packaging--Technological innovations. I. Alavi, Sajid, author, editor II. Thomas, Sabu, editor III. Sandeep, K. P., author, editor IV. Kalarikkal, Nandakumar, editor V. Varghese, Jini, editor VI. Yaragalla, Srinivasarao, editor

TP374.P64 2014 664'.09 C2014-903401-6

Apple Academic Press also publishes its books in a variety of electronic formats. Some content that appears in print may not be available in electronic format. For information about Apple Academic Press products, visit our website at **www.appleacademicpress.com** and the CRC Press website at **www.crcpress.com**

ABOUT THE EDITORS

Sajid Alavi, PhD

Sajid Alavi, PhD, is a Professor of Extrusion Processing and Food Engineering in the Department of Grain Science and Industry at Kansas State University, Manhattan, Kansas, USA. He received his BS degree (1995) in Agricultural Engineering from Indian Institute of Technology (Kharagpur), MS (1997) in Agricultural and Biological Engineering from the Pennsylvania State University, and PhD (2002) in Food Science/ Food Engineering from Cornell University, Ithaca, New York. Dr. Alavi's research activities are focused in the areas of food engineering, extrusion processing for industrial and food applications, nanocomposites for packaging applications, mathematical modeling of flow and structure formation in biopolymer melts during extrusion, food microstructure imaging, structure-texture relationships, and new approaches to global food security and nutrition through processing. He has secured over $6.3 million in extramural funding from various federal, state, and industrial sponsors for his research program. He has supervised seven PhD and 13 Masters level students. Dr. Alavi's received the coveted 2010 Young Research Scientist Award from AACC International, formerly the American Association of Cereal Chemists, which is an important recognition for research accomplishments.

Dr. Alavi designs technology and R&D solutions for numerous food, feed and pet food processors, and is involved in processing and food aid related projects in USA, Africa, India, and other countries around the world. He has been invited to speak at numerous international forums and institutions in USA, Italy, South Africa, Brazil, India, Mozambique, and China. He has provided training and networking opportunities to 800 industry leaders from 30 countries spanning all six continents through his internationally reputed short course "Extrusion Processing: Technology and Commercialization" and similar offerings and workshops in other countries such as India, Brazil and Mozambique.

Sabu Thomas, PhD

Sabu Thomas, PhD, is a Professor of Polymer Science and Engineering at the School of Chemical Sciences and Director of the International and Inter University Centre for Nanoscience and Nanotechnology at Mahatma Gandhi University, Kottayam, Kerala, India. He received his BSc degree (1980) in Chemistry from the University of Kerala, BTech. (1983) in Polymer Science and Rubber Technology from the Cochin University of Science and Technology, and PhD (1987) in Polymer Engineering from the Indian Institute of Technology, Kharagpur. The research activities of Professor Thomas include surfaces and interfaces in multiphase polymer blend and composite systems, phase separation in polymer blends, compatibilization of immiscible polymer blends, thermoplastic elastomers, phase transitions in polymers, nanostructured polymer blends, macro-, micro- and nanocomposites, polymer rheology, recycling, reactive extrusion, processing–morphology–property relationships in multiphase polymer systems, double networking of elastomers, natural fibers and green composites, rubber vulcanization, interpenetrating polymer networks, diffusion and transport and polymer scaffolds for tissue engineering. He has supervised 64 PhD theses, 30 MPhil theses, and 40 Masters theses. He has three patents to his credit. He also received the coveted Sukumar Maithy Award for the best polymer researcher in the country for the year 2008. Very recently Professor Thomas received the MRSI and CRSI medals for his excellent work. With over 600 publications to his credit and over 15,000 citations, with an h-index of 65, Dr. Thomas has been ranked fifth in India as one of the most productive scientists.

K. P. Sandeep, PhD

K. P. Sandeep, PhD, is a Professor of Food Engineering, Research Leader and Associate Head in the Department of Food, Bioprocessing and Nutrition Sciences as well as an associate faculty member in the Department of Biological and Agricultural Engineering at North Carolina State University, Raleigh, North Carolina, USA. He is also Site Director of the Center for Advanced Processing and Packaging Studies. He received his BS degree (1991) in Agricultural Engineering from the Indian Institute of Technology (Kharagpur), his MS (1993), and his PhD (1996) in Agricultural and Biological Engineering from the Pennsylvania State University. His research areas include nanotechnology, thermal and aseptic processing, continuous flow microwave processing, heat exchanger design, development of sensors, mathematical modeling of fluid flow and heat transfer, microelectromechanical systems (MEMS), and nuclear magnetic resonance (NMR). Dr. Sandeep has co-authored scores of books, peer-reviewed publications, and technical abstracts, posters, and presentations on aseptic processing and related topics. He has served as a consultant to several companies and serves as a member of the advisory board to the Southeast Dairy Foods Research Center. He also conducts short courses tailored to meet the specific needs of industry on topics such as nanotechnology, thermal processing and NMR.

Nandakumar Kalarikkal, PhD

Dr. Nandakumar Kalarikkal is a Professor of Physics at the School of Pure and Applied Physics and Joint Director of International and Inter University Centre for Nanoscience and Nanotechnology at Mahatma Gandhi University, Kottayam, Kerala, India. He received his BSc degree (1984) in Physics from Calicut University, and MSc (1986) and PhD (1992) in Physics from Cochin University of Science and Technology, Kerala, India. His research activities involve nanotechnology and nanomaterials, sol-gel synthesis of nanosystems, semiconducting glasses, ferroelectric ceramics, and nonlinear and electro-optic materials. He is the recipient of research fellowships and associateships from prestigious organizations such as the Department of Science and Technology and Council of Scientific and Industrial Research of the Government of India. He has collaborated with national and international scientific institutions in India, South Africa, Slovenia, Canada, and Australia, and is co-author of several books chapters, peer-reviewed publications, and invited presentations in international forums.

Jini Varghese

Ms. Jini Varghese is currently a Research Scholar at the School of Chemical Sciences at Mahatma Gandhi University, Kottayam, Kerala, India. She is engaged in doctoral studies in the area of EPDM rubber-graphene nanocomposites. She received her MSc degree in Analytical Chemistry from Mahatma Gandhi University. Ms. Varghese is a recipient of the Women Scientist Award from the Department of Science and Technology of the Government of India.

Srinivasarao Yaragalla

Mr. Srinivasarao Yaragalla is a Research Scholar at the International and Inter University Centre for Nanoscience and Nanotechnology at Mahatma Gandhi University, Kottayam, Kerala, India. He is engaged in doctoral studies in the area of graphene-based polymer nanocomposites. He has also conducted research work at the Universiti Teknologi MARA in Malaysia. In 2010, Mr. Yaragalla received a prestigious research fellowship administered jointly by the Council of Scientific and Industrial Research and University Grants Commission of the Government of India.

CONTENTS

LIST OF CONTRIBUTORS

Boussad Abbès
University of Reims Champagne-Ardenne, GRESPI/MPSE, Campus Moulin de la Housse, BP 1039, 51687 REIMS Cedex 2, France.
Email: boussad.abbes@univ-reims.fr

Fazilay Abbès
University of Reims Champagne-Ardenne, GRESPI/MPSE, Campus Moulin de la Housse, BP 1039, 51687 REIMS Cedex 2, France.

S. Alavi
Department of Grain Science & Industry, Kansas State University, 201 Shellenberger Hall, Manhattan, KS 66506, U.S.A.
Email: salavi@k-state.edu

C. Anandharamakrishnan
Department of Food Engineering, CSIR - Central Food Technological Research Institute, Mysore – 570 020, India.
Email: anandharamakrishnan@cftri.res.in

D. Saravana Bavan
Department of Mechanical Engineering, National Institute of Technology Karnataka, Mangalore – 575 025, India.
Email: saranbav@gmail.com

Umesh Bhardwaj
Department of Chemical Engineering, Indian Institute of Technology Guwahati–781 039, Assam, India.

Dr. J. Bindu
Fish Processing Division, Central Institute of Fisheries Technology, Cochin 682 029, Kerala, India.
Email: bindujaganath@gmail.com

Sanjaya K. Dash
Orissa University of Agriculture and Technology, Department of Agricultural Processing and Food Engineering, Bhubaneswar, Odisha, India.
Email: sk_dash1006@hotmail.com

S. D. Deshpande
Central Institute of Agricultural Engineering, Nabi Bagh, Berasia Road,Bhopal – 462 016, Madhya Pradesh, India.
Email: sdd1953@gmail.com

Prodyut Dhar
Department of Chemical Engineering, Indian Institute of Technology Guwahati–781 039, Assam, India.

Surendra S. Gaur
Indian Institute of Technology Guwahati, Department of Chemical Engineering, Guwahat – 781 039, Assam, India.

Dr. T. K. Srinivasa Gopal
Central Institute of Fisheries Technology, Cochin 682 029, Kerala, India
Email: tksgopal@gmail.com

Ying-Qiao Guo
University of Reims Champagne-Ardenne, GRESPI/MPSE, Campus Moulin de la Housse, BP 1039, 51687 REIMS Cedex 2, France.

P. P. Kanekar
Microbial Sciences Division, MACS' Agharkar Research Institute, G. G. Agarkar Road, Pune – 411 004, Maharashtra, India.
Email: kanekarpp@gmail.com

S. P. Kanekar
Microbial Sciences Division, MACS' Agharkar Research Institute, G. G. Agarkar Road, Pune – 411 004, Maharashtra, India

Vimal Katiyar
Indian Institute of Technology Guwahati, Department of Chemical Engineering, Guwahati – 781 039, Assam, India.
Email: vkatiyar@iitg.ernet.in

Usha Kiran Kolli
Department of Food Engineering, CSIR - Central Food Technological Research Institute, Mysore, Karnataka – 570 020, India.

S. B. Kondawar
1R.T.M. Nagpur University, Department of Physics, Polymer Nanotech Laboratory, University Campus, Amravati Road, Nagpur-440033, Maharashtra, India.
Email: sbkondawar@yahoo.co.in

Prakash Kotecha
Department of Chemical Engineering, Indian Institute of Technology Guwahati – 781 039, Assam, India

P. R. Kshirsagar
Microbial Sciences Division, MACS' Agharkar Research Institute, G. G. Agarkar Road, Pune – 411 004, Maharashtra, India.

Amit Kumar
Indian Institute of Technology Guwahati, Department of Chemical Engineering, Guwahati – 781 039, Assam, India.

Dr. P. Kumar
Department of Food, Bioprocessing, and Nutrition Sciences, North Carolina State University, Raleigh, NC.

S. O. Kulkarni
Microbial Sciences Division, MACS' Agharkar Research Institute, G. G. Agarkar Road, Pune – 411 004, Maharashtra, India.

G. C. Mohan Kumar
Department of Mechanical Engineering, National Institute of Technology Karnataka Srinivasnagar, Surathkal, Mangalore – 575025, Karnataka, India.

Zhigang Liu
Department of Packaging Engineering, Jiangnan University, Wuxi, 214122, China
Email: liuzg@jiangnan.edu.cn

Lixin Lu
Department of Packaging Engineering, Jiangnan University, Wuxi, 214122, China.

Dr. A. K. Mallick
Export Inspection Agency, 6th Floor CMDA Tower II, No: 1, Gandhi Irwin Road, Egmore, Chennai - 600 008
Email: arunakumarmallick@gmail.com

S. S. Nilegaonkar
Microbial Sciences Division, MACS' Agharkar Research Institute, G. G. Agarkar Road, Pune – 411 004, Maharashtra,

Rahul Patwa
Department of Chemical Engineering, Indian Institute of Technology Guwahati – 781 039, Assam, India.

Akhilesh K. Pal
Department of Chemical Engineering, Indian Institute of Technology, Guwahati – 781 039, Assam, India.

M. Ponraj
Microbial Sciences Division, MACS' Agharkar Research Institute, G. G. Agarkar Road, Pune – 411 004, Maharashtra,

Dr. C. N. Ravishankar
Fish Processing Division, Central Institute of Fisheries Technology, Cochin 682029, Kerala, India
Email: Cnrs2000@gmail.com

K. P. Sandeep
Department of Food, Bioprocessing, and Nutrition Sciences, North Carolina State University, Raleigh, NC.
Email: kp_sandeep@ncsu.edu

S. S. Sarnaik
Microbial Sciences Division, MACS' Agharkar Research Institute, G. G. Agarkar Road, Pune – 411 004, Maharashtra,

Rungsinee Sothornvit
Department of Food Engineering/Faculty of Engineering at Kamphaengsaen/PHTIC
Email: fengrns@ku.ac.th

S. K. Srivastava
School of Biochemical Engineering, Institute of Technology, Banaras Hindu University, Varanasi - 221005, Uttar Pradesh, India.

Panuwat Suppakul
Department of Packaging and Materials Technology, Faculty of Agro-Industry, Kasetsart University, 50 Ngamwongwan Rd., Ladyao, Chatuchak, Bangkok, Thailand 10900.
Email: fagipas@ku.ac.th

X. Z. Tang
College of Food Science and Engineering, Nanjing University of Finance & Economics, Nanjing, Jiangsu Province, 210046, China.

Abhishek Dutt Tripathi
Centre of Food Science and Technology, Institute of Agricultural Sciences, Banaras Hindu University, Varanasi - 221005, Uttar Pradesh, India.
Email: abhi_itbhu80@rediffmail.com

Neelima Tripathi
Department of Chemical Engineering, Indian Institute of Technology Guwahati – 781 039, Assam, India.

V. D. Truong
U.S. Dept. of Agriculture, Agricultural Research Service, South Atlantic Area, Food Science Research Unit, Raleigh, NC 27695, U.S.A.

Zhiwei Wang
Packaging Engineering Institute, Jinan University, Guangzhou, Guangdong 510632, China.

Ajay Yadav
Centre of Food Science and Technology, Institute of Agricultural Sciences, Banaras Hindu University, Varanasi - 221005, Uttar Pradesh, India.

LIST OF ABBREVIATIONS

AAGR	8Average annual growth rate
ADC	Analog to digital convertor
AFM	Atomic force microscope
ATR	Attenuated total reflectance
BC	Bacterial cellulose
BSEs	Back-scattered electrons
CAP	Controlled-atmosphere packaging
CAS	Controlled atmosphere storage
CSIR	Council for Scientific and Industrial Research
CSLM	Confocal scanning laser microscope
DMTA	Dynamic mechanical thermal analysis
DRS	Dielectric relaxation spectroscopy
DSC	Differential scanning calorimeter
DTA	Differential thermal analysis
EELS	Electron energy loss spectroscopy
EM	Elastic modulus
EMI	Electromagnetic interference
EPA	Environmental protection agency
ERH	Equilibrium relative humidity
FDA	Food and drug administration
FDM	Finite difference method
FEA	Finite element analysis
FIB	Focused ion beam
FID	Flame ionization detector
FSI	Food spoilage indicator
FSL	Food-simulating liquids
FTIR	Fourier transforms infra-red spectroscopy
GC-MS	Gas chromatography-mass spectrometry
GFSE	Grape fruit seed extracts
GPC	Gel permeation chromatography
HDPE	High density polyethylene
HIPS	High impact polystyrene
HPLC	High-performance liquid chromatography
ICI	Imperial chemical industry
ICP-MS	Inductively coupled plasma-mass spectrometry
ICPs	Intrinsically conductive polymers
IFT	Institute of Food Technologists
ILT	Ideal laminate theory
LCA	Life cycle assessment

LDH	Layered double hydroxide
LEO	Lemongrass essential oil
LLE	Liquid–liquid extraction
LMIS	liquid metal ion source
LWA	Liquid water absorption
MAP	Modified atmosphere packaging
MWNT	Multi-walled nanotubes
NCF	Nanocellulose fibers
NIAS	Non-intentionally added substances
NMR	Nuclear magnetic resonance
NNI	National nanotech initiative
OML	Overall migration limit
OMLS	Organically modified layered silicate
OP	Oxygen permeability
OTR	Oxygen transmission rate
PCNC	Polymer–clay nanocomposites
PNCs	Polymer nanocomposites
RFID	Radio frequency identification
RH	Relative humidity
SEM	Scanning electron microscope
SML	Specific migration limit
SPI	Soy protein isolate
SPM	Scanning probe microscope
STM	Scanning tunneling microscope
SWNT	Single-wall nanotube
TEM	Transmission electron microscope
TEMAP	Thermally equilibrious modified atmosphere packaging
TEMT	Transmission electron microtomography
TPS	Thermoplastic starch
TS	Tensile strength
UPC	Universal Product Code
VARTM	Vacuum assisted resin transfer molding
VRH	Variable range hopping
WAXS	Wide-angle X-ray scattering
WPI	Whey protein isolate
WVP	Water vapor permeability
WVTR	Water vapor transmission rate
XRD	X-ray diffraction

PREFACE

The world-wide market for plastic films and sheets used in various packaging and non-packaging applications exceeds $100 billion and is growing at an annual rate higher than the global gross domestic product. The polymeric materials that are used include low and high density polyethylene, polyethylene terephthalate, biaxially oriented polypropylene, copolymer polypropylene, poly(vinyl chloride) and ethylene vinyl alcohol. No doubt this massive usage of plastics is driven by several benefits including convenience and economics, but the drawbacks are also becoming apparent.

Plastics do not biodegrade, primarily because they are made of synthetic polymers and no microbe has yet evolved that can feed on them. Disposal of the millions of tons of plastic waste generated every year takes up huge areas in the form of landfills. Plastic polymers may not be toxic themselves but the myriad of chemical monomers added to them for improving their properties can be released to the surroundings and contact materials over time or under conditions such as heat and exposure to sunlight or photodegradation. An example is bisphenol A or BPA that is added as a plasticizer but banned for use in applications involving packaging or containers for infant food due to its potential toxic effects. Waste plastics can also attract and accumulate chemical poisons present in the environment such as water contaminated with DDT and PCB. A striking example of the problem with plastic waste is the 'The Great Pacific Garbage Patch' covering an area roughly the twice the size of France in the Pacific Ocean*. The combined weight of plastic accumulated in this 10-meter deep plastic soup is estimated at three million tons and increasing steadily due to several major sea currents converging to this region that bring flotsam from the Pacific coasts of Southeast Asia, North America, Canada and Mexico. Its toxic effect on marine life is just beginning to be understood.

In this backdrop, the development and use of bio-based and/ or biodegradable polymers is gaining importance. Polylactic acid, polyvinyl alcohol and polybutylene succinate are plastic materials that can be decomposed by bacteria or other living organisms. These materials often lack the performance characteristics, such as strength, flame retardance or barrier properties, of conventional plastics but they can be enhanced by using various nanofillers. Research on such nanocomposites is also gaining widespread attention.

Developments in the above mentioned areas were focus of the International Conference on Polymers for Packaging Applications (ICPPA 2012) organized at Mahatma Gandhi University in Kottayam, India from March 31st to April 2nd, 2012. Scientists from the U.S.A., U.K., France, China, Thailand, Malaysia, Iran and India presented cutting edge research in the areas of food, non-food, and industrial packaging applications of polymers, blends, nanostructured materials, macro, micro and nanocomposites, and renewable and biodegradable materials. This book and its focus have origins in the aforementioned international conference. Several of the speakers at ICPPA 2012 contributed to the various chapters. Due to reasons related to sustainability, recycling and regulatory issues, the topics discussed in the conference and ongoing research has gained even greater urgency in the last two years.

This book emphasizes interdisciplinary research on processing, morphology, structure, and properties as well as applications of polymers in packaging of food and industrial products. It is useful for chemists, materials scientists and food technologists. It details physical, mechanical, electrical and barrier properties of polymers and biopolymers, as well as sustainability, recycling, and regulatory issues. The book contains a good mix of review chapters and experimental studies, and is divided into four major sections. Chapters in the first section provide an overview of traditional plastics in packaging applications including a specific example related to food packaging. Additives used for improving properties of plastics are described along with experimental studies on their migration. The second section focuses on biopolymers and biodegradable plastics, and their synthesis, commercial production, properties and use in packaging of food and industrial products and biomedical applications. Recycling and life cycle analysis for plastics and bioplastics is also discussed. The next section contains chapters related to nanotechnology and bio-nanocomposites in packaging applications. Various nanofillers, including phyllosilicates, metallic nanofillers, carbon nanotubes and graphene, are described and also regulatory issues discussed. Analytical techniques and approaches based on mathematical modeling are presented for understanding the structure, and barrier and mechanical properties of bio-nanocomposites. The final section has chapters describing the state-of-the-art in modified atmosphere packaging for foods, and innovations related to active and intelligent packaging. The last chapter presents an intriguing concept of conductive polymers for functions such as electromagnetic shielding and active packaging.

The editors have made a conscious effort to select authors from various parts of the world representing diverse disciplines including material science, physics, packaging engineering, microbial sciences and food technology. We would like to thank them profusely for their high quality submissions and contributing to this truly multi-disciplinary effort. Special thanks to our readers, and the editorial staff of Apple Academy, Inc. for their assistance and helpful suggestions at every step.

— Sajid Alavi, PhD, Sabu Thomas, PhD, K. P. Sandeep, PhD, Nandakumar Kalarikkal, PhD, Jini Varghese, and Srinivasarao Yaragalla

* Moore, C. J., Moore, S. L., Leecaster, M. K., and Weisberg, S. B. 2001. Marine Pollution Bulletin 42 (12) 1297–300.

Part I: Conventional Plastics in Packaging Applications

CHAPTER 1

PROPERTIES OF PLASTICS FOR PACKAGING APPLICATIONS

VIMAL KATIYAR, SURENDRA S. GAUR, AKHILESH K. PAL, and AMIT KUMAR

ABSTRACT

This chapter discusses the properties of plastics most relevant to packaging applications. General description of thermal, mechanical, and barrier properties of polymers is provided. Properties of most common packaging plastics such as polyethylene and polystyrene are detailed next. Bio-based plastics (PLA, PHA, chitosan, and so on) for packaging applications are discussed as an eco-friendly, sustainable alternative to conventional packaging plastics. In several instances, polymer nanocomposites exhibit superior properties compared to neat, unfilled polymer. Property enhancements associated with incorporation of different nanofillers in the polymer matrix are discussed in the context of packaging applications. Focus is on the improvement of mechanical and barrier properties. Novel packaging techniques such as modified atmosphere packaging and active packaging where the role of packaging goes beyond that of simple containment and protection are reviewed as well. Finally, food-packaging interactions in terms of migration of chemicals from packaging material to foodstuff are discussed.

1.1 INTRODUCTION

Plastics are the most commonly used materials for packaging applications because of low cost, easy process ability and the availability of abundant resources for their production. The world wide annual production of plastics continues to increase at a steady rate every year and is expected to cross the 300million ton mark by the year 2015 (Chen and Patel, 2012). Plastics find applications in a wide variety of segments such as packaging, building and construction, automotive applications, electrical and electronics applications, aerospace applications, and corrosion prevention, and control. Currently the packaging sector accounts for over 40% of the total worldwide plastics consumption (Silvestre et al., 2011).

The properties essential for packaging materials are decided by the physical and chemical characteristics of the product, as well as by the external conditions in which the product is stored/transported. As plastics have a wide range of properties which can be tailored according to the product requirement, they are the most attractive materials for packaging applications. Polyethylene (PE), polypropylene (PP), polyethylene terephthalate (PET), polyvinyl chloride (PVC), and polystyrene (PS) are the most common packaging plastics, accounting for more than 90% of the total volume of plastics used in packaging. Common raw materials for production of plastics are petroleum products obtained from refining processes. Most of the conventional, fossil-fuel based plastics are non-biodegradable and hence can cause environmental pollution. Due to increasing environmental concerns, focus on research in bio-based sustainable plastic packaging materials has increased significantly in the past few years. The most promising bio-based polymers for future packaging applications are cellulose, chitosan, polylactic acid (PLA), and polyhydroxyalkanoates (PHA) (Siracusa, 2012).

The essential functions of packaging can be broadly classified into primary and secondary functions. Primary functions such as those associated with protection, storage, loading, and transport of the product will require the packaging to be strong, leak proof and able to withstand any external conditions that the storage or transportation environment will impose. Secondary functions such as those associated with promoting sales of the product may require the packaging to be transparent or to have good physical appearance (glossy) to attract customer attention. Information such as product ingredients, nutrition content (of food items) and usage instructions may need to be placed on the packaging, which will require the packaging material to be printable. Further, polymer recyclability can be an important criterion in determining its suitability as packaging material especially for volume applications.

This chapter focuses on the general properties of plastics for packaging applications with special emphasis on the most commonly used packaging plastics, and the properties that make them attractive packaging materials. Novel packaging systems such as active packaging and modified atmosphere packaging are briefly discussed as well.

1.2 GENERAL PROPERTIES OFPLASTICS

Several properties are important in determining the suitability of polymers for packaging applications. Such properties depend on the type and structure of the polymer. A list of common polymers used as packaging materials is shown in Figure1. The polymer properties most relevant for packaging applications can be broadly classified into morphology, barrier properties, mechanical properties, thermal properties, and optical properties. These properties strongly influence several key features of the finished packaging material such as strength, transparency, and the ability to seal off oxygen and water vapor:

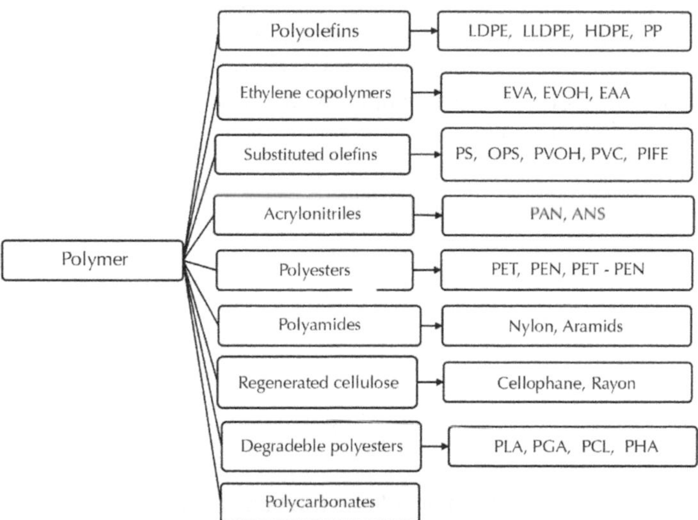

FIGURE 1 Common Polymers for Packaging Applications.

1.2.1 POLYMER MORPHOLOGY

Morphology of polymers refers to the spatial arrangement, orientation, and conformation of polymer chains. One of the most important morphological attributes of a polymer is its degree of crystallinity. A majority of synthetic polymers are semi-crystalline having domains of both crystalline and amorphous regions (Callister and Rethwisch, 2010). A brief description of crystallinity along with the effect of the degree of crystallinity on various properties of plastics is provided next.

CRYSTALLINITY

Polymer crystallinity is characterized by well-ordered regions of parallel, aligned chains. The regular arrangement may be helical, zigzag or planar. Disordered or misaligned polymer chains give rise to amorphous regions in the polymer matrix. Several synthetic polymers are semi-crystallinethat is, they have both amorphous and crystalline regions. Crystallinity decreases with increased branching because the branch points produce irregularities in molecular packing. Similarly, copolymerization introduces asymmetry in the polymer structure, and consequently limits the extent of crystallinity. Chain properties such as tacticity and presence of pendant groups can also have a strong effect on crystallinity. Tacticity is especially important in determining the extent of crystallization in polymers, isotactic, and syndiotactic polymers will generally crystallize whereas atactic polymers will be amorphous (a few exceptions to this rule do exist).

DEGREE OF CRYSTALLINITY

The degree of crystallinity of a polymer is a measure of the relative amounts of crystalline and amorphous regions, and is expressed on either volume basis or mass basis. Degree of crystallinity can be determined using various methods such as density measurement, X-ray scattering, and heat of fusion measurement. Density measurement is based on the two-phase model of polymer behavior (Bower, 2002). If the densities of the crystalline and amorphous components of the polymer are known, the degree of crystallinity of a polymer sample (on mass or volume basis) can be calculated using the following equations:

$$x_{\upsilon} = \frac{\rho - \rho_a}{\rho_c - \rho_a} \tag{1}$$

$$x_m = \frac{\rho_c}{\rho} x_{\upsilon} \tag{2}$$

where x_{υ} and x_m are respectively the volume fraction and mass fraction of crystalline material in the sampler, ρ_a and$\rho_c$are densities of the sample, the amorphous component and the crystalline component respectively.

Wide-angle X-ray scattering (WAXS) can also be used to determine the degree of crystallinity of polymer samples. A typical WAXS curve for a semi-crystalline polymer consists of sharp Bragg peaks corresponding to the crystalline regions and a broad, diffuse halo corresponding to the amorphous regions. The WAXS curve can be resolved into crystalline and amorphous contributions, and the areas under the curve corresponding to amorphous contribution (Aa) and crystalline peaks (Ac) can be used to determine the mass fraction of crystalline component as:

$$X_m = \frac{A_c}{A_a} \tag{3}$$

A comparison between the heat of fusion of the polymer sample and that of 100% crystalline polymer can also provide an estimate of the degree of crystallinity. Measurement of heat of fusion is usually done using a differential scanning calorimeter (DSC).

IMPORTANCE OF POLYMER MORPHOLOGY IN PACKAGING APPLICATIONS

Several important properties of polymers depend on their degree of crystallinity. Table 1 lists the effect of increase in the degree of crystallinity on various mechanical, optical, andbarrier properties of polymers. Crystallinity of polymers affects the tensile strength and transparency of polymer films which are important criteria in some of the packaging applications (Shanks, 2003). As crystallinity and tensile strength of polymer films increases, the transparency decreases. Polymers with a high degree of crystallinity typically exhibit efficient chain packing with low diffusion coefficients and favorable barrier properties. Certain polymers such as PE exhibit high transmission rates for gases despite having considerably high crystallinity. However, even for

PE the transport coefficients (diffusivity and permeability) decrease with increasing degree of crystallinity (Flaconnèche et al., 2001). Properties of semi-crystalline polymers can be tuned for specific packaging applications by adjusting the degree of crystallinity (Rosato et al.,2000; Delpouve, 2012).

TABLE 1 Effect of Increase in Crystallinity on Different Polymer Properties (Selke et al., 2004)

S/N	Properties	Effect of crystallinity
1	Density	↑
2	Tensile strength	↑
3	Clarity	↓
4	Permeability	↓
5	Opacity	↑
6	Compressive strength	↑
7	Impact strength	↓
8	Tear resistance	↓
9	Toughness	↓
10	Ductility	↓
11	Ultimate elongation	↓

[↑ **represents** increase and ↓ represents decrease (with increasing crystallinity]

1.2.2 BARRIER PROPERTIES

In several applications, such as packaging of oxygen sensitive food products, it is desirable for the packaging material to have a high resistance to transmission of gases (such as oxygen and water vapor). Plastics having excellent barrier properties are required to get the desired characteristics in such packaging materials. However, packaging for respiring products such as fruits and vegetables must allow transport of oxygen and carbon dioxide and hence should be permeable to these gases. Barrier property is inversely related to permeability: lower value of permeability implies better barrier property. Permeability of a penetrant in a polymer is dependent on the solubility coefficient and diffusion coefficient.

SOLUBILITY COEFFICIENT AND DIFFUSION COEFFICIENT

Solubility coefficient is a measure of the amount of penetrant absorbed by the polymer from a contacting phase. It represents polymer sorption capacity with respect to a particular sorbate. Diffusion coefficient, on the other hand, is a measure of the mobility of penetrants within the polymer. Gas diffusion rate in crystalline region of polymers is usually negligible compared to that in the amorphous region. Values of solubility and diffusion coefficients depend on various factors such as size of penetrant, polymer-penetrant interactions, polymer morphology, glass transition temperature, and so on. Both solubility and diffusion coefficient typically exhibit Arrhenius-type dependence on temperature:

$$S = S_0 e^{-\Delta H_s / RT} \quad , \quad D = D_0 e^{-E_D / RT} \tag{4}$$

where $S_0$and D_0prefactors, ΔH_s is the heat of solution, $E_D$is the activation energy ofdiffusion, and R is the universal gas constant. For semi-crystalline polymers, the crystalline region does not contribute toward the solubility of gases/vapors in the polymer matrix. Thus, the solubility coefficient is given by the product of the penetrant solubility in the amorphous phase and the volume fraction of the amorphous phase (Michaels and Bixler, 1961; Lin and Freeman, 2004). Presence of impermeable crystalline phase in semi-crystalline polymers not only makes the path of penetrant inside the polymer matrix more tortuous but also restricts the mobility of the amorphous chain segments thus reducing the effective diffusion coefficient (Lin and Freeman, 2004).

PERMEABILITY

Steady-state permeability is defined as the flux of penetrant per unit pressure difference across a sample of unit thickness. Permeation through polymers is generally a three step process: absorption of penetrant into the polymer matrix, diffusion of penetrant through the matrix and desorption of penetrant at the other side (Kirwan and Strawbridge, 2003). Thus, permeability is influenced both by the dissolution and the diffusion of the penetrant in the polymer matrix. In this sorption-diffusion model of penetrant transport across the polymer, permeability (*P*) is given as the product of diffusion coefficient () and solubility coefficient () (Callister and Rethwisch, 2010), that is:

$$P = DS \tag{5}$$

As both D and S show Arrhenius-type temperature dependence, P also depends exponentially onT:

$$P = \left(D_0 e^{-E_D / RT}\right)\left(S_0 e^{-\Delta H_S / RT}\right) = P_0 e^{-E_P / RT} \tag{6}$$

whereP_0=$D_0$and S_0 is a prefactor, and E_P=E_D+ΔH_S is the effective activation energy of permeation. If a fixed pressure difference is maintained across a film of known thickness *i*, the permeability can be experimentally calculated from the measured steady state flux (F):

$$P = \frac{Fl}{\Delta p} \tag{7}$$

The concept of free volume (empty space that is not occupied by polymer molecules) can be used for theoretical analysis of penetrant transport in polymers. The free volume model by Lee (1980) proposes an exponential relation between permeability and specific free volume:

$$P = Ae^{-B/(v-v_0)} \tag{8}$$

where A and B are penetrant-specific constants,ν is the specific volume of the polymer and ν_0 is the volume occupied by the polymer chains. Park and Paul (1997) have proposed a free volume based group contribution approach for the prediction of gas permeability in glassy polymers. Permeability through random copolymers and blends can also be estimated by free volume approach. If the free volume of the multicomponent polymer blend is assumed to be given by additive contributions from the individual polymer components, the blend permeability (P) is given by (Paul, 1984):

$$\ln\left(\frac{P}{A}\right)=\left[\frac{\varphi_1}{\ln\left(P_1/A\right)}+\frac{\varphi_2}{\ln\left(P_2/A\right)}\right]^{-1} \tag{9}$$

whereφ_2 and φ_1 are volume fractions, and P_1 and $P_2$are permeabilities of components 1 and 2 respectively. In case of gas transport through multilayer films, simple models such as the ideal laminate theory (ILT) can be used to estimate the overall permeability using permeabilities of the individual layers (Graff et al., 2004; Jang et al., 2008). Thus, forlayers, the overall permeability, as predicted by ILT, is:

$$P_{overall}=L\left(\sum_{i=1}^{n}\frac{l_i}{P_i}\right)^{-1} \tag{10}$$

where$l_i$are $P_i$the thickness and permeability, respectively, of thei^{th} layer and$L$is the total thickness of the multilayer film, that is, $L=\Sigma i=1 l_i$.

Oxygen permeability, WVP and carbon dioxide permeability are of special interest in packaging applications, especially food packaging where exposure of the product to some of these gases may lead to deterioration in quality. Several factors influence gas permeability of plastics films, such as integrity of the film, crystalline-to-amorphous ratio, mobility of polymeric chains, hydrophobic-hydrophilic ratio, and the presence of plasticizer or other additives (Souza et al., 2009).

OXYGEN PERMEABILITY

This is a key factor in food packaging applications as exposure of food products to oxygen can cause oxidation and undesirable changes in food quality. The changes may include deterioration of odor, color, flavor, and nutrients. Hence, it is necessary to use food packaging films having high resistance to oxygen transmission to ensure quality and maintain long shelf life. The standard methods for determining oxygen transmission rate (OTR) are ASTM D3985 (Film), DIN 53380 (Film), JIS K7126 (Film), ASTM F1307 (Packages) (Johnson, 2009). The OTR is directly related to oxygen permeability, and is an important measure of barrier properties of the packaging film. Representative OTR trend for several polymers is shown in Figure 2:

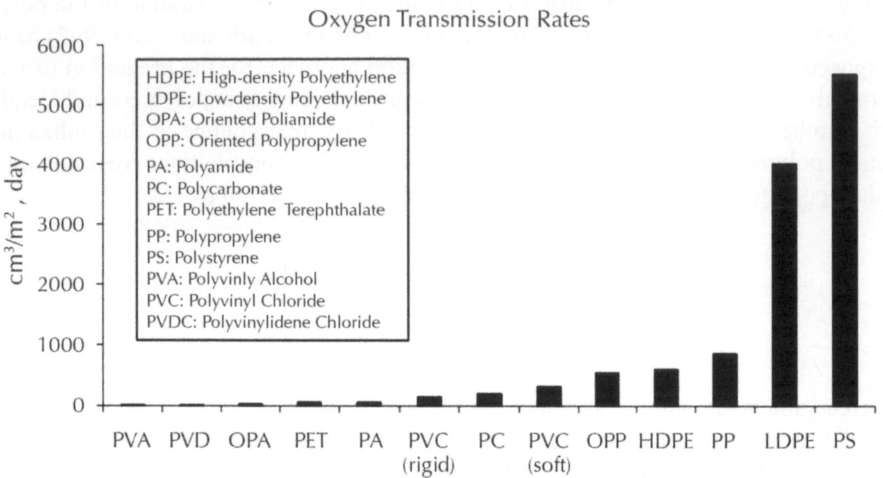

FIGURE 2 Representative Oxygen Transmission Rates (OTR) of Conventional Packaging Materials.

CARBON DIOXIDE PERMEABILITY

Loss of CO_2 is detrimental to the shelf life of products such as soda and carbonated drinks, necessitating a packaging material having low CO_2 transmission rate. Figure 3 shows representative CO_2 transmission rates for several polymers:

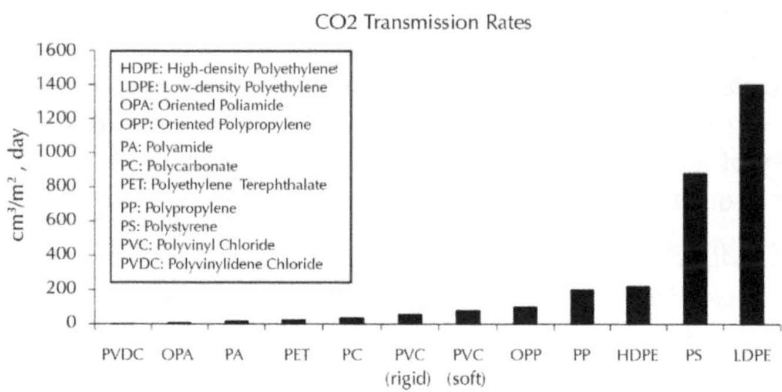

FIGURE 3 Representative CO_2 Transmission Rates of Conventional Packaging Materials.

WATER VAPOR PERMEABILITY

Water vapor permeability (WVP) is a critical parameter in food packaging applications as contact with water vapor may cause certain food items to lose texture. Polymer-water interaction, that is hydrophilicity or hydrophobicity of the polymer, is a crucial factor affecting the WVP. In general, water would permeate preferentially through the hydrophilic portion of the film. Thus, films prepared from hydrophilic polymers are expected to allow a higher transmission rate of moisture than those prepared from hydrophobic polymers. However, increased hydrophobicity does not guarantee a higher resistance to moisture transmission. For instance, water vapor transmission rate (WVTR) for PS is much higher than that for PVC (see Figure 4) although PS is more hydrophobic than PVC. Generally, the measurement of WVP is done by flowing air of controlled relative humidity on one side of polymer film and flowing moisture free nitrogen on the other side. The moisture level in leaving nitrogen stream is measured using a sensor (Kumar and Gupta, 2003). The standard methods for determining the WVTR are ASTM method F1249 and TAPPI method T557 pm-95 (Johnson, 2009):

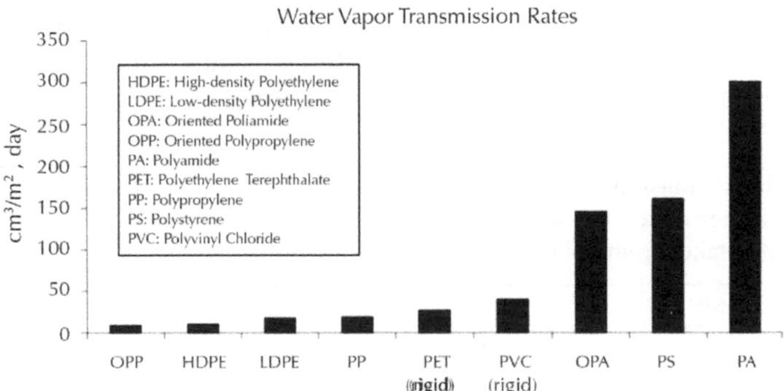

FIGURE 4 Representative Water Vapor Transmission Rates (WVTR) of Conventional Packaging Materials.

IMPORTANCE OF BARRIER PROPERTIES IN PACKAGING APPLICATIONS

Favorable barrier property is an essential requirement for packaging materials as it helps prevent the packaged goods from contamination, oxidative degradation and moisture absorption. Amongst the conventional plastics using in packaging, polyethylene (LDPE and HDPE) has good moisture barrier property (see Figure 4) whereas polyvinyl alcohol (PVA) and polyvinylidene chloride (PVDC) are good oxygen barriers (see Figure2). Polyethylene terephthalate (PET) has superior CO_2 barrier property (see Figure3) and hence, is used extensively in packaging of carbonated drinks. In food product packaging, several layers of films are used in blow molded bottles for improved barrier resistance to gas transmission. Barrier properties can be enhanced significantly by incorporation of nanofillers such as clay, thus reducing the thickness of packaging layers significantly (Mittal, 2010).

1.2.3 THERMAL PROPERTIES

Thermal properties of polymers govern their behavior during heating from solid amorphous or crystalline state to molten state. Polymer materials can undergo several phase transitions upon heating, and each transition determines a specific thermal property. Such, thermal properties associated with phase transitions include glass transition temperature, crystallization temperature, and melting temperature. Thermal conductivity, heat capacity, and heat of fusion determine the thermal behavior of polymers. Thermal stability of plastics defines the temperature up to which they are able to maintain their mechanical properties without degrading. Design of polymer packaging materials generally requires knowledge of the following thermal properties:

- *Glass Transition Temperature*:Amorphous and semi-crystalline polymers exhibit a phenomenon known as glass transition which involves a shift in the dynamics and mobility of chains. Glass transition temperature (T_g) is the temperature at which a polymer changes from glassy to rubbery state accompanied by increase in chain mobility and polymer flexibility. Amorphous polymers are glassy, hard and brittle at temperatures below T_g and become rubbery, soft and elastic above T_g. Table 2 lists T_g values for the most commonly employed conventional packaging plastics. As the T_g of a polymer determines whether it will be hard and brittle or soft and flexible at the temperature of interest, knowledge of T_g of packaging polymers is of utmost importance.

TABLE 2 Glass Transition and Melting Temperatures of Different Polymers and Their Mechanical Characteristics at Room Temperature (compiled from Mark, 1999 and Selkeet al., 2004). ForPVA is taken from Tubbs (1965)

S/N	Polymer	(°C)
1	PS	240
2	PVA	228
3	PET	265
4	PP (isotactic)	185
5	PP (atactic)	—
6	PVDC	195–205
7	PE	105–115 (LDPE) 120–145 (HDPE)

The T_g of a polymer depends on a variety of factors such as flexibility of polymer main chain, nature of side groups, degree of branching and crosslinking, and molar mass. Polymers with flexible main chain such as PE and PP have low T_g whereas those having relatively rigid main chain suchas PET have high T_g. Bulky and polar side groups impede chain rotation and make the chain stiff leading to an increase in T_g of polymers. For example, T_g values for both PS (having bulky phenyl side group) and PVA (having polar hydroxyl side group) are much higher than that for PP (having small and non-polar methyl side group). A small number of branches on a polymer chain lower the T_g due to the increased free volume contribution from larger number of chain ends. However, a high degree of branching and crosslinking restricts chain mobility and leads to increased T_g. Increase in molar mass of the polymer also leads to an increase in T_g. The dependence of T_g on molar mass is approximately given by an expression of the form (Fox and Flory, 1950):

$$T_g = T_g^{\infty} - \frac{K}{M} \tag{11}$$

where$M$ is the molar mass of the polymer, T_g^{∞} is the glass transition temperature of a polymer sample of infinite molar mass and K is a constant. Bicerano (2002) suggested that the constantshould be proportional to $(T_g^{\infty})^3$.

The phenomenon of glass transition in polymers is best analyzed by applying the concept of free volume. It is observed that for a variety of polymers the fractional free volume (defined as the ratio of free volume and total volume) in the glassy state (fg) is relatively constant. Fractional free volume at a temperature T (above T_g) is given by

$$f_V = f_g + \left(T - T_g\right) a_f \tag{12}$$

where α_f is the thermal expansion coefficient of the free volume. Based on the free-volumeconcept of glass transition, several approximate empirical rules have been proposed that relate the T_g of polymers to other polymer characteristics. For example, Simha and Boyer (1962) proposed the following relations between Tg and thermal expansion coefficients:

$$\left(\alpha_l - \alpha_g\right) T_g = 0.113 \quad ; \alpha_l T_g = 0.164 \tag{13}$$

Here $\alpha_l$andα_g are the coefficients of thermal expansion of 100% amorphous polymer above and below T_g respectively. Several relations also exist for determination of T_g of miscible blends and copolymers from knowledge of the individual T_g values of the constituent polymers. Fox equation (Fox, 1956), given below, is one of the simplest relations for estimating blend or copolymer T_g:

$$\frac{1}{T_g} = \frac{\varphi_1}{T_{g1}} + \frac{1 - \varphi_1}{T_{g2}} \tag{14}$$

Here φ_1 is the weight fraction of component *1*, and T_{g1} and T_{g2} are the glass transition temperatures of components *1* and *2*. Another well-known equation forcalculation of polymer blends is the Gordon and Taylor equation (Gordon and Taylor, 1952):

$$T_g = \frac{\varphi_1 T_{g1} + k_{GT}\left(1 - \varphi_1\right) T_{g2}}{\varphi_1 + k_{GT}\left(1 - \varphi_1\right)} \tag{15}$$

where subscript "2"refers to the component having higher Tg value, and the adjustable, empirical parameter K_{GT} accounts for unequal contribution by the two components.

MELTING TEMPERATURE

Melting transition occurs in a crystalline (or semi-crystalline) polymer when the aligned polymer chains become disordered and the material changes from crystalline state to an amorphous, liquid-like state. The temperature at which this transition is observed is called the melting temperature (Tm). Melting transition is typically observed in thermoplastics, thermosets degrade instead of melting. Melting temperatures of a few common thermoplastics are listed in Table 2. The approximate empirical relation $a_cT_m = 0.11$ by Bondi (1968), where a_c is the thermal expansion coefficient of the crystalline polymer, can be used to obtain a rough estimate of T_m.

HEAT DEFLECTION TEMPERATURE

This is the temperature at which a thermoplastic begins to deflect or deform under a specified load. Packaging containers that need to be hot-filled must have a high enough heat deflectiontemperature to avoid any deformation. Heat distortion temperature sets the upper limit on the temperature up to which polymers can be used without undergoing thermal deformation at a given load.

HEAT CAPACITY

Specific heat capacity of any material is defined as the amount of energy required to change the temperature of a unit mass of the material by one degree Celsius (or Kelvin). Heat capacity of plastic materials is temperature dependent, and is different for different phases. For example, in the case of semi-crystalline polymers, the heat capacity of the crystalline phase is typically lower than that of the amorphous phase. The most widespread technique used to measure heat capacity of polymers is differential scanning calorimetry (DSC). Alternatively, differential thermal analysis (DTA) can also be used to determine heat capacity.

HEAT OF FUSION (ΔH_M)

Heat of fusion, denoted by ΔH_m, is defined as the energy required for the transition of polymer from solid, crystalline state to a molten, liquid-like state. For semi-crystalline polymers, $\Delta Hm$is directly proportional to the percentage crystallinity. ΔH_mcan be measured using DSC.

IMPORTANCE OF THERMAL PROPERTIES IN PACKAGING APPLICATIONS

Polymer thermal properties, such as glass transition temperature and heat deflection/distortion temperature, affect the mechanical behavior of polymers and hence, are instrumental in deciding suitability of polymers for certain packaging applications. Glass transition temperature is very important when the packed material is to be stored in a frozen environment. In such cases, one must ensure that the glass transition temperature of packaging material is lower than the freezer temperature. Otherwise, the packaging material will become brittle, and may crack. Heat of fusion is another thermal property that plays an important role in deciding the packaging material. If the heat received from external environment or source is larger than the heat of fusion, the structure of polymer may change and the packaging may fail.

1.2.4 MECHANICAL PROPERTIES

Polymers display a wide spectrum of mechanical properties. The properties that are of importance in plastic packaging are tensile strength, tear strength, impact strength and bursting strength. Standard ASTM, DIN and ISO test methods are commonly used for determining mechanical properties. Based on the form of material (film, bottle, and so on), there are separate standard methods of testing and measuring mechanical properties for packaging materials.

TENSILE PROPERTIES

Stress-strain characteristics under elongation or tensile deformation can be used to understand the mechanical behavior of polymers. Stress is defined as the force per unit area and strain is defined as dimensionless fractional increase in length. Tensile properties of a polymer can be characterized using quantities such modulus of elasticity, stiffness, elastic elongation, ultimate tensile strength, toughness, brittleness, and creep (Tripathi, 2002; Monasse and Haudin, 1995).

TEAR STRENGTH

Tear strength is defined as the energy required to propagate a crack up to tear apart a specimen of standard geometry. In the case of films, tear strength values depend on the stretching ratio, and on whether the measurement is along or across the machine direction. Tear resistance is important in packaging materials such as stretch films because it provides a measure of the protective value of the packaging as well as the strength needed to tear open the packaging.

BURSTING STRENGTH

A large bursting strength in packaging plastics is required to ensure that any external static ordynamic forces experienced during handling and transportation do not destroy the package. Testing of burst strength is performed by increasing air pressure inside the package until it bursts. After several repetitions, an average value of pressure is obtained which determines the burst strength (Johnson, 2009).

IMPACT STRENGTH

Impact resistance of a material refers to its ability to withstand the application of sudden load (Mark, 2007). Packaged products may experience sudden load during handling and transportation, which may cause fracture of the packaging material. Packaging materials of high impact strength are required to ensure safe transportation of goods. Table 3 lists the impact strengths of several conventional packaging plastics. Charpy pendulum impact test and falling weight impact test are commonly used for determining impact strength of materials (Hylton, 2004).

TABLE 3. Tensile Strength and Impact Strength of Conventional Packaging Plastics (from Mark, 1999)

Plastic Name	Tensile Strength (MPa)	Impact Strength (J/m)
LDPE	9–15	No break
HDPE	10–60	30–200

TABLE 3 *(Continued)*

PP	1–2 (atactic)	27 (isotactic)
PS	30–60	~ 20
PET	50	90
PC	~ 66	850
PVDC	73 (machine direction)	
110 (transverse direction)	21–53	

IMPORTANCE OF MECHANICAL PROPERTIES IN PACKAGING APPLICATIONS

Packages are continually subjected to variable mechanical loads from handling, loading and transportation to storage. The packaging material should be able to withstand these mechanical loads, external abrasion and any other changing environmental conditions, such as temperature and pressure, to ensure damage-free product supply. Ideally, the packaging material should undergo a wide variety of mechanical tests such as tensile test, impact test, bursting test, tear strength test, flexural test, compression test, and other product specific tests.

1.2.5 OPTICAL PROPERTIES

Several optical properties, such as transparency and color, are crucial in determining the suitability of plastics for packaging applications. For example, the choice for a plastic alternative to glass or glass-like packaging material may be dictated by properties such as clarity and colorlessness. Visual appearance of materials is usually a result of how the material absorbs/transmits, scatters, reflects or refracts light.

HAZE

When light passes through a sample, the percentage of transmitted light which deviates by more than 2.5° from the direction of the incident parallel beam is known as haze. The deviation or scattering of light takes place due to surface imperfections, in homogeneity and internal scattering from crystallites. Haze can lead to reduced transparency of the material resulting in poor visibility.

GLOSS

Gloss is defined as the percentage of incident light that is reflected at an angle equal and oppositeto that of the incident rays (specular reflection). In case of films, gloss is measured at an angle of 45°. Materials exhibiting a high degree of specular reflection appear glossy and shiny, whereas materials exhibiting low specular reflection (and high diffuse reflection) appear dull and matte. Gloss is reduced by scratches, roughness and irregularities in surface texture.

TRANSPARENCY AND OPACITY

A polymeric material appears transparent because it doesn't absorb or scatter significant amountof visible electromagnetic radiation. Transparency of polymers decreases with increasing crystallinity due to the presence of structural features comparable in size to the wavelength of visible light (Bower, 2002). Transmittance (*T*) is defined as the ratio of the intensity of radiation transmitted by the sample to the radiation intensity incident on the sample. If the transmittance value is higher than 90%, the material is said to be transparent. Higher scattering implies lower transparency and vice versa.

1.2.6 IMPORTANCE OF OPTICAL PROPERTIES IN PACKAGING APPLICATIONS

Glossy packages enhance the aesthetic appeal of the product making it more likely to attract customer attention. Polymers with high transparency are used as an alternative to glass in packaging applications as they provide the added advantage of easier processabilty and lower brittleness.

1.3 PLASTICS IN PACKAGING: PROPERTIES AND APPLICATIONS

The distribution in Figure 5 shows that polyolefins such as LDPE, HDPE and PP, along with PET, PS and PVC, are the most widely used plastics for packaging, accounting for almost 98%of the total packaging plastic consumption worldwide. In this section, the properties and applications of these plastics, as they apply to packaging, are discussed.

1.3.1 POLYETHYLENE

Polyethylene (PE) is currently the most common packaging plastic in use. It is produced from the polymerization of ethylene which in turn is obtained as a petrochemical product by thermal cracking of ethane and propane. Depending on the polymerization process conditions, different types of PE can be obtained as listed in Table 4. The density ranges given in Table 4 differ in value across different literature sources, and are dependent on the synthesis techniques used for polymerization as well as on the polymerization conditions.

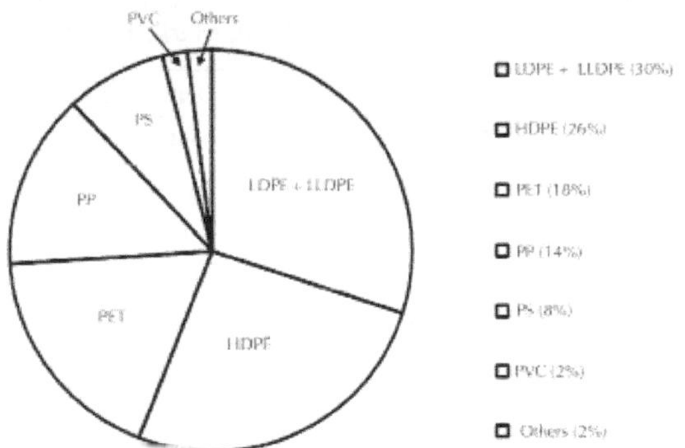

FIGURE 5 Worldwide Distribution of Different Plastics used in Packaging (Twede, 2009).

PROPERTIES

Polyethylene is easy to process, tough, and flexible, has good chemical resistance and good barrier properties for moisture and water vapor, is free from odor and toxicity, has excellent electrical insulation properties, and is easily heat sealable (Brydson 1999; Brandsch and Piringer, 2008; Matar and Hatch, 2000). The PE shows poor barrier properties toward gases, oils, and fats. However, these barrier properties improve with increase in density (Kirwan and Strawbridge, 2003). The PE is flammable and possesses lower thermal stability compared to other plastics used in packaging. It is a semi-crystalline polymer having varying degree of crystallinity. Higher crystallinity in PE leads to improved stiffness, hardness, tensile strength, opacity, barrier properties, and heat and chemical resistance. Drawbacks of increased crystallinity are reduction in impact resistance and permeability (Chanda and Roy, 2009). Properties of different types of PE vary with density, molecular weight, degree and type of branching, and degree of crystallization (Carter, 2009). Thus, different types of PE such as HDPE, MDPE, LDPE, LLDPE, VLDPE, and ULDPE are suited to different types of applications.

The LDPE is a highly branched polymer, and consequently has low crystallinity. The LDPE has higher flexibility and transparency in comparison to other types of PE, and hence it is mostly used in manufacturing sheets and films. The HDPE has higher crystallinity and melting temperature than LDPE because HDPE chains are relatively unbranched (Matar and Hatch, 2000). The LLDPE has similar range of density as LDPE, but has shorter, non-uniformly distributed side chains. The LLDPE has better mechanical properties (Chanda and Roy, 2009), higher melting point, and lower transparency compare to LDPE. Mechanical and thermal properties of LLDPE can be further improved by adding nanofillers such as clay, SiO_2 and TiO_2 (silica and titania nanoparticles). Thermal and mechanical properties of LLDPE can be improved by blending it in proper proportion with LDPE (Liu et al., 2005).

TABLE 4 Types of Polyethylene (PE) and Their Properties (adapted from Firdaus and Tong, 2009; Kirwan and Strawbridge, 2003; Rosato et al., 2000; Mark, 1999)

Type of Polyethylene	Density (g/cm^3)
High density polyethylene (HDPE)	0.94–0.96
Medium density polyethylene (MDPE)	0.93–0.94
Low density polyethylene (LDPE)	0.91–0.93
Linear low density polyethylene (LLDPE)	0.91–0.93
Very low density polyethylene (VLDPE)	0.89–0.91
Ultra low density polyethylene (ULDPE)	<0.89

PROCESSING

The PE is generally melt processed to obtain the final shape for packaging applications. Important techniques used for PE processing are summarized below:

- Injection molding:In this method molten plastic material is cooled under pressure in a mold, solidified and finally taken out of the mold. Injection molding is used to prepare articles with accurate dimension and weight, and desired surface finish (Carraher, 2008). Temperature, pressure, and injection speed are the main parameters which control the injection molding process. Decrease in tensile strength and Young's modulus of HDPE has been observed with increase in injection speed (Jiang et al., 2011).
- Extrusion molding:This is a continuous process in which powdered or granular plastic materials are processed in the molten state using an extruder. An extruder is a screw conveyer in which powder or granules of material are fed through a hopper, and conveyed through a cylindrical barrel. In the barrel the material is squeezed and melted at high temperature and pressure, and forced through a die at a constant rate to form pipes, films, sheets or any other required shape by feeding the melt to the mold (Carraher, 2008).
- Blow molding: Blow molding is used to prepare hollow articles such as bottles. Molten plastic is first formed into a tube-like shape, called parison (or perform), having sealed ends with a hole to allow entry of compressed air. Next, the parison is expanded in the mold of the required shape, cooled and removed. Blow molding can be classified into three types: injection blow molding, extrusion blow molding, and stretch blow molding (extrusion or injection stretch blow molding) (Tadmor and Gogos, 2006).
- Packaging applications:PE is the most commonly used packaging plastic worldwide. Various general and industrial packaging applications of PE include bags for garments and grocery, trash bags, packaging for bakery, meat, poultry and dairy products, frozen food packaging, pharmaceutical and cosmetic products packaging, and packaging for pesticides, insecticides and fertilizers (Vasile and Pascu, 2005; Carter, 2009).

1.3.2 POLYPROPYLENE

Polypropylene (PP) is another common plastic used extensively in packaging applications. Its homopolymer is produced by catalytic addition polymerization of propylene. In the case of PP copolymer, a co-monomer such as ethylene is also used. An organometallic catalyst is used, which attaches to propylene, works as functional group and reacts with the unsaturated bond of propylene (or with other co-monomer in case of copolymerization) to form a long chain polymer (Maier and Calafut, 1998).

PROPERTIES

Polypropylene is a low cost plastic, having a density between LDPE and HDPE, but possessing higher thermal resistance due to the presence of a methyl group in the main chain. It has high chemical resistance, but has poor oxidative resistance due the presence of methyl group. Addition of antioxidants is therefore required for stability in applications involving oxidative environments (Brandsch and Piringer, 2008). PP is nontoxic, easily processable, and has good dielectric and insulation properties, excellent clarity, and good mechanical properties (such as high stiffness and hardness, and good fatigue resistance) (Tripathi, 2002). Due to its high thermal stability and favorable mechanical properties, PP has replaced PE in many applications. Moreover, PP is very lightweight and has high-dimensional stability, making it suitable for replacement of metallic parts in automobiles. Ethylene propylene elastomers, produced by copolymerizing propylene with ethylene, shows improved resistance to heat and oxidation, as well as favorable tensile and tear properties (Monasse and Haudin, 1995).

PROCESSING

Although, the processing techniques for PP are similar to those for PE, the processing parameters are different. Important techniques used for PP processing are injection molding, blow molding and extrusion. Thermoforming and compression molding are also used in small scales for processing of PP.

Thermoforming:In this process thermoplastic sheets are converted into final article shape by heating, shaping, cooling, and trimming in sequence. It is a simple process in which thermoplastic sheet is placed in a mold, heated up to reshaping temperature and subsequently stretched, either by creating a vacuum or by applying pressure or using mechanical tools. When the sheet touches the mold walls, it cools down and the final article is then taken out. As thermoforming requires sheets of thermoplastic material for processing, which must be made by other processing techniques, it incurs higher cost compared to other methods. Thus thermoforming is used only when other processing techniques are not suitable. Heating temperature, a critical parameter in thermoforming, should be maintained in such a way that the material stays within the elastic range (Throne, 1996; Illig, 2001).

PACKAGING APPLICATIONS

Polypropylene is a versatile plastic with applications ranging from packaging to automobile and textile industry. It possesses properties similar to those of PE and hence, competes with PE in several product applications. It is generally used in packaging applications as an alternative to PE. Due to its higher stability, PE is generally preferred over PP in applications where the product is to be used in an oxidative environment. However, the oxidative stability of PP can be improved through addition of antioxidants. Textile industry is another major area of application for PP where it is primarily used to produce synthetic fibers (Maier and Calafut, 1998; Miller, 2009).

1.3.3 POLYVINYL CHLORIDE

Polyvinyl chloride (PVC) is a commercially important thermoplastic. It has a huge demand in the production of flexible goods, and has replaced metal, glass, wood, leather, and rubber in several applications (Wickson and Grossman, 2008). The PVC is synthesized by polymerization of vinyl chloride, which is produced by chemical reaction between chlorine and ethylene. Chlorine is obtained by electrolysis of sodium chloride solution (salt water), and ethylene is obtained as a petrochemical product (Leadbitter et al., 1997).

PROPERTIES

The PVC has some unique characteristic properties that differentiate it from other common plastics. It possesses good combustion resistance due to the presence of chlorine in polymer chain. When burnt, PVC generates hydrogen chloride gas that retards combustion reactions by limiting oxygen supply to the PVC surface (Summers, 2005).

The PVC, in its pure form, has low thermal stability and cannot be processed. Several additives are used in the synthesis of PVC due to which PVC displays a vast range of physical and chemical properties, and has most number of compounded products (Daniels, 2005). The choice of additives depends upon the final application requirement. Different additives used in PVC synthesis are plasticizers, impact modifiers, stabilizers, lubricants, fire retardants, blowing agents, fillers, and viscosity modifiers. Most of the physical properties of PVC are dependent on molecular weight. For example, its tensile strength, abrasion resistance, creep resistance, chemical resistance, and oil resistance increase with increasing molecular weight. Thus, properties of PVC can be tailored according to requirement by adding different additives, and by controlling the molecular weight of the polymer (Wickson and Grossman, 2008).

PROCESSING

The processing techniques used for PVC are injection molding, blow molding, extrusion, thermoforming, casting, rotational molding, and compression molding (Throne, 2005).

PACKAGING APPLICATIONS

Due to its versatile chemical and physical properties PVC is used for a wide range of applications. Applications include piping, building construction, clothing, wire insulation, molded automobile bodies, and toys. Addition of plasticizer to PVC improves its flexibility, toughness, and impact resistance, making it suitable for a variety of packaging applications (Chanda and Roy, 2009).

1.3.4 POLYSTYRENE

Polystyrene (PS) is an addition polymer, which may be produced by bulk, emulsion or suspension polymerization of styrene. Styrene is obtained from ethylbenzene by catalytic dehydrogenation. The PS is a vinyl polymer in which hydrogen is replaced by phenyl group. Its copolymers can be produced using different monomers such as butadiene, acrylonitrile, ethylcneoxide, and divinylbenzene.

PROPERTIES

Properties of PS are highly dependent upon molecular weight and can be controlled by incorporation of different additives. The properties which make it an attractive material for various applications are stiffness, good chemical resistance, absence of odor and toxicity, low water absorption, high transparency, and good electrical insulation property. Foamed or expanded PS, which is obtained by heating polystyrene (PS) and blending with volatile liquid, has good thermal insulation properties. Although, pure PS is brittle in nature, the brittleness can be reduced by blending it with copolymers. Blended PS is known as high impact polystyrene (HIPS) (Kirwan and Strawbridge, 2003).

PROCESSING

The PS is processed using common techniques used for thermoplastics. Generally PS is extruded or injection molded. Expanded PS requires special techniques for extrusion or molding.

PACKAGING APPLICATIONS

The PS is primarily used in packaging of food and nonfood items. Food packaging applications include packaging of meat and fish, egg cartons, fast food packaging, dairy products packaging, fruit packaging, and disposable food containers. Nonfood packaging applications consist of packages for electronic instruments, audio cassettes and compact discs, cosmetics, pharmaceuticals, stationery, various machinery tools and accessories, and shipping packages for delicate equipments (Wunsch, 2000).

1.3.5 POLYETHYLENE TEREPHTHALATE

Polyethylene terephthalate (PET) is a linear thermoplastic polymer, which was initially commercialized for packaging carbonated soft drinks due to its excellent gas barrier properties that allows it to retain CO_2. Raw materials used for production of PET are ethylene glycol and terephthalic acid (or dimethyl terephthalate). Production of PET is a two-step process: in the first step trans-esterification or esterification takes place, depending on whether terephthalic acid or dimethyl terephthalate is used, and in the second step polycondensation of resulting oligomers produces PET (Massa et al., 2011).

PROPERTIES

The PET can exist in either amorphous or crystalline form, which increases its range of applicability to a wide variety of packaging applications. It is a linear chain thermoplastic polymer due to which it can be easily converted from amorphous phase to crystalline phase by annealing or stretching above glass transition temperature. Amorphous PET has high transparency, but is vulnerable to heat degradation. Crystalline PET has good strength, rigidity, dimensional stability, water resistance, and thermal resistance. Crystalline PET also possesses good chemical resistance, but not at the same level as PE or PP (Hough and Dolbey, 1995). The PET has a high glass transition temperature and melting point, and can be easily recycled. The PET possesses excellent gas barrier properties (Mantia, 2002) making it an ideal choice in several packaging applications.

PROCESSING

Common processing techniques used for PET are injection molding, sheet extrusion, and blow molding. The PET is mainly processed as films and bottles because these forms find the most extensive application in packaging (Brady et al., 2002).

PACKAGING APPLICATIONS

Bottles produced from thick PET sheets are primarily used for packaging carbonated drinks, fruit juice, alcoholic drinks, mineral water, perfumes, and so on. The PET films are used for food and pharmaceutical packaging.

1.4 BIO-BASED PLASTICS FOR PACKAGING APPLICATIONS

Most of the synthetic plastics currently used for packaging are non-degradable and hence not friendly to environment (Pilla, 2011). Bioplastics, on the other hand, are biodegradable which resolves the problem of land filling. After disintegration and composting, bioplastics can be used as fertilizer and soil conditioner (Srinivasa and Tharanathan, 2007). Table 5 lists the biodegradation period for several bioplastics.

Starch based films degrade quickly while cellulose based films take a long time for degradation. In comparison, films derived from conventional packing plastics do not degrade at all.

TABLE 5 Biodegradation Periods of Various Packaging Materials (adapted from Srinivasa and Tharanathan, 2007)

Packaging Materials	Duration (days)
Plastic packaging films	Never degraded
Cellulose based films	> 365
PLA based films	150–240 (@ 60°C)
Chitosan based films	~ 90–150
Starch based bioplastics	~ 40–60

Biologically produced materials such as starch, cellulose, fatty acids, sugars, and proteins can be converted into various monomers by microorganisms, which in turn can be used for bio-based polymer production. Some examples of such biologically produced, biodegradable polymers are polyhydroxyalkanoates (PHA), polylactic acid (PLA), poly (butylene succinate) (PBS), and poly (trimethylene terephthalate) (PTT). Tables 6 and 7 list the thermal and mechanical properties of a few microorganism-produced bio-based polymers. Bioplastics can also be chemically synthesized from bio-derived monomers or directly extracted from biomass (Srinivasa and Tharanathan, 2007). A brief introduction to common bioplastics and their properties relevant to packaging applications is provided next.

TABLE 6 Thermal Properties of Microorganism-produced Bio-based Plastics (adapted from Chen and Patel, 2012)

Molecular weight (Mwx10–4)	Glass transition temperature (Tg °C)	Melting temperature (Tm°C)
10–1000	-50–4	60–117
5–50	60	175
3–20	-33–36.6	112–116
5–36 (Mn)	35–36	125–140
3.8	42.6	227.55

Mn: number average molecular weight; PPC: Poly (propylene carbonate)

TABLE 7 Mechanical Properties of Microorganism-produced Bio-based Plastics (adapted from Chen and Patel, 2012)

Molecular weight (Mwx10–4)	Young's modulus (MPa)	Elongation at break (%)
10–1000	flexible	2–1000
5–50	2000–3000	5.2–2.4
3–20	268.0	175.2

TABLE 7 *(Continued)*

5–36 (Mn)	993–6900	3–5
3.8	727.8	159.5

Mn: number average molecular weight; PPC: Poly (propylene carbonate)

1.4.1 STARCH AND DERIVATIVES

Starch is a natural polysaccharide that can be derived from inexpensive and renewable resources. Starch films and coatings are primarily used for food packaging. Natural and modified starch films are also used to change the physical properties of food products such as soups and meat products by modifying the texture, viscosity, adhesion, moisture retention, and gel formation (Thomas and Atwell, 1997). Starch molecules are composed of two macromolecules namely amylose and amylopectin. Amylose has excellent film-formability, and forms odorless, tasteless, and colorless films. The relative amount of amylose and amylopectin depends on the plant source and is a key factor in determining the properties of starch. Generally starch contains 20–25% amylose and 75–80% amylopectin (Jiménez et al., 2012).

Starch can be converted into a thermoplastic material through the application of thermal and mechanical energy in an extruder (Kumar et al., 2011). Starch is a semi-crystalline biopolymer, so thermo-mechanical and gas barrier properties are influenced by both amorphous and crystalline regions (Liu, 2005). Crystallinity, mechanical properties, and barrier properties are dependent on the gelatinization and drying techniques. Lower gelatinization and drying rates render films with the high tensile stress, elastic modulus, and crystallinity. On the other hand, faster gelatinization and drying lead to poor mechanical and water vapor barrier properties (Flores et al., 2007). Although, starch films have low oxygen permeability and low cost compared to non-starch films, they exhibit several drawbacks such as hydrophilic character and poor mechanical properties (Jiménez et al., 2012). Films with adequate mechanical strength can be obtained by the addition of appropriate plasticizer (Campos et al., 2011). Mechanical strength of starch films can also be improved by incorporation of reinforcement agents such as clays or cellulosic fibers (Curvelo et al., 2001; Müller et al., 2009). Addition of lipophilic materials increases the hydrophobicity of starch films and hence can significantly improve the water-vapor barrier properties (García et al., 2000).

1.4.2 CELLULOSE AND DERIVATIVES

Cellulose is the most abundant natural polymer, and is a key structural component of plant cell walls. The main source of cellulose production is wood, which contains 40–50% cellulose by weight (Tang et al., 2012). It is a linear polysaccharide having a hydroglucose as thefundamental repeat unit (Cha and Chinnan, 2004). Cellulose is crystalline, infusible and insoluble in water, and most organic solvents, which makes it unsuitable for film production (Chandra and Rustgi, 1998). Therefore, cellulose is converted to cellophane by dissolving insodium hydroxide and carbon disulphide mixture and extruding into a bath of sulfuric acid (Tang et al., 2012). Cellophane has good mechanical strength, superior oil barrier property, and good gas barrier properties at low relative humidity, but is moisture sensitive. To improve moisture barrier properties cellophane is generally coated with nitrocellulose wax or PVDC (Kumar et al., 2011). Other than cellophane, cellulose can be converted to various derivatives such as cellulose acetate, ethyl cellulose, hydroxyl-ethyl cellulose, and hydroxyl-propyl cellulose, which are commercially available. Films cast from cellulose derivatives are tough, flexible, totally transparent, and highly sensitive to presence of water, but resistant to fats and oils (Campos et al., 2011).

1.4.3 CHITIN AND CHITOSAN

Chitin is second most abundant natural biopolymer (behind cellulose) occurring primarily as a component in the exoskeletons (shells) of insects and crustaceans (such as crab, shrimp, crawfish and so on). Chitin also occurs naturally in fungal cell walls and can be produced by cultivation of fungi (Teng et al., 2001, Vroman and Tighzert, 2009). Chitin can be recovered from crustacean wastes by simple demineralization and deproteinization processing steps (Knorr, 1984). Chitosan, a derivative of chitin, is obtained by deacetylation of chitin. Due to its rigid and crystalline structure and strong intra and intermolecular hydrogen bonding, chitosan is insoluble in water and alkaline medium (Vroman and Tighzert, 2009). Chitosan is biodegradable, biocompatible, and non-toxic. Clear, tough, flexible films having good oxygen barrier properties can be cast from chitosan without the use of any additives (Cutter, 2006). In addition, chitosan shows antimicrobial activity against bacteria, yeasts, and mold, making it an attractive material for food packaging films (Vartiainen et al., 2004). The major drawbacks of chitosan films are low stability and poor water vapor barrier property (Tang et al., 2012; Butler et al., 1996).

1.4.4 PROTEINS

Proteins are biopolymers comprised of long chains of amino acids. The barrier properties of proteins are determined by the polar characteristics of protein films. Protein films have high permeability to polar substances such as water and low permeability to nonpolar substances such as oxygen, oils and several aroma compounds (Kumar et al., 2011). Their moisture sensitivity can be reduced by blending protein with other bio-based or synthetic materials (Morillon et al., 2002). Flexibility and extensibility of protein films can be improved by the use of plasticizers (Gao et al., 2006). Casein and gluten are two major protein based materials, which exhibit properties favorable for food packaging. Due to its excellent mechanical and barrier properties casein has been used in food packaging applications (Kumar et al., 2011). Gluten, commonly found in wheat and corn, is an excellent film forming agent. However, the films formed by gluten are brittle in nature and require addition of plasticizers for mechanical stability (Attenburrow et al., 1990). Wheat gluten films show humidity dependent water vapor and gas permeability, so gas and water vapor composition can be optimized by controlling the humidity level in packages (Gontard et al., 1996). Corn zein, soy protein, whey protein, peanut protein, collagen and gelatin are some of the other protein materials which have been studied and utilized for packaging application. Corn zein possesses good oxygen barrier property and excellent WVP (about 800 times higher than that of a typical shrink-wrap film (Cha and Chinnan, 2004)). It is commercially used in coating preparation for shelled nuts, candy and pharmaceutical tablets.

1.4.5 POLYLACTIC ACID (PLA)

Due to abundantly available feedstock and low cost, poly lactic acid (PLA) is one of the most promising bio-based polymers. PLA is obtained by the controlled polymerization of lactic acid monomers which in turn are obtained from renewable resources such as sugar feedstock, wheat, maize, corn, and waste products from food or agriculture industry by fermentation (Siracusa et al., 2008). Properties of PLA vary according to the *L* - to - *D* lactylenantiomeric ratio. Table 8 lists some important properties of PLA.

TABLE 8 Characteristic Properties of PLA (adapted from Auras et al., 2006, Jamshidian et al., 2010)

Property	Experimental value
Glass transition temperature (Tg °C)	62.1 ± 0.7
Melting temperature (Tm °C)	150.2 ± 0.5
Tensile Strength (MPa)	48–53
Oxygen transmission rate (OTR) at 23°C and 0% relative humidity (RH) (cc / (m2·day))†	56.33 ± 0.12
Oxygen permeability coefficient (OPC) at 23°C and 0% RH (kg·m/(m^2s·Pa))†	4.33 x 10 – 18 ± 1.00 x 10 – 19
Water vapor transmission rate (WVTR) at 37.8°C and 100% RH (g / (m^2·day)) †	15.30 ± 0.04
Water vapor permeability coefficient (WVPC) at 37.8 °C and 100% RH (kg·m / (m^2s·Pa)) †	1.34 x 10 – 14 ± 3.61 x 10 – 17

† Thickness of PLA film for transmission and permeability test: 20.0 ± 0.2 mm

The PLA has good mechanical properties, high thermal plasticity, fabric ability, and biocompatibility (Ahmed and Varshney, 2011). The PLA exhibits tensile strength comparable to other commercially available polymers (Krishnamurthy et al., 2004). The PLA has higher melting temperature (T_m) and moderate glass transition temperature (T_g) compared to other biodegradable plastics. Incorporation of nano-fillers in PLA matrix has been observed to improve the thermo-mechanical and barrier properties of PLA (Sinha et al., 2002). Thus, PLA based nanocomposites are promising biodegradable materials for application in packaging industry.

1.4.6 POLYHYDROXYALKANOATES (PHAS)

The PHA is produced by bacterial fermentation of sugar or lipids. Polymers in PHA family can be produced by 150 different monomers, due which a large variety of PHAs can be synthesized with a wide range of properties. Polyhydroxybutyrate (PHB) is the most common biopolymer in PHA family. PHB is water insoluble and has good resistance to hydrolytic degradation. The T_m and T_g of PHB are approximately 175°C and 15°C respectively (Kumar et al., 2011). PHB exhibits low water permeability which is comparable to that of LDPE, and has similar thermal and mechanical properties as isotactic PP (Savenkova et al., 2000).

1.5 POLYMER NANOCOMPOSITES IN PACKAGING APPLICATIONS

Synthetic polymers generally possess good thermal and mechanical properties for packaging applications. However, several such polymers exhibit inherent permeability to vapors and gases such as oxygen and carbon dioxide making their use in certain food packaging applications undesirable. Further, majority of biopolymers exhibit high WVP. Polymer nanocomposites have been observed to improve the gas barrier properties of several synthetic and bio-based polymers (Arora and Padua, 2010) and hence, are of importance as food packaging materials. Biopolymer nanocomposites are especially important as they provide a route to enhance the often poor thermo-mechanical properties of biopolymers. Nanocomposites are defined on the basis of the dimensions of fillers used. At least one dimension of the fillers should be smaller than 100 nm for a composite to be considered a nanocomposite.

Nanocomposites can display improvement in barrier properties, mechanical strength, thermal stability, durability, chemical stability, flame retardancy, biodegradability, optical, and electrical properties over neat polymers and conventional composites (Sinha and Okamoto 2003, Ray and Bousima, 2005; Ma et al., 2006; Sorrentino et al., 2007). Commonly used fillers to prepare nanocomposites for food packaging applications are montmorillonite (MMT) clays, kaolinite, carbon nanotube, and graphene nanosheet (Arora and Padua, 2010). Small percentage of nanoclay (such as MMT) incorporation into the polymer matrix can lead to significant improvement in elastic modulus, thermal and dimensional stability, and barrier properties to gases and vapors (Lee et al., 2005; Di, et al., 2005; Sanchez-Garcia et al., 2010a). Nano-clay creates a tortuous path for diffusion of gas molecules which generally reduces permeability and improves barrier properties (Pavlidou and Papaspyrides, 2008). Polymer nanocomposites with carbon nanotubes as fillers show improvement in mechanical properties, thermal properties, and electrical conductivity over unfilled polymer. Addition of carbon nanotubes to biopolymers has also been observed to enhance the biodegradation rate (Chen and Wu, 2007; Saeed and Park, 2007). Further, Sanchez-Garcia et al. (2010) have reported that carbon nanotubes can increase the gas and water vapor barrier properties of biopolymers.

Nanocomposites with natural fibers as filler material have also been investigated by researchers. Various sources of natural fibers have been reported in literature such as wood, cotton, bagasse, rice straw, rice husk, wheat straw, flax, hemp, pineapple leaf, coir, oil palm, date palm, doumfruit, ramie, curaua, jowar, kenaf, bamboo, rapeseed waste, sisal, and jute (Majeed et al., 2013). Advantages of natural fibers include light weight, low cost, and biodegradability. Polymer nanocomposites incorporating cellulose nanofibers are relevant to food packaging applications. Two different nanofillers, namely microfibrils and whiskers, can be obtained from cellulose. Microfibrils have diameter in the nanometer range and length in the micrometer range (Samir et al, 2005). The crystalline part of cellulose can be isolated by digestion of amorphous domains of cellulose using methods such as acid hydrolysis. The elongated crystalline cellulosic nanoparticles thus obtained are referred to as cellulose whiskers, and include cellulose nanocrystals (CNC) and nanorods (Dujardin et al., 2003). Siqueira et al.

(2009) studied the effect of using both cellulose microfibrils and cellulose whiskers as nanofillers on the properties of polycaprolactone. Cellulose whiskers were observed to enhance the thermal properties and crystallinity of the polymer whereas chemically functionalized cellulose microfibrils improved the stiffness of the polymer.

1.6 POLYMER PROPERTIES FOR SPECIALIZED PACKAGINGAPPLICATIONS

Specialized food packaging techniques such as active packaging and modified atmosphere packaging are active areas of research presently. Apart from the usual properties expected of good packaging materials, these specialized applications may require certain other unique set of properties.

1.6.1 MODIFIED ATMOSPHERE, CONTROLLED ATMOSPHERE, AND VACUUM PACKAGING

Modified atmosphere, controlled atmosphere, and vacuum packaging are used to maintain conditions which inhibit the growth of microbes and enhance shelf-life of food. Brief description of these packaging techniques is provided next.

MODIFIED ATMOSPHERE PACKAGING (MAP)

In modified atmosphere packaging (MAP) concentrations of atmospheric gases are altered in the food packages to slow down microbial growth, enzymatic deterioration, and respiration of foods. The atmospheric gases whose concentrations are altered are carbon dioxide, oxygen, and nitrogen. Compared to air, lower oxygen concentration and higher carbon dioxide and nitrogen concentrations are maintained in MAP. Concentrations of gases are altered on the basis of dynamic interaction between product, environment and package. Once the concentrations are altered, compositions of gases are not controlled and the composition of different gases will vary inevitably. Nitrogen is used not only to avoid oxidation of fats present in food but also to prevent the package from collapsing. Carbon dioxide is used to suppress the growth of microbes present in food whereas oxygen may be used to slow down anaerobic growth (Yam and Lee, 1995; Sivertsvik et al., 2002; Cutter, 2002; Rao and Sachindra, 2002; McMillin, 2008; Thompson, 2010).

VACUUM PACKAGING

In vacuum packaging air is evacuated from the package without being replaced with any other gas. Thus vacuum packaging requires low air permeability films to maintain the vacuum inside the packages. Vacuum packaging is especially used to create barrier for atmospheric oxygen, which results in prevention of enzymatic reaction and bacterial spoilage (Cánovas, 2003). Due to respiration of packaged food, the trapped oxygen during vacuum packaging will convert to carbon dioxide, which will inhibit the growth of microbes. Flexible packaging materials are suitable for vacuum packaging, as the pressure difference inside and outside the package may cause the package to collapse if rigid packaging material is used. One of the problems associated with vacuum packaging is change in the color of meat due to unavailability of oxygen (Blakistone, 1998). Another problem is that the absence of oxygen may provide favorable conditions for the growth of anaerobic organisms (Paine et al., 1992).

CONTROLLED ATMOSPHERE PACKAGING (CAP)

In controlled atmosphere packaging (CAP), the gaseous environment over a food product is constantly altered to meet the food shelf life demands. The altered gaseous composition is dynamically maintained throughout storage and distribution (Cutter, 2002). In contrast, modified atmosphere packaging involves only an initial alteration of gaseous composition at the time of packaging with no further alterations thereafter. In CAP, continuous monitoring and control of gas compositions, temperature and moisture content are required within the package (Paine et al., 1992). To keep the environment unaltered inside the package, the package should have high barrier properties (Hirsch, 1991).

1.6.2 ACTIVE PACKAGING

Active packaging, or intelligent packaging, are systems that dynamically and invariably alter either permeation properties of the package or concentration of different gases present in the headspace of package during storage. Moreover, such packaging may actively add antimicrobials, antioxidants, or other quality improving agents *via* packaging materials into the packed food in desired amounts during storage. The role of active packaging is to monitor the condition and quality of packaged food during storage and distribution, and improve the shelf life of the packed food item. Table 9 lists a set of active packaging techniques along with the reagents added to achieve the desired effect and representative areas of application. Active packaging may contain several external or internal indicators to actively monitor the state and quality of packaged product. Oxygen indicators, carbon dioxide indicators and time-temperature indicators are used to assess and monitor quality of the product inside. In addition, active packaging may also incorporate leakage indicators, ripening indicators, and freshness indicators (Ahvenainen and Hurme, 1997; Kruijf et al., 2002; Pereira de Abreu et al., 2012). Indicators for active packaging should be easy to understand, should correlate accurately with product quality, and should have reasonable cost.

TABLE 9 Active Packaging Techniques (adapted from Ahvenainen and Hurme, 1997)

Functionality	Reagent	Application
Moisture regulation	Potassium chloride, sodium chloride	Vegetables
Moisture adsorption	Glycerol, clay, silica gel	Meat
O_2 absorption	Metallic and organometallic compounds, glucose oxide, ascorbic acid	No restrictions
CO_2 emission	Ferrous carbonate + metal halide	Vegetables
CO_2 absorption	Calcium hydroxide + sodium/potassium hydroxide	Roasted coffee

Active packaging is designed considering all factors which determine the shelf life of packaged food, such as physiological processes, chemical processes, physical processes, and microbiological aspects (Kruijf et al., 2002). These factors can be controlled using appropriate active packaging system. Active packaging systems can be classified into different categories based on the role they play in preventing deterioration in quality and shelf life of food products. Brief description of these categories is provided next.

OXYGEN SCAVENGERS

The presence of oxygen may have negative effect on the packaged food, making its presence in packaged food an important concern. The 90–95% of the oxygen is removed using vacuum packaging or modified atmosphere packaging technology. Remaining oxygen is controlled using oxygen absorbers to limit the deterioration of packaged food (Pereira de Abreu et al., 2012). Oxygen scavengers are used to actively control the level of residual oxygen inside food packages. Iron powder oxidation, ascorbic acid oxidation, photosensitive dye oxidation, enzymatic oxidation, ferrous salts, unsaturated fatty acids, and combinations of these are typically used for oxygen scavenging (Floros et al., 1997). The advantage of using oxygen scavengers is that the concentration of residual oxygen in the package headspace may be reduced down to 0.01% (Ahvenainen and Hurme, 1997).

ETHYLENE SCAVENGERS

After harvesting, most fruits and vegetables release ethylene, which causes ripening and senescence by increasing the respiration rate. In leafy vegetables, ethylene causes degradation of chlorophyll (Pereira de Abreu et al., 2012). Thus, it is important to control the presence of ethylene in food packages to improve the shelf life of food products. Silica embedded potassium permanganate is the most common ethylene absorbing agent in use currently. Silica absorbs ethylene which in turn is converted to ethylene glycol by potassium permanganate. However, due to its toxic nature potassium permanganate is not integrated with food packaging films (Rubio et al., 2004). Potassium permanganate impregnated zeolite is also used as ethylene absorber after coating with quaternary ammonium ions (Ozdemir and Floros, 2004). Another effective ethylene absorber is activated carbon impregnated with bromine. However, this absorber is not used in food packaging due to release of toxic bromine gas upon reaction between bromine compound and water (Kruijf et al., 2002).

MOISTURE SCAVENGERS

Water is one of the requirements for microbial growth. Hence, to suppress the growth rate, it is important to control excess moisture present in food packages. Moisture scavengers such as silica gel, molecular sieves, natural clays, calcium oxide, calcium chloride, activated carbon, glycerol, and modified starch are commonly used for this purpose (Ozdemir and Floros, 2004; Randenburg, 2009). Moisture scavengers used in food packaging should maintain their properties and should not affect the properties of packaging plastic material as well as the quality of food products (Rubio et al., 2004).

CARBON DIOXIDE (CO_2) SCAVENGERS AND EMITTERS

Carbon dioxide is used to inhibit microbial growth and delay the ripening rate of fruits and vegetables. Due to higher permeability of CO_2 than oxygen through many of the plastic films used in packaging, most of the CO_2 will permeate through the packages (Kerry et al., 2006). To maintain the CO_2 level, a dual function system is generally used which works as oxygen scavenger and CO_2 emitter (Munro et al., 2009).

Ferrous carbonate and a metal halide catalyst are often used for this purpose. The CO_2 scavengers are required for food products such as fresh roasted or ground coffees, which generate significant amount of CO_2. If such products are directly packed to prevent transmission of oxygen and moisture, released CO_2 will build up in the package causing it to burst. A mixture of calcium oxide and activated charcoal is generally used for CO_2 scavenging. In several cases, a dual function system consisting of an oxygen scavenger and a CO_2 scavenger is preferred (Day, 2008). This dual function scavenging system typically contains iron powder for oxygen scavenging and calcium hydroxide for CO_2 scavenging.

FLAVOR AND ODOR ABSORBER AND RELEASER

Presence of aroma components and deterioration reaction during handling and storage may generate undesirable odor and cause loss of flavor of the food products (Rubio et al., 2004). In food packaging, flavor, and odor absorbers are used to remove undesirable volatile gaseous molecules, chemical metabolites of foods, respiration products, or off-flavors in raw food (Rooney, 2005). On the other hand, flavor and odor releaser are added to either make the food more desirable to consumers or improve the flavor of food (Pereira de Abreu et al., 2012).

ANTIMICROBIAL PACKAGING

It is a type of active packaging in which antimicrobial agents are incorporated into the packaging system. The package interacts with food product to prevent microbial growth and extend the shelf-life of food. This system can be divided into two types (Munro et al., 2009):

- Antimicrobial agent on the film migrates to the surface of the food,
- Packaging film achieves antimicrobial effect without migration of any antimicrobial agent to the food.

These two types of antimicrobial packaging systems can be implemented in several ways such as addition of sachets containing antimicrobial agents into packages, incorporation of antimicrobial agents directly into polymers, coating or adsorbing antimicrobials onto polymer surfaces and immobilization of antimicrobials to polymers by ion or covalent linkages (Appendini and Hotchkiss, 2002).

The antimicrobial agents are selected based on their range of activity, mode of action, chemical constitution, and rate of growth of targeted microorganisms (Kenawy et al, 2007). Different classes of antimicrobial agents commonly employed are organic acids (and their salts), acid anhydrides, parabens, chlorides, phosphates, epoxides, sulphites, nitrites, alcohols, ozone, hydrogen peroxide, diethyl pyrocarbonate, bacteriocins, chelators, enzymes, and polysaccharides (Hogan and Kerry, 2008, Ozdemir and Floros, 2004).

Certain polymers are inherently antimicrobial and do not require incorporation of an external antimicrobial agent into the packaging films. Chitosan is one of such polymer, which is cationic in nature and promotes cell adhesion due to presence of charged amines (Kim et al., 2008). Charged amine group in chitosan interacts with negative charges on the cell membrane of microbes causing leakage of intracellular constituents. Hence, chitosan has been used to protect fresh vegetables and fruits from fungal degradation. Poly-*L*-lysine, bactericidal acrylic polymers and polymers containing biguanide substituent also show antimicrobial activity (Appendini and Hotchkiss, 2002).

1.7 FOOD-PACKAGING INTERACTIONS: MIGRATION OF CONTAMINANTS

Additives and fillers used during fabrication of polymeric food packaging material can migrate to foodstuff thereby contaminating and degrading the quality of packaged food. Packaging plastics can release small amounts of chemicals upon coming in contact with foodstuff. If the released chemicals are harmful to humans, migration of such species into food items can raise health concerns. Therefore, knowledge of food-packaging interaction, in terms of the species that can migrate from the packaging to the foodstuff, is essential for assurance of food quality. For example, polycarbonates and phenolic-epoxy resins, used for making plastic packaging containers and as food-can coatings respectively, may release bisphenol A into food which can have adverse health effects. Knowledge of the level at which bisphenol A is released under different conditions to different types of food products is therefore essential to ensure that the intake is under tolerable limits. Several migration studies of contaminants to foodstuff (or food simulants) are reported in literature such as bisphenol A from polycarbonate bottles (Biles et al., 1997) and from epoxy coating on tuna fish cans (Munguía-López, 2005), perfluorochemicals from Teflon and fluorotelomer coated paper bags (Begley, 2005), antioxidants from polypropylene films (Garde, 2001), and so on. Gallart-Ayala et al. (2013) have analyzed several contaminants that can migrate into food from packaging plastics and compiled an extensive list of such contaminants along with typical levels observed in different categories of food. Migration studies on food packaging materials enable the quantitative determination of the amount of undesirable contaminants that can migrate into food brought in contact with the packaging. High performance liquid chromatography (HPLC), gas chromatography-mass spectrometry (GC-MS) and inductively coupled plasma-mass spectrometry (ICP-MS) are the most important analytical equipment used in migration studies. Important parameters that influence the rate and extent of migration of contaminants into food are storage temperature and contact time between packaging and food. Extent of migration of a species also depends on the type of foodstuff in contact: certain species migrate more readily into aqueous food while others show preference towards fatty food. Migration of chemicals from polymer matrix to food is governed not only by kinetic parameters such as diffusivity of the migrant (through polymer or food) but also by thermodynamic equilibrium expressed by the partition coefficient of the migrant between the polymer and the foodstuff. Taking these theoretical considerations into account, models have been proposed for migration of species from polymeric packaging materials to foodstuff. Brandsch et al. (2002) have summarized migration modeling for polyolefin packaging materials (LDPE, HDPE, and PP).

Fasano et al. (2012) studied the migration of phthalates, bisphenol A, alkylphenols, and so on, from common plastic packaging materials such as epoxy-coated food cans, LDPE film, tetrapack, HDPE, polystyrene dish, polycarbonate bottles, and so on, to food simulants representing aqueous food, acidic aqueous food, alcoholic food, and fatty food. They detected the release of several chemical contaminants from most of the packaging material studied. Although the levels detected were lower than the specific migration limit (SML) and overall migration limit (OML) established by the European Union (EU), the authors recommend regular monitoring of levels in frequently ingested packaged foods to ensure low human exposure. Adhesives used to glue together films in multilayer packaging materials can also be a source of potential migrants and other non-intentionally added substances (NIAS). For example, Felix et al. (2012) studied the migration of contaminants from polyurethane adhesives and plastics films (made of PET, PP, PE, polyamide, and so on) to food simulants such as Tenax® and isooctane. They identified more than 63 compounds in the adhesive layer that could be considered as potential migrants.

Polymer nanocomposites contain fillers having characteristic size in the range of a few nanometers. If packaging materials prepared from polymer nanocomposites are brought in contact with food, the nano-filler can migrate from the packaging to the foodstuff. Avella et al. (2005) conducted migration studies on starch/clay nanocomposite films and concluded that nanoclay migration to vegetables and food simulants was within the limits prescribed by EU regulations. Schmidt et al. (2011) performed migration studies on nanocomposite films of PLA and nanosized layered double hydroxide (LDH) platelets using a fatty food simulant. Total migration and specific migration of LDH and laurateorgano modifier (used to improve LDH dispersion in PLA) were found to increase with increased LDH loading in the nanocomposite films. Currently, literature on migration studies of nanoparticles/nanofillers from polymer nanocomposites to food products is limited and hence, more extensive migration studies on polymer nanocomposite packaging materials are required.

1.8 CONCLUSION

Thermal, mechanical, and barrier properties of polymers are critical in determining their suitability as packaging materials. Properties of packaging plastics can vary significantly depending on the field of application. For example, several food packaging applications require polymers having good resistance to oxygen and moisture transmission, that is, good barrier properties. On the other hand, applications such as automobile parts packaging require polymers having excellent mechanical properties. In this chapter a general discussion of polymer properties has been provided in the context of packaging applications. Properties and applications of specific plastics commonly used as packaging materials have also been discussed. Emerging fields of bioplastics and polymer nanocomposites as packaging materials have been highlighted. A brief overview of specialized packaging applications such as active packaging and modified atmosphere packaging, with emphasis on properties required for different functionality, has been provided. In conclusion, knowledge and understanding of polymer properties that dictate their thermo-mechanical behavior and resistance to gas transmission is essential in deciding which polymer is best suited for a particular packaging application.

KEYWORDS

- **Cellulose**
- **Controlled Atmosphere Packaging**
- **Crystallinity**
- **Morphology**
- **Proteins**

REFERENCES

1. Ahmed, J. and Varshney, S. K. *International Journal of Food Properties,* **14**, 37–58 (2011).
2. Ahvenainen, R. and Hurme, E. *Food Additives and Contaminants,* **14**, 753–763 (1997).
3. Appendini, P. and Hotchkiss, J. H. *Innovative Food Science & Emerging Technologies,* **3**, 113–126 (2002).
4. Arora, A. and Padua, G. W. *Journal of Food Science,* **75**, R43–R49 (2010).
5. Attenburrow, G., Barnes, D. J., Davies, A. P., and Ingman, S. J. *Journal of Cereal Science,* **12**, 1–14 (1990).
6. Auras, R., Singh, S. P., and Singh, J. *Journal of Testing and Evaluation,* **34**, 1–7 (2006).
7. Avella, M., de Vlieger, J. J., Errico, M. E., Fischer, S., Vacca, P., and Volpe, M. G. *Food Chemistry,* **93**, 467–474 (2005).
8. Begley, T. H., White, K., Honigfort, P., Twaroski, M. L., Neches, R., and Walker, R. A. *Food Additives and Contaminants,* **22**, 1023–1031 (2005).
9. Bicerano, J. *Prediction of Polymer Properties*, Marcel Dekker, New York (2002).
1. Biles, J. E., McNeal, T. P., Begley, T. H., and Hollifield H. C. *Journal of Agricultural and Food Chemistry,* **45**, 3541–3544 (1997).
2. Blakistone, B. A. *Principles and Applications of Modified Atmosphere Packaging of Foods*, Chapman & Hall, New York (1998).
3. Bondi, A. *Physical Properties of Molecular Crystals, Liquids and Glasses*, Wiley, New York (1968).
4. Bower, D. I. *An Introduction to Polymer Physics*, Cambridge University Press, New York (2002).
5. Brady, G. S., Clauser, H. R., and Vaccari, J. A. Materials Handbook, The McGraw Hill Companies Inc., New York (2002).
6. Brandsch, J., Mercea, P., Rüter, M., Tosa, V., and Piringer, O. *Food Additives and Contaminants,* **19**, 29–41 (2002).
7. Brandsch, J. and Piringer, O. G. *Characteristics of Plastic Materials*. In Plastic Packaging: Interactions with Food and Pharmaceuticals, O. G. Piringer and A. L. Baner (Eds.), Wiley-VCH Verlag GmbH & Co. KGaA: Weinheim, pp. 15–60 (2008).
8. Brydson, J. A. *Plastics Materials*, Butterworth-Heinemann Linacre House, Oxford (1999).
9. Butler, B. L., Vergano, P. J., Testin, J. M., Bunn, J. M., and Wiles J. L. *Journal of Food Science,* **61**, 953–955 (1996).
10. Callister, W. D. and Rethwisch, D. G. *Materials Science and Engineering: An Introduction*, John Wiley & Sons Inc., Hoboken, New Jersey (2010).
11. Campos, C. A., Gerschenson, L. N., and Flores, S. K. *Food Bioprocess Technol*ogy, **4**, 849–875 (2011).
12. Cánovas, G. V. B., Molina, J. J. F., Alzamora, S. M., Tapia, M. S., Malo, A. L., and Chanes, J. W. *Handling and Preservation of Fruits and Vegetables* by Combined Methods for Rural Areas, Food & Agriculture Organization of the United State, Rome (2003).
13. Carraher, C. E. *Seymour/Carraher's Polymer Chemistry*, Taylor & Francis Group, CRC Press, Boca Raton (2008).

14. Carter, S. J. Polyethylene, High Density. In The Wiley Encyclopedia of Packaging Technology, K. L. Yam (Ed.), John Wiley & Sons Inc., Hoboken, New Jersey, pp. 979–982 (2009).
15. Carvalho, A. J. F., Curvelo, A. A. S., and Agnelli, J. A. M. Carbohydrate Polymers, **45**, 189–194 (2001).
16. Cha, D. S. and Chinnan, M. S. *Critical Reviews in Food Science and Nutrition,* **44**, 223–237 (2004).
17. Chanda M. and Roy S. K. *Industrial Polymers, Specialty Polymers, and Their Applications*, CRC Press, Taylor & Francis Group: Boca Raton (2009).
18. Chandra, R. and Rustgi, R. Biodegradable polymers. Progress in Polymer Science, **23**, 1273–1335 (1998).
19. Chen, E. C. and Wu, T. M. *Polymer Degradation and Stability*, **92**, 1009–1015 (2007).
20. Chen, G. Q. and Patel, M. K. Chemical Reviews, **112**, 2082–2099 (2012).
21. Cutter, C. N. *Critical Reviews in Food Science and Nutrition*, **42** (2), 151–161 (2002).
22. Cutter, C. N. Meat Science, **74**, 131–142 (2006).
23. Daniels, C. A. *Physical Properties and Characterization of PVC*. In PVC handbook, C. E. Wilkes, J. W. Summers and C. A. Daniels (Eds.), Carl Hanser Verlag: Munich (2005).
24. Day, B. P. F. *Active Packaging of Food*. Smart Packaging Technologies for Fast Moving Consumer Goods, P. Butler and J. Kerry. (Eds.), John Wiley & Sons Ltd, West Sussex, pp. 1–18 (2008).
25. Debeaufort, F., Quezada-Gallo, J. A., and Voilley, A. *Critical Reviews in Food Science and Nutrition*, **38**, 299–313 (1998).
26. Delpouve, N., Stoclet, G., Saiter, A., Dargent, E., and Marais, S. *Journal of Physical Chemistry B*, **116**, 4615–4625 (2012).
27. Di, Y. W., Iannac, S., Sanguigno, L., and Nicolais, L. *Macromolecular Symposia,* **228**, 115–124 (2005).
28. Dujardin, E., Blaseby, M., and Mann, S. *Journal of Materials Chemistry,* **13**, 696–699 (2003).
29. Fasano, E., Bono-Blay, F., Cirillo, T., Montuori, P., and Lacorte, S. *Food Control,* **27**, 132–138 (2012).
30. Felix, J. S., Isella, F., Bosetti, O., and Nerin, C. *Analytical and Bioanalytical Chemistry,* **403**, 2869–2882 (2012).
31. Firdaus, V. and Tong P. P. *Polyethylene, Linear and Very Low-Density*. In The Wiley Encyclopedia of Packaging Technology, K. L. Yam (Ed.), John Wiley & Sons Inc., Hoboken, New Jersey, pp. 983–987 (2009).
32. Flaconnèche, B., Martin, J., and Klopffer, M. H. *Oil & Gas Science and Technology*, **56**, 261–278 (2001).
33. Flores, S., Famá, L., Rojas, A. M., Goyanes, S., and Gerschenson, L. N. *Food Research International*, **40**, 257–265 (2007).
34. Floros, J. D., Dock, L. L., and Han, J. H. *Food Cosmetics and Drug Packaging,* **20**, 10–17 (1997).
35. Fox, T. G. Bulletin of the American Physical Society, **1**, 123–125 (1956).
36. Fox, T. G. and Flory, P. J. *Journal of Applied Physics,* **21**, 581–591 (1950).
37. Gao, C., Stading, M., Wellner, N., Parker, M. L., Noel, T. R., Mills, E. N. C., and Belton, P. S. *Journal of Agricultural and Food Chemistry,* **54**, 4611–4616 (2006).
38. García, M. A., Martino, M. N., and Zaritzky, N. E. *Journal of Food Science,* **65**, 941–947 (2000).
39. Garde, J. A., Catalá, R., Gavara, R., and Hernandez, R. J. *Food Additives and Contaminants*, **18**, 750–762 (2001).
40. Gontard, N., Thibault, R., Cuq, B., and Guilbert, S. *Journal of Agricultural and Food Chemistry*, **44**, 1064–1069 (1996).
41. Gordon, M. and Taylor, J. S. *Journal of Applied Chemistry*, **2**, 493–500 (1952).
42. Graff, G. L., Williford, R. E., and Burrows, P. E. *Journal of Applied Physics,* **96**, 1840–1849 (2004).
43. Hirsch, A. *Flexible Food Packaging: Questions and Answers*, Van Nostrand Reinhold, New York (1991).

44. Hogan, S. A. and Kerry, J. P. *Smart Packaging of Meat and Poultry Products*, Smart Packaging Technologies for Fast Moving Consumer Goods, P. Butler and J. Kerry (Eds.) John Wiley & Sons Ltd, West Sussex, pp. 33–60 (2008).
45. Hough, M. C. and Dolbey, R. The Plastics Compendium: Key Properties and Sources, Rapra Technology Ltd.: Shropshire (1995).
46. Hylton, D. C. *Understanding Plastic Testing*, Carl Hanser Verlag: Munich (2004).
47. Illig, A. *Thermoforming A Practical Guide*, Carl Hanser Verlag: Munich (2001).
48. Jamshidian, M., Tehrany, E. A., Imran, M., Jacquot, M., and Desobry, S. *Comprehensive Reviews in Food Science and Food Safety*, **9**, 552–571 (2010).
49. Jang, C., Cho, Y.-R. and Han, B. *Applied Physics Letters,* **93**, 133–307 (2008).
50. Jiang, K., Yu, F., Su, R., Yang, J., Zhou, T., Gao, J., Deng, H., Wang, K., Zhang, Q., Chen, F., and Fu, Q. *Chinese Journal of Polymer Science,* **29**, 456–464 (2011).
51. Jiménez, A., Fabra, M. J., Talens, P., and Chiralt, A. *Food and Bioprocess Technology,* **5**, 2058–2076 (2012).
52. Johnson, B. *Testing, Permeation and Leakage*. In The Wiley Encyclopedia of Packaging Technology, K. L. Yam (Ed.) John Wiley & Sons Inc., Hoboken, New Jersey, pp. 1207–1213 (2009).
53. Kenawy, E. R., Worley, S. D., and Broughton, R. Biomacromolecules, **8**, 1359–1384 (2007).
54. Kerry, J. P., O'Grady, M. N., and Hogan, S. A. Meat Science, **74**, 113–130 (2006).
55. Kim, Y. T., Kim, K., Han, J. H., and Kimmel, R. M. *Antimicrobial Active Packaging for Food.* Smart Packaging Technologies for Fast Moving Consumer Goods, P. Butler and J. Kerry (Eds.) John Wiley & Sons Ltd, West Sussex, pp. 99–110 (2008).
56. Kirwan, M. J. and Strawbridge, J. W. *Plastics in Food Packaging.* In Food Packaging Technology, R. Coles, D. Mcdowell, and M. J. Kiewan (Eds.) Blackwell Publishing, CRC Press: Boca Raton, pp. 174–240 (2003).
57. Knorr, D. *Food Technology*, 38, 85–97 (1984).
58. Krishnamurthy, K., Demirci, A., Puri, V., and Cutter, C. N. *Transactions of the American Society of Agricultural Engineers,* **47**, 1141–1149 (2004).
59. Kruijf, N. D., Beesty, M. V., Rijk, R., Malm, T. S., Losada, P. P., and Meulenaer, B. D. *Food Additives and Contaminants*, **19**, 144–162 (2002).
60. Kumar, A. and Gupta, R. K. *Fundamentals of Polymer Engineering.* Marcel Dekker: New-York (2003).
61. Kumar, M. N. S. and Yaakob, Z. *Siddaramaiah Biobased Materials in Food Packaging Applications*, *Handbook of Bioplastics and Biocomposites Engineering Applications*, S. Pilla (Ed.) Scrivener Publishing LLC., Salem, Massachusetts, pp. 121–159 (2011).
62. Leadbitter, J., Day, J. A. and Ryan, J. L. *PVC: Compounds, Processing and Applications*, Rapra Technology Ltd., Shropshire (1997).
63. Lee, J. H., Lee, Y. H., Lee, D. S., Lee, Y. K., and Nam, J. D. Polymer Korea, **29**, 375–379 (2005).
64. Lee, W. M. *Polymer Engineering and Science,* **20**, 65–69 (1980).
65. Lin, H. and Freeman, B. D. *Journal of Membrane Science*, **239**, 105–117 (2004).
66. Liu, G., Li, Y., Yan, F., Zhao, Z., Zhou, L., and Xue, Q. *Journal of Polymer and the Environment*, **13**, 339–348 (2005).
67. Liu, Z. *Edible films and coatings from starch*, Innovations in food packaging, J. H. Han (Ed.) Elsevier Academic Press, London, pp. 318–332 (2005).
68. Ma, H., Xu, Z., Tong, L., Gu, A., and Fang, Z. *Polymer Degradation and Stability,* **91**, 2951–2959 (2006).
69. Maier, C. and Calafut, T. *Polypropylene: The Definitive User's Guide and Databook*, Plastic Design Library, Norwich (1998).
70. Majeed, K., Jawaid, M., Hassan, A., Bakar, A. A., Khalil, H. P. S. A., Salema, A. A., and Inuwa, I. *Materials and Design*, **46**, 391–410 (2013).
71. Mantia, F. L. *Handbook of Plastics Recycling*, Rapra Technology Ltd., Shropshire (2002).
72. Mark, J. E. *Physical Properties of Polymers Handbook*, Springer Science + Business Media, LLC, New York (2007).
73. Mark, J. E. *Polymer Data Handbook*, Oxford University Press, New York (1999).

74. Massa, A., Bugatti, V., Scettri, A., and Contessa, S. *Thermo-Oxidation Stability of Poly (Butylene Terephthalate) and Catalyst Composition*, In Encyclopedia of Polymer Research, C. E. Jones (Ed.), Nova Science Publishers, Inc., New York, pp. 327–342 (2011).
75. Matar, S. and Hatch, L. F. *Chemistry of Petrochemical Processes*, Gulf Publishing Company: Houston, Texas (2000).
76. McMillin, K. W. *Meat Science*, **80**, 43–65 (2008).
77. Michaels, A. S. and Bixler, H. J. *Journal of Polymer Science*. **50**, 393–412 (1961).
78. Miller, R. C. Polypropylene. *The Wiley Encyclopedia of Packaging Technology*, K. L. Yam (Ed.), John Wiley & Sons Inc.: Hoboken, New Jersey, pp. 1005–1009 (2009).
79. Mittal, V. *Barrier Properties of Polymer Clay Nanocomposites*, Nova Science Publishers, Inc., New York (2010).
80. Monasse, B. and Haudin, J. M. *Molecular Structure of Polypropylene Homo and Copolymers*. In Polypropylene: Structure, Blends and Composites, J. K. Kocsis (Ed.) Chapman & Hall: London, pp. 1–30 (1995).
81. Morillon, V., Debeaufort, F., Blond, G., Capelle, M., and Voilley, A. *Critical Reviews in Food Science and Nutrition*, **42**, 67–89 (2002).
82. Müller, C. M. O., Laurindo, J. B., and Yamashita, F. Food Hydrocolloids, **23**, 1328–1333 (2009).
83. Munguía-López, E. M., Gerardo-Lugo, S., Peralta, E., Bolumen, S., and Soto-Valdez, H. *Food Additives and Contaminants,* **22**, 892–898 (2005).
84. Munro, I. C., Haighton, L. A., Lynch, B. S., and Tafazoli, S. *Food Additives and Contaminants*, **26**, 1534–1546. (2009).
85. Munro, I. C., Haighton, L. A., Lynch, B. S., and Tafazoli, S. *Food Additives and Contaminants*, **26**, 1534–1546 (2009).
86. Ozdemir, M. and Floros, J. D. Critical Reviews in Food Science and Nutrition, **44**, 185–193 (2004).
87. Paine, F. A. and Paine, H. Y. *Handbook of Food Packaging*, Blackie Academic & Professional, an Imprint of Chapman & Hall, Glasgow (1992).
88. Park, J. Y. and Paul, D. R. *Journal of Membrane Science*, **125**, 23–39 (1997).
89. Paul, D. R. *Journal of Membrane Science*, **18**, 75–86 (1984).
90. Pavlidou, S. and Papaspyrides, C. D. *Progress in Polymer Science*, **32**, 1119–1198 (2008).
91. Pereira de Abreu, D. A., Cruz, J. M., and PaseiroLosada, P. *Food Reviews International,* **28**, 146–187 (2012).
92. Pilla, S. *Engineering Applications of Bioplastics and Biocomposites - An Overview*. Handbook of Bioplastics and Biocomposites Engineering Applications, S. Pilla (Ed.) Scrivener Publishing LLC., Salem, Massachusetts, pp. 1–15 (2011).
93. Randenburg, J. Modified Atmosphere Packaging. The Wiley Encyclopaedia of Packaging Technology, K. L. Yam (Ed.) John Wiley & Sons Inc., Hoboken, New Jersey, pp. 787–794 (2009).
94. Rao, N. D. and Sachindra, N. M. *Food reviews international*, **18** (4), 263–293 (2002).
95. Ray, S. S. and Bousima, M. *Progress in Materials Science*, **50**, 962–1079 (2005).
96. Rooney, M. L. *Introduction to Active Food Packaging Technologies*. Innovations in Food Packaging, J. H. Han (Ed.) Elsevier Academic Press, Oxford, pp. 63–79 (2005).
97. Rosato, D. V., Rosato, M. G., and Rosato, D. V. *Concise Encyclopedia of Plastics*, Kluwer Academic Publisher: Massachusetts (2000).
98. Rubio, A. L., Almenar, E., Munoz, P. H., Lagaron, J. M., Catala, R. and Gavara, R. *Food Reviews International*, **20**, 357–387 (2004).
99. Saeed, K. and Park, S. Y. *Journal of Applied Polymer Science*, **104**, 1957–1963 (2007).
100. Samir, M. A. S. A., Alloin, F., and Dufresne, A. *Biomacromolecules,* **6**, 612–626 (2005).
101. Sanchez-Garcia, M. D., Lagaron, J. M., and Hoa, S. V. *Composite Science and Technology*, **70**, 1095–1105 (2010).
102. Sanchez-Garcia, M. D., Rubio, A. L., and Lagaron, J. S. *Trends in Food Science & Technology*, **21**, 528–536 (2010a).
103. Savenkova, L., Gercberga, Z., Nikolaeva, V., Dzene, A., Bibers, I., and Kahlnin, M. *Process Biochemistry*, **35**, 573–579 (2000).

104. Selke, S. E. M., Culter, J. D. and, Hernandez, R. J. *Plastics Packaging: Properties, Processing, Applications and Regulations*, Hanser publishers: Munich (2004).
105. Shanks, R. *Technology of Polyolefin Film Production*. In Handbook of Plastic Films, E. M. Abdel-Bary (Ed.) Rapra Technology Ltd: Shropshire, pp. 5–38 (2003).
106. Silvestre, C., Duraccio, D., and Cimmino, S. *Progress in Polymer Science*, **36**, 1766–1782 (2011).
107. Simha, R. and Boyer, R. F. *Journal of Chemical Physics*, **37**, 1003–1007 (1962).
108. Sinha R. S., Maiti, P., Okamoto, M., Yamada, K., and Ueda, K. *Macromolecules*, **35**, 3104–3110 (2002).
109. Sinha, R. S. and Okamoto M. *Progress in Polymer Science*, **28**, 1539–1641 (2003).
110. Siqueira, G., Brass, J., and Dufresne, A. *Biomacromolecules*, **10**, 425–432 (2009).
111. Siracusa, V. *International Journal of Polymer Science*, 1–11 (2012).
112. Siracusa, V., Rocculi, P., Romani, S., and Rosa, M. D. Trends in Food Science & Technology, **19**, 634–643 (2008).
113. Sivertsvik, M., Rosnes, J. T., and Bergslien, H. *Modified Atmosphere Packaging. Minimal Processing Technologies in The Food Industries*, T. Ohlsson and N. Bengtsson (Eds.) Woodhead Publishing Limited, Cambridge, pp. 61–86 (2002).
114. Sorrentino A, Gorrasi G, and Vittoria V. *Trends in Food Science and Technology*, **18**, 84-95 (2007).
115. Souza, B. W. S., Cerqueira, M. A., Casariego, A., Lima, A. M. P., Teixeira, J. A., and Vicente, A. A. *Food Hydrocolloids*, **23**, 2110–2115 (2009).
116. Srinivasa, P. C. and Tharanathan, R. N. *Food Reviews International*, **23**, 53–72 (2007).
117. Summers, J. W. Introduction. In PVC Handbook, C. E. Wilkes, J. W. Summers and C. A. Daniels (Eds.) Carl Hanser Verlag: Munich (2005).
118. Tadmor, Z. and Gogos, C. G. *Principles of Polymer Processing*, John Wiley & Sons, Inc.: New Jersey (2006).
119. Tang, X. Z., Kumar, P., Alavi, S., and Sandeep, K. P. *Critical Reviews in Food Science and Nutrition*, **52**, 426–442 (2012).
120. Teng, W. L., Khor, E., Tan, T. K., Lim, L. Y., and Tan, S. C. *Carbohydrate Research*, **332**, 305–316 (2001).
121. Thomas, D. J. and Atwell, W. A. *Starches*, Eagan Press, St. Paul Minnesota (1997).
122. Thompson, A. K. *Controlled Atmosphere Storage of Fruits and Vegetables*, CAB International, Oxfordshire (2010).
123. Throne, J. L. *Fabrication Processes*. In PVC Handbook, C. E. Wilkes, J. W. Summers and C. A. Daniels (Eds.) Carl Hanser Verlag, Munich (2005).
124. Throne, J. L. *Technology of Thermoforming*, Carl Hanser Verlag: Munich (1996).
125. Tripathi, D. *Practical Guide to Polypropylene*, Rapra Technology Ltd., Shropshire (2002).
126. Tubbs, R. K. *Journal of Polymer Science Part A*, **3**, 4181–4189 (1965).
127. Twede, D. *Economics of Packaging*. In The Wiley Encyclopedia of Packaging Technology, K. L. Yam (Ed.) John Wiley & Sons Inc., Hoboken, New Jersey, pp. 383–389 (2009).
128. Vartiainen, J., Motion, R., Kulonen, K., Ratto, M., Skytta, E., and Advenainen, R. *Journal of Applied Polymer Science*, **94**, 986–993 (2004).
129. Vasile, C. and Pascu, M. *Practical Guide to Polyethylene*, Rapra Technology Ltd: Shropshire (2005).
130. Vroman, I. and Tighzert, L. *Materials*, **2**, 307–344 (2009).
131. Wickson, E. J. and Grossman, R. F. *Formulation Development*. In Handbook of Vinyl Formulating, R. F. Grossman (Ed.) John Wiley & Sons, Inc., Hoboken, New Jersey, pp. 1–12 (2008).
132. Wunsch, J. R. *Polystyrene - Synthesis, Production and Applications*, Rapra Technology Ltd., Shropshire (2000).
133. Yam, K. L. and Lee, D. S. *Design of Modified Atmosphere Packaging for Fresh Produce*. Active Food Packaging, M. L. Rooney (Ed.) Blackie Academic & Professional, an Imprint of Chapman & Hall, Glasgow, pp. 55–73 (1995).

CHAPTER 2

INTERACTION PHENOMENA BETWEEN PACKAGING AND PRODUCT

BOUSSAD ABBÈS, FAZILAY ABBÈS, and YING-QIAO GUO

ABSTRACT

This chapter provides a review of interactions phenomena occurring between polymer packaging and packed products. The migration of species from primary packaging to contained product and the sorption of the product in the polymer could have many consequences. Migration may alter the quality, the shelf-life, the organoleptic or therapeutic efficacy of the packed product. Sorption may cause stress-cracking, squeeze or paneling of the container.

A general description of the main thermoplastics used together with usual additives is first given. The second part concerns a presentation of experimental methodologies set up for the interaction phenomena characterization. The characterization of the sorption phenomenon of amyl acetate in polypropylene packaging by FTIR and gravimetric methods is then detailed. The effects of amyl acetate concentration and interaction temperature are studied by means of sorption kinetics. We show that the sorption kinetics is of sigmoïdal-type, highlighting an asymptotic behavior of the sorption kinetics at the polymer surface. The mathematical modeling of this mass transfer phenomenon leads to the introduction of the surface mass transfer coefficient H, in addition to the diffusion coefficient D. These parameters are determined from the sorption curves showing that the diffusion coefficient D remains constant, while the surface mass transfer coefficient H increases with increasing amyl acetate concentration and temperature of the solution.

The second part of this chapter concerns the setup of an experimental methodology and the resulting empirical models for the prediction of the mechanical behavior of aged bottles. Tensile tests have been performed on strip-shaped samples cut from virgin and aged PP bottles to determine elastic modulus and peak load. Empirical models correlating these properties to the amyl acetate concentration are proposed. Furthermore, top-load tests are performed on several bottles to evaluate the evolution of the resistance to vertical compression as a function of aging. A model correlating the maximum top-load to the concentration of amyl acetate is drawn. All experimental results show that the measured mechanical properties decrease with increasing concentration of amyl acetate until reaching a threshold for high concentration levels. Furthermore, thermal studies by DSC show a decrease in the polymer cristallinity with the ester concentration of the packed solution. This result is correlated with the plasticizing effect of the amyl acetate on one hand, and with the migration of the incorporated additives from the polymer to the solution on the other hand. A model relating the decrease of the polymer cristallinity with the increasing concentration of amyl acetate is also derived from the experimental results.

2.1 INTRODUCTION

Polymer packaging is widely used in all industry sectors: food, cosmetics, health and care, chemicals, transportation, distribution, industrial and agricultural products, and so on. A key role of packaging is to ensure the protection of the product it contains. The packaging prevents the product from external contamination and it maintains its quality. But when a product is brought into contact with a thermoplastic material, transfer of compounds occur from the material to the product (migration) or from the product to the material (sorption) (Safa and Bourelle, 1999; Auras et al., 2006).

The mass transfer between packaging and products has been subject of numerous studies. Currently, bibliographic data on this phenomena show considerable effects on the polymer packaging including: a change of crystallinity, a loss of mechanical strength, stress cracking, squeeze, and collapse, and so on.

The deterioration of the packaging and (or) its contents during aging result in a loss of quality, product returns, and then a decline in brand image of the company. The companies require increasing the quality of their packaging, and additional costs in manufacturing and or transport.

Furthermore, due to mechanical stresses rising during storage, transport or handling, the packaging undergo irreversible deformations. Aging can also affect the appearance (brightness, color, and deep degradation of the surface) and the mechanical properties (resistance to vertical compression, impact resistance, and so on) of the packaging.

To solve this problem, two approaches are currently used in the industry: the first one is completely empirical and preventive, leading to random results; the second approach, rather corrective, consists on modifying the packaging by changing its constitutive material, or its shape, or by increasing its thickness, or even by modifying the product formulation. Both approaches, however, reveal unsatisfactory, for various reasons, being expensive, and not ecofriendly.

Thus, a better understanding of the container/content interactions is essential. This can be achieved by coupling thermo mechanical and physicochemical studies of the interaction between polymer packaging and product. The feedback of such studies will allow constructive and sustainable industrial development to protect the packaging from all constraints and preserve the packed product quality and integrity.

This chapter provides an overview of experimental methodologies set up for the interaction phenomena characterization, and presents analytical models generally used for its numerical simulation.

2.2 GENERAL DESCRIPTION OF MAIN THERMOPLASTICS USED IN PACKAGING

Within packaging, polyolefin cover the largest part of the demand (74% of all the plastic packaging in Europe), followed by Polyethylene Terephthalate (PET), Polystyrene (PS), and Polyvinyl Chloride (PVC). Packaging is by far the biggest application segment for thermoplastics in general, reaching 40.1% of the European plastic demand. Low-Density Polyethylene (LDPE) and Linear Low-Density Polyethylene (LLDPE) plastics have the largest plastic packaging consumption (31%), followed by HDPE (22%), and PP (21%).

The Polypropylene (PP) is one of the most versatile polymers available with a wide variety of packaging applications. The PP can be produced either as a homopolymer or as a copolymer. A homopolymer molecule is derived from a single monomer, which in the case of PP is derived from propylene. Alternatively, a copolymer molecule is derived from two or more monomers, which in the case of PP is derived from a combination of propylene and ethylene. Homopolymers are more crystalline than copolymers and have the highest strength and rigidity. Copolymers are classified into block and random variants. The properties and applications of PP are given in Table 1.

2.2.1 ADDITIVES

An additive is a substance which is incorporated into plastics to achieve a technical effect in the finished product. Additives in plastics provide the means whereby processing problems, property performance limitations and restricted environmental stability are overcome (Lau and Wong, 2000). We can categorize additives based on their primary purpose, which will be either to make easier a manufacturing process, to enhance thermo mechanical and physicochemical properties, or to satisfy more stringent environmental and health regulations. As the scope of plastics has increased, so has the range of additives: better mechanical properties, heat and thermal resistance, light and weathering, flame retardancy, electrical conductivity, and so on. Some examples of additives are antioxidants, antistatic agents, emulsifiers, plasticizers, UV absorbers, and so on.

They are classified into two main categories:

- Stabilizers and,
- Technological adjuvants.

TABLE 1 Properties and applications of polypropylene (PP) in packaging

Properties	Applications
Rigidity (sterilization resistance)	Containers
Cold resistance	Caps
Water vapor barrier	Tubes
Good at freezing (- 40°C)	Boxes
Good at microwave (+ 120°C)	Films
Transparency	Flasks
Low density	Pumps
Shininess	Thermoforming sheets
PP clarified	Reusable packages
OPP	Reheat able dishes
BOPP	Valves
EPP (expanded): repeated shocks resistance	Pill-boxes
	medicine packages

STABILIZERS

These additives help maintaining the physicochemical properties of polymers over time by reacting instead of the polymer. They inhibit or delay the process responsible for the polymer structure alterations during the manufacturing process or the end use of the product.

The stabilizers can be classified into two main categories:

- *Antioxidants*: These additives are used to prevent degradation of polymers that can result in loss of strength, flexibility, thermal stability, and color. Antioxidants perform by inhibiting atmospheric oxidation during processing and usage. The oxidation is catalyzed by the presence of catalyst residue, defects in the polymer chains and also by light and temperature. While primary antioxidants inhibit oxidation by scavenging free radicals (molecules cleaved from the polymer chain) to give hydroperoxyde groups, secondary antioxidants stop oxygenated radical molecule propagation by decomposing them into stable products (alcohol groups). The two types of antioxidants are complementary and are generally used simultaneously in polymers.
- *Anti-lights*: These additives, which absorb light instead of the polymer, help materials withstand UV light. Ultraviolet radiation damages chemical bonds of polymeric materials, therefore addition of UV stabilizers is crucial and essential to produce materials that will provide good performance even when exposed to UV radiation over extended periods. Generally, these stabilizers function by absorbing high energy UV radiation and then releasing it at lower energy level, less harmful to the polymer.

TECHNOLOGICAL ADJUVANTS

Technological adjuvants are used to modify the physical and/or chemical properties of polymers.

There is a wide panel of adjuvants, which can be classified according to their final effect:

- *Mechanical properties modifiers*: Which make the polymer more resistant to mechanical stresses (fillers, anti-shock, plasticizers, and so on).
- *Processing additives*: Plasticizers, stabilizers, nucleating agents, crosslinking agents, and so on.
- *Surface properties modifiers*: Antistatic lubricants, and so on.

2.3 INTERACTIONS IN PACKAGING

When content is brought into contact with a thermoplastic container, several interactions can take place from the container to the content or from the product to the packaging. These interaction phenomena between the packaging material and the product can manifest themselves in different ways (Figure 1) either during the manufacturing, conditioning process, or during storage:

- Migration of constitutive substances of the packaging material to the product (Nir and Ram, 1996).

- Gas permeation: Generally oxygen (O_2) diffusion into food, or carbon dioxide (CO_2) diffusion out of the packaging (Tawfik et al., 1998; Hernández-Muñoz et al., 1999).
- Sorption (or absorption) of the product components by the packaging (For example, flavors) (Hernández-Muñoz et al., 2001; Auras et al., 2006).

These interactions may have several consequences, visual or not, even under normal conditions of use: organoleptic modification of the formula or of the primary packaging, chemical modification of the formulation, and so on.

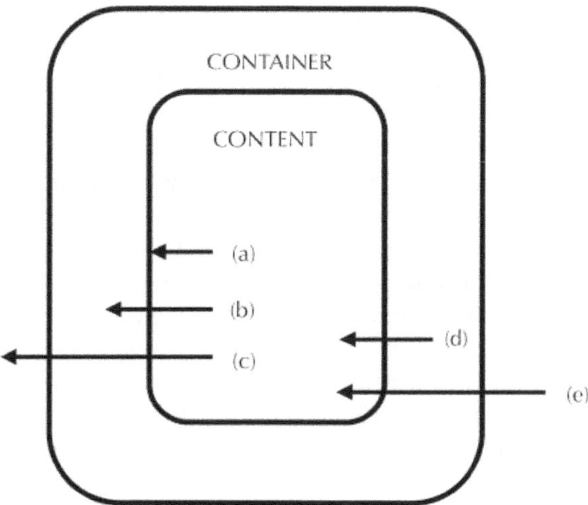

FIGURE 1 Schematic diagram for container/content interaction phenomena: (a) adsorption, (b) absorption, (c) permeation, (d) migration, and (e) permeability.

2.3.1 MIGRATION

Migration is the transfer of the thermoplastic packaging constituents to the packed product (Lickly et al., 1995). The main migrant products can be classified as:

- The constituents of synthetic polymers: Residual monomers (styrene, terephthalic acid, and so on), pre-polymers (mono or di-hydroxyethyl terephthalate, and so on), oligomers (low molecular weight polystyrene, and so on).
- The degradation products of synthetic polymers due to photo-oxidation of polyolefins, hydrolysis of polyesters, and so on.
- The adjuvants of natural or synthetic polymers: surfactants, catalysts, antistatics, colorants, plasticizers, antioxidant, and so on.

European and American regulations in particular give two definitions to the migration phenomenon: The global migration, which estimates the total mass of migrants lost by the package, without distinguishing between the nature and the specificity of these components, and the specific migration which tries to identify and quantify each of those elements.

Measuring the specific migration, especially with a given food for example, is an approach that has been questioned by several authors due to the difficulties of setting up the testing procedures. To overcome this, several liquid simulants were used to study this phenomenon. The Shwope et al. (1987) have studied the migration of antioxidants into food stored in LDPE wraps. Two typical antioxidants, BHT and Irganox 1010, were radiolabelled to allow accurate analytical measurement of the extent of their migration into foods and food-simulating liquids (FSL). The results show that BHT, a much smaller and more volatile molecule than Irganox 1010, migrates more rapidly into foods, but the differences are less for FSL. Hernández-Muñoz et al. (2002) have characterized the partition equilibrium of food aroma compounds between plastic films and foods or food simulants. Two polymers (LLDPE and PET), three organic compounds (ethyl caproate, hexanal and 2-phenylethanol), four food products with varying fat content (milk cream, mayonnaise, margarine, and oil) and three simulants (ethanol 95%, n -heptane and isooctane) were selected for study. The results show the effect of the aroma compound volatility and polarity, as well as its compatibility with the polymer and the food or food simulant. They showed that the measurement of liquid/vapors equilibrium can be regarded as a powerful tool to compare the effectiveness of food simulants as substitutes of a particular food product and can be used as a guide for the selection of the appropriate simulant. Dopico Garcia et al. (2003) have developed an analytical method for the determination of specific migration levels of phenolic antioxidants from low-density polyethylene (LDPE) into food simulant. The screening and response surface experimental designs to optimize the liquid–liquid extraction (LLE) of these antioxidants have been tested and the analyses have been carried out by reversed-phase high-performance liquid chromatography (HPLC) coupled with ultraviolet diode-array detector. The procedure developed has been applied to specific migration tests in different commercial LDPE films. The considered antioxidants Ethanox 330 and Irgafos 168 have been found at trace level. Quinto-Fernandezy et al. (2003) have reported different methods of analysis for four additives (two antioxidants, IRGANOX 245 and 1035; an ultraviolet absorber, CHIMMASORB 81; and an optical brightening agent, UVITEX OB) in olive oil.

The studied methods allowed establishment of additive stability and measurement of migration of the selected additives into olive oil at different time-temperature conditions. Begley et al. (2004) have measured the migration characteristics of the UV stabilizer Tinuvin 234 (2-(2H-benzotriazol-2-yl)-4, 6-bis (1-methyl-1-phenylethyl) phenol) into food simulants from polyethylene terephthalate (PET) using HPLC with UV detection. Ethanol/water, isooctane and a fractionated coconut oil simulant (Miglyol®) were used as food simulating solvents. Isooctane is determined to be a good fatty food simulant that provides similar results for PET to those of fatty foods. Stoffers et al. (2004) selected six most suitable certified reference materials (LDPE/Irganox 1076/Irgafos 168, LDPE/1, 4-diphenyl-1, 3-butadiene (DPBD), HDPE/Chimassorb 81/Uvitex OB, PP homo/Irganox 1076/Irgafos 168, HIPS, 1% mineral oil/styrene, and PA 6/caprolactam) as candidate for specific migration testing. Several other researchers have focused their work on prediction and simulation of the migration of packaging constituents to assess health risks and to predict the retention time of the products. O'Brien and Cooper (2001) have calculated the predicted migration values using the Piringer "Migratest Lite" model by entering the measured initial concentration of additive in the polymers into the equations together with known variables such as additive molecular weight, temperature, and exposure time. The results indicate that the Piringer migration model, using the "exact" calculations of the Migratest Lite program, predicted migration values into olive oil close to, or in excess of, the experimental results for > 97% of the migration values generated in their study. Brandsh et al. (2002) have predicted for polyolefin packaging materials, an upper limit of migration into foodstuffs with a high degree of statistical confidence. For easy handling of both the experimental results and the diffusion model, user-friendly software has been developed. For migration modeling of packaging materials with multilayer structures, a numerical solution of the diffusion equation is described and the procedure has been also applied for modeling the migration into solid or high viscous foodstuffs. Reynier et al. (2002) have presented a computing program describing precisely the migration of additives from a polymer into a food simulant. The model has been validated with a UV absorber in polypropylene migrating into glyceryl tripelargonate. Six parameters were used to fit the simulant sorption and additive extraction kinetics, and these were determined by independent experiments. The Han et al. (2003) have developed a simulation model computer program, which accounts for not only the diffusion process inside the polymer but also partitioning of the contaminant between the polymer and the contacting phase, based on a numerical treatment, the finite element method, to quantify migration through multilayer structures. For their study, three-layer co-extruded high density polyethylene (HDPE) film samples, having a symmetrical structure with a contaminated core layer and virgin outer layers as the functional barriers, were fabricated with varying thickness of the outer layers and with a known amount of selected contaminant simulant, 3, 5-di-t-butyl-4-hydroxytoluene (BHT), in the core layer. Migration of the contaminant simulant from the core layer to the liquid food simulants was determined experimentally as a function of the thickness of the outer layer at different temperatures and compared successfully with the model's results. Roduit et al. (2005) have developed a diffusion model to simulate the migration from multilayer packaging and under non-isothermal temperature conditions. Finite Element Analysis (FEA) is used as a numerical approximation method to solve the diffusion equations describing such processes. Correlations between experimental and computed results are presented for benzophenone in LDPE/LDPE/PP films. Kanavouras and Courtelieris (2006) have obtained a reliable estimation of the shelf life of packaged olive oil for various combinations of storage conditions, close to "real-life" situations, by using a theoretical model supported by experimental results.

2.3.2 PERMEABILITY

Permeation refers to the penetration and transfer of a component through a wall under the influence of a concentration gradient (Wang et al., 1998). The permeability is the mass (or molar) flux of that particular component or species per unit driving force.

The permeability aims at measuring over a time-period the amount of a species crossing a polymer film subjected to a difference in potency of the species (Soares et al., 1999; Ulsten and Hedenqvist, 2003). Permeability properties of a polymer depend on many parameters and in particular on the kinetics of diffusion and the solubility of the permeant:

$$P = D \, . \, S \tag{1}$$

where P is the permeability coefficient, D is the diffusion coefficient and S is the solubility coefficient.

Permeability is greater for linear molecules than those of spherical geometry. It is more important for polymers in the rubbery state than for glassy polymers and for amorphous polymers than semi-crystalline ones. Some authors have attempted to characterize and model the permeability of thermoplastics to gases (O_2, CO_2, N_2) and water vapor. Jasenka et al. (2000) have measured the permeability's of polyolefins to oxygen and nitrogen at different temperatures using the manometric method. Permeability, diffusion and solubility constants were calculated while temperatures and enthalpies of the phase change were determined by the DSC method. Permeability of gases increases with temperature for all investigated samples. Permeability data of all investigated samples showed good agreement with the Arrhenius equation. Kata and Nada (2001) have reported the permeability of gases for some monofilms as well as laminated polymers at temperatures from 20 to 60°C. In order to simulate the effect of food/packaging interaction on permeability of polymeric materials, all investigated samples were subjected to 15% (v/v) ethanol and 3% (w/v) acetic acid aqueous solution prior to permeability measurements. The permeability of the selected polymers treated with food simulants increased significantly compared to untreated ones, mainly due to the swelling effect of the water molecules. Del Nobile et al. (2007) have studied the packaging of three different fresh processed fruits: prickly pear, banana and kiwifruit; with two different films, a laminated PE/aluminum/PET film, and a co-extruded polyolefin film, and storing the packages at 5°C. The package headspace composition, in terms of oxygen and carbon dioxide concentration, was monitored for a period ranging between 4 and 12 days, depending on film type. A simple mathematical model was used for describing and predicting the respiration behavior of packed fresh processed fruits during storage. It was found that the predictive ability of the proposed model depends on both the permeability of the package and the fresh processed fruit packed. The results of optimization suggest that the optimal film thickness strongly depends on packed fruits. Techavises and Nikida (2008) have developed a mathematical model based on Fick's law for predicting O_2, CO_2, N_2, and water vapor exchanges in modified atmosphere packaging (MAP) films with macro perforations. The temperature and film thickness had no significant effect on the effective permeability ($P > 0.05$). For most conditions, the effective permeability did not differ between gas types (O_2, CO_2, N_2, and water vapor). An empirical equation of the effective permeability of a macro perforation in a thin film as a function of perforation diameter was developed. The transmission rate of LDPE film was determined for temperatures between 5 and 25°C. The use of the proposed MAP model coupled with an effective permeability model was found to yield a good prediction of gas concentration and RH when compared to experimental results for MAP of "Kiyomi" fruit.

2.3.3 SORPTION

The "sorption" term is generally used to describe any process including penetration and dispersion of a species in a polymer. This process therefore includes adsorption, absorption, diffusion and dispersion of the species in a free volume. The species transport depends on their ability to move and on the mobility of the polymer chains.

The content's molecules can be adsorbed on the walls of the package, and then can penetrate through the polymer if they have low mass and low steric hindrance.

EXPERIMENTAL CHARACTERIZATION OF SORPTION

Experimental characterization consists in putting into contact a polymer with a diffusing species under specific conditions and following its concentration evolution with time. It is often tempting to simulate real situations of products sorption (liquid or solid food, perfume, cream and so on) in polymers. However, the large number of diffusing species involved in such phenomena makes too complex the analysis of those products.

Moreover, if the diffusing species is solid at the temperature of the experiment it may have a bad contact with the polymer. It is then necessary to put it in a solution by adjusting its concentration to achieve the desired levels of sorption.

Several experimental techniques can be used to study transport phenomena in polymers: nuclear magnetic resonance imaging (NMR), UV spectrophotometer, gas chromatography/flame ionization detector (GC/FID), high-performance liquid chromatography (HPLC), laser interferometry, gravimetric method and Fourier Transform Infra-Red spectroscopy (FTIR).

We have collected in Table 2 some data for diffusion coefficient (D) obtained in the literature for some couples of stimulant/PP determined by different experimental techniques. The diffusivity ranges from 10^{-14} to 10^{-09} depending on the temperature and the couple stimulant/PP.

TABLE 2 Diffusion coefficients for some couples of stimulant and PP determined by different experimental techniques

Simulant	D (cm²/s)	Temperature (°C)	Experimental Technique	Reference
Olive oil	7.5×10^{-11}	23	FTIR	Riquet et al. (1998)
Dichloromethane	1.4 to 9.78×10^{-09}	25	Gravimetric method	D'Aniello et al. (2000)
Irganox 1010	$(0.7 \pm 0.1) \times 10^{-12}$	40	UV spectrophotometer	Ferrara et al. (2001)
	$(18.7 \pm 0.1) \times 10^{-12}$	70		
	$(49.2 \pm 2.3) \times 10^{-12}$	80		
Hexadecane	1.3×10^{-09}	40	GC/FID	Reynier et al. (2001)
Octadecane	8.7×10^{-10}			
Hexatriacontane	2.0×10^{-11}			
Octadecanol	2.1×10^{-10}			
DEHP	3.8×10^{-11}			

TABLE 2 *(Continued)*

γ-decanolactone	5.79×10^{-11}	45	FTIR	Safa and Abbès (2002)
Octyl acetate	2.40×10^{-09}			
Dodecyl acetate	7.28×10^{-10}			
Cetyl acetate	3.79×10^{-11}			
Trichloroethane	0.55×10^{-9}	40	GC/FID	Dole et al. (2006)
Toluene	1.50×10^{-9}			
Chlorobenzene	1.20×10^{-9}			
Phenylcyclohexane	0.4×10^{-9}			
Benzophenone	0.47×10^{-9}			
IRGANOX 1010	0.10 to 51.0×10^{-14}	40	HPLC	Begley et al. (2008)
Limonen	0.30 to 98.0×10^{-11}	80		
	0.95 to 310.0×10^{-10}	40		

PARAMETERS INFLUENCING SORPTION OR DIFFUSION

Molecule sorption and diffusion in a polymeric material are influenced by several factors, like the molecule structure, its concentration and its temperature, and the polymer molecule morphology.

- Influence of the shape and structure of diffusing molecule

The shape of the diffusing molecule strongly influences its diffusion in the polymer. Thus, for a given molecular volume, linear molecules diffuse more rapidly than those with branches; which diffuse more rapidly than spherical molecules (Siddaramaia and Mallu, 1998). Safa and Abbès (2002) showed that a linear molecule of 10–14 carbons favors the sorption in a polypropylene film. While the presence of a double bond, or a branch or a cycle is unfavorable to the sorption. The rate of sorption decreases in general: from ketone to ester to aldehyde. Aminabhavi and Phayde (1995) also showed that in the case of alkane, the addition of a linear carbon chain has a significant effect on the diffusion coefficient. Al-Malaika et al. (1991) tried to correlate the values of the diffusion coefficient (D) as a function of molar mass (M) of the diffusing molecule by using the following equation:

$$D = K.M^{-\alpha} \tag{2}$$

where K and a are the model parameters.

- Influence of the concentration of the diffusing molecule

Mohney et al. (1988) reported that low concentrations of the diffusing molecule affect the polymer but only at a very limited degree, and the amount of absorbed compounds is directly proportional to its concentration. At higher concentrations, the absorbed quantity can even modify the polymer matrix (Saddler and Braddock, 1990; Charara et al., 1992).

Dhoot et al. (2001) showed that the sorption of n-butane and n-pentane in biaxially oriented polyethylene terephthalate increases with the concentration of these alkanes.

- Influence of temperature: Diffusion phenomena of a molecule in a polymer follow an Arrhenius type activation process (Chiang et al., 2002; Kulkami et al., 2003; Patzlaff et al. 2006). The diffusion rate is a function of temperature according to the equation:

$$D = D_0.\exp(-\frac{E_a}{RT}) \quad (3)$$

where D_0 is the diffusion coefficient for infinite temperature, T is the absolute temperature (K), E_a is the activation energy ($J.mol^{-1}$) and R is the gas constant (R = 8.314 $J.mol^{-1}.K^{-1}$).

- Influence of the microstructure of polymers: The crystalline regions in a polymer have higher density than the amorphous ones. They are then virtually impermeable. Consequently, diffusion takes place mainly in the amorphous regions of the polymer, where small vibratory movements occur along the polymer chains. These Micro-Brownian movements may result in the formation of holes while the macromolecular chains are moving away from each other. These resulting spaces become then active sites for the diffusing molecules.

The Johannson (1993) and Wesselingh and Krishna (2000) noted that polymers having more amorphous areas absorb more volatile compounds than polymers with high crystallinity. Escobal et al. (1999), for example, measured larger amounts of amyl acetate and ethanol in LDPE than in HDPE or PP. They also noted that PP absorb more than OPP. Indeed, the chains orientation obtained by mechanical stretching, causes a preferential orientation of the individual chains in both the amorphous and crystalline phases. Moisan (1980) found that the kinetics of sorption in a semi-crystalline polymer depends strongly on the orientation of the polymer chains in the diffusion direction.

2.3.4 EFFECTS OF MASS TRANSFER

EFFECTS ON THE PACKED PRODUCT

In case of migration process, constituents of packaging materials (additives and other low molecular weight compounds…) can diffuse into packaging content, like food or drug for example. The tendency of food to interact with its packaging is a significant factor that can affect quality, appearance and shelf life. Adherence of food residues to the packaging may decrease product acceptability, enhance oxidation and off flavors, increase waste, and result in lowering overall product quality. Significant concentrations of packaging substances with potential health concern were mainly identified in foods packed in plasticized PVC cling-films (Harison, 1988; Sharman et al., 1994; Petersen and Breindahl, 2000). Modifications in connection with container/content interactions therefore potentially have an impact on the quality, efficacy and safety of the packed product.

On the other hand, the sorption of some molecules of food or drug in the polymer can cause a loss of flavor and aromatic imbalance (Moisan, 1980). Several studies (Arora et al., 1991; Paik, 1992; Fayoux et al., 1997; Berlinet at al., 2005; Simko et al., 2005) have shown that significant amounts of aromatic compounds can be absorbed by thermoplastic packaging.

These interactions may also influence the sensorial quality and shelf life of the product and even lead to a total loss of smell or change the aroma nature.

EFFECTS ON THE POLYMER

Transport phenomena can cause physical aging of polymers which results in irreversible alteration of its properties. This alteration can affect its chemical structure, its composition or its physical state (Kausch, 2005). Physical aging is reflected in several ways that we will discuss hereafter.

Plasticization occurs when the diffusing molecules penetrate into the macromolecular network. This produces disorders that weaken or even destroy the secondary bonds responsible for the cohesion of the material.

Theoretically, plasticization is reversible. In fact, it induces internal rearrangements and may also facilitate the relaxation of internal stresses. These phenomena are often found when the material has chains of low molecular weight or when it has a low degree of crystallinity. Plasticization is characterized by a change in the mechanical properties of the material, resulting in a decrease of the glass transition temperature (T_g).

Diffusion of solvent in the polymeric material may induce swelling causing changes in its internal structure if there are heterogeneities which induce stresses between more or less swollen areas. These areas may be amorphous (relatively accessible) or crystalline (relatively inaccessible). This swelling can also occur when the kinetics of diffusion of the solvent create concentration gradients.

Damage under stress can produce cracks in the material. Crazing concerns the areas constituted of vacuum and highly oriented fibrils in the polymer. These can lead to the formation of a micro-cracks and cracks.

2.4 MATHEMATICAL MODELLING OF DIFFUSION

2.4.1 DIFFUSION EQUATION

Diffusion is the process by which the transfer of the material of one part of the system to another. It results from random movements of molecules in the system. Consider two areas of a system with different concentrations where the molecules move randomly. A large number of molecules will move from high concentration area to low concentration, which leads to smooth out spatially varying concentrations.

FICK'S LAWS OF DIFFUSION

In 1855 Adolf Fick drew an analogy between the heat transfer phenomena by conduction and those of mass transfer. Assuming a steady state, the first diffusion theory postulates that the flux of the diffusing substance through a unit area is directly proportional to the concentration gradient measured in a direction normal to this surface unit. In one spatial dimension, this relationship can be written as:

$$F = -D.\frac{\partial C}{\partial x} \tag{4}$$

where F is the flow of material along the x-axis corresponding to the diffusion direction, C is the concentration at position x and D is the diffusion coefficient.

In some cases, the diffusion coefficient can be considered as constant. However, in the case of polymers, this coefficient may depend strongly on parameters like concentration and temperature.

Equation (4) is called the Fick's first law. The negative sign in the above expression indicates that the diffusion occurs in the opposite direction to that of the concentration gradient.

If only one space direction is considered, there is an accumulation of material over time in a volume bounded by two planes perpendicular to the flow direction of diffusion. When the thickness of this volume tends to zero, the increase in concentration over time is given by Fick's second law:

$$\frac{\partial C}{\partial t} = \frac{\partial}{\partial x}\left(D.\frac{\partial C}{\partial x}\right) \tag{5}$$

where t is the time.

When D is constant, the Fick's second law can be written as:

$$\frac{\partial C}{\partial t} = D.\frac{\partial^2 C}{\partial x^2} \tag{6}$$

A diffusion problem is given by the geometry of the medium and a set of boundary conditions. Crank (1975) established mathematical solutions to these equations for given geometries and different boundary conditions.

By plotting $M_t/M_\infty = f(t^{1/2})$ (Figure 2), for Fickean sorption kinetics the slope of the linear portion of the obtained curve provides direct access to the diffusion coefficient. But no direct access to the diffusion coefficient is possible for sigmoïd-type and two-steps sorption kinetics. The diffusion coefficient can be determined by fitting the experimental curves.

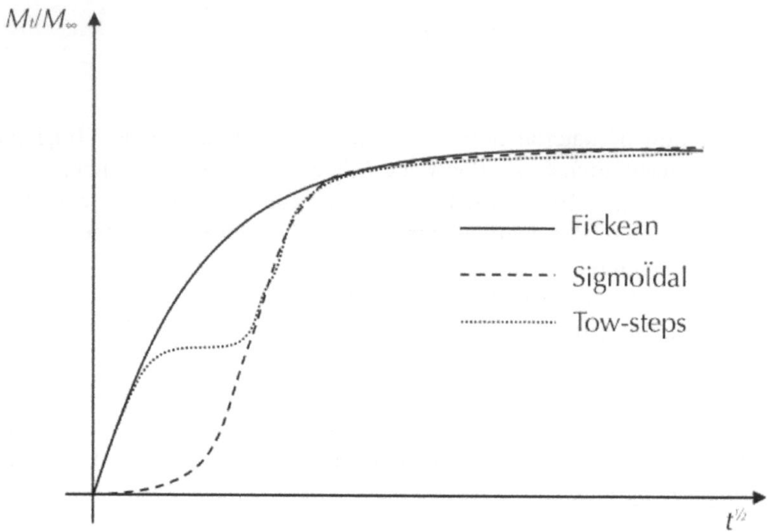

FIGURE 2 Schematic representation of different sorption kinetics.

DIFFUSION COEFFICIENT

The diffusion coefficient D is the key parameter for modeling sorption experiments. When a compound is found in the polymer with a significant concentration, it can induce a plasticization. This can especially:

- Increase the polymer's diffusion coefficient. This is the case of "self plasticization" occurring during the sorption of a solvent,
- Increase the diffusion coefficient of the other compounds in the material. This is the case of migration experiments, where the diffusion coefficients of migrants in the polymer depend on the concentration of sorbed molecules.

This dependence of the diffusion coefficient as a function of concentration is generally expressed as (Reynier et al., 2002; Dole et al., 2006):

$$D(C) = D_0 + A.C \tag{7}$$

$$D(C) = D_0 \exp(A.C) \tag{8}$$

$$D(C) = D_0 \exp\left(A + \frac{C}{1+C}\right) \tag{9}$$

where C is the concentration, D_0 is the diffusion coefficient in the virgin material and A is a constant.

2.5 SORPTION EXPERIMENTAL RESULTS

2.5.1 SAMPLES PREPARATION AND CONTACT CONDITIONS

In the present study, we have used samples from commercial extruded blown polypropylene (PP) bottles cut into strips of 1 mm thickness. This polymer is commonly used as a food or cosmetic packaging because of its low weight, flexibility, strength, good moisture barrier and low cost. We had no information about the polypropylene structure and forming conditions. FTIR analysis showed that the polymer is a copolymer polypropylene containing 83% of propylene and 17% of methylene monomers.

The amyl acetate (n-pentyl acetate) used is an analytical grade reagent obtained from Aldrich Chemical Company (Lyon, France), with about 99% purity. The product properties are as follows: molar weight, 130.2 g/mol, density at 20°C, 0.876g/cm^3, boiling point, 146°C, and fusion point, −71°C. The enthalpy of vaporization of amyl acetate calculated from data in the Handbook of Chemistry and Physics is 41.9 kJ/mol. This molecule is used as a food aroma and as a test odorant in studies of olfactory function.

From pure amyl acetate, we have prepared aqueous solutions of 5000ppm, 10000ppm, 20000ppm and 50000ppm. These solutions were prepared from pure amyl acetate solution with 2.5% ethanol to improve solubility in some cases.

Samples were cut in rectangular shape (9 cm x 1 cm) using a warm scalpel. All of the samples had an aspect ratio (length over thickness) greater than 10, which ensured application of a one-dimensional diffusion equation for analysis of the transport data.

2.5.2 GRAVIMETRICALLY MEASURED DIFFUSION EXPERIMENTS

The gravimetric method is one of the most used methods for the study of sorption of liquids in solid materials. It consists in following the evolution of the mass of a sample immersed in a solution. This method can be used only when one compound is studied, but it remains universal with respect to the substance to quantify. The polymer samples are soaked in screw-tight test glass vials containing the solution of interest. The immersion experiments can be conducted at different temperatures. These samples are removed periodically, the solvent drops adhering to the surface are wiped off using soft filter paper wraps, and are weighed immediately on a balance with minimum accuracy of ± 0.01mg. After contact with the solution, the mass of the samples is determined immediately. Equation (10) is used to compute the relative mass gain of the samples at time *t*:

$$\Delta M = \frac{M_t - M_0}{M_0} \tag{10}$$

where M_0 is the initial mass of the sample and M_t is the mass of the sample after a contact time *t*.

Figure 3 shows the sorption kinetics of amyl acetate in polypropylene at 23°C. We can observe that the sorption equilibrium is reached after 10 days. The largest sorption is observed with the solution with the highest concentration. The sorption curves at 40°C (Figure 4) show that the equilibrium is reached after 7 days whatever the concentration of amyl acetate. When the contact temperature increases, the amount of the sorbed ester into the polypropylene samples increases and the time to reach the equilibrium state decreases. These effects are essentially due to the increase in the mobility of the macromolecules in the amorphous phase of the polypropylene accompanied by the increase in the polymer free volume.

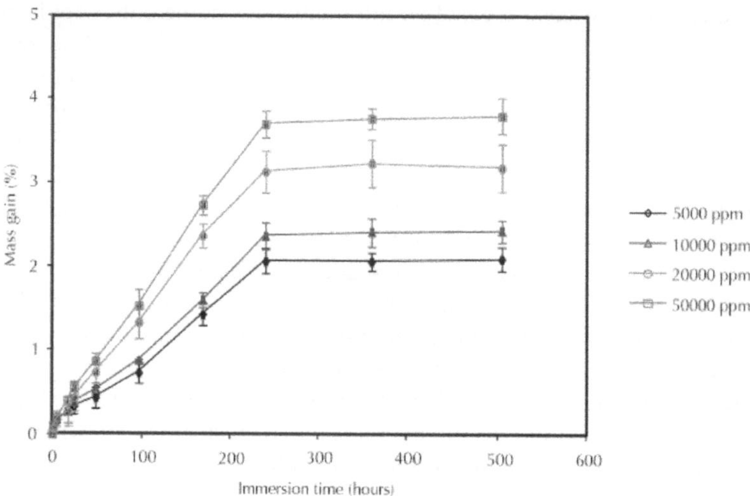

FIGURE 3 Relative mass gain in polypropylene samples for different concentrations of amyl acetate at 23°C.

2.5.3 FTIR MEASURED DIFFUSION EXPERIMENTS

The Fourier Transform Infra-Red spectroscopy (FTIR) technique is a valuable method for investigating the kinetics of sorption of organic molecules in polymers. In further works (Safa and Abbès, 2002; Safa et al., 2007; Zaki et al. 2009), we have used such method to study the sorption rate of different esters by polypropylene. The ester function is characterized by absorption band in regions where the polypropylene does not absorb. The sorption can be evaluated by the ratio of the carbonyl ester (CO) band area at 1747 cm^{-1} to the area of a characteristic peak of the PP at 841 cm^{-1} used as a reference band. The polymer samples are cut-out in thin films of 50 μm thickness by using a microtome. These films are then put on a support adapted for the spectroscopic analysis. Figure 5 shows FTIR spectra of a polypropylene sample after contact with 5000 ppm of amyl acetate solution at 23°C after 5 hours, 4 days and 15 days of immersion. The absorption characteristic band of esters at 1747 cm^{-1} and its growth is clearly observed.

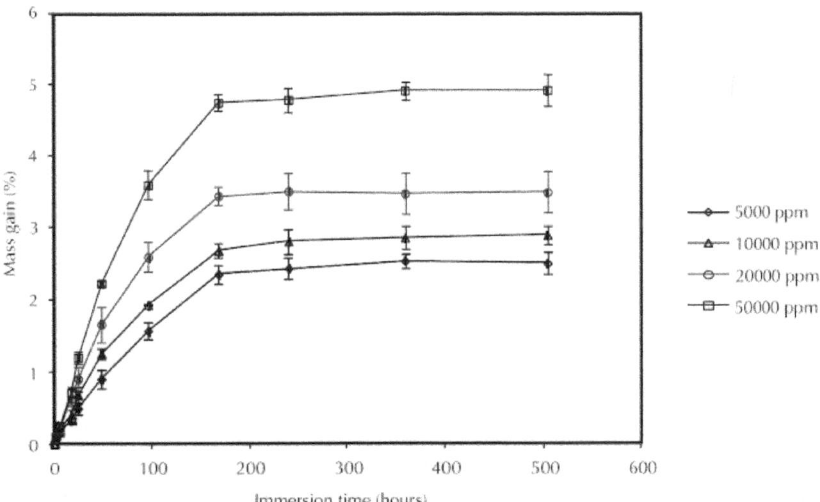

FIGURE 4 Relative mass gain in polypropylene samples for different concentrations of amyl acetate at 40°C.

The quantification of the sorbed product is achieved by comparing the ratio A_t of the area of the absorption band of amyl acetate A_{1747} at 1747 cm^{-1} and the area of the characteristic band of polypropylene A_{841} at 841 cm^{-1}:

$$A_t = \frac{A_{1747}}{A_{841}} \tag{11}$$

On Figure 6 and Figure 7, we have illustrated a comparison of the evolution of the A_t ratio for different concentrations of amyl acetate according to ageing duration at 23°C and 40°C respectively.

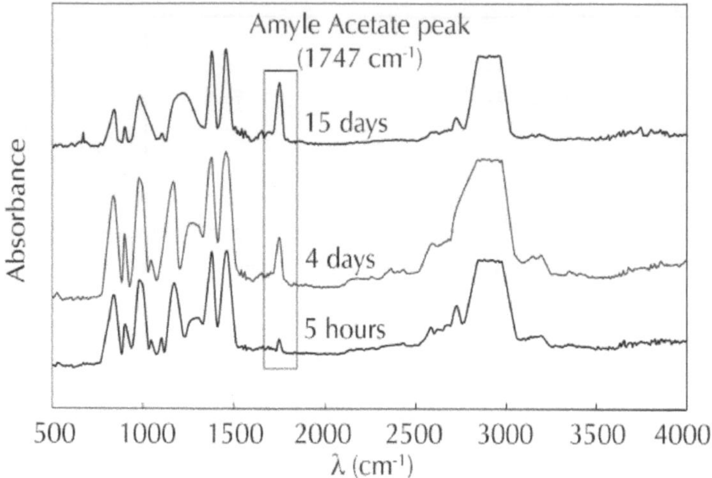

FIGURE 5 FTIR spectra after immersion of polypropylene in 5000 ppm amyl acetate solution at 23°C.

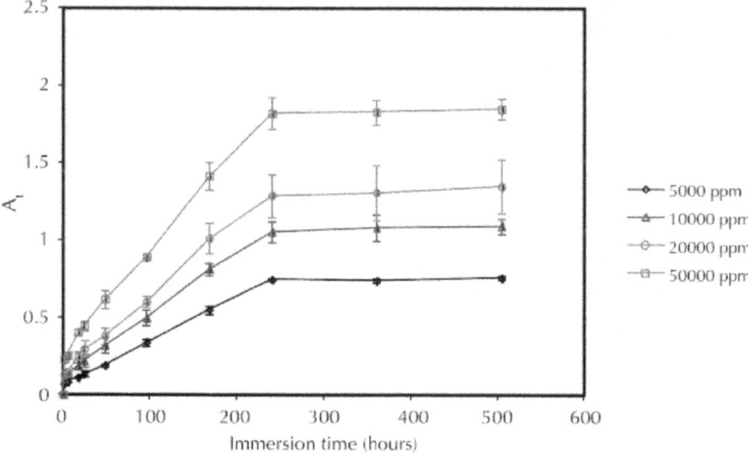

FIGURE 6 Evolution of the integrated intensity ratio At during diffusion of amyl acetate into polypropylene for different concentrations at 23°C.

2.6 SORPTION MODELLING

To identify the sorption type, we have adopted a more adequate representation which consists in plotting the evolution of M_t/M_∞ and A_t/A_∞ versus the square root of the time of immersion. M_∞ and A_∞ are the mass gain and the area ratio at the equilibrium state respectively. This representation shows that the kinetics of sorption of the amyl acetate in polypropylene is sigmoid-type for both ageing temperatures considered in this study. Such non-Fickian kinetics has been observed experimentally in many systems. (Park and Crank, 1968; Vergnaud, 1991) have proposed a model where the total amount of diffusing substance M_t *(or* A_t*)* penetrating into the sample up to time *t* can be expressed as a fraction of M_∞ (or A_∞) corresponding to the quantity after infinite time:

$$\frac{M_t}{M_\infty} = 1 - \sum_{n=1}^{\infty} \frac{2R^2}{\beta_n^2\left(\beta_n^2 + R^2 + R\right)} \exp\left(\frac{-\beta_n D_t}{L^2}\right) \tag{12}$$

where β_n are positive roots of the following equation:

$$\beta \tan \beta = R \tag{13}$$

and $R = LH/D$ is a dimensionless parameter.

L denotes here the thickness of the sample, *H* is the surface mass transfer coefficient and *D* is the diffusion coefficient. Recall that *H* and *D* are both unknown parameters, determined by a fitting procedure using adequate software.

Figure 8 and Figure 9 show two examples of experimental sorption kinetics data fitted with sigmoid-shape curves at 23°C. Figure 10 and Figure 11 show two examples of experimental sorption kinetics data fitted with sigmoid-shape curves at 40°C. These figures show that sigmoid sorption model fits well the experimental results obtained by gravimetric and FTIR techniques.

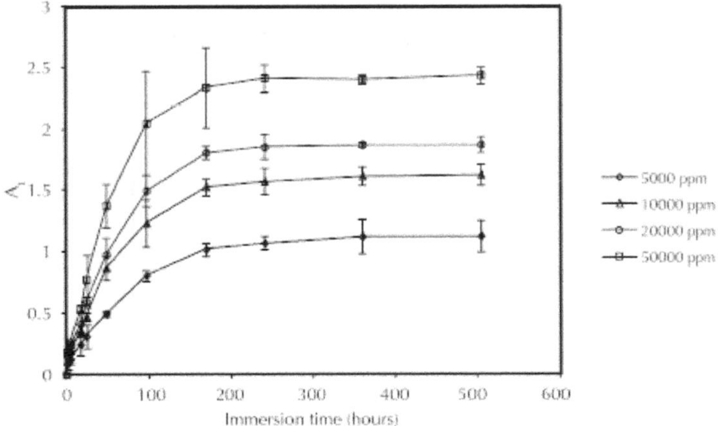

FIGURE 7 Evolution of the integrated intensity ratio At during diffusion of amyl acetate into polypropylene for different concentrations at 40°C.

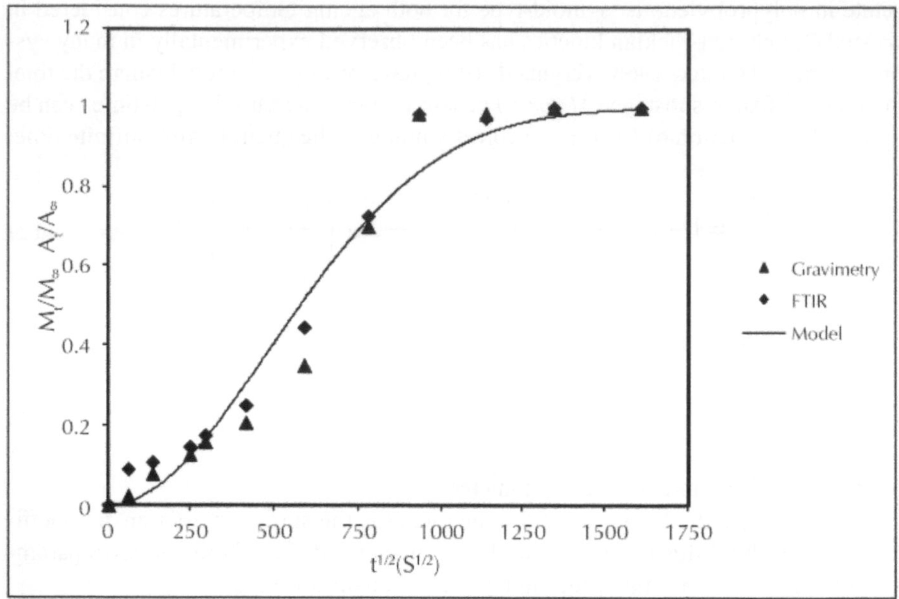

FIGURE 8 Comparison of experimental sorption kinetics data with sigmoid-shape fitted curves for immersed samples in 5000 ppm solution at 23°C.

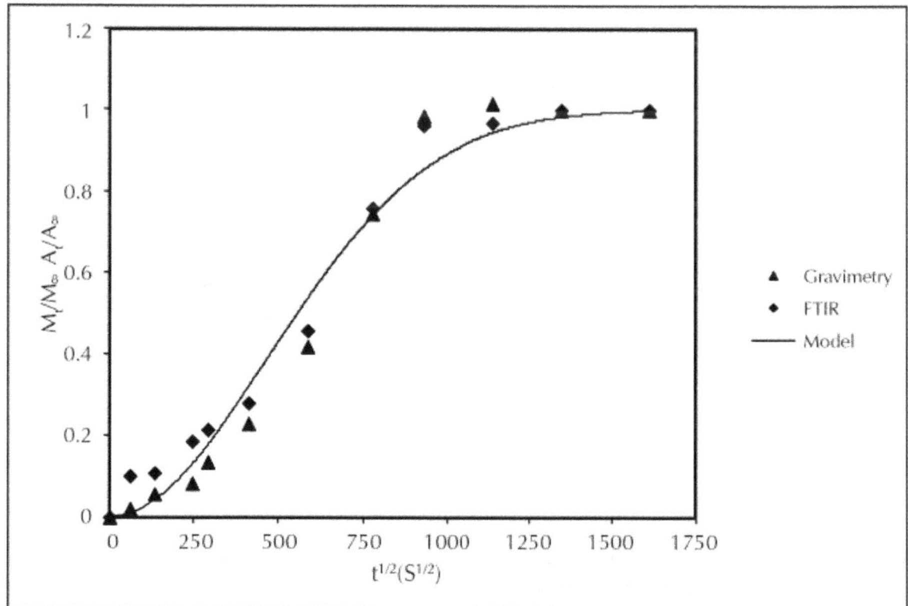

FIGURE 9 Comparison of experimental sorption kinetics data with sigmoid-shape fitted curves for immersed samples in 20000 ppm solution at 23°C.

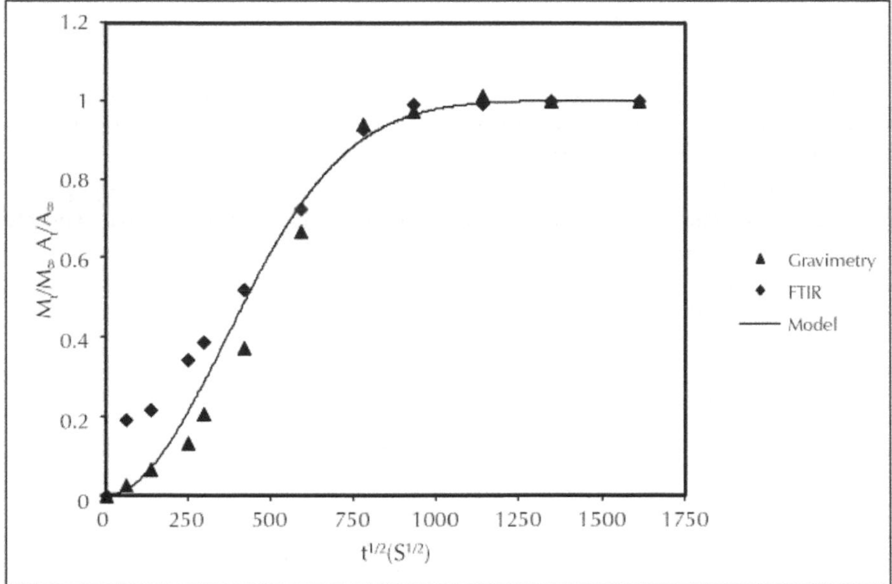

FIGURE 10 Comparison of experimental sorption kinetics data with sigmoid-shape fitted curves for immersed samples in 5000 ppm solution at 40°C.

2.7 MECHANICAL PROPERTIES: EXPERIMENTS AND MODELLING

2.7.1 EXPERIMENTAL RESULTS ON SAMPLES

In this section the effect of amyl acetate concentration and ageing temperature on the mechanical properties of polypropylene samples are presented.

After ageing, the test samples are wiped and dried. Each sample is then gripped at each extremity by suitable grips in a testing machine. The latter exerts an axial tension at a displacement rate of 10 mm/min so that the polymer is stretched. In order to quantify the ageing conditions effect, the elastic modulus (Young's modulus) and the peak tensile load are determined for each sample. Before testing, the crystallinity of the samples is determined by using a modulated Differential Scanning Calorimetry (DSC) analysis. The enthalpy of fusion ΔH_f and the enthalpy of crystallization ΔH_c of the samples are determined from the DSC thermograms of the virgin and aged polypropylene. The crystallinity (in %) of the polymer is then calculated by using the following equation:

$$X_C(\%) = \frac{\Delta H_f - \Delta H_C}{\Delta H_0} \times 100 \tag{14}$$

where $\Delta H_0 = 209 J / g$ is the enthalpy of fusion of the polymer counterpart 100% crystalline (Belofsky, 1995).

The tensile curves for virgin and aged polypropylene samples at 23°C and at 40°C are shown in Figure 12 and Figure 13 respectively. One can notice that the load increases quickly with a slight increase of the displacement before the yielding point. After peak load, it can be observed that all materials undergo necking propagation. Nevertheless, we can see differences in the slopes of the linear-elastic part and the maximum tensile loads between the virgin and aged specimens. These confirm that the sorption of the amyl acetate in polypropylene as well as the temperature contact condition influence the mechanical behavior of the polymer.

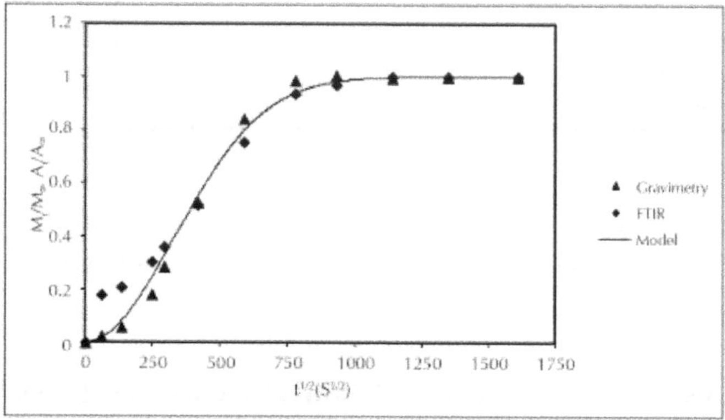

FIGURE 11 Comparison of experimental sorption kinetics data with sigmoid-shape fitted curves for immersed samples in 20000 ppm solution at 40°C.

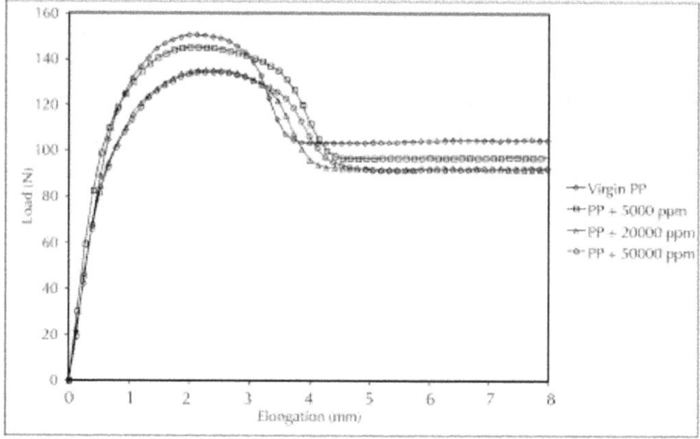

FIGURE 12 Load vs. elongation curves of virgin and aged polypropylene in various amyl acetate solutions at 23°C.

2.7.2 MODELLING MECHANICAL RESULTS

Experimental data issued from tensile tests on samples with varying amyl acetate concentration and ageing temperature have evidenced a decrease of mechanical properties (elastic modulus, peak load, cristallinity) over time. These experimental results match the amyl acetate molecules penetration effect in the polymeric material, inducing a decrease of the intermolecular interactions in the polymer, like a plasticizer should do.

The decrease of Young's modulus, maximum tensile load and cristallinity can be related to the concentration by using the following empirical models respectively (Abbès et al., 2010):

$$E = E_0 \exp\left(-\frac{a_E C}{b_F + C}\right) \tag{15}$$

where E_0 is the elastic modulus of the virgin polypropylene, C is the concentration of amyl acetate, a_E and b_E are the model's parameters.

$$F_{\max} = F_0 \exp\left(-\frac{a_E C}{b_F + C}\right) \tag{16}$$

where F_0 is the maximum tensile load of the virgin polypropylene, a_F and b_F are the model's parameters.

$$X_C = X_0 \exp\left(-\frac{a_X C}{b_X + C}\right) \tag{17}$$

where X_0 is the cristallinity of the virgin polypropylene, a_X and b_X are the model's parameters.

The unknown parameters (a_E and b_E, a_F and b_F, a_X and b_X) are determined by fitting experimental results.

The identified numerical results are given in Table 3–5 respectively. The experimental data and the fitted curves are depicted on Figure 14–16 for Young's modulus, maximum tensile loading and cristallinity respectively. One can notice the good agreement between experimental data and analytical models proposed for the evolution of mechanical properties like elastic modulus, peak load and cristallinity with varying concentration, whatever is the ageing temperature.

TABLE 3 Model parameters identified for the elastic modulus

Model parameters	T = 23°C	T = 40°C
E_0 (MPa)	843.5	843.5
a_E	0.41	0.51
b_E (ppm)	23286	15042

TABLE 4 Model parameters identified for the maximum tensile load

Model parameters	T = 23°C	T = 40°C
F_0 (N)	150.8	150.8
a_F	0.15	0.18
b_F (ppm)	11131	6395

TABLE 5 Model parameters identified for the cristallinity

Model parameters	T = 23°C	T = 40°C
X_0 (%)	35.53	35.53
a_X	0.12	0.20
b_X (ppm)	38592	30132

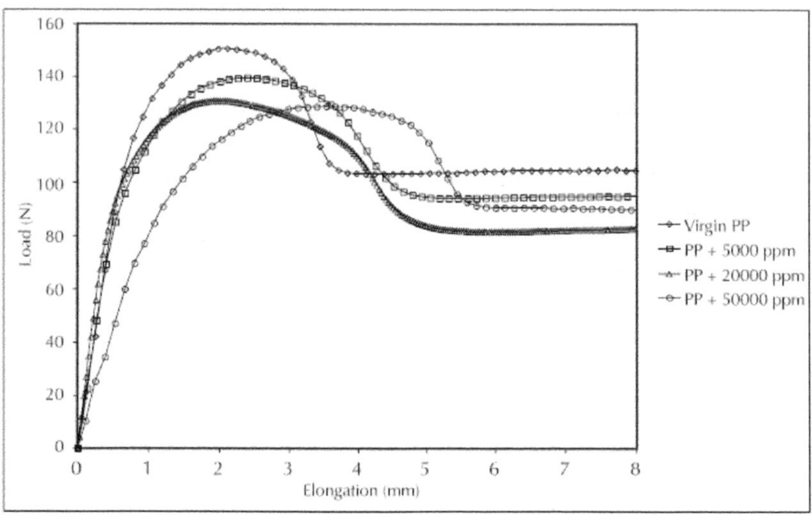

FIGURE 13 Load vs. elongation curves of virgin and aged polypropylene in various amyl acetate solutions at 40°C.

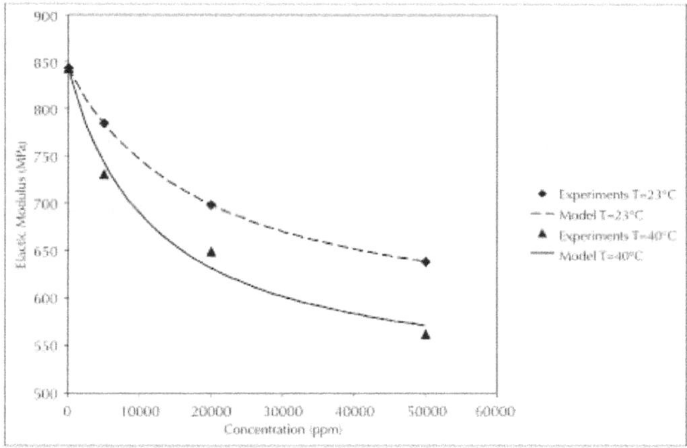

FIGURE 14 Comparison between experimental data and numerical results issued from modeling the elastic modulus evolution.

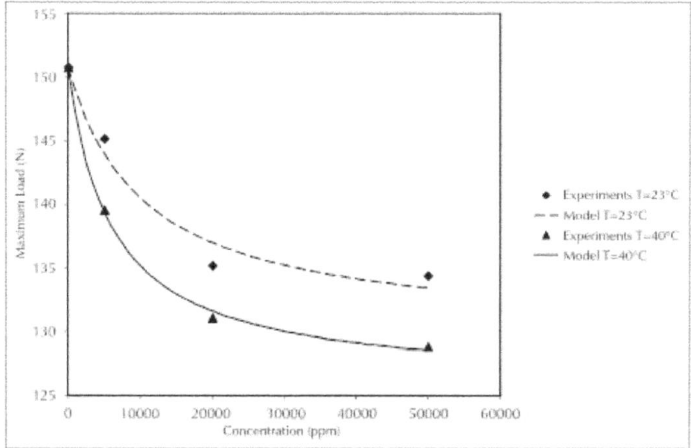

Figure 15 Comparison between experimental data and numerical results issued from modeling the peak load evolution.

2.7.3 MECHANICAL EXPERIMENTAL RESULTS ON BOTTLES

The top-load of crushable test is used to assess the axial load capacity of beverage and household product bottles. This is a critical parameter in guaranteeing the bottles will not buckle during filling and capping, or when stacked for transportation and storage.

To perform the test, a bottle is set-up beneath a compression plate and a compressive load is applied at a displacement rate of 10 mm/min until the bottle has visibly buckled.

Figure 17 and Figure 18 show an example of experimental curves for virgin and aged polypropylene bottles at 23°C and 40°C. We observe that during the top loading of the bottles, the load increases until reaching a maximum value at which occurs the buckling of the bottles. Similar results were obtained by van Dijk et al. (1998) in the case of PET and PVC bottles. The compressive strength decreases by increasing the concentration of amyl acetate and by increasing the ageing temperature. These results are coherent with the ones obtained in the previous section, where we have shown that the mechanical properties of polypropylene are strongly influenced by the amyl acetate sorption. The influence of amyl acetate concentration on the top-load (*TL*) of the bottles is modelled by using the following expression:

$$TL = TL_0 \exp\left(-\frac{a_{TL}C}{b_{TL}+C}\right) \tag{18}$$

where TL_0 is the top-load of the virgin polypropylene, a_{TL} and b_{TL} are the model's parameters.

The latter are determined by fitting the experimental results and are given in Table 6. The experimental results and the fitted curves are shown in Figure 19. There is a very good correlation between the experimental results and the proposed model whatever is the ageing temperature.

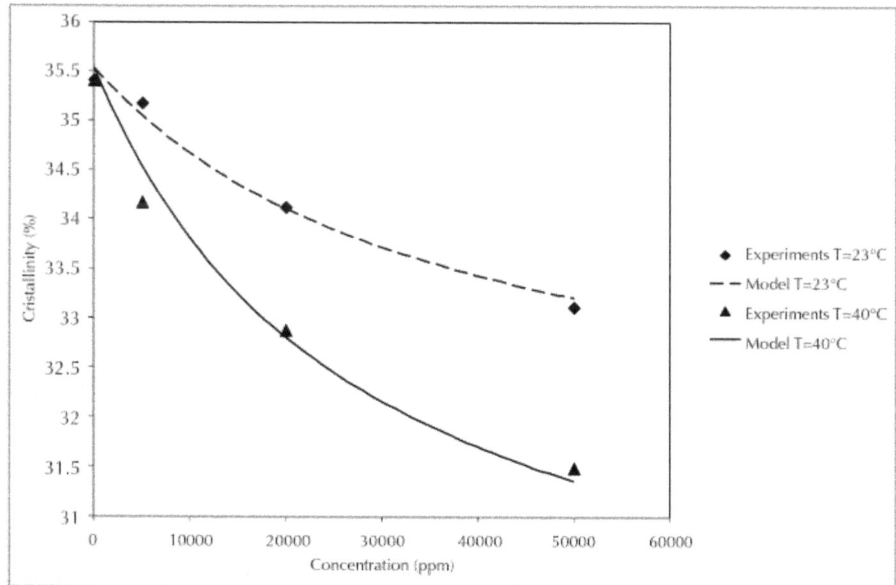

FIGURE 16 Comparison between experimental data and numerical results issued from modeling the cristallinity evolution.

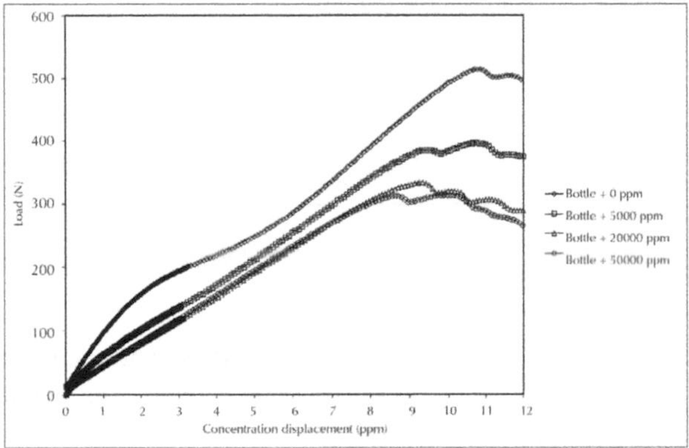

FIGURE 17 Load vs. compressive displacement curves of virgin and aged PP bottles at 23°C.

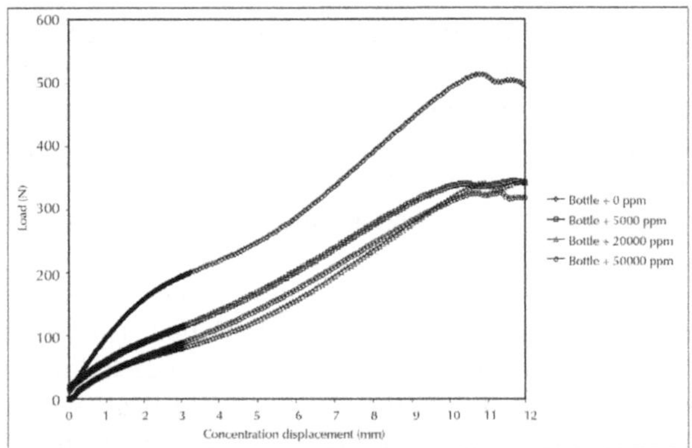

FIGURE 18 Load vs. compressive displacement curves of virgin and aged PP bottles at 40°C.

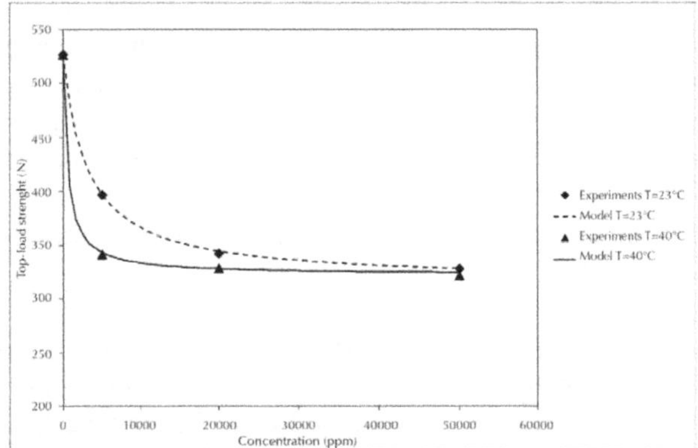

FIGURE 19 Comparison between experimental data and numerical results issued from modeling the top load evolution.

TABLE 6 Model parameters identified for the top-load

Model parameters	T = 23°C	T = 40°C
TL_0 (N)	526.6	35.53
a_{TL}	0.51	0.49
b_{TL} (ppm)	4043	728

2.8 CONCLUSION

This chapter provided a non-exhaustive review of studies concerning content/container interactions generally encountered in polymer packaging. The interaction is called migration process if substances pass from primary packaging to packed product, and sorption in the other case. Both could have many consequences on quality, shelf-life, organoleptic or therapeutic efficacy properties concerning the product, or strength, stress-cracking-resistance, resilience and more generally the mechanical properties of the packaging.

After a brief overview of experimental methodologies set up for the interaction phenomena characterization, we have focused on amyl acetate sorption in polypropylene (PP) packaging. FTIR and gravimetric techniques used here evidenced that the sorbed quantity of amyl acetate increases with the solution concentration and with the solute temperature. They also allowed the determination of the diffusion coefficient, which is a crucial parameter when a better understanding of the diffusion process is needed for predicting properties over time.

Furthermore, thermal studies by DSC have shown a decrease in the polymer cristallinity with the ester concentration of the packed solution. This result has been correlated with the plasticizing effect of the amyl acetate on one hand, and with the migration of the incorporated additives from the polymer to the solution on the other hand.

For the amyl acetate/PP system considered here, we have found that the sorption kinetics is of sigmoïdal-type. This indicates an asymptotic behavior of the sorption kinetics at the polymer surface.

To model this mass transfer, we have introduced, in addition to the diffusion coefficient D, the surface mass transfer coefficient H representing the above asymptotic sorption-kinetics at content/container interface. Both parameters have been determined from sorption tests. Experimental results show that the diffusion coefficient D remains constant, while the surface mass transfer coefficient H increases with amyl acetate concentration and temperature of the solution.

Another part of this work concerned the setup of an experimental methodology and the associated analytical modeling strategy for the prediction of the mechanical behavior of aged bottles. Tensile tests have been performed on strip-shaped samples cut from virgin and aged PP bottles. In particular, we have determined the experimental time-evolution of mechanical properties like elastic modulus and peak load. Evolution models of those properties as a function of the amyl acetate concentration have been proposed. Furthermore, top-load tests performed on several bottles have been performed to evaluate the evolution of the resistance to vertical compression as a function of aging. The comparison between experimental data from top-load tests and analytical model results of such compressive test has shown good agreement.

In this chapter, we have proposed a new approach for the experimental characterization and the analytical modeling of the mechanical behavior of polymer packaging taking into account mass transfer phenomena occurring in content/container interactions. This approach has shown its efficiency and can then be extended to the behavior analysis of any product/packaging system.

KEYWORDS

- **Additive**
- **Permeation**
- **Polymer packaging**
- **Polypropylene (PP)**
- **Stablizers**

REFERENCES

1. Abbès, B., Zaki, O., and Safa, L. Experimental and numerical study of the aging effects of sorption conditions on the mechanical behaviour of polypropylene bottles under columnar crush conditions. *Polymer Testing*. **29**, 902–909 (2010).
2. Al-Malaika, S., Goonetileka M. D., and Scott, G. Migration of 4-substituted 2-hydroxy benzophenones in low density polyethylene: part I – Diffusion characteristics. *Polymer Degradation and Stability*, **32**, 231–247 (1991).
3. Aminabhavi, T. M., and Phayde, H. T. S. Molecular transport characteristics of Santoprene thermoplastic rubber in the presence of aliphatic alkanes over the temperature interval of 25 to 70°C. *European Polymer Journal*, **36**(5), 1023–1033 (1995).
4. Arora, D. K., Hansen, A. P., and Armagost, M. S. Sorption of flavor compounds by low-density polyethylene film. *Journal of Food Science*, **56**, 1421–1423 (1991).
5. Auras, R., Harte, B., and Selke, S. Sorption of ethyl acetate and d-limonene in poly (lactide) polymers. *Journal of the Science of Food and Agriculture*, **86**, 648–656 (2006).
6. Begley, T. H., Biles, J. E., Cunningham, C., and Pringer, O. Migration of a UV stabilizer from polyethylene terephthalate (PET) into food simulants. *Food Additives and Contaminants*, **21**(10), 1007–1014 (2004).
7. Begley, T. H., Brandsch, J., Limm, W., Siebert, H., and Piringer, O. Diffusion behaviour of additives in polypropylene in correlation with polymer properties. *Food Additives and Contaminants*, **25**, 1409–1415 (2008).
8. Belofsky, H. *Plastics: Product design and process engineering.* Hanser Publisher (1995).
9. Berlinet, C., Ducruet, V., Brillouet, J. M., Reynes, M., and Brat, P. Evolution of aroma compounds from orange juice stored in polyethylene terephthalate (PET). *Food Additives and Contaminants*, **22**(2), 185–195 (2005).
10. Brandsh, J., Mercea, P., Ruter, M., Tosa, V., and Piringer, O. Migration modelling as a tool for quality assurance of food packaging. *Food Additives and Contaminants*, **19**(Supplement), 29–41 (2002).
11. Charara, Z. N., Williams, J. W., Sshmidt, R. H., and Marshall, M. R. Orange flavor absorption into various polymeric packaging materials. *Journal of Food Science*, **57**, 963–966 (1992).
12. Chiang, I. J., Chau, C. C., and Lee, S. The mass transport of ethyl acetate in syndiotactic polystyrene. *Polymer Engineering and Science*, **42**(4), 724–732 (2002).

13. Crank, J. *The mathematics of diffusion*. Clarendon Press, Oxford, 2nd Edition (1975).
14. D'Aniello, C., Guadagno, L., Gorrasi, G., and Vittoria, V. Influence of the crystallinity on the transport properties of isotactic polypropylene. *Polymer*, **41**, 2515–2519 (2000).
15. Del Nobile, M. A., Licciardello, F., Scrocco, C., Muratore, G., and Zappa, M. Design of plastic packages for minimally processed fruits. *Journal of Food Engineering*, **79**, 217–224 (2007).
16. Dhoot, S. N., Freeman, B. D., Stewart, M. E., and Hill, A. J. Sorption and transport of linear alkane hydrocarbons in biaxially oriented polyethylene terephthalate. *Journal of Polymer Science: Part B: Polymer Physics*, **39**, 1160–1172 (2001).
17. Dole, P., Feigenbaum, A., De La Cruz, C., Pastorelli, S., Paseiro, P., Hankemeier, T., Voulzatis, Y., Aucejo, S., Saillard, P., and Papaspyrides, C. Typical diffusion behaviour in packaging polymers – application to functional barriers. *Food Additives and Contaminants*, **23**(2), 202–211 (2006).
18. Dole, P., Voulzatis, Y., Vitrac, O., Reynier, A., Hankemeier, T., Auceio, S., and Feigenbaum, A. (2006). Modelling of migration from multi-layers and functional barriers: Estimation of parameters. *Food Additives and Contaminants*, **23**(10), 1038–1052.
19. Dopico-Garcia, M. S., Lopez-Vilaño, J. M., and Gonzalez-Rodriguez, M. V. Determination of antioxidant migration levels from low-density polyethylene films into food simulants. *Journal of Chromatography A*, **1018**, 53–62 (2003).
20. Escobal, A., Iriondo, C., and Katimed, I. Organic solvents adsorbed in polymeric films used in food packaging: Determination by head-space gas chromatography. *Polymer Testing*, **18**, 249–255 (1999).
21. Fayoux, S. C., Seuvre, A. M., and Voilley, A. J. Aroma transfers in and through Plastic packagings: Orange juice and d-limonene. A review. Part I: Orange juice aroma sorption. *Packaging Technology and Science*, **10**, 69–82 (1997).
22. Ferrara, G., Bertoldo, M., Scoponi, M., and Ciardelli, F. Diffusion coefficient and activation energy of Irganox 1010 in poly (propylene-co-ethylene) copolymers. *Polymer Degradation and Stability*, **73**, 411–416 (2001).
23. Han, J. K., Selke, S. E., Downes, T. W., and Harte, B. R. Application of a computer model to evaluate the ability of plastics to act as functional barriers. *Packaging Technology and Science*, **16**, 107–118 (2003).
24. Harison, H. Migration of plasticizers from cling-film. *Food Additives and Contaminants*, **5**, 493–499 (1988).
25. Hernández-Muñoz, P., Catala, R., and Gavara, R. Effect of sorbed oil on food aroma loss through packaging materials. *Journal of Agricultural and Food Chemistry*, **47**, 4370–4374 (1999).
26. Hernández-Muñoz, P., Catala, R., and Gavara, R. Food aroma partition between packaging materials and fatty food simulants. *Food Additives and Contaminants*, **18**(7), 673–682 (2001).
27. Hernández-Muñoz, P., Catala, R., and Gavara, R. Simple method for the selection of the appropriate food simulant for the evaluation of a specific food/packaging interaction. *Food Additives and Contaminants*, **19**(Supplement), 192–200 (2002).
28. Jasenka, G., Kata, G., Zelimir, K., and Nada, C. Gas permeability and DSC characteristics of polymers used in food packaging. *Polymer Testing*, **20**, 49–57 (2000).
29. Johansson, F. Packages for food - materials, concepts and interactions, a literature review. *SIK - The Swedish Institute for Food and Biotechnology*, Göteborg (1993).
30. Kanavouras, A. and Courtelieris, F. A. Shelf-life predictions for packaged olive oil based on simulations. *Food chemistry*, **96**, 48–55 (2006).
31. Kata, G. and Nada, C. Permeability characterisation of solvent treated polymer materials. *Polymer Testing*, **20**, 599–606 (2001).
32. Kaushc, H. H. The effect of degradation and stabilization on the mechanical properties of polymers using polypropylene blends as the main example. *Macromolecular Symposia*, **225**, 165–178 (2005).
33. Kulkami, S. B., Kariduraganavar, M. Y., and Aminabhavi, T. M. Molecular migration of aromatic liquids into a commercial fluoroelastomeric membrane at 30, 40, and 50°C. *Journal of Applied Polymer Science*, **90**, 3100–3106 (2003).

34. Lau, O. and Wong, S. Contamination in food from packaging material. *Journal of Chromatography A*, **882**, 255–270 (2000).
35. Lickly, L. D., Lehr, K. M., and Welsh, G. C. Migration of styrene from polystyrene foam food-contact articles. *Food and Chemical Toxicology,* **33**(6), 475–481 (1995).
36. Mohney, S. M., Hernandez, R. J., Giacin, J. R., Harte, B. R., and Miltz, J. Permeability and solubility of d-limonene vapor in cereal package liners. *Journal of Food Science*, **53**, 253–257 (1988).
37. Moisan, J. Y. Les additifs du polyéthylène – II: Influence de l'orientation. *European Polymer Journal*, **16**, 997–1002 (1980).
38. Nir, M. M. and Ram, A. Sorption and migration of organic liquids in Poly (Ethylene Terephthalate). *Polymer Engineering and Science*, **36**(6), 862–868 (1996).
39. O'Brien, A. and Cooper, I. Polymer additive migration to foods-a direct comparison of experimental data and values calculated from migration models for polypropylene. *Food Additives and Contaminants*, **18**(4), 343–355 (2001).
40. Paik, J. S. Comparison of sorption in orange flavor components by packaging films using the headspace technique. *Journal of Agricultural and Food Chemistry*, **22**, 1822–1825 (1992).
41. Park, G. S. and Crank, J. *Diffusion in polymers*. Academic Press (1968).
42. Patzlaff, M., Wittebrock, A., and Reichert, K. H. Sorption studies of propylene in polypropylene. Diffusivity in polymer particles formed by different polymerization processes. *Journal of Applied Polymer Science*, **100**, 2642–2648 (2006).
43. Petersen, J. H. and Breidahl, T. Plasticizers in total diet samples, baby food and infant formulae. *Food Additives and Contaminants*, **17**, 133–141 (2000).
44. Quinto-Fernandezy, J., Perez-Lamelaz, C., and Simal-Gandaraz, J. Analytical methods for food-contact materials additives in olive oil simulant at sub-mg kg-1 level. *Food Additives and Contaminants*, **20**(7), 678–683 (2003).
45. Reynier, A., Dole, P., Humbel, S., and Feigenbaum, A. Diffusion coefficients of additives in Polymers. I. Correlation with geometric parameters. *Journal of Applied Polymer Science*, **82**, 2422–2433 (2001).
46. Reynier, A., Dole, P., and Feigenbaum, A. Integrated approach of migration prediction using numerical modelling associated to experimental determination of key parameters. *Food Additives and Contaminants*, **19**(Supplement), 42–55 (2002).
47. Riquet, A. M., Wolff, N., Laoubi, S., Vergnaud, J. M., and Feigenbaum, A. Food and packaging interactions: determination of the kinetic parameters of olive oil diffusion in polypropylene using concentration profiles. *Food Additives and Contaminants*, **15**(6), 690–700 (1998).
48. Roduit, B., Borgeat, C. H., Cavin, S., Frangnière, C., and Dudler, V. Application of Finite Element Analysis (FEA) for the simulation of release of additives from multilayer polymeric packaging structures. *Food Additives and Contaminants*, **22**(10), 945–955 (2005).
49. Sadler, G. D. and Braddock, R. J. Oxygen permeability of low density polyethylene as a function of limonene absorption: An approach to modeling flavor scalping. *Journal of Food Science*, **55**, 587–588 (1990).
50. Safa, H. L. and Bourelle, F. Sorption-desorption of aromas on multi-use PET bottles: A test procedure. *Packaging Technology and Science*, **12**, 37–44 (1999).
51. Safa, L. and Abbès, B. Experimental and numerical study of sorption/diffusion of esters into polypropylene packaging films. *Packaging Technology and Science*, **15**, 55–64 (2002).
52. Safa, L., Abbès, B., and Zaki, O. Effect of amyl acetate sorption on mechanical and thermal properties of polypropylene packaging. *Packaging Technology and Science*, **20**, 403–411 (2007).
53. Safa, L., Zaki, O., Le Prince-Wang, Y., and Feigenbaum, A. Evaluation of model compounds – polypropylene film interactions by Fourier Transformed Infrared spectroscopy (FTIR) method. *Packaging Technology and Science*, **21**, 149–157 (2008).
54. Schwope, A. D., Till, D. E., Ehntholt, D. J., Sidman, K. R., Whelan, R. H., Schwartz, P. S., and Reid, R. C. Migration of BHT and ingranox 1010 from low-density polyethylene (LDPE) to foods and food-simulating liquids. *Food and chemical Toxicology*, **25**, 317–326 (1987).

55. Sharman, M., Readn, W. A., Castle, L., and Gilbert, J. Levels of di-(2-ethylhexyl) phthalate and total phthalate esters in milk, cream, butter and cheese. *Food Additives and Contaminants*, **11**, 375–385 (1994).
56. Siddaramaia, H. and Mallu, P. Sorption and diffusion of aldehydes and ketones through castor oil-based interpenetrating polymer networks of PU–PS. *Journal of Applied Polymer Science*, **67**, 2047–2055 (1998).
57. Simko, P., Sklarsova, B., Simon, P., and Belajova, E. Decreased benzo (a) pyrene concentration in rapeseed oil packed in polyethylene terephtalate. *European journal of lipid science and technology*, **107**, 187–192 (2005).
58. Soares, N. F. F. and Hotchkiss, J. H. Comparative effects of de-aeration and package permeability on ascorbic acid loss in refrigerated orange juice. *Packaging Technology and Science*, **12**, 111–118 (1999).
59. Stoffers, N. H., Stormer, A., Bradley, E. L., Brandsch, R., Cooper, I., Linssenk, J. P. H., and Franz, R. Feasibility study for the development of certified reference materials for specific migration testing. Part 1: Initial migrant concentration and specific migration. *Food Additives and Contaminants*, **21**(12), 1203–1216 (2004).
60. Tawfik, M. S., Devliegherea, F., and Huyghebaert, A. Influence of D-limonene absorption on the physical properties of refillable PET. *Food chemistry*, **61**(1–2), 157–162 (1998).
61. Techavises, N. and Hikida, Y. Development of a mathematical model for simulating gas and water vapor exchanges in modified atmosphere packaging with macroscopic perforations. *Journal of Food Engineering*, **85**, 94–104 (2008).
62. Ullsten, N. H. and Hedenqvist, M. S. A new test method based on head space analysis to determine permeability to oxygen and carbon dioxide of flexible packaging. *Polymer Testing*, **22**, 291–295 (2003).
63. van Dijk, R., Sterk, J. C., Sgorbani, D., and van Keulen, F. Lateral deformation of plastic bottles: Experiments, simulations and prevention. *Packaging Technology and Science*, **11**, 91–117 (1998).
64. Vergnaud, J. M. *Liquid transport processes in polymeric materials: Modelling and industrial applications.* Prentice Hall (1991).
65. Wang, Y., Allan, J. E., and Chen, X. D. Ethylene and oxygen permeability through polyethylene packaging films. *Packaging Technology and Science*, **11**, 169–178 (1998).
66. Wesselingh, J. A. and Krishna, R. *Mass transfer in multicomponent mixtures*. Delft University Press, Netherlands (2000).
67. Zaki, O., Abbès, B., and Safa, L. Non-Fickian diffusion of amyl acetate in polypropylene packaging: Experiments and modelling. *Polymer Testing*, **28**, 315–323 (2009).

CHAPTER 3

SPECIFIC MIGRATION OF ANTIOXIDANTS BHT, IRGANOX 1076, AND IRGAFOS 168 INTO TYPICAL EDIBLE OILS UNDER MICROWAVE HEATING CONDITIONS

ZHIGANG LIU, ZHIWEI WANG, and LIXIN LU

ABSTRACT

Three groups of PE films added with antioxidants BHT, Irganox 1076, and Irgafos 168 were separately and purposely made for specific migration testing under microwave heating conditions with different microwave power and time. The initial concentration of contaminant simulant in the plastic films was tested both with Soxhlet extraction method and migration cell extraction method and determined by RP-HPLC method. Six types of edible oils were used as fatty food stimulants, whose temperature-time profiles were made by using MWS OSR-8 type fiber optic thermometer system (made by FISO Technology, Canada). There is a linear relationship between temperature and time under testing conditions. Migration of BHT, Irganox 1076, and Irgafos 168 separately from PE films into six types of edible oils was tested under microwave heating conditions (microwave power from 100 to 900W), whose diffusivity were calculated. Results show that the relationship between migration and microwave energy (product of microwave power and time) is also almost linear.

3.1 INTRODUCTION

Migration of chemical substances, such as plastic additives, monomers, oligomers, and their degradation products, and so on, from plastic packaging materials into foods, which could lead to food safety problems for consumers has been widely studied for many years by researchers from FDA, EC, and other organizations all over the world. In China, several affairs on migration problems of food contact materials led to great concern of China government especially the General Administration of Quality Supervision, Inspection, and Quarantine of the People's Republic of China(AQSIQ) since 2005 including "PVC film affair" (General Administration of Quality Supervision, Inspection and Quarantine of the People's Republic of China) and "PVC gasket phthalate affair" (Zur Druckversion dieser Aktualität) on about migration problems. Then plastic and paper materials for food contact should make QS Certification since 2006 and 2007 as required by AQSIQ. (General Administration of Quality Supervision, Inspection and Quarantine of the People's Republic of China)

Wang and Liu et al. started researches on migration problems of food contact materials especially polyolefin films since early 2004. Wang et al. firstly presented to Chinese consumers about migration testing technique, migrants' analysis technology, and migration models. (Wang, Sun, Liu, 2004) Liu reviewed the advancing on analytical mathematical models for migration prediction of chemical substances from plastic packaging materials and evaluates by non-dimensional method the influence of parameters on efficiency of analytical migration models. (Liu, Wang, 2007) Liu also numerically simulated migration of chemical substances from plastic food packaging materials into foods (simulants) and instability of migrants in foods (simulants) by Finite Difference Method (FDM) (Liu, Wang, 2007) and experimented migration of three antioxidants BHT, Irganox 1076, and Irgafos 168 into 100% ethanol and made comparison between experimental and numerical simulation results.(Liu, Hu, Wang, 2007)

Over the last few years, more and more microwave ovens have been put into work in kitchens as microwave heating is a kind of fast, clean, and convenient food heating method, which is so popular in ordinary family in China now. As microwave heating is a kind of special heating method, usually only plastic materials and ceramic containers should be used as food contact materials from which chemicals (such as antioxidants, plasticizers, and so on) and heavy metals (such as Pb, Cd, and so on) might migrate into foods. Migration of chemical substances from PET, CPET, PVC, PP, PVDC (Startin, Sharman, Rose, Parker, Mercer, Castle, Gilbert, 1987; Begley, Hollifield, 1990; Begley, Dennison, Hollifield, 1990; Castle, Jickells, Gilbert, Harrison, 1990; Henry, Bouquant, Scholler, Klinck, Feigenbaum, 1992; Rijk, Kruijf, 1993; Risch, 1993; Lau, Wong, 1996; Valdez, Gramshaw, Vandenburg, 1997; Gramshaw, Valdez, 1998; Badeka, Pappa, Kontominas, 1998; Galotto, Guarda, 1999; Nerin, Acosta, 2000; Nerin, Acosta, Rubio, 2002; Nerin, Fernandez, Domeno, Salafranca, 2003; Galotto, Guarda, 2004; Alin, Hakkarainen, 2010), and cartonboard (Johns, Jickells, Read, Castle, 2010) into several types of foods or food simulants under microwave heating conditions has been widely studied, however, seldom on PE stretch wrap film. Since, PVC film is forbidden to be used as stretch wrap film for foods under microwave heating conditions in China since 2006, PE film begin to be widely used as a new kind of stretch wrap film from which migration is a new problem. This chapter presents some results of specific migration of BHT, Irganox 1076, and Irgafos 168 from specially made PE film.

3.2 EXPERIMENTAL

3.2.1 MATERIALS

Three Groups of PE films were produced added with three different antioxidants respectively by Key Laboratory of Heat Transfer and Energy Conservation of The State Education Ministry (South China Univ. of Tech) and Guangzhou Feida Plastic Packaging Co., Ltd.

Three antioxidants used were:

- 2,6-di-tert-butyl-4-methylphenol (BHT, CAS:128-37-0, pure quality from Bayer, Germany),
- Octadecyl 3,5-di(tert)-butyl- 4-hydroxyhydrocinnamate (Irganox 1076, CAS:2082-79-3, pure quality from Ciba, Switzerland), and
- Tris(2,4-di-tert-butylphenyl)phosphate(Irgafos 168, CAS:31570-04-4, pure quality from Ciba, Switzerland).

Six edible oils used as fatty food simulants were bought from local supermarket:

- Soya bean oil,
- Peanut oil,
- Sunflowerseed oil,
- Maize oil,
- Rapeseed oil, and
- Reconciled edible oil.

3.2.2 CHEMICAL REAGENTS

Main chemical reagents used were: Methanol, acetonitrile, tetrahydrofuran(THF) and methylene chloride, HPLC grade, Merck, Germany; Ethanol, HPLC grade, Tianjin Shield Company, China.

3.2.3 TEMPERATURE MEASUREMENT UNDER MICROWAVE HEATING CONDITIONS

Traditional method as by spirit thermometer, mercury thermometer or thermocouple thermometer cannot be used to measure temperature profile of foods under microwave heating conditions. However, fiber optic thermometer is able to complete the task. In this chapter, MWS OSR-8 type fiber optic thermometer system made by FISO Technology (Canada) is used to obtain temperature-time profile of edible oils under different microwave power conditions.

3.2.4 RP-HPLC ANALYSIS

Concentration of antioxidants in chemical reagents was analyzed by RP-HPLC method. The RP-HPLC system consisted of a Waters 2695 Separations Module (Alliance Waters, Milford, USA), Waters 2996 PDA Detector, Waters 2487 Dual λ Absorbance Detector, and a column of length 150 mm × 4.6 mm ID, 5 μm particle size. RP-HPLC analysis conditions (See Table 1).

TABLE 1 RP-HPLC analysis conditions for antioxidants BHT, Irganox 1076, and Irgafos 168

Antioxidants	Mobile phase(v/v)	UV wavelength (nm)	Column temperature (°C)	Flow rate(mL/min)	Injection (μL)	Retention time(min)
BHT	88ethanol + 12% water	280	25	1	10	5.12
Irganox 1076	100% methanol	230	30	1.2	10	7.95
Irgafos 168	100% methanol	230	30	1.2	10	8.13

3.2.5 STANDARD CALIBRATION CURVE

Standard calibration curves were made for BHT, Irganox 1076, and Irgafos 168 in methanol respectively (See Figure 1).

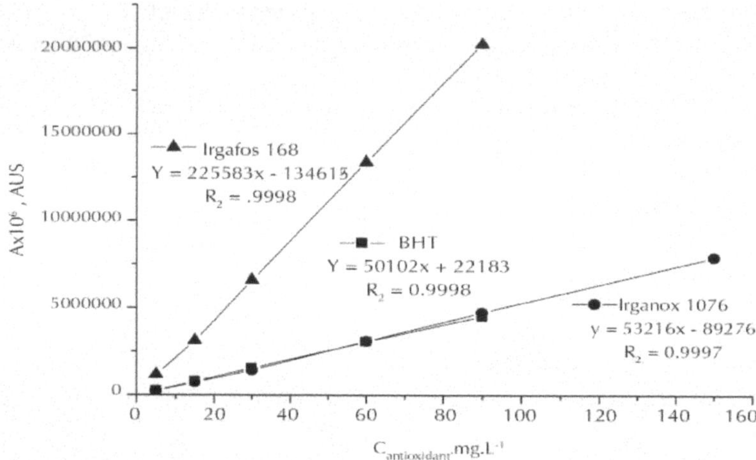

FIGURE 1 Standard calibration curve of BHT, Irganox 1076, and Irgafos 168 in methanol.

3.2.6 INITIAL CONCENTRATIONS

According to Han et al., (Han, Selke, Downes, 2003) the initial concentration of contaminant simulant in the plastic samples could be determined either by Soxhlet extraction method (for samples with thickness of 2.4 mil (0.006 cm) and lower) or migration cell extraction method (for thicker samples with thickness of 3.6 mil (0.009 cm) and higher). The PE films used in our tests were about 50 μm, so we applied both methods (See Table 2).

TABLE 2 Initial concentration of antioxidants BHT, 1076, and 168 in PE film

Antioxidants	Film thickness(μm)	Sample mass(g)	Initial concentration(g/g)		Maximum concentration in simulants(mg/L)
			Soxhlet	Migration Cell	
BHT	49.9	0.40736	0.00504	0.00490	49.893
Irganox 1076	50.2	0.40981	0.01073	0.01052	117.85
Irgafos 168	50.1	0.40434	0.00732	0.00746	75.459

3.3 DISCUSSION AND RESULTS

Migration of three typical antioxidants BHT, Irganox 1076, and Irgafos 168 purposely added into three PE films into six types of edible oils (fat food simulants) under different microwave power conditions is studied.

3.3.1 TEMPERATURE-TIME PROFILES

Temperature-time profiles of six types of edible oils are almost the same, thus, Figure 2 gives the average results. We can see there is almost the linear relationship between temperature and time from under 100W microwave power to 900W (See Table 3). The higher the microwave power, the faster the edible oil to reach the same temperature. The slope of temperature-time profile represents the speed to reach a temperature (See Figure 3). As PE film is usually used below 90°C, sampling time of migration testing under different microwave power conditions is determined in Table 4.

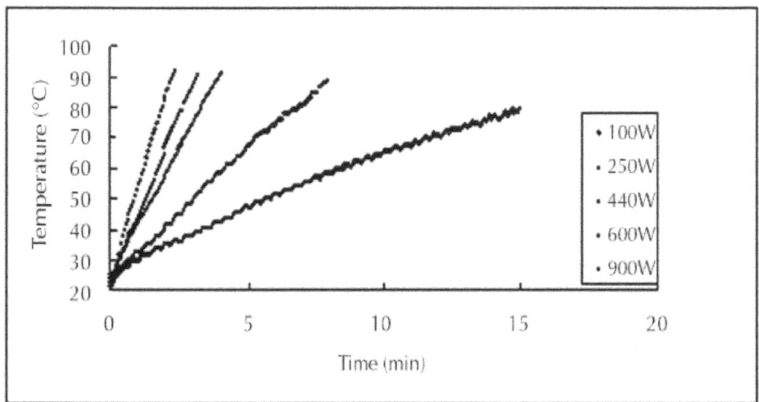

FIGURE 2 Temperature-time profiles of edible oil under different microwave power conditions.

TABLE 3 Equations of heating-up curve of edible oil under different microwave power conditions

Number	Microwave power (W)	Temperature-time equation (least squares for linear fitting)
1	100	y = 3.593x + 28.151
2	250	y = 8.2979x + 24.689
3	440	y = 16.664x + 25.175
4	600	y = 22.287x + 22.062
5	900	y = 30.527x + 23.138

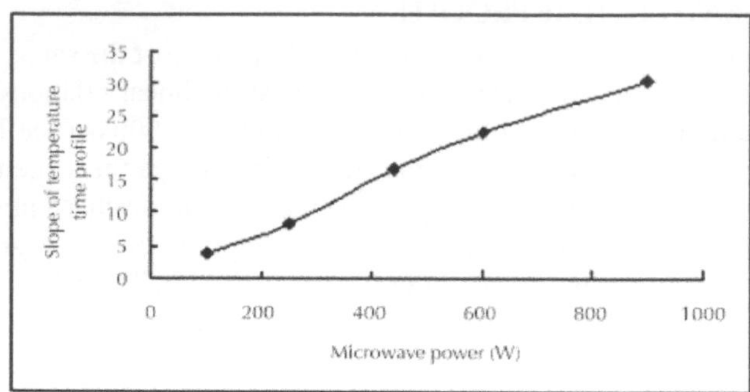

FIGURE 3 Relationship between microwave power and slope of temperature-time profile of edible oils.

TABLE 4 Sampling time of migration under different microwave power conditions

Time(min)	Microwave power(W)				
	100w	250w	440w	600w	900w
2	Δ	Δ	Δ	Δ	Δ
2.5				Δ	
3	Δ	Δ			
4			Δ		
5	Δ	Δ			
10	Δ				
15	Δ				

3.3.2 MIGRATION OF BHT, IRGANOX 1076, AND IRGAFOS 168 INTO SOYA BEAN OIL

Since, specific migration of BHT, Irganox 1076, and Irgafos 168 into six types of edibles is almost the same according to the results, only migration into soya bean oil is presented here. Because test condition is set below 90°C, the higher the microwave power, the less the testing points. Specific migration increases with time, when microwave power is constant (for example 100W) and with microwave power, when time is constant (for example 2 min), (See Figure 4–6). As we supposed, migration is quite related to microwave energy (product of microwave power and time). Migration increases with microwave energy input and there may be a linear relationship between migration and microwave energy (See Figure 7).

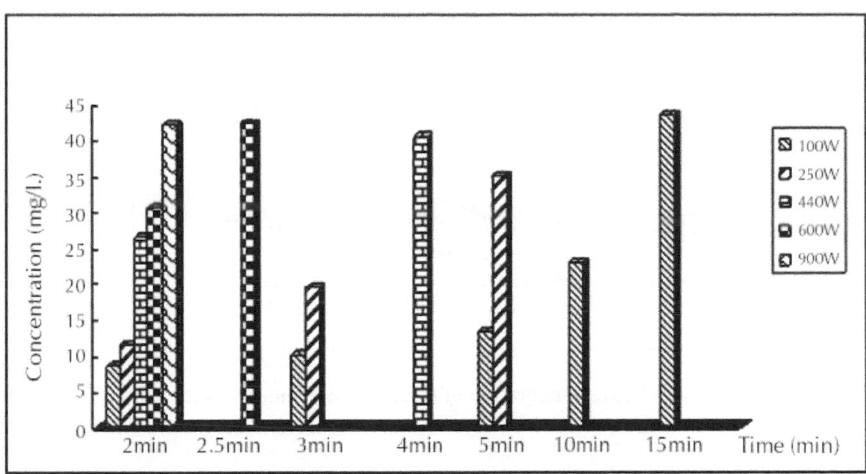

FIGURE 4 Migration of BHT into Soya bean oil under different microwave power conditions.

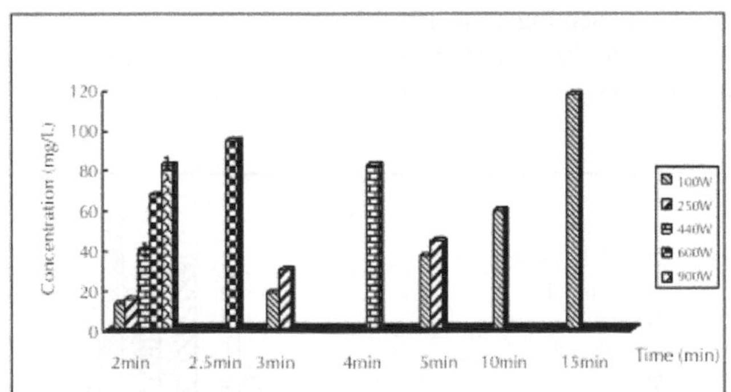

FIGURE 5 Migration of Irganox 1076 into Soya bean oil under different microwave power conditions.

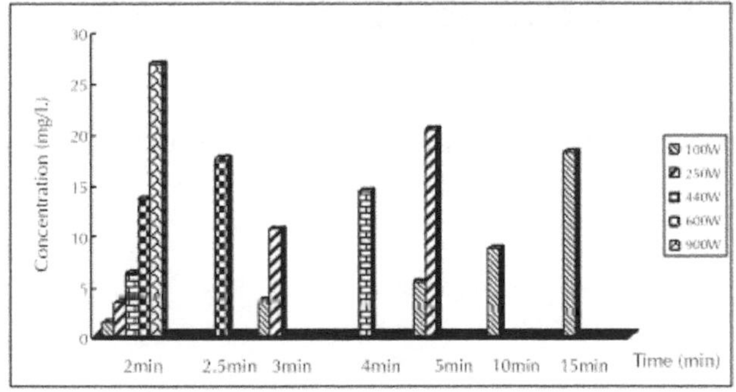

FIGURE 6 Migration of Irgafos 168 into Soya bean oil under different microwave power conditions.

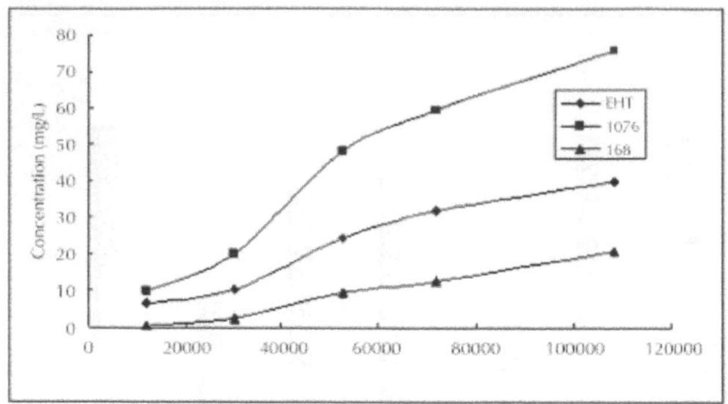

FIGURE 7 Relationship between migration of three antioxidants and microwave power.

3.3.3 EFFECT OF OIL TYPE ON MIGRATION

As stated in the previous chapter according to migration test results, oil type has little effect on migration of BHT, Irganox 1076, and Irgafos 168 from PE films, which is proved to be true here (See Figure 8–10). When microwave power is 100W, five test points with time 2, 3, 5, 10, and 15 min are studied. At the same time, migration is almost the same, whichever edible oil it is.

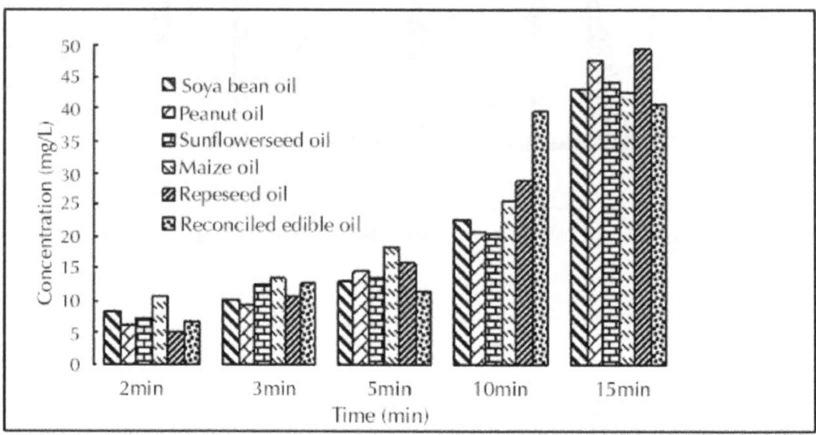

FIGURE 8 Migration of BHT into six types of edible oils under microwave power of 100 watt.

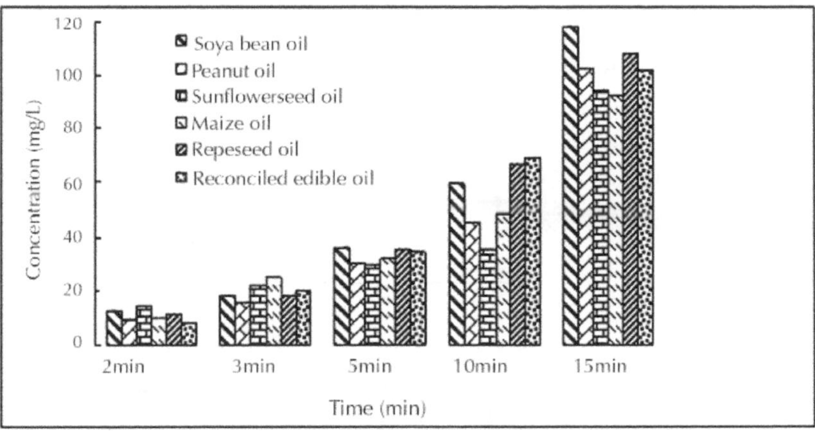

FIGURE 9 Migration of Irganox 1076 into six types of edible oils under microwave power of 100 watt.

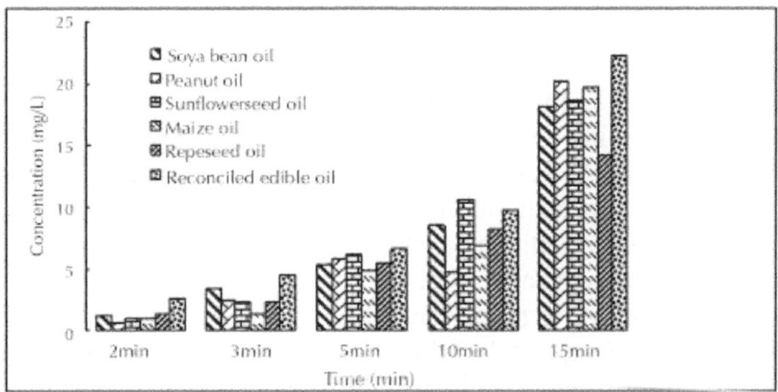

FIGURE 10 Migration of Irgafos 168 into six types of edible oils under microwave power of 100 watt.

3.3.4 DIFFUSIVITY OF BHT, IRGANOX 1076, AND IRGAFOS 168 UNDER MICROWAVE HEATING CONDITIONS

Four types of traditional diffusivity models (by Baner–Piringer, Limm–Hollifield, Brandsch, and Helmroth) are compared and discussed, which all are based on migrants' molecular weight (M) and temperature (T).(Liu, Wang, 2008) Brandsch's diffusivity model is a "Worst-Case" model, which has been the most popular one accepted by many researchers all over the world. However, the four diffusivity models are all under traditional conditions but not under microwave heating conditions. There are few chapters on migration of PE under microwave heating conditions and thus no diffusivity model is presented. Diffusivity of BHT, Irganox 1076, and Irgafos 168 is obtained according to migration results in this chapter, which is shown in Figure 11.

For BHT and Irganox 1076, diffusivity increases with microwave power and great increment is found, when microwave power is over 400W. For Irgafos 168, diffusivity increases slowly with microwave power, which mostly relates to its molecular weight and space structure.

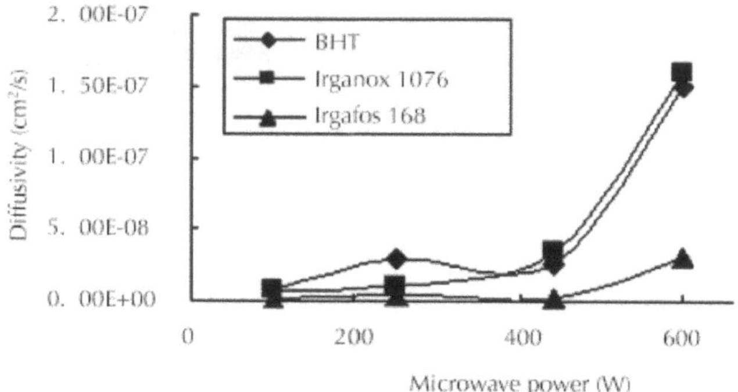

FIGURE 11 Relationship between diffusivity of three antioxidants and microwave power.

3.4 CONCLUSION

As PVC film is no longer permitted to be used as stretch wrap film for food packaging since 2006, PE film is widely used now from supermarket to home in China. BHT, Irganox 1076, and Irgafos 168 are typical antioxidants usually added in several plastic packaging materials especially in PE films. Migration of additives from PP, PET, PVDC, PVC, and PE under traditional conditions is studied in the last a few years but less under microwave heating conditions.

Three groups of special PE films were made on purpose added with BHT, Irganox 1076, and Irgafos 168 separately, whose initial concentration was determined by Soxhlet extraction method and migration cell extraction method. Migration of BHT, Irganox 1076, and Irgafos 168 separately from PE films into six types of edible oils, that is soya bean oil, peanut oil, sunflower seed oil, maize oil, rapeseed oil, and reconciled edible oil, were experimented under different microwave heating conditions with RP-HPLC analysis. Results show that six different types of edible oils had the same migratory behavior and it could be due to their components, which were almost the same as unsaturated and saturated fatty acids. MWS OSR-8 is a fiber optic thermometer system, which is made by FISO Technology (Canada) and it is quite fit to be used to obtain temperature-time profile of edible oils under different microwave power conditions. There is an obvious linear relationship between temperature and microwave time, whichever the edible oil type is. Migration obviously increases with microwave power and time and there is nearly a linear relationship between migrant concentration and microwave energy input (product of microwave power and time). Diffusivity of BHT, Irganox 1076, and Irgafos 168 is obtained according to migration results, under microwave heating conditions, which are quite larger than those obtained under room temperature conditions.

ACKNOWLEDGMENT

The authors acknowledge the Fundamental Research Funds for the Central Universities (JUSRP20901), the Key Projects in the National Science & Technology Pillar Program in the Eleventh Five-year Plan Period (2009BADB9B04) and the Opening Fund of Key Laboratory of Product Packaging and Logistics of Guangdong Higher Education Institutes, Jinan University.

KEYWORDS

- **Antioxidants**
- **Calibration curves**
- **Fatty food stimulants**
- **Microwave heating**
- **Specific migration**

REFERENCES

1. General Administration of Quality Supervision, Inspection and Quarantine of the People's Republic of China. AQSIQ Announcement No.155, 2006 [EB/OL]. http://www.aqsiq.gov.cn/zwgk/jlgg/zjgg/ zjgg2005/ 200610/t20061027_10885.htm, 2005-10-25
2. Zur Druckversion dieser Aktualität. Verunreinigung von öligen Lebensmitteln asiatischer Herkunft mit bedenklichen Weichmachern aus Schraubdeckel von Gläsern [EB/OL]. http://www.klzh.ch/aktuelles/ detail.cfm? id=81&archiv=set, 2007-06-07
3. General Administration of Quality Supervision, Inspection and Quarantine of the People's Republic of China. AQSIQ Announcement No.334, 2006 [EB/OL]. http://spscjgs.aqsiq.gov.cn/xzzq/xgcpc/ 200701/t20070105_23709.htm, 2006-07-25
4. General Administration of Quality Supervision, Inspection and Quarantine of the People's Republic of China. AQSIQ Announcement No.76, 2007 [EB/OL]. http://spscjgs.aqsiq.gov.cn/spxgcpgl/ spxgcpjhzpzxjdxx/200707/t20070720 _33901.htm, 2007-07-13
5. Wang, Z. W., Sun, B. Q., and Liu, Z. G. Packaging Engineering. China, 5, 1 (2004).
6. Liu, Z. G. and Wang, Z. W. Polymer Materials Science and Engineering. China, 5, 19 (2007).
7. Liu, Z. G. and Wang, Z. W. Journal of Chemical Industry and Engineering. China, 8, 2125 (2007).
8. Liu, Z. G., Hu, C. Y., and Wang, Z. W. Packaging Engineering. (China). 2007 ; 28(1) : 1–3, 9
9. Startin, J. R., Sharman, M., Rose, M., Parker, I., Mercer, A. J., Castle, L., and Gilbert, J. Food Addit Contam. 4, 385 (1987).
10. Begley, T. H. and Hollifield, H. C. Food Addit Contam. 3, 339 (1990).
11. Begley, T. H., Dennison, J. L., and Hollifield, H. C. Food Addit Contam. 6, 797 (1990).
12. Castle, L., Jickells, S. M., Gilbert, J., and Harrison, N. Food Addit Contam. 6, 779 (1990).
13. Henry, J. E., Bouquant, J., Scholler, D., Klinck, R., and Feigenbaum, A. Food Addit Contam. 4, 303 (1992).
14. Rijk, R. and Kruijf, N. D. Food Addit Contam. 6, 631 (1993).
15. Risch, S. Food Addit Contam. 6, 655 (1993).
16. Lau, O. W. and Wong, S. K. Packag Technol Sci. 9, 19 (1996).
17. Valdez, S. H., Gramshaw, J. W., and Vandenburg, H. J. Food Addit Contam. 3, 309 (1997).
18. Gramshaw, J. W. and Valdez, S. H. Food Addit Contam. 3, 329 (1998).
19. Badeka, A. B., Pappa, K., and Kontominas, M. G. Z Lebensm Unters Forsch A. 208, 429 (1999).
20. Badeka, A. B., Pappa, K., and Kontominas, M. G. Z Lebensm Unters Forsch A. 208, 69 (1999).

21. Galotto, M. J. and Guarda, A. Packag Technol Sci. 12, 277 (1999).
22. Nerin, C. and Acosta, D. J Agric Food Chem. 50, 7488 (2002).
23. Nerin, C., Acosta, D., and Rubio, C. Food Addit Contam. 6, 594 (2002).
24. Nerin, C., Fernandez, C., Domeno, C., and Salafranca, J. J Agric Food Chem. 51, 5647 (2003).
25. Galotto, M. J. and Guarda, A. Packag Technol Sci. 17, 219 (2004).
26. Alin, J. and Hakkarainen, M. J Appl Polym Sci. 118, 1084 (2010).
27. Johns, S. M., Jickells, S. M., Read, W. A., and Castle, L. Packag Technol Sci. 13, 99 (2000).
28. Han, J. K., Selke, S. E., and Downes, T. W. Packag. Technol. Sci. 16, 107 (2003).
29. Liu, Z. G. and Wang, Z. W. Polymer Materials Science and Engineering. China, 12, 25 (2008).

CHAPTER 4

SUITABILITY OF POLYMER-BASED RETORTABLE POUCHES FOR PACKAGING OF FISH PRODUCTS

J. BINDU, A. K. MALLICK, C. N. RAVISHANKAR, and T. K. S. GOPAL

ABSTRACT

Commercially available six opaque and three clear retortable pouches were evaluated based on physical parameters like thickness, oxygen, and water vapor transmission rates, tensile strength, elongation at break, bond strength, bursting strength, process resistance, migration residual air, and so on for their suitability for thermal processing. All pouches were found suitable for thermal processing, opaque pouches had lower oxygen and water vapor transmission rates compared to clear pouches. Imported clear pouch was superior than indigenously developed ones. Further analysis was conducted by using one opaque and three clear pouches for packaging of smoked tuna in natural pack. The heat penetration characteristics for processing of smoked tuna to a Fo value of 10 in the four pouces were recorded separately using CTF 9008 (Ellab, Roedovre, Denmark) temperature scanner recorder cum Fo and cook value integrator. The total process time was 42.01 min for opaque pouches (OP-E) and in clear pouches it was 33.83 min for CP-A, 38.04 min for CP-B, and 37.3 min for CP-C pouches respectively. Opaque pouches containing aluminum foil offer greater protection and shelf life than clear pouches, which allow visibility of the product packed in it.

4.1 INTRODUCTION

The retortable packaging has advanced tremendously due to the developing of films from the petroleum industry and improvements in the thermo processing techniques (Yamaguchi, 1990). The concept of pouch as a food container was developed by the US Army Natick Laboratories and a consortium of food packaging companies in the early 1960s (Herbert, Bettison, 1987). In the year 1968–1969 curry in foil free and aluminum foil containing pouches was commercialized in Japan (Tsutsumi, 1972). A typical pouch had an outer layer of polyethylene terephthalate for heat resistance, aluminum foil for oxygen/light barrier, bi-axial oriented nylon for resilience, and an inner cast polypropylene for pack sealing (Lampi, 1977). Comprehensive work on flexible packaging for thermal processed foods was done by (Lampi, 1980). Heat sterilized low acid solid foods in pouches created a new segment within the canned foods category (Brody, 2002).

The shelf life of a retorted food product is about 2 years for an aluminum foil pouch (Szczeblowski, 1971; Thorpe, Atherton 1972) 2–3 months for a barrier type non foil pouch and 1–2 months for a non foil retortable pouch (Yamaguchi, 1990). Studies on hamburgers packed in retort pouches showed that sensory characteristic depends on the oxygen permeability of the packaging material (Ishitani, Hirata, Matsushita, Hirose, Kodani, Ueda,, Yyanai, Kimura, 1980). Komatsu et al. (Komatsu, Yamaguchi, Kishimoto, 1970) have studied the effect of headspace gas on thermal processing and shelf life of processed products. Dymit (Dymit, 1973) reported that after 8 years shrimp in a foil laminate was superior in flavor and color to the canned item. Pouches using aluminum foil have a longer shelf life with respect to quality. Several workers have reported that thermal processing of fish curry in retort pouches resulted in products with good sensory attributes and also gave a shelf life of more than one year at room temperature (Vijayan, Gopal, Balachandran, Madhavan, 1998; Ravishankar, Gopal, Vijayan, 2002; Bindu, Gopal, Nair, 2004; Gopal, 2005l; Manju, Sonaji, Leema, Gopal, Ravishankar, Vijayan, 2004).

4.2 MATERIALS AND METHODS

Nine types of locally available opaque and clear (foil free) retortable pouches were evaluated based on physicochemical properties for their suitability for thermal processing. The details of the pouches as to the type of laminations and their layers are given in Table 1.

TABLE 1 Retort pouches and their specifications

	SOURCES OF RETORT POUCHES	LAYERS	CODE
1.	Pradeep Lamination, Pune, India	PEST/Al foil/ CPP	[OP-A]
2.	Premier Plastic Pack, Coimbatore, India	PEST/Al foil/Nylon/ CPP	[OP-B]
3.	Paper Products Ltd, Mumbai, India	PEST/Al foil/ CPP	[OP-C]
4.	Packaging India Ltd, Pondicherry	PEST/Al foil/ CPP	[OP-D]
5.	MH Packaging Ltd, Ahmedabad, India	PEST/Al foil/ CPP	[OP-E]
6.	Korean Retort Pouch (Imported)	PEST/Al foil/Nylon/ CPP	[OP-F]
7.	Korean Retort Pouch (Imported)	PEST coated with Al oxide /Nylon/ CPP (aluminum coated clear)	[CP-A]
8.	Pradeep Lamination, Pune, India	PEST coated with Silicon dioxide /Nylon/ CPP (silicon dioxide Clear)	[CP-B]
9.	MH Packaging Ltd, India	PEST/ CPP (clear two layer)	[CP-C]

Where, PEST-polyester, Al-aluminum, and CPP-cast polypropylene.

4.2.1 PHYSICAL PROPERTIES OF RETORT POUCHES

The thickness of each layer was measured with a handheld micrometer screw guage by recording the reading on scale as per (American Society for Testing and Materials Determination of thickness of paper and paperboard except electrical insulating papers). Tensile strength and elongation at break were determined using universal testing machine (Lloyd instruments LRX plus, UK) as per (Indian Standard Institute, Specification for low density polyethylene films New Delhi, India). For determining the tensile strength a sample of size 5 × 1.5 cm size was clamped on the seperating jaws of the equipment and pulled apart at a speed of 500 mm/min. The force required to make a break was measured alongwith the elongation as percentage at break. The test was repeated with 10 samples in both directions separately and the average value was calculated and expressed as kg/cm^2. The bond strength of the films was measured as per (American Society for Testing and Materials Standard method of test for peel or stripping strength of adhesive boards). The test specimens were cut into strips of size 25 × 2.5 cm in both directions. Initial separation of one half of the different layers of the specimen was done mechanically using solvents. The separated ends were then clamped in the jaws of the equipment and pulled apart at a test speed of 300 mm/min. The value is expressed as N/m or kg/2.54 cm or as psi. Heat seal strength of sealed seams were determined in a universal testing machine (American Society for Testing and Materials 1434) with samples cut 2.54 cm in width through the heat sealed area and by clamping each leg of the specimen in the jaws of the equipment. The sealed area should be equidistant between the clamps with at least 5 cm between the seal and clamp. The maximum stress applied on the seal at break (300 mm/min) is recorded and expressed as kg/cm^2. Bursting strength was measured using a bursting strength measuring equipment (Sevana Electrical Appliances Ltd., Kizhakkambalam) as per (Lampi, 1977). The conditioned samples of 12 × 12 cm size was clamped with the instrument, and subjected to hydrostatic pressure until the specimen ruptures. The maximum registered reading of pressure guage was noted down. Ten such replications was performed, and the average of the values was reported in psig. Water vapor transmission rate (WVTR) was measured as per (Indian Standard Institute, Methods of sampling and testing for paper and allied products, New Delhi, India). The test specimen of known area was fixed at the rim of the shallow aluminum dishes containing fused calcium chloride with wax. The set up was kept in a dessicator containing saturated solution of potassium nitrate so as to expose to 94% Relative Humidity (RH) and 37°C after weighing for a known period. Measurement of oxygen transmission rate (OTR) was carried out using gas permeability apparatus Gas and steam permeability, Ats Faar, Societa' Per Azioni, Milano, and Italia as per (American Society for Testing and Materials, Standard method for determining gas permeability characteristics of plastic film and sheeting). For conducting the tests samples were clamped in the equipment. The upper side of the specimen was provided with test gas at one atmosphere pressure and the lower portion was evacuated to create a vacuum below, which a mercury column was maintained in the graduated capillary tube. The flow of gas was measured with time noting the displacement in the capillary tube for a known period of time. The quantity of gas transmitted was calculated and expressed as cc/

m^2/24 hrs at one atmosphere pressure difference. Overall migration test was performed by using the food simulants such as distilled water, 3% acetic acid and n-heptane (Indian Standard Institute, Determination of overall migration of constituents of plastics materials and articles intended to come in contact with foodstuffs – methods of analysis). For process resistance of pouches (Gopakumar, 1993). One crumpled and another uncrumpled pouches were placed in a retort containing some water and heated to 121°C, 15 psi steam pressure, for 30 min. The pouches were then cooled, taken out, and examined for delamination if any.

Residual air in the pouch was determined as per (Shappee, Stanley, Werkowski, 1972). The test was performed by holding the pouch inverted below water under a funnel attached to a graduated cylinder filled with water. A corner of the pouch was cut open under the funnel and the air is squeezed out. The amount of residual air in the pouch was measured as the water displacement in the cylinder. The volumetric measurements of air were corrected to atmospheric pressure by Boyle's law:

$$V1 = (Pa - Wn)\ Vm/Pa$$

where, V1 = residual air in pouch at atmospheric pressure (ml)
Pa = atmospheric pressure (inches of mercury)
Vm = volume of measured air (ml)
Wh = pressure of water in graduated cylinder (inches of mercury)
where, $W = \rho gh$
where, ρ = density of water (kg/m^3)
g = acceleration due to gravity (m/s^2 or N/kg)
h = height of water in graduated cylinder (m)

4.2.2 PREPARATION OF SMOKED TUNA IN RETORTABLE POUCHES

Yellow fin tuna (*Thunnus albacares*) was purchased from the Cochin fisheries harbor, India. The tuna was washed, bled in chilled water and loined. The dark meat was removed from the loins and loins cut into steaks of 1.5 mm thickness. The steaks were then brined in 5% brine solution (w/v) for one hr. The drained steaks were then smoked for one hr using dried coconut husks at 75°C in a smoke kiln, Kerres–Germany (Model No.CS 350 "G" EL) to develop the smoke flavor.

4.2.3 THERMAL PROCESSING OF TUNA

The smoked tuna (100 g) was vacuum packed in four different pouches (150 × 200 mm) of 100 g capacity. The pouches used were opaque pouches (OP-E), aluminum oxide coated clear pouches (CP-A), silicon dioxide coated clear pouches (CP-B), and two layer clear pouches (CP-C). Tuna was packaged as natural or dry pack without any addition of filling medium. The heat penetration characteristics like Fo and cook value of tuna processed to Fo value of 10 in the above mentioned pouches were determined separately. For determining the data pouches were fixed with thermocouple glands through, which thermocouples were inserted. The tips of the thermocouples were inserted into the tuna for recording the core temperature during heat processing in a still over pressure retort (Model No 5682: John Fraser & Sons Ltd, Newcastle-upon-Tyne, UK). The thermocouple outputs were recorded using a CTF 9008 (Ellab, Roedovre, Denmark) temperature scanner recorder cum Fo and cook value integrator. The retort temperature (RT) was maintained at 121°C and air pressure was maintained at 28 psig throughout the heating and cooling period. After processing the pouches to the required Fo value, the pouches were rapidly cooled to 55°C (Tc) by pumping water into the retort and recirculating it. The lag factor for heating (Jh), slope of the heating curve (fh), time in min for sterilisation (U), and lag factor for cooling (Jc) were calculated by plotting temperature deficit (RT - Tc) against time on semi log paper. The process time (B) was calculated by the mathematical method (Indian Standard Institute, Determination of overall migration of constituents of plastics materials and articles intended to come in contact with foodstuffs – methods of analysis). The total process time was calculated by adding 58% of the come up time to (CUT) to B. Cook value (Cg), a measure of heat treatment with respect to nutrient degradation and textural changes that occur during processing, was also determined by measuring the extent of cooking and nutritional loss during processing in a manner similar to the D value, except that the reference temperature is 100° C instead of 121°C, and the Z value is 33°C, which is required for the denaturation of thiamine (Ranganna, 2000).

4.2.4 COMMERCIAL STERILITY

The processed pouches were subjected to sterility test as per (Indian Standard Institute: 2168, Specification for Pomfret canned in oil. New Delhi, India). Commercial sterility is ensured so that there is no growth of microorganisms after thermal processing. Four pouches were incubated at 55°C for 4 days and another four were incubated at 37°C for 14 days before conducting the tests. The incubated pouches were asceptically opened and 1–2 g of the samples were taken with a sterilized forceps and inoculated into the sterilized fluid thioglycolate broth in test tubes. In order to create anaerobic condition, a little sterilized liquid paraffin wax was put on the top of the broth and kept for incubation. The absence of any turbidity was observed after incubation, which indicated that the pouches were sterile.

4.3 DISCUSSIONS AND RESULTS

4.3.1 THICKNESS OF THE POUCHES

The total thickness of the pouch and individual thickness of the different layers of all the retort pouches are presented in the Table 2. It has been observed that there is not much variation of thickness of pouches of individual layers from one laminate to another. The thickness of different pouches ranged from 135–110 μm for the opaque retort pouches and 150–120 μm for the clear pouches. Both imported and indigenous pouches were having uniform thickness throughout the same specimen. The OP-B pouch was having one extra nylon layer apart from polyester, aluminum foil, and cast polypropylene layer. In case of clear pouches, CP-C had only two layers that is polyester and CPP, whereas in CP-A, there were three layers, where the outer polyester layer was coated with aluminum oxide and nylon was also present. In the case of CP-B, the polyester layer was quoted with silicon dioxide. The thickness of the pouch had a direct influence on the heat penetration characteristics and product quality. Non uniform thickness can affect the machine performance, product protection, and integrity of the packages (Hemavathi, Anupama, Vani, Asha, Vijayalakshmi, Kumar, 2002). The acceptable limit for variation in thickness of individual layers is ± 2 μm (inner ply) or 10% of the value (4–5). The thin profile helps in rapid transfer of heat to the inner regions unlike in cans and glass bottles. About 20–30% reduction in process time was observed by (Simpson, Almonacid, Mitchell, 2004) for vacuum packed mackerel in retortable pouch and processed in steam/air mixture at 116.8°C.

TABLE 2 Physicochemical properties of different types of opaque and see through retortable pouches

Properties		OP-A	OP-B	OP-C	OP-D	OP-E	OP-F	CP-A	CP-B	CP-C
Thickness (μm)	Total	115.00	135.00	110.00	115.00	115.00	120.00	120.00	150.00	131 ± 0.01
	Polyester	15.00	15.00	15.00	15.00	15.00	15.00	15.00	15.00	15.00
	Al Foil	30.00	15.00	15.00	30.00	15.00	25.00**			
	Nylon	-	20.00	-	-	-	-	15.00	20.00	
	CPP	70.00	85.00	80.0	70.00	85.00	80.00	90.00	125.00	115
*Tensile strength (Kg/cm²)	MD	462.82 ± 2.34	454.77 ± 8.63	460.93 ± 8.36	521.98 ± 6.39	451.25 ± 7.25	441.85 ± 9.51	816 ± 50.10	717 ± 0.01	292 ± 0.01
	CD	434.53 ± 6.49	436.90 ± 2.79	328.22 ± 11.0	506.19 ± 5.54	425.54 ± 8.21	415.62 ± 6.53	488 ± 0.07	592 ± 0.01	316 ± 0.01
*Elongation at break (%)	MD	48.83 ± 3.95	46.04 ± 4.25	36.42 ± 1.75	45.89 ± 2.40	31.24 ± 3.21	43.45 ± 4.83	95 ± 2.23	78 ± 2.22	53 ± 0.02
	CD	34.38 ± 3.56	36.83 ± 8.06	34.29 ± 2.17	39.07 ± 1.99	28.34 ± 2.48	35.77 ± 3.35	76 ± 3.42	68 ± 3.85	44 ± 0.01
*Heat seal strength (N/25 mm)	Machine direction	392.28 ± 1.06	253.75 ± 2.75	193.85 ± 2.45	289.63 ± 1.13	310.68 ± 3.25	381.50 ± 3.98	504 ± 1.31	538 ± 1.25	303 ± 0.01
	Lab seal	308.45 ± 2.32	205.69 ± 1.59	150.73 ± 2.52	212.63 ± 3.32	296.77 ± 2.31	324.47 ± 2.11	179.47 ± 2.24	189.04 ± 2.34	191 ± 1.14
	Cross direction	384.10 ± 4.14	246.76 ± 3.82	181.42 ± 3.13	210.68 ± 2.34	237.95 ± 2.11	317.43 ± 2.19	224.37 ± 2.21	412 ± 2.41	286 ± 0.02
*Bond strength (g/10 mm)		182.11 ± 0.11	138.21 ± 0.35	102.32 ± 0.29	189.44 ± 0.16	184.34 ± 0.21	109.52 ± 0.33	149.13 ± 0.04	123 ± 0.11	100.4 ± 0.11
Bursting strength (psig)		30.00	30.00	30.00	30.00	30.00	30.00	29.00	29.00	30.00
*OTR (ml/m²/ 24 hrs at 1 atm. at 21°C)		0.35 ± 0.01	0.32 ± 0.02	0.42 ± 0.04	0.34 ± 0.03	0.39 ± 0.02	0.30 ± 0.01	0.06 ± 0.01	2.0 ± 0.03	55 ± 0.03
*WVTR (g/m²/24 hrs at 37°C at 90% RH)		0.021 ± 0.005	0.04 ± 0.001	0.095 ± 0.004	0.05 ± 0.001	0.032 ± 0.006	0.087 ± 0.004	0.21 ± 0.003	0.87 ± 0.012	1.99 ± 0.03

TABLE 2 *(Continued)*

	D/D (121°C/2h)	0.066 ± 0.002	0.105 ± 0.002	0.103 ± 0.011	0.032 ± 0.021	0.091 ± 0.013	0.096 ± 0.012	4.8 ± 0.015	3.2 ± 0.022	6.8 ± 0.11
*Global migration (mg/dm^2)	3% AA (121°C/2h)	0.252 ± 0.045	0.126 ± 0.064	0.268 ± 0.074	0.252 ± 0.065	0.124 ± 0.084	0.137 ± 0.054	0.140 ± 0.028	0.124 ± 0.021	0.14 ± 0.11
	n-heptane (66°C/2h)	1.722 ± 0.032	0.545 ± 0.051	0.783 ± 0.041	2.120 ± 0.012	1.580 ± 0.036	1.708 ± 0.024	2.45 ± 0.021	1.65 ± 0.02	1.98 ± 0.01
Residual Air (%)		1.821	1.88	1.854	1.854	1.90	1.82	1.73	1.89	1.64

4.3.2 TENSILE STRENGTH AND ELONGATION AT BREAK

The tensile strength and elongation at break of the packaging materials determines the resistance of rupture and breakage, when subjected to tensile force. It has been observed from the Table 2 that the different laminates possessed good tensile strength and elongation at break for both machine direction and cross/transverse direction. In the case of opaque pouches the highest tensile strength was observed in OP-D and the lowest in OP-F. In machine direction, it was 521.25 kg/cm^2 and in cross direction, it was 506.19 kg/cm^2, the lowest values were 441.85 kg/cm^2 in MD, and 415.62 kg/cm^2 in CD. In the clear pouches the tensile strength was high in both the three layer pouches, whereas in the two layer pouches, it was comparatively less. The higher tensile strength in the clear pouches (CP-A and CP-B) over opaque pouches was perhaps due to the absence of the aluminum layer. The tensile strength of container is required to protect the product during processing, distribution, and retail handling. Polyester and nylon are commonly used as oriented films since, they are generally stronger than unoriented films (Ghazala,1994). These properties of packaging materials are of utmost importance to perform well with form fill seal operations (Vijayalakshmi, Sathish, Rangarao, 2003). The elongation at break in percentage also depended upon the tensile strength of the films.

4.3.3 HEAT SEAL STRENGTH

The heat seal strength, which is important for package integrity and prolonged shelf life is shown in Table 2. Heat seal strength of the three side sealed by manufacturer was measured for machine direction and cross direction, whereas for processor seal, it was measured in one direction only. It was observed from the results that the heat seal strength of manufacturer seal was higher than the processor seal. However, in all the cases the values were well above the prescribed limits (4–5). Polypropylene has good heat seal strength and as such it is used as the inner layer in pouches for heat sealing. The material can tolerate temperatures up to 130°C (Forshaw, 1990) and has necessary properties for food contact application.

4.3.4 BOND STRENGTH

Bond strength is one of the most important requirements for retort pouch, which contributes to prevent delamination of the films during thermal processing. The bond strength of all the pouches were higher than the prescribe requirements (Table 2).

The highest bond strength was for OP-D (189.44 g/10 mm) and the lowest for OP-C (102.32g/10 mm) for opaque pouches. In the case of clear pouches CP-C showed the lowest (100.4g/10 mm). Lower value indicates easy delamination of the layers during thermal processing, which results in physical destruction of pouches and reduction of barrier properties of retorts pouches. Bond strength was carried out between CPP layer and aluminum foil, whereas in other layers it is practically not feasible due to less strength of the layers, which break during the experiment.

4.3.5 BURSTING STRENGTH

In this study it is observed that all the pouches passed the bursting test by holding at above 25 psi for one minute in the bursting strength testing machine (Table 2). Low values of bursting strength indicate easy delamination of the layers during thermal processing, which results in physical destruction of the pouch and reduction in barrier properties (Vijayalakshmi, Sathish, Rangarao, 2003). Bursting strength combines tensile strength and tear resistance and serves as a rough guide to compare the packaging materials (Vijayalakshmi, Sathish, Rangarao, 2003).

4.3.6 OXYGEN TRANSMISSION RATE

The OTR values observed in all opaque retort pouches were very low, when compared to clear retort pouches (Table 2). The OTR is an important parameter to check the barrier properties of the retort pouches. The higher OTR value in case of clear pouches, when compared to opaque pouches is due to the lack of aluminum foil layer. Similar results have been observed for imported and indigenous three layer pouches (Vijayalakshmi, Sathish, Rangarao, 2003). For CP-A the OTR values were 0.6 ml/m^2/24 hrs at one atmosphere at 21°C and CP-B it was 2.0 ml/m^2/24 hrs at one atmosphere at 21°C. For CP-C, which is a two layer pouch, the OTR values were 55 ml/m^2/24 hrs at one atmosphere at 21°C. The polyester outer layer of the CP-A and CP-B pouches were coated with aluminum oxide and silicon dioxide and had a nylon layer, which would have given the desired barrier properties. The OTR is a very important parameter to know the headspace oxygen of the packaged retort pouch. Low oxygen barrier properties for retort pouches allow the penetration of oxygen into the pouch, which can lead to rancidity in the food product present inside the processed pouch.

4.3.7 WATER VAPOR TRANSMISSION RATE (WVTR)

The WVTR values for all opaque retort pouches were very low, which may be due to the presence of aluminum foil layer in the pouches (Table 2). In case of clear pouches the WVTR values for imported pouches were similar to opaque pouch. This may be due to good water vapor barrier property of aluminum oxide layer, which is present in the imported see through pouch. For indigenous see through pouches the WVTR rates were higher. The highest WVTR rate (1.99 g/m^2/24 hrs at 37°C at 90% RH) was reported in CP-C pouch, which is made of two layers. Plastic films may have pinholes, which under normal conditions are unlikely to occur, when they are laminated (Yamaguchi, 1990). It has been confirmed that even a single layer film does not allow the microorganisms to pass (Lampi, 1980). Hence the problem lies in careless handling during the manufacture of the products and during transportation and storage.

Contamination by penetration of water contaminated with microorganisms is of major concern during the cooling process. The cooling water should be clean for maintaining a quality product (Yamaguchi, 1990). In retort pouched products the wide sealing width does not allow the entry of water into the pouch from the seal area.

4.3.8 GLOBAL MIGRATION TEST

The results show that migration into n-heptane stimulants was higher than in distilled water and 3% acetic acid in all the pouches (Table 2). Most plastics in finished form contains components that may leach out from plastic to foods, whenever direct contact occurs between food and plastic, thereby contaminating the food product with the consequent risk of toxic hazard to the consumer (Gopal, Antony, Govindan,1981). Since, the migration of food is inevitable, various countries have formulated standards, which specify maximum limit of migration. Bureau of Indian Standard specify 10 mg/dm^2 or 60 ppm for finished materials (Indian Standard Institute, Determination of overall migration of constituents of plastics materials and articles intended to come in contact with foodstuffs – methods of analysis). Similar results were also reported by (Vijayalakshmi, Sathish, Rangarao, 2003). The selection of suitable packaging material for food contact application is decided on the basis of the physical, mechanical, barrier, and performance properties of the films (Iyer, 1992). However, the results of migration in our study indicated that the extractives were below the specified limit.

PROCESS RESISTANCE OF POUCHES

All retort pouches subjected to process resistance test were found to be suitable for thermal processing and no delamination and wrinkles were observed after treating the pouches at 121.1°C for 30 min.

4.3.9 THERMAL PROCESSING AND HEAT PENETRATION CHARACTERISTICS OF SMOKED TUNA IN RETORTABLE POUCHES

The heat penetration characteristics of smoked tuna in four different pouches are given in Table 3 and Figures 1–4. The Fo recommended for fish and fish products ranges from 5–20 (Frott, Lewis, 1994). An Fo value of 10 gave a good tuna product with desired sensory and textural characteristics (Vanloey, Francis, Hendrickx, Maesmans, Norohna, Tobback, 1994). The tuna packed in different pouches were processed separately to a Fo value of 10. The heating rate index (fh) was 17 min for OP-E pouches, 14 min CP-A pouches, 16.5 min for CP-B pouches, and 16 min for CP-C pouches. The lack of any filling medium in the dry pack decreased the heat penetration rate since, heat transfer took place by conduction only. The total process time (T_B) for smoked tuna by adding 58% of come up time for tuna dry pack was 42.01 min, for OP-E pouches, 33.83 min for CP-A pouches, 38.04 min for CP-B pouches, and 37.53 min for CP-C pouches. The other parameters like lag factor of heating (Jh), lag factor of cooling (Jc), time in minutes for sterilization at RT (U), and final temperature deficit (g) are also given in the Table 3. The graphs depicting the Fo value, cook value, RT, and core temperature for tuna dry pack are given in Figure 1–4 respectively. When we analyze the process time, it is seen that the total process time was highest for the opaque pouches (OP-E), followed by clear pouches (CP-B, CP-C, and CP-A). The higher values in opaque pouches may be due to the thick aluminum foil layer, whereas in clear pouches the process time increased with increasing thickness. Heat penetration characteristics at the slowest heating point should be specific, if parameters such as filled weight, head space, type of container, dimension of container, come up time of the retort, heating media, and initial temperature were uniform for a given product (39).

Sterility tests conducted for tuna dry pack showed that there was no turbidity in any of the test tubes, indicating the absence of bacterial growth. From this we conclude that all the pouches were commercially sterile at both the temperature of 37°C and 55°C.

TABLE 3 Heat penetration characteristics of smoked tuna dry packed in different pouches

Types of pouches	f_h(min)	J_h	J_c	f_h/U	g (°C)	B (min)	Total process time T_B (min)	Cook value (Cg) (min)
Opaque (OP-E)	17.00 ± 0.25	1.37 ± 0.45	1.05 ± 0.14	1.51 ± 0.19	0.72 ± 0.17	37.72 ± 0.15	42.01 ± 0.54	89.31 ± 0.23
Aluminum oxide coated clear (CP-A)	14 ± 0.23	1.02 ± 0.21	0.96 ± 0.2	1.17 ± 0.12	0.39 ± 0.01	28.35 ± 0.76	33.83 ± 0.67	77.80 ± 0.56
Silicon dioxide coated clear (CP-B)	16.5 ± 0.12	0.67 ± 0.21	1.00 ± 0.01	1.6 ± 0.12	0.67 ± 0.01	34.24 ± 0.1	39.1 ± 0.12	84.05 ± 0.65
Two layered clear (CP-C)	16 ± 0.34	0.77 ± 0.11	0.10 ± 0.02	1.01 ± 0.01	0.63 ± 0.11	32.80 ± 0.9	37.53 ± 0.98	82.45 ± 0.19

Each value is represented by the average ± standard deviation of at least 3 determinants. Where fh = slope of heating curve, Jh = lag factor of heating, Jc = lag factor of cooling, U = time in minutes for sterilization at RT, g = final temperature deficit, B = Ball's process time, T_B = total process time, and Cg = cook value.

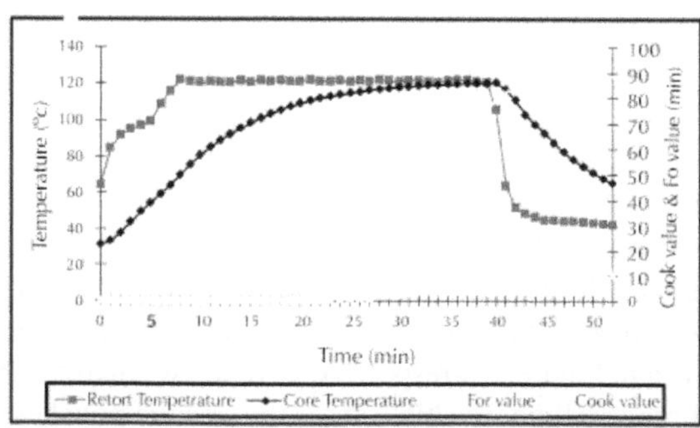

FIGURE 1 Heat Penetration characteristics of smoked tuna dry pack in opaque (OP-E) pouches.

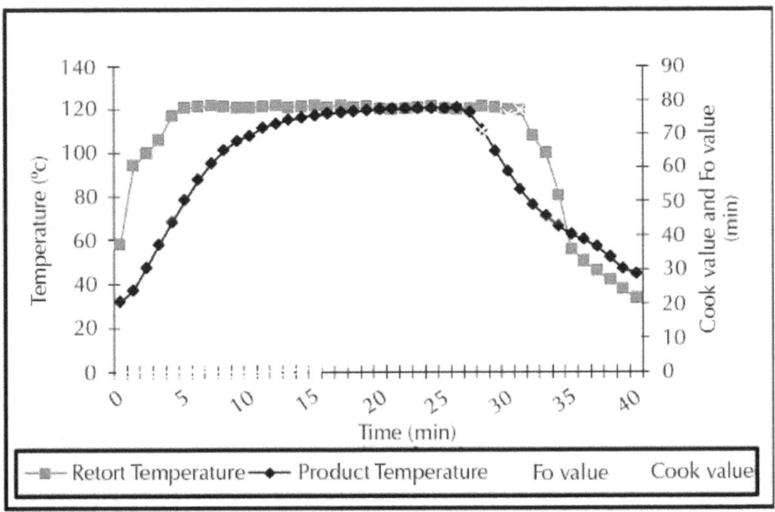

FIGURE 2 Heat Penetration characteristics of smoked tuna dry pack in aluminum oxide coated clear (CP-A) pouches.

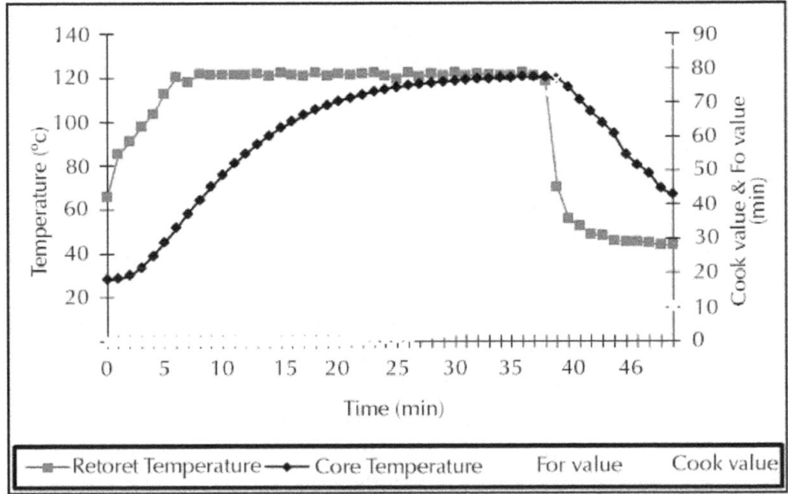

FIGURE 3 Heat Penetration characteristics of smoked tuna dry pack in silicon dioxide coated clear (CP-B) pouches.

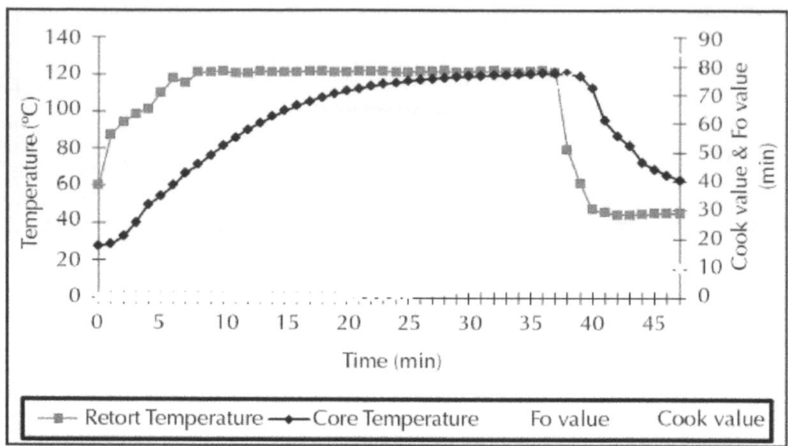

FIGURE 4 Heat Penetration characteristics of smoked tuna as dry pack in two layer clear (CP-C) pouches

4.3.10 RESIDUAL AIR TEST AFTER PROCESSING

The residual air test after the thermal processing of tuna showed that the value was well below the prescribed limit of 2% as described by (Shappee, Stanley, Werkowski, 1972). If the limit is exceeding, it will affect the shelf life of the product leading to rancidity. The excess air present in the pouch expands during thermal processing and also affects the seal integrityof the pouches.

4.4 CONCLUSION

The six opaque pouches and three clear pouches were found suitable for thermal processing since, they exhibited the required physical properties. The inner most layer of the laminates, which was in contact with the food and exhibited high heat seal strength was cast polypropylene for all the pouches. The high non permeability of the opaque pouches was mainly due to the aluminum foil, which was found sandwiched in between the polypropylene and polyester layer. The outer most layers in all the retortable pouches was the polyester, which had high mechanical strength and was abrasion resistant and can be effectively utilized for reverse color printing. Three layer clear pouches had nylon instead of the aluminum foil and were found to be coated either with aluminum oxide and silicon dioxide on the polyester layer. These pouches were found to be as good as opaque pouches during processing and storage of the products. Two layered clear pouches were found suitable for thermal processing but had a high oxygen transmission rate, which could affect the sterility of the product at ambient storage (28 ± 2°C).

KEYWORDS

- **Aluminum foil**
- **Opaque**
- **Physicochemical**
- **Retortable pouches**
- **Thermal processing**

REFERENCES

1. Yamaguchi, K. Retortable packaging. T. Kadoya (Ed.), *Food packaging*. Academic Press, Inc., New York, pp. 185–211 (1990).
2. Herbert, D. A. and Bettison, J. Packaging for thermally sterilized foods. S. Thorne (Ed), *Developments in food preservation*, **4**, 87–122 (1987).
3. Tsutsumi, Y. Retort Pouch. Its development and application to foodstuffs in Japan. *J. Plastics*, **6**, 24–30 (1972).
4. Lampi, R. A. Flexible packaging for thermo-processed foods. *Adv. Food Res.*, **23**, 306–428 (1977).
5. Lampi, R. A. Retort pouch. The development of a basic packaging concept in today's high technology era. *J. Food Proc. Engg.*, **4**, 1–18 (1980).
6. Brody, A. L. The return of the retort pouch. *Food Technol.*, **57**(2), 76 (2002).
7. Szczeblowski, J. W. An Assessment of the Flexible Packaging System for Heat-processed Foods. Tech. Rep. 71-57-GP. U.S. Army Natick Laboratories, Natick, Massachusetts (1971).
8. Thorpe, R. H. and Atherton, D. Sterilized food in flexible packages. Tech. Bull., No-21., Fruits and vegetable preservation research association, Chipping Campden, Gloucestershire, England (1972).
9. Ishitani, T., Hirata, T., Matsushita, K., Hirose, K., Kodani, N., Ueda, K., Yyanai, S., and Kimura, S. The effects of oxygen permeability of pouch, storage temperature, and light on the quality change of a retortable pouched food during storage. *J. Food Sci. Technol.*, **27**(3), 118–124 (1980).
10. Komatsu, Y., Yamaguchi, K., and Kishimoto, A. Effect of oxygen transmission rate of retort pouches on the browning of food-simulated systems during the heat-processing and storage: 17th Annual Meeting. *Society Food Science & Technology* (1970).
11. Dymit, J. M. Today's food service systems. *Food Packag. Syst.*, **25**(2), 56–63 (1973).
12. Vijayan, P. K., Gopal, T. K. S., Balachandran, K. K., and Madhavan, P. K. Fish curry in retort pouches. K. K. Balachandran, T. S. G. Iyer, J. Joseph, P. A. Perigreen, M. R. Raghunath, and M. D. Varghese (Eds.), *Advances and Priorities in Fisheries Technology,* Society of Fisheries Technologists(I), Cochin, India, pp. 232–235 (1998).
13. Ravishankar, C. N., Gopal, T. K. S., and Vijayan, P. K. Studies on heat processing and storage of seer fish curry in retort pouches. *Packag. Technol. Sci.,* **15**, 3–7 (2002).
14. Bindu, J., Gopal, T. K. S., and Nair, T. S. U. Ready to Eat Mussel Meat Processed in Retort Pouches for Retail and Export Market. *Packag. Technol. Sci.,* **17**, 113–117 (2004).
15. Gopal, T. K. S. Safety aspects of packaging materials. J. Joseph, T. K. S. Gopal, C. N. Ravishankar, J. Bindu, and G. Ninan (Eds.), Current trends in packaging of fish and fishery products. Central Institute of Fisheries Technology, Cochin, India, pp. 173–179 (2005).
16. Manju, S., Sonaji, E. R., Leema, J., Gopal, T. K. S., Ravishankar, C. N., and Vijayan, P. K. Heat penetration characteristics and shelf life studies of seer fish moilee packed in retort pouch. *Fish. Technol.*, **41**(1), 37–44 (2004).

17. American Society for Testing and Materials Determination of thickness of paper and paperboard except electrical insulating papers. ASTM. D 645, Mc Graw-Hill Book Co. Inc., New York (1964).
18. Indian Standard Institute, Specification for low density polyethylene films New Delhi, India, IS: 2508 (1984).
19. American Society for Testing and Materials Standard method of test for peel or stripping strength of adhesive boards. ASTM D 90349. Mc Graw-Hill Book Co. Inc., New York. (1972).
20. American Society for Testing and Materials 1434. Standard methods of test for seal strength of flexible barrier materials, ASTM 68–88. Philadelphia (1973).
21. Indian Standard Institute, Methods of sampling and testing for paper and allied products, New Delhi, India. IS: 1060 Part II (1960).
22. American Society for Testing and Materials, Standard method for determining gas permeability characteristics of plastic film and sheeting. ASTM. D1434Mc Graw-Hill Book Co. Inc., New York (1982).
23. Indian Standard Institute, Determination of overall migration of constituents of plastics materials and articles intended to come in contact with foodstuffs – methods of analysis, I S: 9845 New Delhi (1998).
24. Gopakumar, K. Retortable pouch processing. K. Gopakumar (Ed.), *Fish packaging technology, materials, and methods*, Concept publishing company, New Delhi, pp. 113–132 (1993).
25. Shappee, J., Stanley, J., and Werkowski, S. J. Study of a non-destructive test for determining the volume of air in flexible food packages. Technical Report 73-4-GP, US Army Natick Laboratories, Natick, Massachusetts (1972).
26. Stumbo, C. R. *Thermobacteriology in Food Processing. Second Edition*. Academic Press, New York, pp. 93–120 (1973).
27. Ranganna, S. *Handbook of Canning and Aseptic Packaging*. Tata McGraw-Hill Publishing Co. Ltd., New Delhi, India, pp. 507–508 (2000).
28. Indian Standard Institute: 2168, Specification for Pomfret canned in oil. New Delhi, India (1971).
29. Hemavathi, A. B., Anupama, G., Vani, K. V., Asha, P., Vijayalakshmi, N. S., and Kumar, K. R. Physico-chemical and mechanical properties evaluation of polyolefin films and composite. *Popular Plastics Packag.*, **47**, 65–70 (2002).
30. Simpson, R., Almonacid, S., and Mitchell, M. Mathematical model development, experimental validation, and process optimization: Retortable pouches packed with seafood in cone frustum shape. *J. Food Eng.*, **63**(2), pp. 153–162 (2004).
31. Ghazala, S. New packaging technology for seafood preservation-shelf life extension and pathogen control. A. M. Martin (Ed.), *Fisheries Processing: Biotechnological applications*, Chapman and Hall, London, pp. 83–110 (1994).
32. Vijayalakshmi, N. S., Sathish, H. S., and Rangarao, G. C. P. Physico-chemical studies on indigenous aluminum foil based retort pouches vis-a -vis their suitability for thermal processing. *Popular plastics & Packaging*, **48**, 71–74 (2003).
33. Forshaw, T. Prepared meals choices for packaging in plastics. *Food Rev.,* **17**, 23–25 (1990).
34. Gopal, T. K. S., Antony, K. P., and Govindan, T. K. Packaging of fish and fishery products, present status, and future prospects. *Seafood Export J.*, **13**(1), 15–22 (1981).
35. Iyer, H. R. Need for quality control of packaging materials. Packaging India, **24**(6), 23–29 (1992).
36. Frott, R. and Lewis, A. S. *Canning of Meat and Fish Products*. R. Frott and A. S. Lewis (Eds.), Chapman and Hall, London, UK, pp. 200–202 (1994).
37. Ali, A. A., Sudhir, B., and Gopal, T. K. S. Effect of rotation on the heat penetration of thermally processed tuna (*Thunnus albacares*) in oil in aluminum cans. *Fish Technol.*, **42**(1), 55–60 (2005).
38. Vanloey, A., Francis, A., Hendrickx, M., Maesmans, G., Norohna, J., and Tobback, P. Optimizing thermal process for canned white beans in water cascading retorts. *J. Food Sci.*, **59**(4), 828–832 (1994).

Part II: Bio-Based and Biodegradable Materials for Packaging

CHAPTER 5

A VIEW ON ECO-FRIENDLY NATURAL FIBERS FOR PACKAGING

D. SARAVANABAVAN and G. C. MOHAN KUMAR

ABSTRACT

Plastics are the fore most materials for packaging. In the present days, plastic materials obtained are of two kinds namely thermoset plastic and thermoform plastic. But, these plastic materials are hazardous to mankind and leave lot of carbon emissions to the atmosphere. Alternative plastic material called green plastics can be used to reduce the foot print of carbon emission and these materials are obtained from natural resources such as plant fibers which are biodegradable and renewable. The present work is focused on natural fibers for packaging applications. Survey is concentrated on environmental and packaging issues, packaging innovation techniques, reducing the packaging waste, sustainable, flexible and eco-packaging, recycling of waste plastics and biodegradable plastic materials. Experimental set up is also carried for natural fibers for packaging applications.

5.1 INTRODUCTION

Natural fibers are the immense sources of materials for packaging, structural, automotive and composite applications. From the history, plant fibers were used for packaging and building applications. Even today, it can be seen in rural part of India, plantain leaves are used for wrapping food products and spices. Today's materials used for packaging include paper, metals (aluminium and steel), glass, earthen ware materials and plastics (petroleum- derived polymers). Paper boards used for packing include corrugated boxes, bags and wrapping paper. They are usually thicker than paper, made in multiple form layers and used for transport packaging. These packaging's are low cost, compared to other packing materials. Fiber board and chip board along with small quantity of paper materials are used for packing food materials, dry food products, and other materials (Al-Salem, Lettieri, and Baeyens, 2009).

Metallic packing provides more barrier protection and storage for the product, but they are expensive, should spend lot of time in manufacturing and availability source are limited. Aluminium used for packing applications includes foils, beverages, cans and used for packing food, beverages, and paints. Glass packing and containers are suited for gaseous products and they do not react with content of the product and safe for handling containers. They can be re-used, reformed and recycled easily with minimum effort but manufacturing of these materials are time consuming and need more care for packing. Earthen ware materials are also used for storing liquid and solid foods and the materials are porous, which suits for cooling. Petroleum derived polymers are called plastics; they are more flexible and can easily moulded to any form and shape. These materials are light in weight, easily fabricated, and processed.

The present article is focussed on different plastic and bio-plastic materials for packaging applications and their impact on the environment; conversion and recycling of waste plastics are also briefed. Use of friendly packaging material (natural fibers) and their outcomes are explained. Ecological packaging with minimal waste and packaging innovation techniques are also added. For practice, experimental work is conducted on using maize plant fibers and thermoset polymeric resin as matrix processed by Vacuum Assisted Resin Transfer Molding (VARTM) to obtain product in the form of packing box.

5.2 ENVIRONMENTAL AND FRIENDLY PACKAGING

For obtaining an environmental and friendly packing minimum roles should be performed, which are shown below:

- It should provide full physical and chemical protection to the products,
- Must be free from damage and protect from moisture, dust and other contaminant factors,
- Convenient in handling, easy distribution and transportation,
- Design factors such as shape, size and weight should be considered,
- Disposal of the packing material must be bio degradable,
- The material should be feasible, flexible, and recycled.

5.2.1 NATURAL PLANT FIBERS IN PACKAGING

Natural fibers are divided in to plant fibers, animal fibers, and mineral fibers (Andrady, 2011; Anne, 2011). Plant fibers are playing a major role in establishing a green environment for composites, structural applications, automotive, and packaging sectors. Uses of these fibers are increasing day by day because of added advantages such as low density, less cost, renewability, bio degradable, carbon neutral, and environmental friendly (Andrady, 2011; Anne, 2011; Anshuman, Khardenavis, Sandeep, and Tapan, 2007; Averous, and Pollet, 2012). At the end of their life cycle, they are 100% biodegradable and they reduce carbon emissions. Natural fibers are bio degradable through the help of microorganisms such as bacteria and fungi. Plant fibers used for packing are light weight, and hence mainly used for packing food items and dry food products. Some of the fibers are cotton, kenaf, sisal, plantain leaves, coconut palm, and bamboo. The properties of natural fibers are comparably good when compared to the properties of synthetic fibers as shown in table 1.

Cotton fabrics are inexpensive and can be used for wrapping grains, sugars, and dry foods. They are made in large quantities and processing of these materials is easy and the availability of the material is abundant. These materials are suitable for holding light products and also protect from environment. Banana/ plantain leaves are used for wrapping food and spicy items. These leaves provide pleasant flavor over the food and thereby it may also create some positive medicinal effect to the food items. Plantain and green leaves are commonly used for eating plates and the pseudo stem sheath provides wrapping material for food and fruits.

TABLE 1 Properties of Glass and Natural Fibers

Fibres	Density (g/cm³)	Tensile Strength (MPa)	E-Modulus (GPa)	Moisture Absorption (%)
Flax	1.4	800–1500	60–80	7
Hemp	1.4	550–900	70	8
Jute	1.46	400–800	10–30	12
Jute	1.46	400–800	10–30	12
Ramie	1.5	500	44	12–17
Coir	1.25	220	6	10
Sisal	1.33	600–700	38	11
Cotton	1.51	400	12	8–25
E-glass	2.55	2400	73	-

Coconut palm leaves are used for shelters for home and the fibers can be used as yarn/string and the dried leaves are used for roof tops. They also play a major role in packing dry foods and fruits. Hence forth, in India, banana and coconut plants are said to be Kalpatharu or Kalpavriksha (full filling divine tree) because the usage of these sources are plenty. Kenaf fibers and sisal fibers are used as rope and string material for packing. Bamboo fibers are used for transportation and packaging heavy products. Palm leaves can be weaved and used as basket for carrying vegetables and fruits. But these natural fibers have the problem of moisture absorption and rotting effect, which in turn create problems for the food items. Kraft papers, parchment paper, grease proof paper, sulphite paper, are some of the paper items used for packaging. The paper items should be chemically treated, coated and impregnated to have better functional properties. Paper packaging is commonly used for corrugated boxes, wrapping paper, folding bags, paper plates, and cups. Papers are obtained from network of cellulose fibers which are derived from wood sources. Paper board are thicker often processed in multiple layers and are used for transportation and shipping. White board, solid board, chip board, fiber board are some of the common boards used for boxes and trays (Al-Salem, Lettieri, and Baeyens, 2009).

The problems in natural fibers are low strength properties such as impact strength; moisture absorption, swelling of fibers, processing temperature is restricted and poor fire resistance. But these problems can be tackled by chemical treatment of fibers (Azeredo, 2009) and changing the fibers from hydrophilic to hydrophobic. Adhesion properties between the natural fibers and matrix can be improved by treating the fibers with chemicals. Always, good impregnation makes the composite material to carry more loads.

5.2.2 CONVENTIONAL PLASTICS IN PACKAGING

The conventional types of materials used for packing are plastic, paper, glass, fiber board and metallic (steel and aluminium). Synthetic polymers (Plastics) are derived from petroleum fossils and are made-up of long chain of hydrocarbons with additives. They offer many advantages over other materials in terms of light weight, appreciable strength, ease process and are easily moulded into finished product. Traditional plastics are usually used for packing food items, films, packing bottles, confectionary items and they are non-biodegradable and recyclable. Table 2 shows traditional plastic materials and their applications in packaging.

TABLE 2 Conventional Plastics and their Applications in Packaging (Błędzki, Jaszkiewicz, Urbaniak, and Walczak, 2012)

Plastics	Applications
Low density polyethylene (LDPE) Poly vinyl chloride (PVC)	Films and packaging
Polyethylene terephthalate (PET) PVC High Density Polyethylene (HDPE)	Bottles, tubes, and pipes
Polystyrene Polypropylene (PP) PVC	Tanks, containers, and jugs
Polyurethane, LDPE, HDPE	Bags, paints, Insulation, and coating

The two types of plastics are thermoplastics and thermosetting plastics. Thermoplastics can be softened; melt and recyclable. Recyclable plastics include Polyethylene Terephthalate (PET), Low Density Polyethylene (LDPE), High Density Polyethylene (HDPE), and Polypropylene (PP). Thermosets can melt and take the shape only once and not used for repeated work. Sometimes, they are termed as non-recyclable plastics, and include polycarbonate, bakelite, nylon, and so on. Table 3shows synthetic polymer properties and their uses in packaging.

TABLE 3 Synthetic Polymer Properties and their Uses in Packaging

Sl No	Polymer	Properties	Uses in Packaging
1	Poly Ethylene Terephthalate (PET)	- Light Weight and low density - Prevents gas from escaping package - Very tough	- High flavoured food - Carbonated drinks - Water bottles

TABLE 3 *(Continued)*

2	Poly Vinyl Chloride (PVC)	- Chemical resistant - Weather resistant - Strong and abrasive resistant	- Pharmaceutical products - Food, confectionary and juices
3	Poly Propylene (PP)	- Light weight, rigid and flexible - Low moisture absorption - Chemical and impact Resistance	- Food packaging - Laminating paper and Board - Medicine bottles
4	Polystyrene	- Rigid and light weight - Water absorption - Good heat insulation	- Food packaging - Electrical and fragile products
5	Others		

5.2.3 ECO- FRIENDLY AND ECO-PACKAGING

The main aim of packaging is to produce the packing material with low investment cost and accessibility of the raw material should be convenient. Packing the product should not create any irritation to the eyes or skin, should be a friendly packing with ease of design, ease processing features and thus making the packing product an interesting one. For creating an eco- design packing, several factors are considered such as environmental condition, market demand, economical consideration, social element and laws with regulation as shown in figure 1. The idea of eco-design is to produce green and environmental friendly system. Laws and regulations makes manufacturers to create good product and free from defects and hazardous elements.

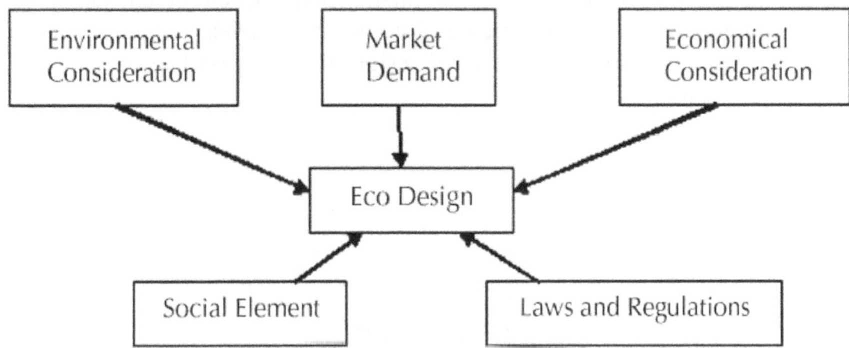

FIGURE 1 Environmentally Friendly Design.

Market demand and social elements play an important role in creating an eco-product. Preservation, protection, distribution and movement of the product should be ecological and these products should be renewable, biodegradable and sustainable (Choi, Park, Han, Kim, Goh, Kim, and Park, 2006; Clark, Kosoris, Hong, and Crul, 2009; Environmental Association for Universities and Colleges Waste Management Guide; Environmental benefits of recycling, Pub: Dept. of Envrn, 2010). The raw products can be from natural resources such as corn starch, card board, fiber board and paper and more necessary to perform the packing according to the international standards. Eco- labels should be pasted on the product mentioning the features of the packing material such as physical, chemical, and biological protection. The material should be easy to identify and create value in the market. The parameters for eco-packing are shown in table 4.

TABLE 4 Various Parameters for Eco-Pacakaging (Fontaine, 2008)

SlNo	Parameters	Responsibility
1	Physical protection	To protect from physical damage
2	Chemical protection	In protection from chemical substances
3	Biological protection	Protect from against biological contaminants
4	Barrier protection	To protect against moisture, light and temperature
5	Identification	In easy identification of the product
6	Marketing	Brand image in high lighting and differntiating the product
7	Convience	Consumer access and distribution

5.2.4 FLEXIBLE AND SUSTAINABLE PACKAGING

Flexible packaging term is used for several purposes:

- Must contain the needed materials and products,
- In protecting the products from contamination and other environmental damage,
- To facilitate ease in transportation and storage of materials,
- In carrying information and colourful designs that make packaging more attractive with beautiful displays.

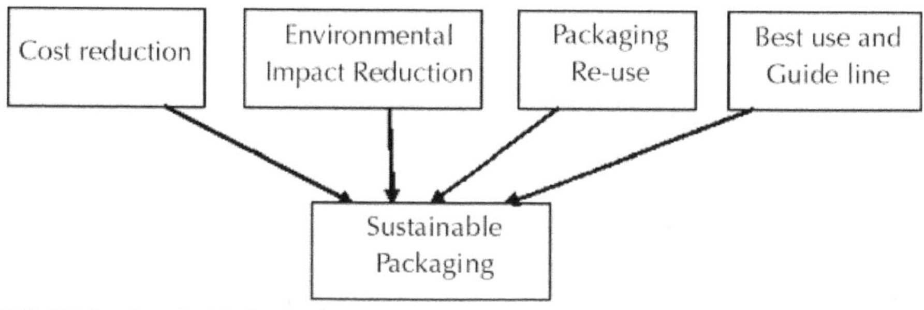

FIGURE 2 Sustainable Packaging.

Some of the functions for flexible packaging are: protect the product from environmental damage, protect from contamination, ease to carry and should have enough space to contain the product. Sustainable materials are those materials that can be continued for further use and still available for future generations. Sustainability design and packaging is an eco design concept and includes environmental, economic and social components of production (Gaceva, Avella, Malinconico, Buzarovska, Grozdanov, Gentile, and Errico, 2007). The sources can be renewable resource of energies or biological resources. In future, innovation and design of product for sustainability should be created and must be imposed in a society to create a good supply chain in the material (Huang, Shetty, and Wang, 1990; James, Fitzpatrick, Sonneveld, and Lewis, 2005). Cost reduction, environmental impact reduction, packaging re-use, and moral guide lines are some of the elements of sustainable packaging, as shown in Figure 2.

5.2.5 BIO BASED POLYMERS IN PACKAGING

Bio polymeric materials can be sorted out into three classes (John and Thomas, 2008):

- Polymeric materials extracted directly from biomass such as starch, cellulose, proteins, chitin and lipids,
- Polymeric material produced by chemical synthesis using renewable bio based monomers such as poly lactic acid,
- Polymers produced by micro-organisms.

Biopolymers are biodegradable, compostable polymers, economically sustainable, and easily processed. They do not produce any harmful gases to the atmosphere and hence environmental friendly (Khoo, Tan, and Chng, 2010; Lee and Xu, 2005). Biopolymers are obtained from microbial systems, extracted from organisms and also obtained from the seasonal crops and plants such as corn, wheat, potato and sugarcane as shown in Figure 3. They are chemically synthesized from basic biological materials like amino acids, sugars, oils, and so on. These polymers have wide applications for production of food, clothing fabrics, cosmetics, medicinal materials, industrial plastics, and so on as shown in table 5 (Leja and Lewandowicz, 2010; Li, Tabil, and Panigrahi, 2007).

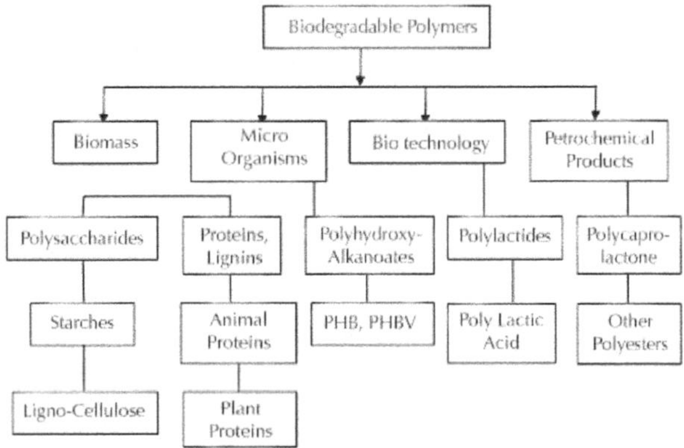

FIGURE 3 Classification of Bio-polymers (Majeed, Jawaid, Hassan, Abu Bakar, Abdul Khalil, Salema, and Inuwa, 2013; Marsh and Bugusu, 2007).

TABLE 5 Few Biopolymers with Source and Applications (Mitrus, Wojtowicz, and Moscicki, 2009)

Biopolymer	Source	Applications
Starch	Plants– corn, potatoes, rice, barley, wheat	Adhesives, stabilizers, thickeners, soil conditioners, packaging.
Plant Cellulose	Plant - cell wall	Thickening solutions, gels, food, textiles, ceramics, paper coatings, liquid fuels
Lignin	Plants- herbaceous and woody	Non-energy, industrial packaging, tape, binding, and thinning agents, chemicals.
Chitin	Plants, insects, shells	Medicine, manufacturing, agriculture, contact-lenses, cosmetics, shopping-bags.
Protein	Plants (soy, zein from corn), animal (collagen)	Films, gels, wetting agent, fibers.

Biopolymers are also termed as natural polymers because they form under natural conditions during the growth cycles of organisms. Some of the biopolymers and their manufacturers are listed in table 6 (Mohanty, Misra, and Drzal, 2002; Mohanty, Misra, and Hinrichsen, 2000). Polymers of natural origin (for example, starch and cellulose) are modified either by physical or chemical methods in order to make them suitable for processing (Mohee, Unmar, Mudhoo, and Khadoo, 2008; Muller, Kleeberg, and Deckwer, 2001). Availability of thermoplastic biopolymers includes Poly Lactic acid (PLA), cellulosic plastic, starch plastic, soybean and corn based polymer resins. Poly lactic acid is derived from corn starch via commercial fermentation and polymerization technologies. They are the most common biopolymer in use (Marsh and Bugusu, 2007; Rapa, Popa, Cinelli, Lazzeri, Burnichi, Mitelut, and Grosu, 2011; Riedel, 1999).

TABLE 6 Some of the Bio-Based Polymers and their Manufacturers (Risch, 2009)

Class	Bio -Based Polymers	Manufacturers
Bio Synthetic Polymers	Polyhydroxyalkanoates (PHAs)	- Biomer, Germany - Telles, USA - Kaneka, Japan
Bio Chemo Synthetic Polymers	Poly (lactic acid)(PLA) and Poly (butylene succinate) (PBS)	- Nature works, USA - Hycail, Netherlands, - Mitsui chemicals, Japan - Showa High Polymer, Japan
Modified Natural Polymers	Starch Polymers andCellulose derivatives	- Novamont, Italy - BIOP, Germany - Sanstar, India - Japan corn starch, Japan

Degradation action takes place with the help of enzymes (biological), chemical, thermal and photo (light) sources. Bio degradation takes place with the help of living organisms and the process occurs in two steps. The first step is to break down the polymers into lower molecular mass species (Leja and Lewandowicz, 2010) and the action takes place with the aid of oxidation or hydrolysis. Second step is by, bio assimilation of polymer fragments by microorganisms. The conditions for biodegradation depend on other factors such as processing characteristics, ageing conditions, storage facilities and chemical compositions (Roy, Nei, Orikasa, Xu, Okadome, Nakamura, and Shiina,2009; Salmoral, Gonzalez, and Mariscal, 2000; Sannino, Demitri, and Madaghiele, 2009). Bio-degradation of few packaging materials are shown in table 7.

Degradation can be from several types (Sarasa, Gracia, and Javierre, 2009).

1. Biodegradation- action from living organisms,
2. Thermal degradation- action from high temperatures,
3. Thermo oxidative degradation- slow oxidation with moderate temperatures,
4. Photo degradation- action from light sources,
5. Hydrolysis- break down reaction with water.

TABLE 7 Biodegradation of Few Packaging Materials (Scheller and Conrad, 2005)

Bio plastics	Duration (Days)
Starch based bioplastics	Approx 40–60
Polymer based films	Approx 90–150
Chitosan based films	Approx 90–150
Cellulose based films	More than 365

5.2.6 USE OF BIO PLASTICS FOR PACKAGING

Bio plastics are plastics made of renewable resources, biodegradable and in some cases they may not be compostable or biodegradable. Bio plastics are made from sugar cane, starch, corn and contain little amount of petroleum or oil contents (Singh, Sharma, and Malviya, 2011; Srinivasa and Tharanathan, 2007). Table 8 shows the comparison between conventional plastic to a bio plastic and the energy necessities for bio bag product as shown in table 9.

TABLE 8 Comparison between petroleum based conventional plastics to bio based plastics

Sl No	Factors	Petroleum based plastics	Bio - plastics
1	Sustainable	No	Yes
2	Renewable	No	Yes
3	Degradation	Few degradable by polymer oxidation	Bio degradable
4	Emissions	High	Low
5	Fossil fuel usage	High	Low

TABLE 9 Energy Necessities for Bio-Bag Product (Sudesh and Iwata, 2008)

Agriculture farming - Corn production	2.5 MJ/kg Corn
Wet Milling - Corn to glucose	4.9 MJ/kg glucose
Fermentation and recovery - PHA	52.53 MJ/kg PHA

5.3 PACKAGING INNOVATION TECHNIQUES

The innovation techniques should make the consumer to feel satisfy by the product. It is the duty of a packaging engineer to create the design that must be flexible, active and intelligent and innovative. Intelligent packaging are set as an integral part of any business, including retailers, food service providers, and general merchandisers. The main design parameters include the product protection, product transport, product storage, product information, product utilization, product marketing and product satisfaction as shown in Figure 4.

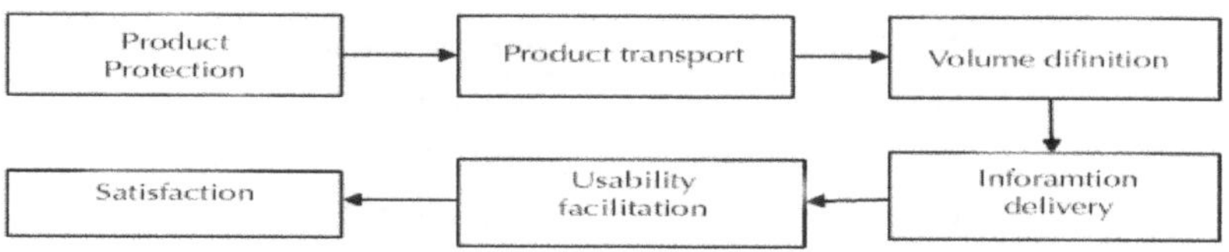

FIGURE 4 Package Design (Thamae, Marien, Chong, Wu, and Baillie, 2008).

The three types of packaging are: Primary package- which refers to the goods bought by the consumers (interior container contact with the contents), Secondary packaging that relates to the materials which are not in direct contact with the product, but are sold to the consumer (containing one or more containers also termed as intermediate package), Tertiary/ Transit packaging includeboxes, boards, wooden pallets and plastic wrapping papers (exterior package used to protect the goods during process of handling and transportation) and they are used to bear and deliver the load (Tokiwa, Calabia, Ugwu, and Aiba, 2009). The basic role of packing is fivefold: to protect the product, to provide easy transportation, to identify product specific factors, to deliver important information about the product, and to facilitate the use of the product. Cute packing, convenience in packaging, rigid packaging, thin film packing and environmental design packing is some of the innovation packing techniques.

5.4 RECYCLING AND REDUCING WASTE PLASTIC MATERIALS

Waste is defined as a substance that is discarded and thrown after use and they are classified namely: controlled waste and noncontrolled waste as shown in Figure 5. The waste objects can be re utilized and can be beneficially regained to its own value; if the system is allowed to follow the 4 R's namely reduce, reuse, recycle and replace (Velde and Kiekens, 2002).

- Reduce–eliminate and minimization of waste materials and reduction of source,
- Reuse–create and use of materials in their original state, facilitate retention and repair it,
- Recycle–reprocessing into useful materials and making new products,
- Replace– consumption of environmental friendly materials and try to replace it.

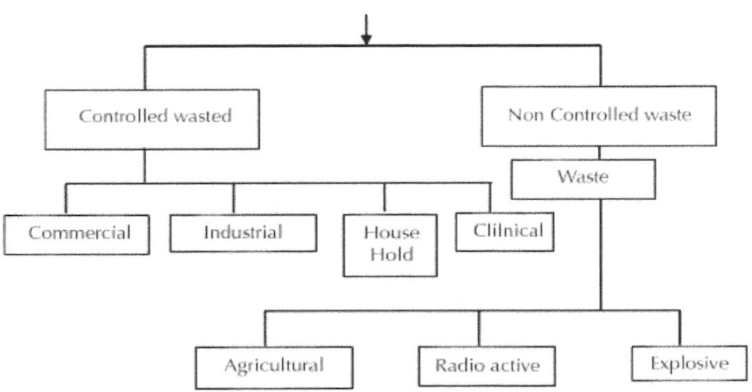

FIGURE 5 Waste classification (Vroman and Tighzert, 2009).

Recycling of materials can be achieved through physical, chemical and thermal recycling methods (Wool and Sun, 2005). The process of recycle includes collecting the waste product materials and transforming the new product with enforced design, appearance and look. Physical recycling includes change in size; shape of the material and to obtain appearance. Chemical recycling includes breaking the molecular structure of polymers using chemical reactions and are used to obtain the same polymer. Thermal recycling undergoes recycling with the aid of thermal (heat) and they break down the structure of polymer to obtain a new polymeric material.

Some of the reasons for using plastics in packing are: they are light weight materials, non toxic, resistance to chemicals, easily processed, excellent mechanical and chemical properties, good barrier properties, not affected by microorganisms, low cost, ease in storage and transportation, availability in sachet packets and reusability (Yu (Ed.), 2008; Yu, Dean, and Li, 2006). During recycling, code numbers for the polymeric materials are assigned for sorting and for ease identification, as shown in table 10.

TABLE 10 Polymer with their Code number

Code No	Polymer Name
1	Polyethylene Terephthalate (PET)
2	High Density Polyethylene (HDPE)

TABLE 10 *(Continued)*

3	Polyvinyl Chloride (PVC)
4	Low Density Polyethylene (LDPE)
5	PolyPropylene (PP)
6	PolyStyrene (PS)
7	Others

The production and disposal of conventional plastic and bio-plastic materials can be investigated using Life Cycle Assessment (LCA). The LCA is defined as the procedure of evaluating the potential effects that a product can process and has service on the environment over the entire period of its life cycle (Tokiwa, Calabia, Ugwu, and Aiba, 2009; Zheng, Yanful, and Bassi, 2005). Life Cycle Assessment consists of four factors namely: Definition of goal and scope, Life cycle inventory analysis, Life cycle impact assessment, Life cycle interpretation. The beneficial of LCA are: the results of LCA can be highlighted and whether the products are environmental friendly and also comparison can be made to the existing one.

Some of the common methods for reducing packing waste are: Eliminate the non essential components of packing, reduce weight and thickness, transfer the waste in to the container, buy the product in bulk, replace single use with multi use, maximum utilization of the product, and use the product for longer life. Life cycle analysis of product is shown in Figure 6.

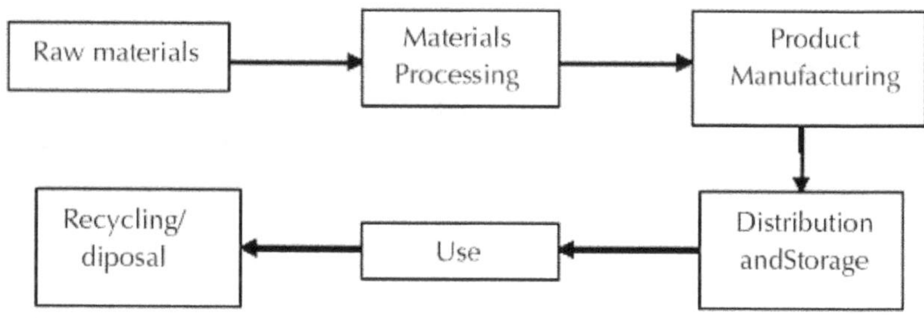

FIGURE 6 Life cycle analysis of product.

5.5 NATURAL FIBERS FOR PACKAGING APPLICATIONS- A PRACTICAL LOOK

An experimental work is carried out for use of natural fibers for packaging application. Natural fiber of maize stalk and synthetic polymer of epoxy resin are used for obtaining a packing material and are processed through Vacuum Assisted Resin Transfer Molding (VARTM). The setup for the flow is identical as shown in Figure 7. Maize stalk fibers were selected as shown in figure 8 (a) and cleaned thoroughly in running water and sun dried. These fibers were collected in local farm regions of Karnataka,

India. The fibers are chemically treated with alkaline method and finely powdered using a pulverizer. Thermoset polymer of epoxy resin with suitable hardener was used as a matrix material. The resin inlet is prepared using the distribution spiral connected at one end of the reinforcement material so that it avoids the amount of resin being transferred. The mold is placed on the table and covered by vacuum bag.

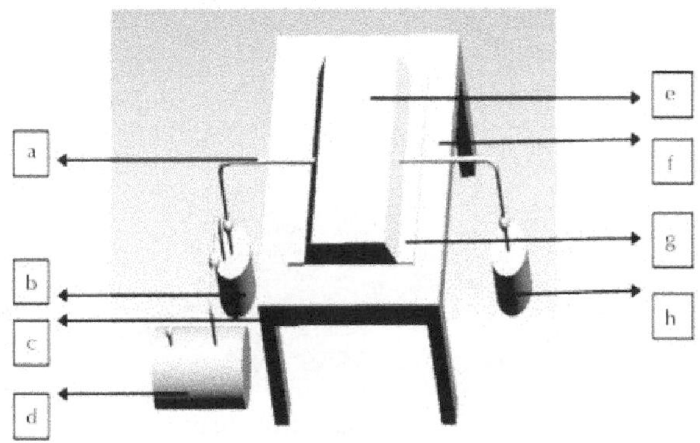

FIGURE 7 Vacuum Assisted Resin Transfer Molding.

The vacuum is applied through a layer of peel ply and the mold is made air tight using a vacuum bag. Sealant tape is applied along the edges to remove the possible high permeable flow path. During the experiment the vacuum level is controlled by a regulator and finally the resin is prepared to inject. The vacuum source is left on until the resin system begins to gel. Once the vacuum is turned off, the part is left to cure at room temperature. The product of pack box is shown in Figure 8 (b).

FIGURE 8 *(Continued)*

FIGURE 8 (a) Maize Stalk (b) Packing box made from maize fiber.

5.6 CONCLUSIONS

The present work was focused to study the nature of plastic and bio plastic material with application refereeing to a packaging sector. Plastic materials have lot of problems such as toxic emissions, oil spills and global working and the only solutions to these problems are by using natural fibers and a sustainable material like bio plastic which can be easily renewed. Experimental work of bio plastic box made from natural fiber and polymeric resin showed a future sign for packing materials. In future, to have little damage to the environment and for green packing, the bio- packing materials must be fabricated, processed, consumed and re-used. Natural fibers are one among the fibers that can create a green environment mode in packaging sectors.

KEYWORDS

- **Bio Based Polymers**
- **Biopolymers**
- **Degradation**
- **Sustainable Packaging**
- **Vacuum**

REFERENCES

1. Al-Salem, S. M., Lettieri, P., and Baeyens, J. *Waste Mgmt.***29**, 2625–2643 (2009).

2. Andrady, A. L. *Marine Poll. Bull.*, **62**, 1596–1605 (2011).
3. Anne, B. Environmental-Friendly Biodegradable Polymers and Composites. Sunil Kumar (Ed.), *Integrated Waste Management - Volume I*, InTech, (2011).
4. Anshuman, A., Khardenavis, Kumar, M. S., Sandeep, N. M., and Tapan, C. *Biores Tech.*, **98**, 3579–3584 (2007).
5. Averous, L. and Pollet, E. Biodegradable Polymers, *Environmental silicate nano – biocomposites*, Springer -Verlag, **6** (2012).
6. Azeredo, H. M. C. *Food. R. Intl.* 42, 1240–1253 (2009).
7. Błędzki, A. K., Jaszkiewicz, A., Urbaniak, M., and Walczak, D. S. *Fibres& Textiles in Eastern Europe.*, **20**(6B), 15–22 (2012).
8. Choi, D. H., Park, K. W., Han, B. J., Kim, J. H., Goh, B. S., Kim, E. H., and Park, Y. J. *Ahn Graphics*, (2006).
9. Clark, G., Kosoris, J., Hong, L. N., and Crul, M. *Sustainability.*, **1**, 409–424 (2009).
10. Environmental Association for Universities and Colleges Waste Management Guidehttp://www.eaucwasteguide.org.uk/
11. Environmental benefits of recycling, Pub: Dept. of Envrn. Climate Change and Water NSW, (2010).
12. Fontaine, K. A. *A Practical Approach to Sustainability for Flexible Packaging Buyers*, Amgraph Packaging, Inc., (2008).
13. Gaceva, G. B., Avella, M., Malinconico, M., Buzarovska, A., Grozdanov, A., Gentile, G., and Errico, M. E. *Poly Compos*, 98–107 (2007).
14. Huang, J., Shetty, A. S., and Wang, M. *Adv. in Poly. Tech.*, **10**, 23–30 (1990).
15. James, K., Fitzpatrick, L., Sonneveld, K., and Lewis, H. Sustainable packaging systemsdevelopment, in Handbook on Sustainability Research, W. L., Filho (Ed.) Peter LangScientific Publishers: Frankfurt, (2005).
16. John, M. J. and Thomas, S. *Carboh. Poly.*, **71**, 343–364 (2008).
17. Khoo, H. H., Tan, R. B. H., and Chng, K. W. L. *Int. Journal of Life Cycle Assess.*, **15**, 284–293 (2010).
18. Lee, S. G. and Xu, X. *Int. Journal of Envt. Tech. Mgmt.*, **5**, 14–41 (2005).
19. Leja, K. and Lewandowicz, G. *Polish Journal of Envt. Stud.*, **19**, 255–266 (2010).
20. Li, X., Tabil, L. G., and Panigrahi, S. J. *Polym. and Envt.*, **15**, 25–33 (2007).
21. Majeed, K., Jawaid, M., Hassan, A., Abu Bakar, A., Abdul Khalil, H. P. S. Salema, A. A., and Inuwa, I. *Materials and Design*, **46**, 391–410 (2013).
22. Marsh, K. and Bugusu, B. *Journal of Food Science.*, **72**, 39–55 (2007).
23. Mitrus, M., Wojtowicz, A., and Moscicki, L. *Thermoplastic Starch in Biodegradable Polymers and their Practical Utility*; Janssen, L. P. B. M, L. Moscicki (Eds), Wiley-VchVerlag GmbH & Co., Weinheim, pp 1–33 (2009).
24. Mohanty, A. K., Misra, M., and Drzal, L. T. *J. Polym. and Envt.*, **10**, 19–26 (2002).
25. Mohanty, A. K., Misra, M., and Hinrichsen, G., *Macromol. Mat. Engg.*, **276/277**, 1–24 (2000).
26. Mohee, R., Unmar, G. D., Mudhoo, A., and Khadoo, P. *Waste Mgmt.*, **28**, 1624–1629 (2008).
27. Muller, R. J., Kleeberg, I., and Deckwer, W. D. *Journal of Biotech*, **86**, 87–95 (2001).
28. Rapa, M., Popa, M. E., Cinelli, P., Lazzeri, A., Burnichi, R., Mitelut, A., and Grosu, E. *Romanian Biotech.* Letters., **16**, 59–64 (2011).
29. Riedel, U. *Natural fiber-reinforced biopolymers as construction materials - new discoveries*, 2nd *International Wood and Natural Fiber Composites Symposium*, Kassel/Germany, (1999).
30. Risch, S. J. *Journal Agric. Food Chem.* **57**, 8089–8092 (2009).
31. Roy, P., Nei, D., Orikasa, T., Xu, Q., Okadome, H., Nakamura, N., and Shiina, T. *Journal of Food Engg.*, **90**, 1–10 (2009).
32. Salmoral, E. M., Gonzalez, M. E., and Mariscal, M. P. *Indl. Crops and Prd.*, **11**, 217–225 (2000).
33. Sannino, A., Demitri, C., and Madaghiele, M. *Materials.*, **2**, 353–373 (2009).
34. Sarasa, J., Gracia, J. M., and Javierre, C. *Biores. Techn.*,**100**, 3764–3768 (2009).
35. Scheller, J. and Conrad, U. *Current Opn. Plant Bio.*,**8**, 188–196 (2005).
36. Singh, A., Sharma, P. K., and Malviya, R. W. *Appl. Sci. Journal***14**, 1703–1716 (2011).

37. Srinivasa, P. C. and Tharanathan, R. N. *Food R. Intl.*, **23**, 53–72 (2007).
38. Sudesh, K. and Iwata, T. *Clean.*, **36**, 433–442 (2008).
39. Thamae, T., Marien, R., Chong, L., Wu, C., and Baillie, C. *Journal Mater Sci.,* **43**, 4057–4068 (2008).
40. Tokiwa, Y., Calabia, B. P., Ugwu, C. U, and Aiba, S. *Int. Journal Mol. Sci.,* **10**, 3722–3742 (2009).
41. Velde, V. D. K. and Kiekens, P. *Poly. Testing*, **21**, 433–442 (2002).
42. Vroman, I. and Tighzert, L. Materials., **2**, 307–344 (2009).
43. Wool, R. and Sun, X. S. *Bio-Based Polymers and Composites*, Elsevier, (2005).
44. Yu, L. (Ed.) *Biodegradable Polymer Blends and Composites from Renewable* *Resources*, John Wiley & Sons, Inc, (2008).
45. Yu, L., Dean, K., and Li, L., *Prog. Polym. Sci.* **31** 576–602 (2006).
46. Zheng, Y., Yanful, E. K., and Bassi, A. S. *C. R. in Biotech*. **25**, 243–250 (2005).

CHAPTER 6

ENVIRONMENT FRIENDLY PACKAGING PLASTICS

VIMAL KATIYAR, NEELIMA TRIPATHI, RAHUL PATWA, and PRAKASH KOTECHAO

ABSTRACT

Polymer based packaging forms an integral part of human life, however, most of the currently used packaging plastics are produced from fossil feedstock and are not usually biodegradable, thereby jeopardizing the long term sustainability of the planet. Hence, there is an immediate need to explore alternative packaging materials that are eco-friendly. This chapter reviews the current state of the art on various commercially viable environmental friendly, biodegradable packaging plastics that can be obtained from plants, animals, and other synthetic biodegradable plastics such as poly lactic acid (PLA) and poly hydroxyalkonates. An overview of various aspects of these environment friendly packaging plastics including their synthesis, commercial production, processing, properties, applications, advantages, and limitations have been discussed in considerable detail.

6.1 INTRODUCTION

Packaging has become an integral part of almost every product for a variety of reasons. These include protection of packaged product from degradation to enhancing the sales appeal of the product. A wide variety of packaging is available to accommodate product specific requirements such as inertness, conditioned environment, and specific barrier properties. Packaging is usually made of plastics, glass, wood, metal, paper,and their products, whereas plastic packaging is commonly made of polyethylene (PE), polypropylene (PP), polystyrene (PS), polyvinyl chloride (PVC), or polyethylene terephthalate (PET), which are predominantly produced from fossil feedstocks. The favorable properties of plastics include its cost, lightness, non-corrosiveness, chemical resistance, and comparatively low toxicity. After their usage, thermo-plastic based packaging are mostly recycled and reused for limited number of times with its own set of disadvantages. Thereafter, these non-recyclable plastics are usually discarded as solid wastes and are either disposed as landfill or are incinerated. Despite their current wide usage, these plastics are non-degradable in nature and lead to waste disposal problems (Panda et al., 2010; Punčochářet al., 2012) and continue to cause irrevocable damage to the environment (Yang et al., 2012; Kumar et al., 2004). To overcome the disadvantages of the conventional plastics, the scientific community has been exploring a large number of alternatives including bio-based plastics. These have obtained increased attention both due to its ability to be degraded into fundamental components by micro-organisms and enzymes and also due to the fact that it can be prepared from renewable resources. Significant effort has been made to produce biodegradable plastic materials with similar functionalities as their conventional counterparts. At present, the environment friendly plastic based packaging has established their acceptability to be used on commercial scale. In this chapter, we will be discussing on many such environmental friendly plastics that are under active scientific investigation.

This chapter is organized as follows: In the next section, we will be describing about functions of packaging, advancement in packaging, industrial classification of packaging along with the properties of packaging. Subsequently, a description of some of the bio-based packaging materials is presented along with their properties, advantages, and limitations. Finally, we conclude this chapter by summarizing the developments in this article.

6.2 PACKAGING

In this section, we will discuss the functions, advancements, classification, and properties of various materials that have been conventionally used for packaging. This will help to provide broader perspective on the requirements of an environmental friendly plastic.

6.2.1 FUNCTIONS OF PACKAGING

Packaging serves a wide range of functions, which can be broadly divided into two types namely:

1. Protection and
2. Enhancement of product value.

Both of these can be further subdivided into subcategories (Hernandez et al., 2000) as discussed below:

BARRIER PROTECTION

This involves protection of the packaged item from unfavorable external influences such as oxygen, temperature variations, moisture, and light.

BIOLOGICAL PROTECTION

This involves protection of the packaged item from biological contaminants such as micro-organisms.

PHYSICAL PROTECTION

This involves protection of the packaged item from physical damage that is the shape and size for solid or semi-solid contents.

INFORMATION

Packaging helps in providing authentic information about the product such as brand name, ingredients, expiry period, and other essential information required for the benefit of the manufacturer and consumer.

IDENTIFICATION

Packaging helps inrecognizing the product from the detailed information that is mentioned on it. Recent trends include transparent smart packaging, which can also enable the aesthetic value of the product.

6.2.2 ADVANCEMENT IN PACKAGING

Materials such as metals, glass, paper, and its products, plastics along with wood have long been the preferred choices for packaging. The choice of packaging material has been primarily governed by the scientific development of such materials. A brief description of these conventional packaging is presented below:

METALS

Being considered as non-toxic, gold and silver are used since ancient times for various purposes including preparation and subsequent processing and packaging of food. However, their widespread use was restricted due to its disadvantages like cost and weight. Subsequently, various low cost metals were used for packaging but their success was limited in specialized food packaging due to issues related to toxicity. In the latter half of twentieth century, tin and aluminum started to find widespread usage in packaging because of its various unique properties including high ductility, low cost, and density. Its relative non-toxic nature also enabled its usage in widespread food packaging applications. Despite the development of various other alternatives, these materials, especially aluminum still continues to find widespread usage (Cui andRoven, 2012) primarily due to significant developments in material processing.

GLASS

Glass has also been used for various types of packaging. The use of glass was restricted primarily to rigid packaging. Due to its inertness and non-toxic nature, it was preferred material of choice especially for chemicals and food packaging. Moreover, it has very high barrier properties and hence acts as an excellent material for preserving the aroma of its contents. However, its density and brittle nature led to shedding its market share to various alternatives such as metal and plastics. Despite this, glass bottles are still used in packaging for products that require brandishing a high quality image (Marcinkowskiand Kowalski, 2012).

PAPER

Paper is also widely used for packaging of solid materials that do not require very high barrier properties. It is usually characterized by low strength but its comparatively lower cost to metals and glass has made it a preferred alternative for packaging of many products. Additionally, its light weight helps in easier handling of the packaged material. Paper based composite packaging has been a significant development in this field and contributed towardsits use in additional areas like packaging of liquid products and also in addressing issues related to its barrier properties. Paper based composite laminates (Landge et al., 2009; Siracusa et al., 2008), which include polymers and metal foils like aluminum have gained usage as they increase the shelf life of the products. Hence, these are preferred choices for modified atmospheric packaging and are commonly used for products such as milk, juices, jams, ketchups, and so on.

PLASTIC

Plastic are long chain polymer made of small units known as monomer. Today, plastics are one of the most commonly used packaging materials (Siracusa et al., 2008). It has been shown that if plastics were replaced with other materials, the weight of packaging would increase by a factor of four and volume by two and half (CallisterandRethwisch, 2009). In the beginning, plastic originated from bio feedstock and were predominantly derived from cellulose. Subsequently, cellophane was developed, which was used as flexible packaging and had the added advantage of being resistant to moisture. Further developments included utilizing its heat sealable property as one of the packaging requirements especially in case of MAP. The use of synthetic flexible plastics can be attributed to the discovery of polyvinylidene chloride known as Saran™(Wiley and Reilly, 1939) and was initially used in protecting military equipments and thereafter widely used in food packaging and spray bottles. Another significant development in the area of plastics was the invention of low density polyethylene (LDPE), which has found applications in flexible packaging (Oliveira et al., 2009).Subsequently, high density polyethylene (HDPE) was developed that possessed considerably high thermal, thermo-mechanical and barrier properties (Hernandez et al., 2000). The discovery ofPP enabled the use of plastics for semi-flexible packaging and ensured the use of plastics for rigid to flexible packaging.The lifecycle of these conventional packaging materials can be divided into six stages (Hernandez et al., 2000) and their environmental impacts in each of the stages are shown in Figure 1. In the next section, we will discuss about the industrial classification of packaging, which is also practiced for environmental friendly plastics.

6.2.3 INDUSTRIAL CLASSIFICATION OF PACKAGING

Packaging is usually classified into two categories namely:

1. Rigid packaging and
2. Flexible packaging, based on its ability to change its shape.

As the name indicates, rigid packaging does not change its shape and is more reliable in transporting the materials without changing the aesthetic appearance of the product. In general, when compared to flexible packaging, rigid packaging has greater impact strength, stiffness, and barrier properties making it ideal for more shape sensitive contents. Examples of rigid packaging include cartons, crates, tanks, cans, bottles, thermoformed cups, trays, and so on. In many instances, rigid packaging is used as a protection to flexible packaged materials and imparts additional strength towards its transportation and storage. Usually, rigid packaging is costlier but can be reused multiple times. However, rigid packaging increases the size and can thereby substantially increase the requirement of the storage space.

Flexible packaging is the most “source-reduced” form of packaging, which means that a flexible package requires least amount of material compared to other forms of packages that would hold the product. This also means that flexible packaging adds very little weight to the overall product. They require 60–70% less space and also save around 30–40% of energy than rigid packaging material. However, it cannot be reused and have environmental issues related to its disposability.There are many different types of flexible packaging (Wooster, 2002) depending on its processing and use. Some of them are described below:

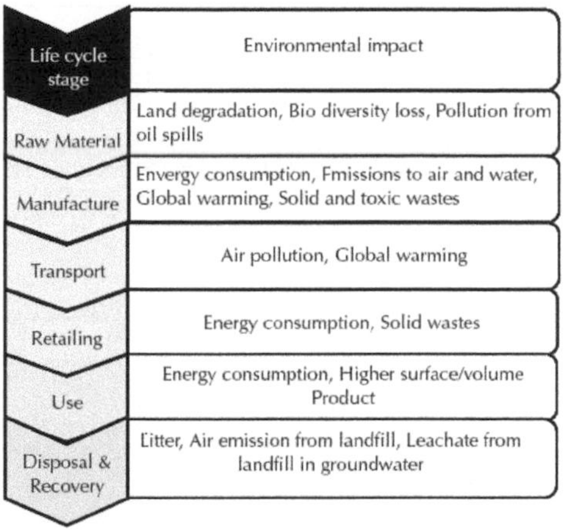

Life cycle stage	Environmental impact
Raw Material	Land degradation, Bio diversity loss, Pollution from oil spills
Manufacture	Envergy consumption, Fmissions to air and water, Global warming, Solid and toxic wastes
Transport	Air pollution, Global warming
Retailing	Energy consumption, Solid wastes
Use	Energy consumption, Higher surface/volume Product
Disposal & Recovery	Litter, Air emission from landfill, Leachate from landfill in groundwater

FIGURE 1 Life cycle stages of packaging and their associated environmental impacts.

SHRINK WRAP PACKAGING

These are films, which are used to shrink tightly over the covering material either by stretching the plastic film or by the application of an external agent such as pressure, heat and/or an adhesive. Shrink wrap are usually used for tight bundling of goods. Polyolefin is commonly used for shrink wrap.

BUBBLE WRAP PACKAGING

This is a pliable transparent plastic wrap with regularly spaced, protruding air filled bubbles that are intended to act as cushion for fragile items primarily during transportation and handling.

CLING WRAP PACKAGING

This is a very thin plastic film in the order of 10 microns and is commonly used for covering food items. It does not possess very high barrier properties and often used for covering the items for avoiding their direct contact with foreign substances for short durations especially for food items.

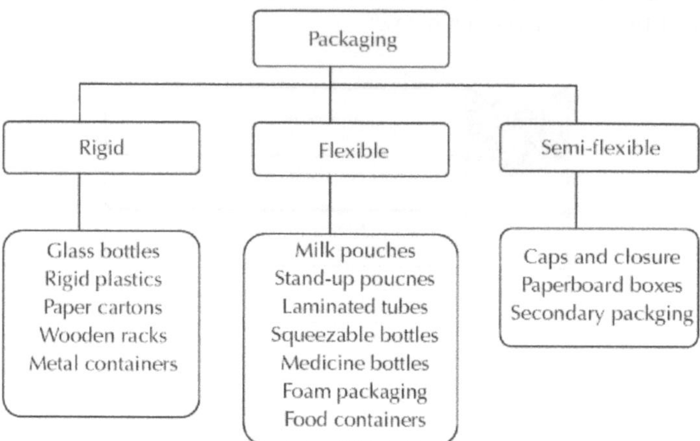

FIGURE 2 Types of packaging.

BREATHABLE FILM PACKAGING

These are new generation films used in food packaging. These films are used either to absorb or transmit moisture from the environment to food product or vice versa depending on the requirement of the food that is packaged using such films. Another feature is that these films can be designed not to allow liquid water transport across the film, while allowing water vapor.

RETORT PACKAGING

This is a multi-layered laminated packaging, wherein the layers are selected to provide properties such as sealing, puncture resistance, gas barrier, printing, and gloss for aesthetic appearance. The layers can be composed of only polymers or can be of hybrid nature involving the use of multi polymers and metallic materials.

BLISTER PACKAGING

These packaging have a cavity or pocket made by vacuum forming of thermo-plastics such as PP, PVC, PET, and so on. These are generally used in packaging medicines and are usually packed aseptically in a sterile environment and small consumable items.

GAS FLUSH PACKAGING

This is primarily used for Modified Atmosphere Packaging (MAP) and Controlled Atmosphere Packaging (CAP) and helps to maintain a conditioned atmosphere inside the pack to preserve the quality and increase the shelf life of the packaged material. Examples include bottling of beverages and other packaging for organic food such as different types of meat.

VACUUM PACKAGING

This is practiced to store frozen food products, wherein a vacuum is created during the packaging and thus enable it to be frozen. It also serves in the removal of atmospheric oxygen to prevent the rotting of the food items.

6.2.4 PROPERTIES OF PACKAGING

The packaging properties play a dominant role in the selection of an appropriate polymer and can be broadly classified into four categories as shown in the Figure 3.

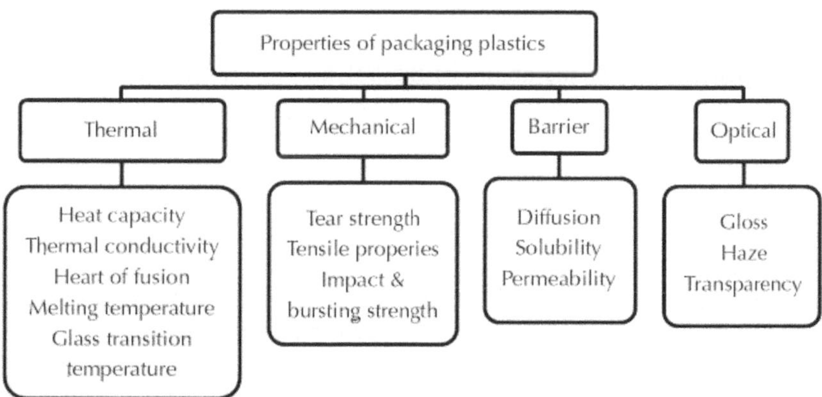

FIGURE 3 Properties of packaging.

THERMAL PROPERTIES

Unlike metal packaging, plastics are characterized with high heat capacity and lower thermal conductivity thereby, requiring significantly higher exposure to heat for transmitting the undesired heat to the packaged material. For example, aluminum has a specific heat capacity of around 0.9 kJ/Kg K whereas the specific heat capacity of PE is around 1.8 kJ/Kg K. For a given polymer, heat of fusion determines the crystallinity of the packaging, which in turn governs the gas barrier and mechanical strength of the packaging. A higher heat of fusion will imply a better gas barrier and increased mechanical strength. Different grades of the same polymer may possess different melting points depending on the size of the crystals. A higher melting point would indicate the presence of bigger crystal size, which in turn would increase the barrier and tensile strength of the polymer at the cost of decreasing the tear and impact strength. However, a higher melting point would require higher energy during the processing of the polymer.Glass transition temperature (T_g) of polymer is an important selection criteria for packaging under a specified temperature application. For example, for packaging of frozen products, it is recommended to select a polymer, which has a glass transition temperature lower than the freezing temperature. As a case, T_g for PE is around 120°C and can be used to freeze materials, whereas the T_g for PET is 70°C and would not be preferred for sub-zero freezing applications.

MECHANICAL PROPERTIES

Mechanical property is one of the primary selection criteria of the packaging polymer. Various structural and morphological characteristics of polymers such as molecular weight, crystallinity and tacticity influence the mechanical properties. Tensile strength determines the strength required for the rupture of the packaging and hence an ideal packaging polymer should be characterized by a higher tensile strength. Further, tear strength is also an important property for flexible polymer packaging as it determines the energy required for tear initiation. Thus, a polymer with higher strength would be preferred for preventing the spillage of its content. Most packaging films are biaxial as they possess higher tear strength. Similarly, the polymer should also possess a high impact and bursting strength (TekadeandGattani, 2010).

BARRIER PROPERTIES

Barrier properties refer to the ability of the polymer to prevent the entry of undesired gas or vapor and are characterized by solubility, permeability, diffusivity across the barrier, and also the packaging material's affinity towards moisture. Additionally, barrier properties are also dependent on the morphological properties of the material such as crystallinity, chain conformation, and so on. In general, crystallinity increases the barrier properties particularly towards moisture, light and grease. However, the increase in crystallinity should not lead to increase in channeling as it can restrict the barrier properties. A higher barrier property ensures longer shelf-life for packaging and also protects flavors or aromas that might be lost by permeation.

OPTICAL PROPERTIES

Optical properties of polymers are characterized by haze, transparency, and gloss (Villalobos et al., 2005). Transparency increases the aesthetic value of the packaging and is usually desirable as it helps in the identification of the packaged material. However, there are light sensitive products, which do not require transparency to have a longer shelf life. Hence, the level of transparency varies from product to product. Some of transparent packaging plastics are PET, PS, and polycarbonate (PC).

6.3 ENVIRONMENT FRIENDLY PLASTICS

There has been an increasing awareness of the anthropogenic effects of the non-biodegradable plastic disposal to the environment. In addition to this, the depletion of the fossil feedstocks is bound to cause a significant demand and supply gap in the conventional fossil based packaging products. Both these factors have been a driving force for exploring sustainable alternatives to conventional plastics. Some of the bio-based polymers have been shown to be potential alternatives (Gross and Kalra, 2002; Demirbas, 2007; Cha et al., 2004) in addressing both these problems. In certain instances, the properties of environmental friendly plastics need to be enhanced by adding some of the fossil based polymers. For example, plasticizers and compatibilizers are used to reduce the brittleness and increase the miscibility of some incompatible polymers.Environmental friendly plastic based nanocomposites (Rhimand Perry, 2007) comprise of polymer matrix reinforced with particles and have high surface area and aspect ratio. Production capacity ofbiopolymers is shown in Figure 4. It can be seen that the Europe and America produces the maximum amount of biopolymers followed by Asia.

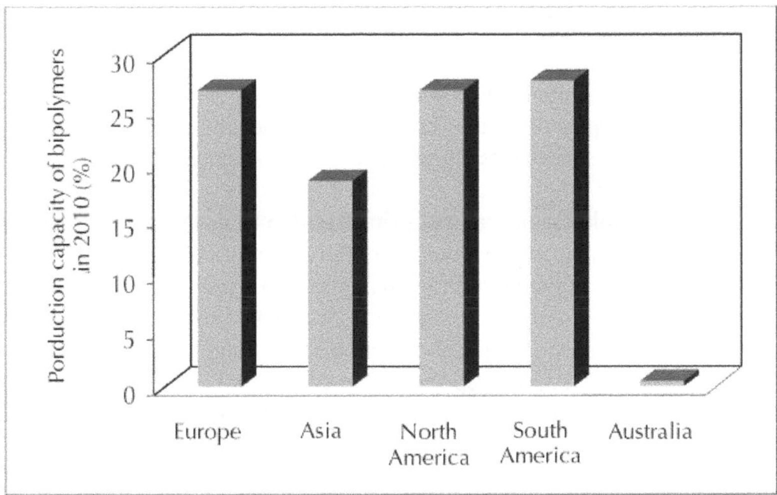

FIGURE 4 Production capacity of biopolymers in 2010.

6.3.1 CLASSIFICATION OF ENVIRONMENTAL FRIENDLY PACKAGING PLASTICS

The environment friendly packaging plastics can be broadly divided into biopolymers and synthetic polymers derived from bio feedstock as shown in Figure 5. Biopolymers are those polymers, which are directly obtained from plants and animals. Synthetic polymers are further subdivided into two categories based on the origin of the precursor. Precursors that are derived from renewable resources constitute one category, whereas precursors that do not originate from bioresource form the other category.

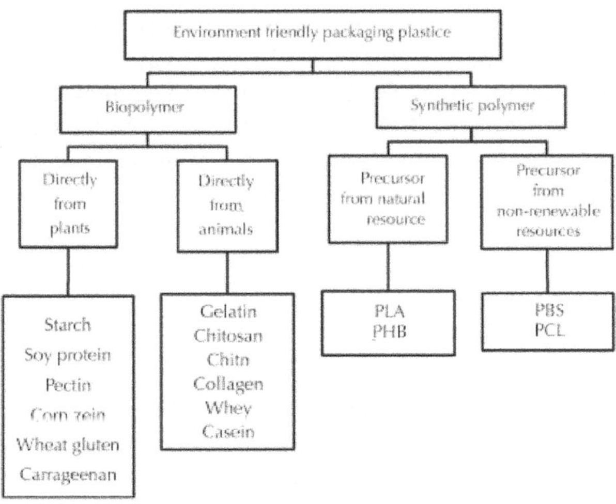

FIGURE 5 Classification of environment friendly polymers.

6.3.2 BIOPOLYMER: DIRECTLY FROM PLANTS

STARCH

Starch is a naturally and abundantly available linear polysaccharide. It essentially consists of glucose based repeat units, amylose, and amylopectin linked by glucosidic linkages in the 1–4 carbon positions as shown in Figure 6. Average chain length of starch is in the range of 500–2000 glucose units.Despite its wide availability and its low cost, neat starch is not suitable for packaging, primarily due to its poor tensile and flexural strength along with low percentage elongation. Further, it has a poor water vapor transmission property despite being hydrophilic in nature. Nevertheless, starch modified with plasticizer such as glycine, sorbitol, amino acids, fatty acids, and surfactants can be used for film packaging with adequate gas barrier properties, toughness, film flexibility, and transparency. It has been recently reported (Cao et al., 2009; Huang et al., 2005) that ethanolamine can be used as a plasticizer to destroy the pristine starch granules and provide a uniform continuous phase as similar to thermo-plastic.

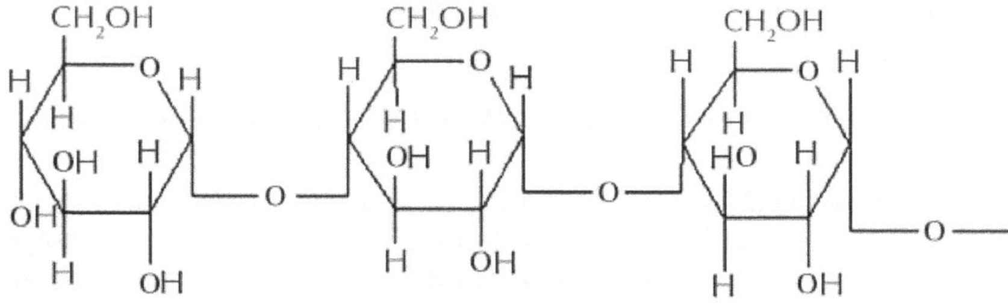

FIGURE 6 Structure of amylose.

Starch-based coatings not only reduce the microbial count but also extend the storage life of the food product (Chinnan, 2004). Starch is also used as a precursor to derive a variety of environmental friendly synthetic polymers such as PLA, polyhydroyalkanoates (PHA), and its derivatives like polyhydroxybutyrate (PHB), polylcaprolactone (PCL), and so on. The biodegradation of starch products recycles atmospheric CO_2 trapped by starch producing plants. Biodegradation of starch-based polymers occur between the sugar groups leading to a reduction in chain length and the splitting off of mono-, di-, and oligosaccharide units as a result of enzymatic attack at the glucosidic linkage. Starch can also be used as biodegradable filler in synthetic fossil polymers like PE and PS, which make them semi-degradable in nature.

CELLULOSE

Cellulose is one of the most abundant natural polymer on Earth and is made up of glucose monomer units, which are joined together by β-1,4glycosidic linkages as shown in Figure 7(a). The packing of monomers is close and leads to a network structure as compared to starch due to the formation of inter-chain hydrogen bonding as shown in Figure 7(b) and has a semi-crystalline morphology. Because of its hydrophilic nature and semi-crystalline structure, it is less suitable for packaging purpose. In cellulose based packaging materials, the alternating hydroxyl side-chains along the cellulose backbone causes poor moisture barrier properties. These bonds also cause the high crystalline structure, which results in brittle nature, poor flexibility, and tensile strength.

However, cellulose based derivatives such as hydroxy or carboxymethyl cellulose, cellulose acetate, and so on have been demonstrated to be commercially viable due to its easy processing and excellent film forming properties making it suitable for packaging applications. In everyday life, cellulose based packaging are commonly used as primary packaging such as PE coated papers and bulk paper products as secondary packaging.

Cellulosic materials, such as cotton and wood pulp, when subjected to acid hydrolysis yield defect-free, rod-like crystalline residues called as cellulose nanocrystals. These are often referred to as microcrystals, whiskers, nanocrystals, nanoparticles, microcrystallites, or nanofibres. Their nanoscale dimension, high surface area, high specific strength, thermal, and mechanical strength along with gas barrier properties increase its merit towards its usage in advanced applications like reinforcement in polymeric matrix to form nanocomposite materials, chemical transformations and so on.

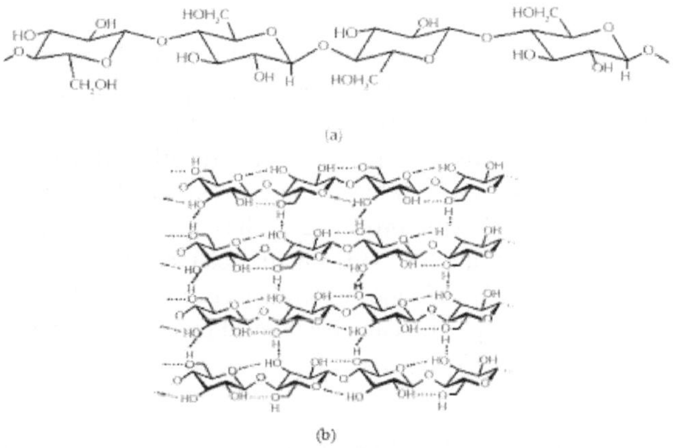

(a)

(b)

FIGURE 7 (a) Structure of cellulose (b) cellulose with matrix showing hydrogen bonding.

Polymer matrix gets transformed, when they are pooled with cellulose nanocrystals and thereafter utilized to prepare biocomposites having enhanced mechanical and electrical properties along with ease in degradability. Cellulose is also formed as slender rod like crystalline microfibrils. The crystal nature (monoclinic sphenoidal) of naturally occurring cellulose is known as cellulose I. Cellulose is resistant to strong alkali (17.5 wt%) but is easily hydrolyzed by acid into water-soluble sugars. Cellulose is relatively resistant to oxidizing agents. Cellulose has been used as a model system for understanding the biodegradation process. The studies in simulated compost environments revealed that cellulose acetates with degree of substitution of up to 2.5 are biodegradable. A decrease in the degree of substitution of cellulose acetate from 2.5 to 1.7 results in a large increase in the rate of their biodegradation (Demirbas, 2007).

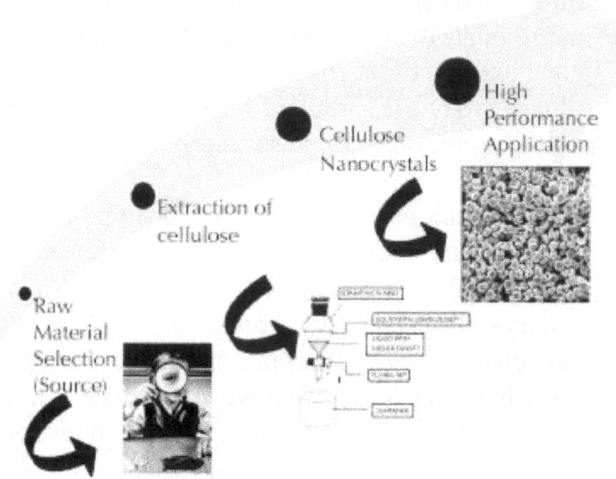

FIGURE 8 Cellulose nanocrystals from plant feedstocks.

Some of the properties of nanocrystalline cellulose (NCC) are high density compared to other biopolymers, high surface area, high aspect ratio, biodegradability in aqueous environment, and modifiable surface properties due to reactive -OH side groups. The crystalline cellulose has high tensile strength and Young's Modulus of around 7 and 150 GPa respectively. It has been reported that the dispersion of NCC in PLA matrix can lead to high barrier properties. It has also been found that the dispersion of silver nanoparticles in the NCC-PLA matrix can lead to enhanced antibacterial properties against *Escherichia coli* and *Staphylococcus aureus*. Further, 74% reduction in water transmission properties can be achieved by the addition of 10%NCC into based films (Brinchi et al., 2013).

PECTIN

Pectin is a branched anionic polysaccharide, which consists of methylated ester of-polygalacturonic acid having segments with rhamnose residues in between, some of which may be linked to short neutral sugar side chains (Gorrasi et al., 2012). It can be used for eco-friendly packaging as it is available in large amounts in the form of agricultural wastes. This can be used to form edible films after demethylation. Depending on the degree of methoxylation, pectins can be classified into high (methoxylation of 50 or greater) and low methoxyl pectin (methoxylation less than 50) (Tripathi et al., 2010). Main drawback of pectin packaging is its water vapor permeability (WVP), which limits its use in food packaging applications. Efforts are being made to combine pectin with other biodegradable materials to modify its properties.Oxygen and water vapor barrier properties of pectin films can be improved by the addition of nanoclays. An 80% decrease in the oxygen transmission rates has been reported for pectin films with ~30% nanoclays, whereas 15–88% reduction has been reported with PET and PLA based layered silicate nanocomposites (Vartiainen et al., 2010).

PROTEINS

Various kinds of proteins have been used to produce bio-based polymers. Some of the examples include soy protein, corn zein (CZ), and wheat gluten. In this section, we will restrict the discussion to a brief description and additional material can be obtained from the cited literature.

SOY PROTEIN

Proteins are natural polymers made up of amino acids and are found in living organisms as enzymes and as a component of tissues, bones, and so on. They have complex structures that contain many sites that allow for modification using plasticizers and other polymers. Bioplastics made from proteins may use corrosives and other toxic chemicals during production such as formaldehyde and gluteraldehyde. However, these may act as cross-linking agents to improve barrier properties, whereas alcohols like glycerol may be used as plasticizers.

Proteins have wide range of functional properties. They have high intermolecular binding due to the Van der Waals forces and also have very high cross-linking potential due to the formation of hydrogen bonds at different positions. Proteins have much better barrier and mechanical properties than polysaccharides. Edible films have been made from lipid compounds for excellent moisture barrier properties. However, textural and organoleptical problems may occur due to the usage of such lipid compounds in the edible films(Vieria, 2011). Some of the advantages of soy protein based films over other plant protein sources are lower brittleness, water solubility and higher transparency. In addition, these films have excellent film forming abilities, low cost and high barrier properties against oxygen permeation under varying humidity conditions.However, these films have poor mechanical properties and heat sealability, when compared to synthetic polymer films such as LDPE and polyester film. In general, the tensile strength of soy protein films is lower than the conventional polymers such as LDPE and polyester films (Cho et al., 2010).

CORN ZEIN

The CZ is a group of alcohol soluble prolamin protein found in the endosperm of corn with a molecular weight in the range of 15–50 kDa. CZ is produced commercially by extraction using aqueous alcohol and subsequently drying to a granular powder. Zein films are formed by dissolving zein into aqueous ethanol or isopropanol at 60–80°C and cast using standard solution technique. Plasticizers like glycerol and polyethylene glycol (PEG) are used to reduce brittleness of zein films but these tend to migrate out of the films as there is weak interaction with protein, which leads to loss in flexibility of the film over a period of time. The WVP of zein films is lower than or similar to those of other protein films. The WVP of zein is still higher than that of many commonly used conventional fossil based polymers like LDPE. Such protein based films find commercial applications in coating formulations for shelled nuts, candy, pharmaceutical tablets, and so on. When the heat sealable CZ layer is laminated on soy protein isolate (SPI) film, an oxygen barrier film with improved heat sealability can be obtained. The CZ/SPI film has up to four times lower oxygen permeability and improved heat sealibility even at higher temperatures, when compared to some of the non-degradable polymer blends based films such as nylon-metalocene catalyzed linear low-density PE (NY/mLLDPE) (Cho et al., 2010).

WHEAT GLUTEN

It is an enriched protein (70–80%) complex containing (proladins) gliadins and glutenins in combination with small amounts of wheat oils, starch, and insoluble hemicellulose. Both of these are water insoluble, gliadins are soluble in ethanol, while glutelins are insoluble in ethanol. The gliadins are mainly monomeric single chain polypeptides, whereas the glutenins are polymeric and disulfide-linked polymeric chains. Edible films prepared by combing unmodified and modified polypeptides from wheat gluten hydrolysates exhibit high tensile strength and are practically impermeable to oil but are readily water-soluble and very brittle.Wheat gluten has high film-forming capacities due to distinctive elastic and cohesive properties of gluten proteins (Türe et al., 2012). These films are selectively permeable to carbon dioxide versus oxygen with permselective value up to 28 depending on the relative humidity, which is significantly higher than some of the conventional synthetic films with permselectivity of around 4(Paz et al., 2005).

GUM AGAR

Agar is fibrous polysaccharide extracted from marine algae such as *Gelidiumsp.* and*Gracilariasp*(red seaweeds).It consists of slightly branched and sulfonated, agarose, and agaropectin.It forms gels that have melting points above the initial gelation temperature. Agar coatings containing antibiotics are effective in prolonging the shelf life of poultry meat stored at 2°C. Agar gel edible films containing bacteriocin like Nisin inhibited microbial activity in food articles, when stored for long time under refrigeration. Agar can be used as packaging material because it has high mechanical strength, biodegradability, and biocompatibility (Giménez et al., 2013).

ALGINATE

Alginates are the salts of alginic acid, a linear copolymer of D-mannuronic and L-guluronic acid (Figure 9) monomers that are extracted from brown seaweeds of the phaephyceae class (Schettini et al., 2013). These react with divalent and trivalent cations, which initiate gelling that leads to film formation. Calcium ions, which are more effective than magnesium, manganese, aluminum, ferrous, and ferric ions, has been used to form an edible calcium alginate film, which prevents dehydration of raw fish.

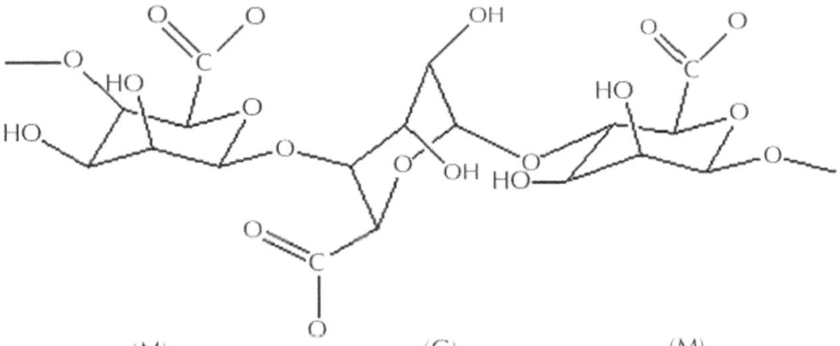

FIGURE 9 Structure of alginate: α-L-guluronic acid (G) β-D-mannuronic acid (M), MG block copolymer.

Sodium alginate (SA) is a water-soluble polysaccharide mainly composed of (1–4)-linked β-D-mannuronic acid units and α-L-guluronic acid units. It consists of carboxyl groups in each constituent residue and possesses various functionalities for different applications. The SA has great potential in making biodegradable or edible films due to its biocompatibility, biodegradability, non-toxicity, and reproducibility. However, it has poor mechanical and gas barrier properties and high water sensitivity (Abdollahia et al., 2013).

CARRAGEENAN

Carrageenans are structural polysaccharides obtained from *Chondruscrispus* (red seaweed) and is water-soluble galactose polymer.It is a complex mixture of five types namely ι-, κ-, λ-, μ-, and ν-carrageenan. Of these, the mixture of ι-, κ-, and λ-carrageenan is being used in food applications. Gelation of ι- and κ-carrageenan occurs with both monovalent and divalent cations, whereas λ-carrageenan is prevalent as a non-gelling and thickening agent. Carrageenan can be used as edible protective coatings for extending the shelf life of poultry products. κ-carrageenan also has good film forming properties along with an adequate WVP. In comparison to PE films, κ-carrageenan has higher tensile strength (22–32MPa) (Choi et al., 2005).

6.3.3 BIOPOLYMER: DIRECTLY FROM ANIMALS

GELATIN

Gelatin is prepared by thermal denaturation or partial hydrolysis of collagen that is found in animal skins and bones, in the presence of dilute acids (Cha, 2004, Wu et al., 2013). Gelatin is primarily used as a gelling agent forming transparent elastic thermoreversible gels on cooling below 35°C. However, its poor mechanical properties limit its application as a packaging material. Many techniques such as vapor cross-linking, orientation technique, and use of fillers such as hydroxylapatite and tricalcium phosphate are used to reinforce gelatin-based films, which are strong, flexible, transparent, and impermeable to oxygen (Giménez et al., 2013, Cha, 2004). Film-forming applications of gelatin, particularly in blend with other polymers like chitosan, have been used in the pharmaceutical and food industry

(Cha, 2004) in order to maintain the conditioned atmosphere. The physical properties of edible composite films can be improved by preparing films from fish gelatin and different polysaccharides such as chitosan or pectin(Giménez et al., 2013). Gelatin based composite films can be produced from beef gelatin and glycerol, where glycerol acts as a hydrophilic plasticizer. It has been reported that glycerol increases the film solubility, sealing strength, and gas permeability due to the reduction in the intermolecular forces, which in turn increases the chain mobility inside the protein network (Hanani et al., 2013).

CHITOSAN

Chitosan is deacetylated form of chitin, which is the second most abundant natural polymer after cellulose. As shown in Figure 10, chitosan structurally, is a long chain of repeating units of (1, 4)-linked 2-deoxy-2-aminoglucose. Chitosan films are clear, tough, and flexible and act as good oxygen barrier that can be formed by film casting from 1–2% acetic acid based aqueous solution. Moreover, the non-toxic nature along with biodegradability, biocompatibility, and antimicrobial activity makes it a strong candidate for the eco-friendly plastics (Kanatt et al., 2012, Tang et al., 2012). Chitosan films can be used to protect foods from fungal decay and modify atmosphere for fresh fruits (Tang et al., 2012).Unlike many other bio-derived polymers such as PLA, PCL, andPHB, chitosan is not a thermo-plastic and therefore its film processing requires solution casting approach, which restricts its usability for mass production application. Moreover, chitosan films also lack long term stability, possess lower water vapor barrier characteristics and have selective permeability to carbon dioxide and oxygen (Elsabee et al., 2013). Properties of chitosan can be modified by the addition of various compounds. For example, The WVP decreases by 40–45% by the addition of 1–3 wt% montmorillonite (MMT) and by 58% on the addition of 3 wt% ofclay and 1–1.5% rosemary essential oil (REO) (Abdollahi et al., 2012). Tensile strength increases to 20% and 15%, when 3 and 5 wt%, MMT is added respectively.

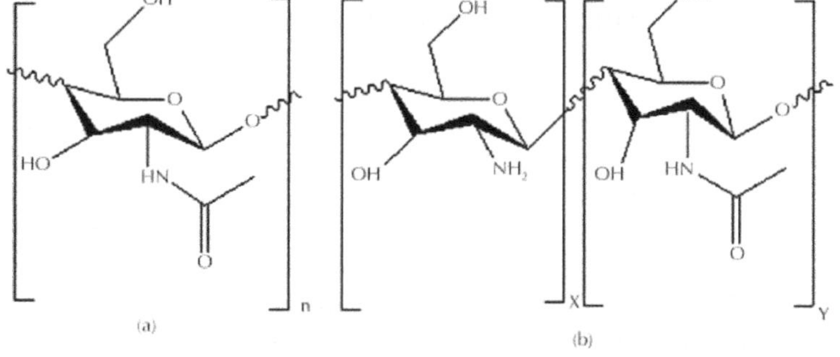

FIGURE 10 (a) Structure of chitin and (b) Structure of chitosan.

Chitosan films find wide usage in food, formation of edible biodegradable films, and pharmaceutical packaging due to its inherent antimicrobial activity. By using enzymes, chitosan can also be functionalized with antioxidant molecules (Zemljic et al., 2013). There are many factors, which influence the antimicrobial activity of chitosan films. The factors are pH, intrinsic factors, chitosan metal complex, sorption, and bacterial properties. Due to the hurdle effect of the acid stress on the bacterial cells, the antimicrobial activity of chitosan based films decreases with increasing pH. Chitosan has the ability to bind to the bacterial cell wall with electrostatic interaction. Antimicrobial character of chitosan is due to its positively charged amino group, which interacts with negatively charged microbial cell membranes and causes the leakage of proteinaceous and other intracellular constituents of the micro-organisms (Dutta, 2009).

6.3.4 SYNTHETIC POLYMER: PRECURSOR FROM NON-RENEWABLE RESOURCE

POLYCAPROLACTONE (PCL)

The PCL is a biodegradable thermo-plastic polymer obtained by chemical conversion of crude oil followed by ring opening polymerization (ROP) of ε-caprolactone. The structure of the PCL can be understood by the Figure 11. The PCL is mainly produced by the company Solvay under the trade name Capa®. It is prepared from renewable resources by chemical treatment of saccharides. Firstly, the saccharides are converted to ethanol and acetic acid by fermentation. In the second step, ethanol is converted to cyclohexanone by use of chromic acid. Naturally, such product is expensive and is therefore mixed with large amount of other natural materials to obtain a good biodegradable material at a low price.

The PCL has good water, oil, solvent, and chlorine resistance, low melting point (58–60°C) and low viscosity and hence, is easy to process. Due to the low melting point it can degrade at lower temperature, which is a favorable condition for composting. For example, PCL can biodegrade at 50°C by thermotolerant *Aspergillus sp.* within 6 days into succinic, butyric, valeric, and caproic acids (Sanchez et al., 2000). Otherwise,the PCL without additives degrades after 6 weeks in compost with activated sludge.The rate of hydrolysis and biodegradation of PCL depends on its molecular weight and degree of crystallinity. These related polyesters are water resistant and may be melt-extruded into sheets, molded bottles, and various shaped articles.

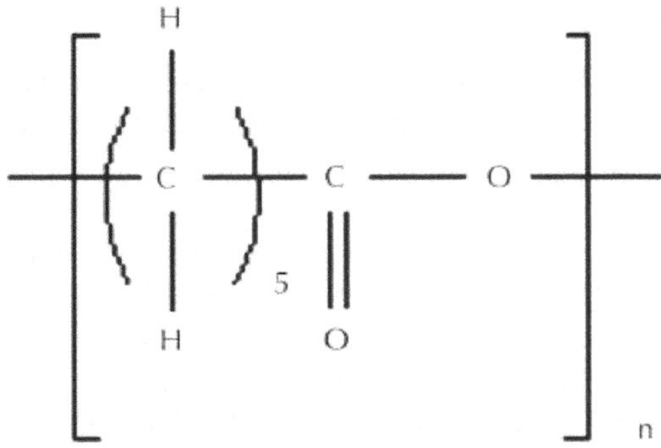

FIGURE 11 Structure of polycaprolactone.

The PCL is used mainly in surface coatings, adhesives and synthetic leather, and fabrics. It also serves to make stiffeners for shoes and orthopedic splints, fully biodegradable compostable bags, sutures, and fibers. Recent studies show that it has many biomedical applications too.

POLY TRIMETHYLENE TEREPHTHALATE (PTT)

The PTT is a copolymer of 1,3-propanediol and aromatic terephthalic acid or dimethyl terephthalate, which can be synthesized by condensation polymerization. The above precursors for PTT can be derived from both renewable resources like corn and non-renewable resources like conventional petroleum based products. The structure of PTT is shown in Figure 12. In its class, it has been observed to be degradable by enzymatic hydrolysis in addition to the thermal degradation. So far pristine PTT does not seem to have found applications in packaging but its blend with LDPE, PP, or PET can be used for packaging applications.

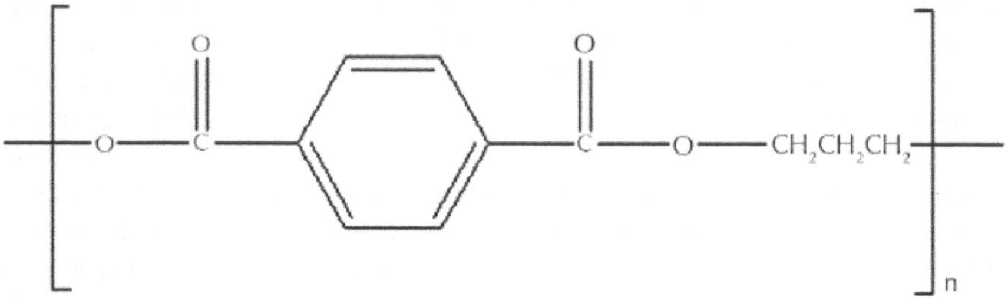

$$\left[-O-\overset{O}{\overset{\|}{C}}-C_6H_4-\overset{O}{\overset{\|}{C}}-O-CH_2CH_2CH_2- \right]_n$$

FIGURE 12 Structure of Poly (trimethylene terephthalate).

The PTT has properties similar to PET and has processing characteristics similar to PBT. The PTT has comparable tensile strength close to PLA, which is around 60MPa(Supaphol et al., 2004), whereas the density of PTT is 1.34 g/cc as compared to PLA (1.24 g/cc).

6.3.5 SYNTHETIC ENVIRONMENT FRIENDLY POLYMER: PRECURSOR FROM NATURAL RESOURCE

POLYLACTIC ACID (PLA)

Synthetic environment friendly polymers with precursors from natural resource such as polylactic acid have shown enormous potential to substitute a wide variety of conventional fossil based packaging plastics and have been demonstrated to be even commercially viable. In view of this, a detailed description about various aspects of PLA such as synthesis, properties, and processing have been discussed as below.

Lactic acid based polymer such as PLA belongs to the family of aliphatic polyesters. The PLA is formed by polycondensation of lactic acid (2-hydroxy propionic acid). It is a biodegradable polymer with a reasonable shelf life, for a wide variety of consumer products, such as paper coatings, films, moulded articles, and fiber applications (Datta et al., 1995). It degrades slowly by simple hydrolysis of the ester bond to convert into harmless, natural products like CO_2 and H_2O (Drumright et al., 2000).

Hence, it could be a functional and economical replacement to the large amount of existing plastics, used in the world. The PLA has degradation time in the environment of the order of six months to two years, which compares very favorably with 500–1000 years for conventional plastics such as PS and PE(Ohara et al., 2003).

As mentioned above, the basic building block for PLA is lactic acid, which exists in two optically active configurations L(+) lactic acid and D(-) lactic acid. Another intermediate monomer for PLA synthesis is a lactide and obtained by the depolymerization of low molecular weight PLA under reduced pressure to give a mixture of L-lactide, D-lactide, or meso-lactide as shown in Figure 13.

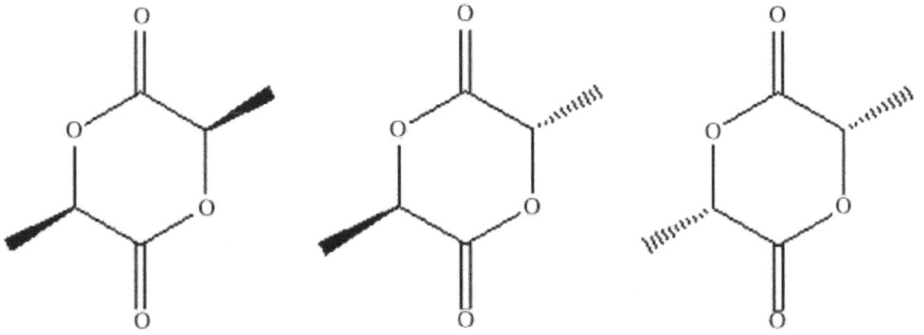

D-la ctide meso-lactide L-lactide

FIGURE 13 Structure of isomeric lactides as precursor for PLA.

POLYMERIZATION PROCESS OF POLY LACTIC ACID

The PLA can be synthesized by various synthesis routes such as direct polycondensation, azeotropicdehydrative condensation, ROP, melt polycondensation (MP), and solid state polymerization (SSP) of low molecular weight (MW) PLA (Lunt, 1998).

Lactic acid is polymerized by direct polycondensation to yield a viscous to brittle glassy material with MW up to 10,000 Da, depending on the polymerization conditions. The reason for the low MW is the presence of water. The statistical presence (low concentration) of reactive end groups causes various unwanted side reactions such as transesterification, ester exchange, and backbiting equilibrium reactions, which favor the formation of lactide as byproduct (Garlotta, 2002). Subsequently, in order to avoid this unwanted equilibrium, the PLA synthesis by lactic acid is performed in the presence of a catalyst and organic solvents *via* azeotropicdehydrative condensation process. In this process, the water is removed from the reaction system *via*azeotropic distillation with the organic solvent (Ajioka et al., 1995). The average MW of the PLA depends on the type and moisture content of the organic solvent. When the solvent has high water content, ~ 400–500 ppm, the average MW of the PLA obtained is 15,000–50,000 (Ohta et al., 1995). However, at optimum conditions, this process yields MW > 10^5 Da, although the polymerization time is also high (~50 h) (Enomoto et al., 1994).

In order to achieve high MW PLA from lactic acid, polycondensation synthesis of moderate MW PLA is done through the MP of lactic acid in the presence of an appropriate catalyst (Moon et al., 2000). Subsequently, MW is enhanced by lowering the temperature below the melting point of the PLA and carrying out SSP of the moderate MW PLA pre-polymer. In SSP, the polymerization reaction can be favored over depolymerization and other side reactions because of restricted mobility of the PLA chain and their end groups and due to the high activation energy required for the unwanted side reactions. Particularly, in the process of crystallization of the resultant pre-polymer, both, reactive end groups, and catalyst, are concentrated in the amorphous region leading to the preferential condensation between the reactive end groups of pre-polymer, yielding PLA of high MW, upto 6,00,000 Da (Moon et al., 2001). Figure 14 illustrates the entire process of the MP of lactic acid followed by SSP.

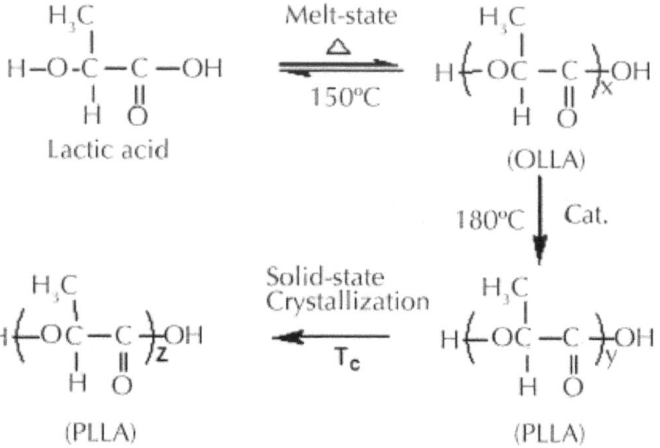

FIGURE 14 MP/SSP reaction for lactic acid polymerization (adopted from Moon et al., 2001).

The increase in MW can continue until the crystallinity exceeds 45% (Moon et al., 2001). This inference has been reasonably supported by the observed decrease in MW and crystallinity of the polycondensates obtained, after heating times exceed 20 hrs. However, increase in MW has been observed for longer pre heat treatment time.

Thus, one can tailor the MW of PLA *via* the polycondensation process, by adjusting the reaction time. Addition polymerization of lactide (dimer of lactic acid) is capable of yielding high MW PLA in short polymerization times (minutes to a few hours)*via* ROP. This process involves the synthesis oligomer of lactic acid (OLLA) by direct polycondensation followed by depolymerization of OLLA in the temperature range 190–240°C, which yields the lactide precursor for ROP. The different percentages of the lactide isomers (D-, L-, meso-, and D, L-racemic stereocomplex) formed *via* the backbiting reaction of OLLA, depend on the lactic acid isomer feedstock, temperature, and catalyst (Lunt, 1998). The racemic stereocomplex melts at 126–127°C, significantly higher than pure isomer (97°C). Purification of lactide isomers is very critical in order to obtain high MW PLA. The ROP of L-lactide requires presence of a suitable catalyst or initiator. Several organometallic compounds such as oxides, carboxylates, and alkoxides, have been reported as effective initiators for the synthesis of PLA by ROP (Stridsberg et al., 2002). The most effective and versatile catalyst among many, which produces PLA by coordination-insertion mechanism is stannous octoate(Dechy-Cabaret et al., 2004). Recently, Creatinine, a non-toxic metabolite in human body and a derivative of guanidine has also been examined for the ROP of lactide and has been found to initiate the reaction (Wang et al., 2004). The PLA obtained by using this catalyst has M_n ~ 15,600 Da and PDI ~ 1.28.

Enzymes as biocatalysts are also examined due to their non-toxic, natural catalysts and therefore, are superior candidates for ROP of lactides.

Advantages of enzyme catalyzed polymerizations over conventional methods include

- Mild reaction conditions, that is, temperature, pressure, pH, and absence of organic solvents,
- High enantio and regioselectivity, and
- Recyclability of catalyst (Garlotta, 2002, Dechy-Cabaret et al., 2004).

In case of PLA and other polyester synthesis, lipase catalyst has been used and found to be efficient. Lipase is usually an enzyme that catalyzes the hydrolysis of fatty acid esters in living systems and sometimes, can also be used as catalyst for esterification and transesterifications. Lipase enzyme can be derived from *Candida cylindracea, Pseudomonas cepacia, Pseudomonas fluorescens, and Porcine pancreas*, which are powdery and commercially available crude enzymes (Albertsson and Varma, 2003).

PROPERTIES OF POLY L-LACTIC ACID

The High MW PLA is a glossy, colorless, stiff, and thermo-plastic polymer with properties similar to PS. In this section, the thermophysical, mechanical, electrical, rheological, solubility properties, and degradation characteristics of poly L-lactic acid are presented. The PLA is a semi-crystalline polymer with higher glass transition, equilibrium crystalline, and melting temperatures, which decrease with increase in content of D-isomer. Polymer prepared from meso- or racemic lactide is generally amorphous. The melt enthalpy estimated for a pure PLA of 100% crystallinity reported by various authors is in the range between 93 J/g and 148 J/g(Lunt, 1998, Miyata, 1998). The melting temperature and degree of crystallinity, are dependent on the molar mass, purity, and thermal history of the polymer. However, the PLA is a slow crystallizing material, similar to PET. The fastest rates of crystallization for pure PLA occur in the temperature range of 110–130°C, yielding spherulitic crystalline morphology.Spherulitic growth rate for PLA has been determined to be 4 μm/min at 125°C.

The mechanical properties of PLA are significantly influenced by its MW and degree of crystallinity. By controlling these two parameters, the mechanical properties of PLA can be varied, ranging from soft and elastic plastics to stiff and high strength materials. The tensile strength and modulus of PLA increase by a factor of two, when its MW is raised from 50,000–100,000 Da (Hartmann, 1998). The impact strength and softening temperature increase with crystallinity and MW. However, after annealing of the same PLA, the impact resistance increases due to the crosslinking effects in the crystalline region, while the tensile strength increases, presumably due to the stereoregularity of the chain. However, electrical properties such as volume resistivity, dielectric constant, and dielectric loss tangent of PLA at room temperature are similar to that of cross linked polyethylene (XLPE), which is currently used as the insulating material for cables and electric wires. The mean value of the impulse breakdown strength of PLA is about 1.3 times that of PE. Thus, there is a possibility for the insulation thickness reduction of electric wire and cable (Nakagawa et al., 2003). Further, solubility of PLA depends on its MW and degree of crystallinity. The Amorphous PLA is soluble in most organic solvents such as chlorinated solvents, tetrahydrofuran, benzene, acetonitrile, and dioxane. Crystalline PLA is soluble only in chlorinated solvents and benzene at elevated temperatures. The most superior solvent till date is 1,1,1,3,3,3-hexafluoro-2-propanol (HFIP). Typical non-solvents for lactic acid based polymers are water, alcohols (for example, methanol, ethanol, and propylene glycol), and unsubstitutedhydrocarbons (for example, hexane and heptane). The PLA melt is viscoelastic in nature, exhibiting a flow behavior that is combination of irreversible viscous flow due to the polymer chain slippage as well as reversible elastic deformation due to molecular entanglement.

MELT PROCESSING

Extrusion based melt-processing of lactic acid based polymers is required for thermoforming, injection molding, fiber drawing, and film blowing. The properties of polymer depend on the processing conditions such as shear rate and temperature. A major concern in the manufacturing of PLA products is the limited thermal stability during the melt processing, because the MW decreases by 14–40% (Garlotta, 2002). The stress-strain curve of poly (D,L-lactide) oriented to $\lambda = 2.5$ is shows more brittle behavior comparison to unorientedPLA(Nijenhuis et al., 1992).As far as PLA fiber production is concerned, manufacturing of biodegradable fibers of aliphatic polyesters has been extensively investigated for potential use in medical applications (Garlotta, 2002). One of the first commercially available bioreabsorbable products for medical use has been poly(glycolide) fibers (Dexon), for use as sutures. Other fiber-form bioabsorbable medical products introduced in the market are based on copolymers of glycolide in combination with *L*-lactide (Vicryl) and copolymers of ε-caprolactone (Monocryl).

APPLICATIONS OF POLY L-LACTIC ACID

Advances in the manufacturing processes of PLA together with improvements in the material properties have realized its applications in various fields including packaging. The growing waste disposal problems throughout the world require ban of some petroleum based polymer products such as plastic bags. In this respect, biodegradable PLA can prove to be a viable alternative to petrochemical based plastics for several applications (Lunt, 1998). Its unique physical properties make it useful in diverse applications including paper coating, fibers, films, and packaging. The PLA can also be used in fibers for woven and non-woven fabrics starting from cord and rope to mattresses carpeting and clothing (Oksman et al., 2006). The PLA blended with cotton, wool, and silk can be used to make exercise clothing, suits, and even a 100% corn fiber wedding dress.Other applications in this field include PLA carpet tiles and action wear (Gupta et al., 2007).

LIMITATIONS OF PLA

In recent years, the demand for biodegradable polymer with excellent material properties has been growing at an enormous rate. Virgin polymers may not meet the requirements for several food packaging applications due to medium gas barrier properties and also require to enhance their resistance to fire and ignition or sometimes to simply reduce the cost.However, The PLA is considered not suitable for high strength, high performance, and high temperature applications, due to its weak mechanical properties and low heat deflection temperature. In order to improve its mechanical properties such as storage modulus, flexural modulus, ultimate tensile strength, and distortion at break, the PLA is mixed with particles of other materials, so as to yield composite materials for enhanced properties. The PLA is associated with layered silicates. In literature, PLACN with MMT, synthetic mica and smectite clays, have been reported (Chang et al., 2003).

POLYHYDROXYALKANOATES (PHA)

The PHAs are a family of polyhydroxyesters of 3-, 4-, 5-, and 6-hydroxyalkanoic acids (Figure15), produced by bacterial fermentation under nutrient limiting conditions with excess carbon in the form of lipids or sugars. These are an important class of water insoluble, biocompatible, and biodegradable thermo-plastic, which can be produced from renewable carbon sources. Thus, there has been considerable interest in the commercial utilization of these biodegradable polyesters (Suriyamongkol et al., 2007).In the late 1980s, Imperial Chemical Industries commercialized the PHAs under the trade name Biopol using bacterium *Ralstoniaeutropha.* Usually, the PHA naturally accumulates within the microbes as granules that can constitute up to 90% of a single cell mass. It has excellent mechanical properties similar to synthetically produced degradable polyester. Poly(3-hydroxybutyrate) (P3HB) is a member of PHA family and was discovered by the French microbiologist Maurice Lemoigne, in the form of intracellular granules in Gram-positive bacterium *Bacillus megaterium.* It has a higher glass transition, melting temperature, and degree of crystallinity relative to PHAs. Thermal and mechanical properties of PHA and its copolymers can be enhanced by dispersion of biodegradable plasticizers such as soybean oil (SO), epoxidized soybean oil (ESO), triethyl citrate (TEC), and dibutyl phthalate (DBP) (Vieira, 2011).

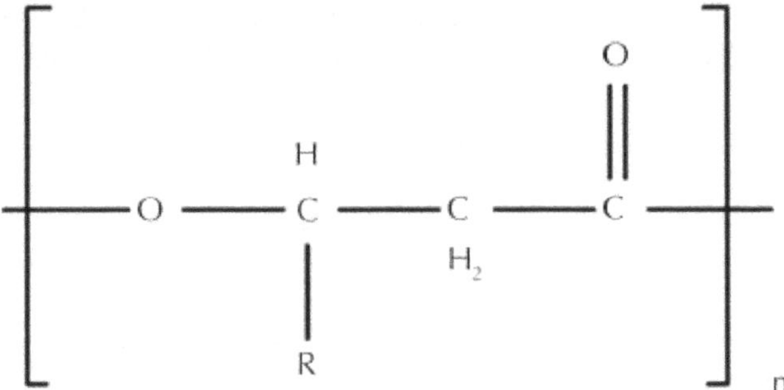

FIGURE 15 Structure of PHA [R-: $-CH_3$ (PHB) and $-CH_2CH_3$(PHBV)].

TABLE 1 Physical, mechanicalW, and thermal properties of some commercialized PHAs (adapted from Chanprateep, 2010)

Properties Biomer 240		PHB		PHB co-Polymers		PHBV		PHBH
		Biomer P226	Mirel P1001	Mirel P1002	Bio-cycle 100	Bio-cycle 24005	Kane-ka	
Physical	Melt flow rate (g/10 min)	5–7	9–13	-	-	10–12	15–25	5–10
	Density (g/cm^3)	1.17	1.25	1.39	1.30	1.22	1.20	1.2
	Crystallinity (%)	60–70	60–70	-	-	50–60	-	-
Mechani-cal	Tensile strength (MPa)	18–20	24–27	28	26	30–40	25–30	10–20
	Elongation (%)	10–17	6–9	6	13	2.5–6	20–30	10–100
	Flexural strength (MPa)	17	35	46	35	-	-	-
	Flexural modulus (GPa)	-	-	3.2	1.9	-	-	0.8–1.8
Thermal	Melting temperature (°C)	-	-	-	-	170–175	-	-
	VICAT softening point (°C)	53	96	148	137	-	-	120–125

The physical properties and biodegradability of the PHAs can be regulated by blending with synthetic or natural polymers. The PHAs with short side chains behave similarly to PP, whereas the PHAs with longer side chains are elastomeric with crystals acting as physical crosslinks. It can be extruded into films, molded/coated onto other substrates using conventional processing equipment. Various properties of different commercialized PHAs are shown in Table 1. In addition, the PHA can either be in a latex form, or as dry powders ready to use that can be suitable for melt processing. In addition, the PHAs offer a low moisture vapor permeability that is comparable to that of LDPE, which is very useful in packaging applications. Recent application developments based on medium chain length PHAs include biodegradable cheese coatings (Nair and Laurencin, 2007), lawn and leaf bags, disposable diapers, fast food service ware, single use medical devices, paints, as well as performance materials, which take advantage of superior properties inherent to the PHA. Table 2 shows the various PHA products that are commercially available in the market.

TABLE 2 Commercial production of polyhydroxyalkanoates (adapted from Chanprateep, 2010)

Type	Manufacturer	Trade Name	Capacity (Tonnes)
PHB	Mitsubishi Gas Chemical Company Inc.(Japan)	Biogreen®	10,000
	Telles (US)	Mirel™	50,000
	PHB Industrial Company (Brazil)	Biocycle®	50
	Meredian (US)	Meredian	272,000
PHBH	P&G (US)	Nodax™	20,000–50,000
	Lianyi Biotech (China)	Nodax™	2000
	Kaneka Corporation (Japan)	Kaneka	1000
PHBV+ PHB	Biomer Inc. (Germany)	Biomer®	50
P(3HB-co-4HB)	Tianjin Gree Bio-Science Co/DSM	Green Bio	10,000
PHBV + Ecoflex blend	Tianan Biologic, Ningbo (China)	Enmat®	10,000

The PHAs are produced *via* challenging microbial biosynthesis having low productivity and high costs compared to the conventional fossil based plastics. However, relatively high productivity has been achieved for P(3HB), poly (3-hydroxybutyrateco-3-hydroxyvalerate), P(3HB-co-3HV) and poly (3- hydroxyhexanoate-co-3-hydroxyoctanoate), and P(3HHx-co- 3HO) (Keshavarzand Roy, 2010). Despite low productivity in majority of cases, the versatility of these biopolymers has made them good candidates for high value low volume based packaging products. As such, properties of PHAs depend upon the composition of the monomer unit, the microorganism used in fermentation (for example, Gram-negative or Gram-positive), nature of the carbon source used during the fermentation process, fermentation conditions, and modes of fermentation (batch, fed-batch, and continuous modes). Biopolymers have found an increasing use in packaging. Representative applications of biopolymers in food packaging are given in Table 6.3. (Avena-Bustillos et al., 1997; Balasubramaniam et al., 1997; Banks, 1985; KamperandFennena, 1985; Makino et al., 1997; Park et al., 1996; Ayranci et al., 1997).

6.3.6 PROPERTIES OF BIOPOLYMERS

Conventional polymers can be replaced by biopolymers if the required properties are comparable for a given application. In this section, we will compare some of the critical properties of environmental friendly polymers with various conventionally used fossil based polymers.

THERMAL PROPERTIES

As discussed earlier, glass transition temperature and melting point are two important properties that determine the usability and processabiltiy of a particular bio-based polymer for a targeted application. Figure 16 shows the melting and glass transition temperatures of a wide variety of bio and fossil based polymers. It can be seen that bio-based alternatives are available for both stiff and flexible polymers along with intermediate to higher temperatures. Hence, the thermal properties of packaging are no longer a severe limitation for the replacement of conventional fossil based polymers. It should be noted that the glass transition and melting temperature of a polymer can widely vary depending on various factors primarily MW. From Figure 16, it can be seen that the set of environmental friendly plastics have wide range of melting and glass transition temperatures that can provide the required flexibility to select an appropriate environmental friendly polymer for a specific application.

MECHANICAL PROPERTIES

Mechanical properties depend largely on the polymer architecture and on the process utilized for fabrication (injection molding, sheet extrusion, blow molding, thermoforming, and filmforming). Tensile test analyses are made to determine the tensile strength (MPa), the percent elongation at yield (%), the percent elongation at break (%), and the elastic modulus (GPa) of the polymer packaging material. Table 6.4 shows the various physical and mechanical properties of different biopolymers (Luecha et al., 2010).

TABLE 3 Representative applications of biopolymers in food packaging

Biopolymer	Representative applications	Improvement in properties
	Beef and Chicken	Absorb moisture
	French fries	Containment
Starch based	Bread	Moisture barrier
	Beverages	Acid resistant, high barrier properties
	Snacks	Light barrier, protection against crushing
	Potato Chips	Light, oxygen & moisture barrier
	Soft Fruits	moisture barrier
Cellulose Based	Bananas	Oxygen andCO_2 barrier
	Beans	Moisture barrier
	Meat	High moisture barrier and absorption

TABLE 3 *(Continued)*

Protein(Casein, Alginate, Whey, Carrageenan, Gelatin, Soy, Zein)	Bell Peppers/Cucumbers	Moisture barrier
	Tomatoes	Moisture barrier and gas barrier
	Pears	Moisture, oxygen and CO_2 barrier
	Potato Chips	Light, moisture and oxygen barrier
	Meat	Inhibit microbial growth, frying oil barrier, adhesionandbarrier properties
	Pizza base/sauce	Moisture barrier
	Bread	Moisturebarrier
	Yoghurt	Mechanical protection, moisture and CO_2 barrier
	Butter	Moisture, light and grease barrier
	Hard Cheese	Light, moisture and gas barrier
PLA	Milk	Light and gas barrier and high moisture
	Mushrooms	High mechanical strengths, flexible not opaque
	Frozen Products	High moisture, light and oxygen barrier
	Beverages	Acid resistant, high barrier, inert towards micro-organisms
	Dry Products	High moisture and oxygen barrier
Chitosan	Tomatoes	Moisture and gas barrier
	Fresh Vegetables	Moisture, oxygen and CO_2 barrier, mechanical protection
PHB/V	Milk	Light and gas barrier and high moisture
	Beverages	Acid resistant, high barrier, inert towards micro-organisms
	Snacks	Light barrier, protection against crushing
	Dry Products	High moisture and oxygen barrier
PCL	Fresh Meat	High moisture barrier and absorption
	Fresh Vegetables	Moisture, oxygen and CO_2 barrier, mechanical protection
	Butter	Moisture, light and grease barrier

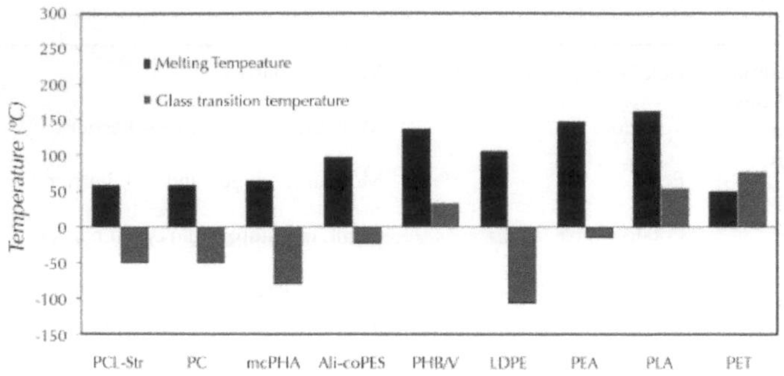

FIGURE 16 Representative melting and glass transition temperatures of bio and fossil based polymers (PCL-Str – Polycaprolactone/Starch blends, PC – Polycaprolactone, mcPHA – medium chain Polyhydroxyalkanoates, Ali-coPES – Aliphatic (co)Polyesters, PHB/V – Polyhydroxy Butyrate/valerate, and PEA – Polyester Amides).

TABLE 4 Representative physical and mechanical properties of various biopolymers

Properties	Type of biopolymer			
	PLA	**PGA**	**PCL**	**PHB**
Density, ρ (g/cm³)	1.2–1.3	1.5–1.7	1.11–1.15	1.18–1.26
Tensile strength, σ (MPa)	21–60	60–100	21–42	40
Tensile modulus, *E* (GPa)	0.4–4	6–7	0.2–0.5	3.5–4.0
Ultimate strain, ε (%)	2.5-6.0	1.5-20	300-100	5-8
Specific tensile strength, σ*(Nm/g)	16.8–48.0	40.0–45.1	18.6–36.7	32.0–33.9
Specific Tensile modulus, *E (kNm/g)**	0.28–2.80	4.0–4.51	0.19–0.38	2.80–2.97
Glass transition temperature, T_g (°C)	45–60	35–45	-60–(-65)	5-15
Melting point, T_m (°C)	150–162	220–233	58–65	168-182

BARRIER PROPERTIES

The determination of the barrier properties of a polymer is crucial to estimate and predict the product-package shelf-life.Oxygen permeability and water vapor transmission are two of the most important barrier properties particularly for food packaging. It has been reported that the oxygen permeability of a substance is closely interrelated to the permeability of other gases making it as a general measure of gas barrier. Representative values of oxygen permeability for various bio and conventional plastics are shown in Figure 17. Moreover, like their conventional counterparts, bio-based polymers also have a constant ratio of oxygen to CO_2 permeability. Figure 18 shows the relative comparison of water vapor transmittance of bio and fossil based polymers. It can be seen that a considerable number of bio alternatives to fossil based polymers are available for applications requiring low water vapor barrier. Carbon dioxide is now important for the packaging in MAP technology because it can potentially reduce the problems associated with processed fresh product, leading to a significantly longer shelf-life (Siracusa et al., 2008). However, for applications that require higher water barrier, bio-based polymers have limitations of substituting the use of fossil based materials. Increasing the water barrier properties of biopolymers is a major area of research in biopolymers and thus futuristic biopolymers may have higher water vapor barrier properties.

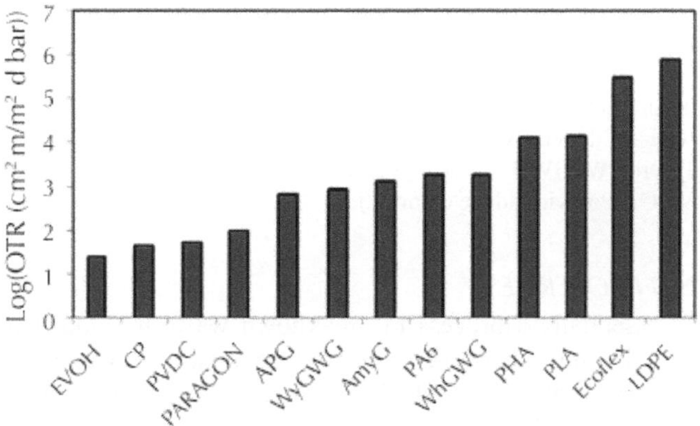

FIGURE 17 Representative comparison of oxygen permeability of bio and fossil based polymers, (EVOH–Ethylene Vinyl Alcohol, CP –Chitosan Plasticized, PVDC –Polyvinylidene chloride, APG–Amylopectin/glycerol, WyGWG–Whey/gluten/water/glycerol, AmyG–Amylose/glycerol, and WhWG–Wheat gluten/water/glycerol).

Barrier properties of biopolymers for use in packaging applications can be improved by:

- Use of coating with materials to add hydrophobicity to the packaging material,
- Lamination of two or more biopolymers (co-extrusion),
- Use of an edible coating with the required barrier properties for the food and subsequently use of biopolymers as primary packaging,
- Development of blends of biopolymers with different properties,
- Chemical and/or physical modification of biopolymers, and
- Development of micro and nanocomposites based on biopolymers. (Mensitieri et al., 2011).

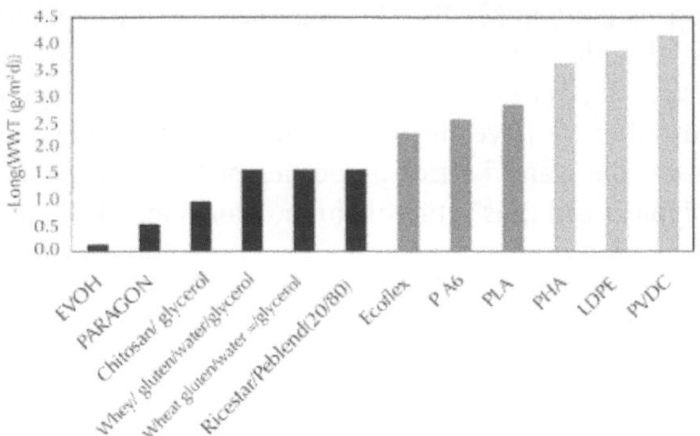

FIGURE 18 Representative Comparison of water vapor transmittance of bio and fossil based polymers, (EVOH – Ethylene Vinyl Alcohol, Ch/Gl – Chitosan/glycerol, WyGWG – Whey/ gluten/water/glycerol, WhGWG – Wheat gluten/water/glycerol, RS + PE – Ricestarch/Pe blend (20/80), and PVDC – Polyvinylidene chloride).

COMPOSTING PROPERTIES

Composting is essentially a process to breakdown waste by biodegradation and is considered to be the most attractive route for treatment of biodegradable packaging waste. Following disposal, biodegradable polymers are expected to undergo complete mineralization to carbon dioxide under aerobic conditions and methane under anaerobic conditions together with other elements that may be present in the materials. Composting time (months) for various fossil based and bio-based environment friendly plastics are shown in Figure 19. It can be seen that the composting time can range up to 6 months for materials such as PEA, cellulosediacetate, and wood. The values are representative, as the actual values depend on a large number of factors including the technology employed for composting.

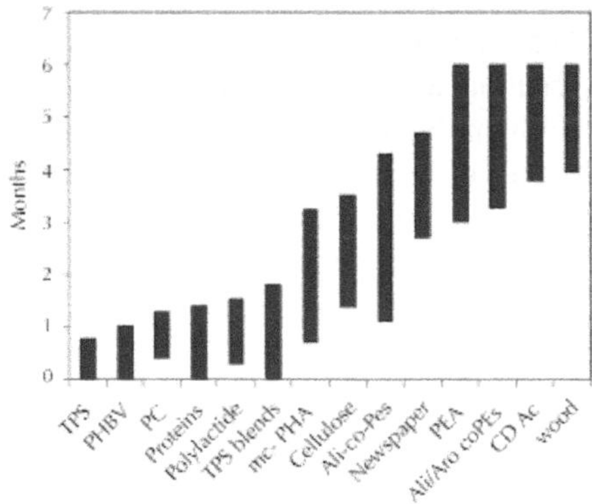

FIGURE 19 Representative composting time (in months) of biopolymers (TPS – Thermoplastic starch, PC – Polycaprolactone, mc-PHA – medium chain length PHA, Al-co-PEs – Aliphatic co-polyester, PEA – Polyester amides, Ali/arocoPEs – Aliphatic/aromatic copolyester, and CDAc – Cellulose diacetate).

Advanced technology can reduce the time required for composting, whereas primitive composting techniques may require longer time to achieve a desired level of degradation. Additionally, the values represented in the Figure 19, do not include the time required for the final biodegradation of the polymer but is the duration for breaking down of the polymer.

Composting properties of these biodegradable polymers are affected by two factors:

1. Exposure conditions and
2. Polymer characteristics.

Exposure conditions can be further categorized as:

1. Abioticfactorsand
2. Biotic factors (Kale et al., 2007), which can be seen in Figure 20.

It has been reported that the surface whitening can be a signal that the process of degradation has started, inducing a change in the refraction index of the sample as a consequence of water absorption and/or presence of products formed by the hydrolytic process. The visual degradation is accompanied by a loss in physical weight and formation of holes (Fukushima et al., 2011).

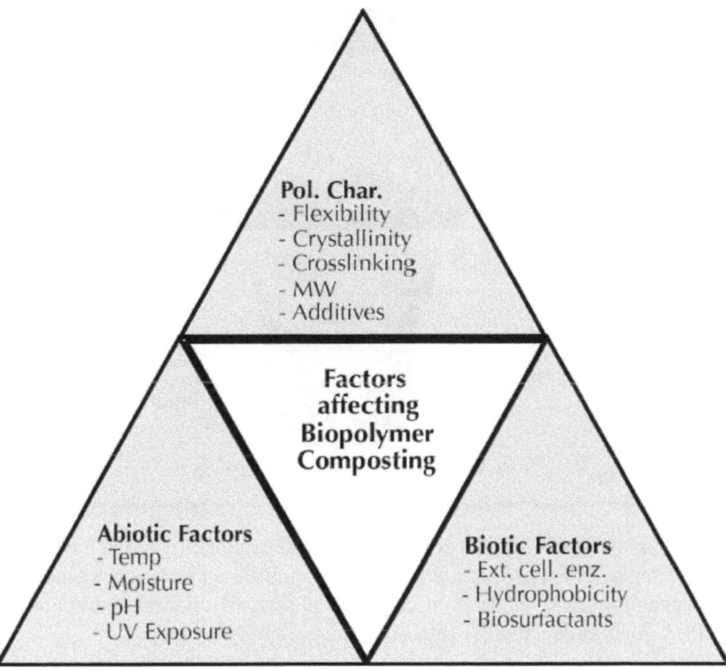

FIGURE 20 Factors affecting biopolymer composting (Pol. Char. – Polymer characteristics, Ext. cell. enz. – Extracellular enzymes,and Temp. – Temperature).

6.3.7 ADVANTAGES AND CURRENT COMMERCIAL STATUS

In addition to the fact that most of the biopolymers are produced from renewable feedstocks, there are various additional benefits, which can be categorized broadly into:

i. Reduction in CO_2 emissions,
ii. Reduction in the fossil energy required for production,
iii. Biodegradability and compostability, and
iv. Lower levels of toxicity.

For example, bio-based polymers like PTT, PHAs, PLA, TPS, cellulose, lignin, CZ, and soy protein are some of the biopolymers that are largely compostable. Similarly, PLA and PTT reduce the CO_2 emissions by more than 30%, whereas polymers like PLA, TPS, and PTT save at least 25% of the energy required, when compared to the production of fossil based polymers.

However, the current market for packaging is dominated by conventional fossil based synthetic polymers but the use of biopolymers has been continuously increasing. Biopolymers are more readily accepted in medicinal applications, where the cost of the product is not as important as its highly specialized function. Figure 21 shows some of the famous brands that have adopted the use of biopolymers and Figure 22 shows the current and projected demand of biopolymers.

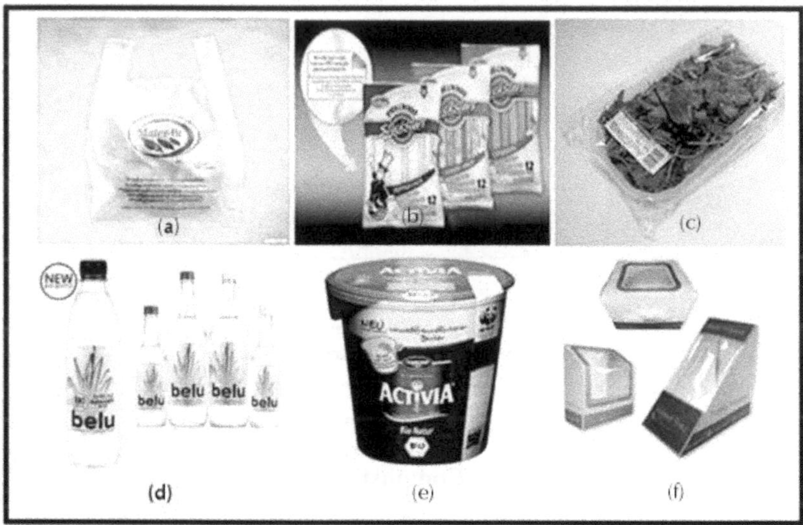

FIGURE 21 Commercial bio-based polymers (a) starch copolymer based biodegradable carry bag, (b) 50–50 film blend of polylactic acid and petroleum-based materials for packaging cheese snacks, (c) packaging blister made from cellulose acetate, (d) biodegradable bottle on sale in Britain, (e) PLA yoghurt cup, and (f) 100% degradable sandwich box.

Bioplastics development is in growing stage, now it covers approximately 5–10% of the current plastic market. The European countries mainly France and Germany have the highest utilization of bioplastics of about 50,000 tons per year (Siracusa et al., 2008). There are a large number of manufacturers spread in USA, Germany, Japan, UK, and The Netherlands among others, who commercially produce bio-based polymers for packaging. Ecofoam, Ecoplast, Novon, and Biopur are some of the commericialtradenames for starch based biodegradable polymer. NatureWorks, Biomer, Ecoflex, Biomax, Bionelle, and Bioplast are some of the tradenames of biopolymers that are based on PLA. EnviroPlastic, ACEPLAST, and Bioceta are bio-based polymers that are based on cellulose acetate, whereas GreenFill, POVAL, and Hydrolene are based on PVA. Some of the commercial tradenames for PCL based bioprodcuts include TONE, CAPA, and Aqua-Novon. It should be noted that the tradenames discussed here are representative in nature.

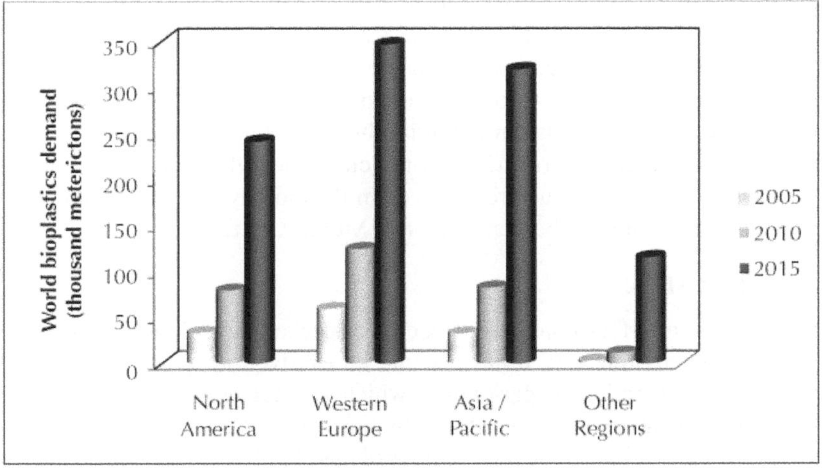

FIGURE 22 World bioplastics demand.

6.3.8 ISSUES RELATED TO ENVIRONMENTAL FRIENDLY PACKAGING

One of the major stumbling blocks in the large scale adoption of biopolymers is that their current production costs are higher than the conventional fossil based polymers. A part of the increased cost is due to the different polymer processing conditions that are used in processing of conventional polymer.However, it can be significantly reduced with the advancement in biopolymer based production technology and novel materials to enhance the properties of the environmental friendly plastic. However, there have been certain sections of the society, which has raised the issue of environmental, occupational, and safety hazards of biopolymers. The environmental issues include the fact that a considerable amount of feedstock is harvested by industrial agricultural production that depends substantially on Genetically Modified Organisms, which have been a subject of controversy particularly related to the data on energy requirements. Additionally, the processing of some biopolymers like cellulose and lignin require very large amount of water and also are associated with reasonably high emissions. Occupational and safety hazards are usually associated with the exposure to pesticides and other chemicals like sulfuric acid, tin octoate, glycerol, methylene di-isocynate, and other volatile, flammable products.The disposal of some of the bioplastics can be challenging as these may be biodegradable only under some specific conditions.

For example, the PLA is found to be biodegradable by industrial composting but is unsuitable for garden composting as the temperatures required for the degradation of the PLA are not sufficiently high in small-scale compost heaps. Hence, procedures for disposal need to be strictly certified, clearly classified, and labeled so that they can be readily identified by both the householder and the waste contractor for composting purposes. On the other hand, there are biopolymer based packagings, which need to be discarded within a particular timeframe to prevent the onset of biodegradation of the packaging before the expiry of the product in it. Segregation of biodegradable plastic litters from conventional plastic litter is necessary if the plastic litter is to be recycled failing, which the durability of the recycled product may be compromised. Moreover, instances of incomplete biodegradation can lead to toxic intermediates. Overall, bioplastics show environmental advantages over petrochemical polymers in many impact categories (abiotic depletion, Ozone depletion Potential, Global Warming Potential 100, and Photochemical Oxidation Potential) but incur higher acidification, eutrophication, and fresh water andterrestrial eco-toxicity, where the emissions from the agriculture systems and infrastructure involved in the biopolymer production processes dominate the environmental burdens(Guo and Murphy, 2012).

6.4 CONCLUSION

With the depletion in fossil based feedstocks for packaging and increasing anthropogenic impact on environment, it has become imperative to explore alternative feedstocks for the preparation of packaging, which are not only environment friendly but are also sustainable over a long term. In many instances, biopolymers have been shown to perform similar to conventional fossil based polymers and are commercially used thereby establishing their economic viability. With continuing research efforts, the limitations of bio-based environmental friendly feedstocks are bound to shrink and along with a wider social acceptability, environmental friendly plastics can aid in enhancing long term sustainability of the environment.

KEYWORDS

- **Carrageenan**
- **Polycaprolactone**
- **Poly lactic acid**
- **Polyhydroxyalkanoate**
- **Polycarbonate**

REFERENCES

1. Abdollahi, M., Rezaei, M., and Farzi, G. *Journal of Food Engineering*. **111**, 343–350 (2012).
2. Abdollahia, M., Alboofetileha, M., Behroozb, R., Rezaeia, M., and Mirakib, R. *International Journal of Biological Macromolecules*. **54**, 166–173 (2013).
3. Ajioka, M., Enomoto, K., Suzuki, K., and Yamaguchi, A. *Bulletin of the Chemical Society of Japan*. **68**, 2125–2131 (1995).

4. Albertsson, A. C. and Varma, I. K. *Biomacromolecules*. **4**, 1466–1486 (2003).
5. Avena-Bustillos, R., Krochta, J. M., and Saltveit, M. E. *J. Food Science*. **62**, 351–354 (1997).
6. Ayranci, E. and Tunc, S. *Z Lebensm Unters Forsch A*. **205**, 470–473 (1997).
7. Balasubramaniam, V. M., Chinnan, M. S., Mallikarjunan, P., and Phillips, R. D. *J. Food Proc. Eng*. **20**, 17–29 (1997).
8. Banks, N. H. *Scientia Horticulturae*. **26**, 149–157 (1985).
9. Brinchi, L., Cotana, F., Fortunati, E., and Kenny, J. M. *Carbohydrate Polymers*. doi:10.1016/j.carbpol.2013.01.033.
10. Callister, W. D. and Rethwisch, D. G. *Materials Science and Engineering: An Introduction*. Wiley, John Wiley & Sons, Inc. (2009).
11. Cao, N., Yang, X., and Fu, Y. *Food Hydrocolloids*. **23**, 729–735 (2009).
12. Cha, D. S. and Chinnan, M. S. *Critical Reviews in Food Science and Nutrition*. **44**, 223–237 (2004).
13. Chang, J. H., An, Y. U., Cho, D., and Giannelis, E. P. *Polymer*. **44**, 3715–3720 (2003).
14. Chanprateep, S. *Journal of Bioscience and Bioengineering*. **6**, 621–632 (2010).
15. Cho, S. Y., Lee, S. Y., and Rhee, C. *LWT-Food Science and Technology*. **43**, 1234–1239 (2010).
16. Choi, J. H., Choi, W. Y., Cha, D. S., Chinnan, M. J., Park, H. J., Lee, D. S., and Park, J. M. *LWT-Food Science and Technology*. **38**, 417–423 (2005).
17. Cui, J. and Roven, H. J. *Trans. Nonferrous Met. Soc. China*. **20**, 2057–2063 (2012).
18. Datta, R., Tsai, S., Bonsignore, P., Moon, S., and Frank, J. *FEMS Micro-biology Reviews*. **16**, 221–231 (1995).
19. Dechy-Cabaret, O., Martin-Vaca, B., and Bourissou, D. *Chem. Rev*. **104**, 6147–6176 (2004).
20. Demirbas, A. *Energy Sources: Part A: Recovery, Utilization, and Environmental Effects*. **29**, 419–424 (2007).
21. Drumright, R. E., Gruber, P. R., and Henton, D. *Advanced Materials*. **12**, 1841–1846 (2000).
22. Dutta, P. K., Tripathi, S., Mehrotra, G. K., and Dutta, J. *Food Chemistry*. **114**, 1173–1182 (2009).
23. Elsabee, M. Z. and Abdou, E. S. *Materials Science & Engineering* (2013) doi:10.1016/j.msec.2013.01.010.
24. Enomoto, K., Ajioka, M., and Yamaguchi, A. US Patent 5310865A (May 10, 1994).
25. Fukushima, K., Tabuani, D., Abbate, C., Arena, M., and Rizzarelli, P. *EurPolym J*. **47**, 14 (2011).
26. Garlotta, D. J. *J. Polym. Environt*. **9**, 63–84 (2002).
27. Giménez, B., Lacey, A. L., Pérez-Santín, E., López-Caballero, M. E., and Montero, P. *Food Hydrocolloids*. **30**, 264–271 (2013).
28. Gorrasi, G., Bugatti, V., and Vittoria, V. *Carbohydrate Polymers*. **89**, 132–137 (2012).
29. Gross, R. A. and Kalra, B. *Science*. **297**, 803–807 (2002).
30. Guo, M. and Murphy, R. J. *J. Polym Environ*. **20**, 976–990 (2012).
31. Gupta, B., Revagade, N., and Hilborn, J. *Prog. Polym. Sci*. **32**, 455–482 (2007).
32. Hanani, Z. A. N., McNamara, J., Roos, Y. H., and Kerry, J. P. *Food Hydrocolloids*. **31**, 264–269 (2013).
33. Hartmann, M. H. *Biopolymers from Renewable Resources*. D. L. Kaplan (Ed.), Springer-Verlag, Berlin, Germany, Chapter 15, pp. 367–411 (1998).
34. Hernandez, R. J., Selke, S. E. M., and Culter, J. D. *Plastics Packaging: Properties, Processing, Applications, and Regulations*, Hanser Gardner Publications, Inc., Cincinnati (2000).
35. Huang, M., Yu, J., and Ma, X. *Polymer Degradation and Stability*. **90**, 501–507 (2005).
36. Kakuta, M., Hirata, M., and Kimura, Y. *Journal of Macromolecular Science: Part C: Polymer Reviews*. **49**,107–140 (2009).
37. Kale, G., Kijchavengkul, T., Auras, R., Rubino, M., Selke, S., and Singh, S. P. *Macromol. Biosci*. **7**, 255–277 (2007).
38. Kamper, S. L. and Fennema, O. *J. Food Sci*. **50**, 382–384 (1985).
39. Kanatt, S. R., Rao, M. S., Chawla, S. P., and Sharma, A. *Food Hydrocolloids*. **29**, 290–297 (2012).
40. Keshavarz, T. and Roy, I. *Curr. Opin. Microbiol*. **13**, 321–326 (2010).

41. Kumar, S. M., Mudliar, S. N., Reddy, K. M. K., and Chakrabarti, T. *Bioresource Technology*. **95**, 327–330 (2004).
42. Landge, S. N., Chavan, B. R., Kulkarni, D. N., and Khedkar, C. D. *J. Dairying, Foods & H. S.***28**, 20–25 (2009).
43. Luecha, J., Sozer, N., and Kokini, J. L. *Journal of Materials Science*. **45**, 3529–3537 (2010).
44. Lunt, J. *Polym. Degrad. Stabil.* **59**, 145–152 (1998).
45. Makino, Y. and Hirata, T. *Postharvest Biology and Technology*. **10**, 247–254 (1997).
46. Marcinkowski, A. and Kowalski, A. M. *Resources, Conservation, and Recycling*. **69**, 10–16 (2012).
47. Mensitieri, G., Maiob, E. D., Buonocorea, G. G., Nedi, I., Olivieroa, M., Sansoneb, L., and Iannace, S. *Trends in Food Science & Technology*. **22**, 72–80 (2011).
48. Miyata, T. and Masuko, T. *Polymer*. **39**, 5515–5521 (1998).
49. Moon, S. I., Lee, C. W., and Kimura, Y. *J. Polymer Science: Part A: Polymer Chemistry*. **38**, 1673 (2000).
50. Moon, S. I., Lee, C. W., and Kimura, Y. *Polymer*. **42**, 5059 (2001).
51. Nair, L. S. and Laurencin, C. T. *Prog. Polym. Sci.* **32**, 762–798 (2007).
52. Nakagawa, T., Nakiri, T., Hosoya, R., and Tajitsu, Y. *Proceedings of 7th International Conference on Properties and Applications of Dielectric Materials*. **2**, 499–502 (2003).
53. Nijenhuis, A. J., Grijpma, D. W., and Pennings, A. J. *Macromolecules*. **25**, 6419–6424 (1992).
54. Ohara, H. and Ito, M. US Patent 6569989B2 (May 27, 2003).
55. Ohta, M., Obuchi, S., and Yoshida, Y. US Patent. 5,444,143 (Aug 22, 1995).
56. Oksman, K., Mathew, A. P., and Kvien, B. I. *Compos. Sci. Tech.* **66**, 2776–2784 (2006).
57. Oliveira, V. M., Ortiz, A. V., Mastro, N. L. D., and Moura, E. A. B. *Radiation Physics and Chemistry*. **78**, 553–555 (2009).
58. Panda, A. K., Singh, R. K., and Mishra, D. K. *Renewable and Sustainable Energy Reviews*. **14**, 233–248 (2010).
59. Park, J. W., Testin, R. F., Vergano, P. J., Park, H. J., and Weller, C. L. *Journal of Food Science*. **61**, 766–768 (1996).
60. Paz, H. M., Guillard, V., Reynes, M., and Gontard, N. *Journal of Membrane Science*. **256**, 108–115 (2005).
61. Punčochář, M., Ruj, B., and Chatterjee, P. K. *ProcediaEngineering*. **42**, 420–430 (2012).
62. Rhim, J. W. and Perry, K. W. *Critical Reviews in Food Science and Nutrition*. **47**, 411–433 (2007).
63. Sanchez, J. G., Tsuchii, A., and Tokiwa, Y. *Biotechnology Letters*. **22**, 849–853 (2000).
64. Schettinia, E., Santagatab, G., Malinconicob, M., Immirzib, B., Mugnozzaa, G. S., and Vox, G. *Resources, Conservation, and Recycling*. **70**, 9–19 (2013).
65. Siracusa, V., Rocculi, P., Romani, S., and Rosa, M. D. *Trends in Food Science & Technology*. **19**, 634–643 (2008).
66. Stridsberg, K. S., Maria, R., and Albertsson, A. C. *Adv. Polym. Sci.* **157**, 41–43 (2002).
67. Supaphol, P., Dangseeyun, N., Thanomiat, P., and Nithitanakul, M. *J. Polym. Sci. Part B: Polym. Phy*. **42**, 676–686 (2004).
68. Suriyamongkol, P., Weselake, R., Narine, S., Moloney, M., and Shah, S. *Biotechnology Advances*.**25**, 148–175 (2007).
69. Tang, X. Z., Kumar, P., Alavi, S., and Sandeep, K. P. *Critical Reviews in Food Science and Nutrition*. **52**, 426–442 (2012).
70. Tekade, A. R. and Gattani, S. G. *International Journal of Pharm. Tech. Research*. **2**, 106–112 (2010).
71. Tripathi, S., Mehrotra, G. K., and Dutta, P. K. *Carbohydrate Polymers*. **79**, 711–716 (2010).
72. Türe, H., Gällstedt, M., and Hedenqvist, M. S. *Food Research International*. **45**, 109–115 (2012).
73. Vartiainen, J., Tammelin, T., Pere, J., Tapper, U., and Harlin, A. *Carbohydrate Polymers*. **82**, 989–996 (2010).

74. Vieira, M. G. A., da Silva, M. A., dos Santos, L. O., and Beppu, M. M. *European Polymer Journal*. **47**, 254–263 (2011).
75. Villalobos, R., Chanona, J., Hernandez, P., Gutierrez, G., and Chiralt, A. *Food Hydrocolloids*. **19**, 53–61 (2005).
76. Wang, C., Li, H., and Zhao, X. *Biomaterials*. **25**, 5797–5801 (2004).
77. Wiley, R. M. and Reilly, R. M. J. H. Canadian Patent 385753 (Dec 10, 1939).
78. Wooster, J. J. *Encyclopedia of Polymer Science and Technology*. John Wiley & Sons, Inc.(2002).
79. Wu, J., Chen, S., Ge, S., Miao, J., Li, J., and Zhang, Q. *Food Hydrocolloids*. **32**, 42–51 (2013).
80. Yang, X., Sun, L., Xiang, J., Hu, S., and Su, S. *Waste Management* (2012).
81. Zemljič, L. F., Tkavc, T., Vesel, A., and Šauperl, O. *Applied Surface Science*. **265**, 697–703 (2013).

CHAPTER 7

BIOPOLYMERS: POTENTIAL BIODEGRADABLE PACKAGING MATERIAL FOR FOOD INDUSTRY

ABHISHEK DUTT TRIPATHI, S. K. SRIVASTAVA, and AJAY YADAV

ABSTRACT

Food packaging is an integral and essential part of modern food processing and it will play an increasingly significant role in food industry as the use of new and alternative food processing operations expands. It is estimated that 41% of all plastics are being employed for packaging applications, and moreover, almost half of these are used for food packaging. Today, more than 99% of plastics are petroleum-based because of their easy availability and cheaper cost. These conventional petrochemical plastics are recalcitrant to microbial degradation. However, increased use of synthetic packaging films has led to serious ecological problems due to their total non-biodegradability. Hence, recent research developments have demonstrated the feasibility, utilization, and commercial application of a variety of bio-based polymers or bio-polymers made from a variety of materials including renewable/sustainable agricultural commodities. The aim of this review is to offer a complete view on different biodegradable polymer packages for food application.

7.1 INTRODUCTION

In today's modern era of science and technology, plastics have become one of the most widely used materials all over the world. Plastic literally means "changeable" and it refers to any natural and synthetic materials that can be shaped when soft and then hardened. Plastics are typically polymers of high molecular mass. A polymer is a large molecule (macromolecule) composed of repeating structural units. These subunits are typically connected by covalent chemical bonds. Plastic applications are nearly universal components in automobiles, home appliances, computer equipment's, packages, and even medical applications are areas, where plastics clearly have become indispensable. Approximately 25 million tons of plastics are produced by the plastics industry every year (Wong, Chua, Lo, Lawford, Yu, 2002). Today, the plastic industry is heavily integrated with the oil industry. It emphasizes how dependent the plastic industry is on oil and consequently and how the increasing demand of crude oil and natural gas price can have an economical influence on the plastic market. Plastics are produced from non-renewable resources such as petrochemical and are not compatible with natural carbon cycles because of their non-degradable characteristics. The development of biodegradable plastics has become one of the major concerns in the present society because the disposal of the plastics has pointed out their major weaknesses. Biopolymers are polymers that can be synthesized from living organism. Examples of input materials that can be used to produce biopolymer are starch, sugar, cellulose, or other synthetic materials. Biopolymers may be defined as products which are based on renewable agricultural or biomass feedstock, capable of behaving like conventional plastics in production and utilization but degradable through microbial processes upon disposal. It is this progressive development of biopolymers which has led to a surging interest of a plastic and composite industry based on biological materials. Previously, biodegradable polymers found limited applications in food packaging. In comparison to the usual wrap, labels, films, and laminates which comes from fossil fuel resources, biodegradable polymers seems to be an appropriate substitute from environmental concern. Currently, Countries like Italy, Brazil, Finland, Greece, and USA are utilizing these bio-based materials for development of biodegradable packaging material in food industries, for example, NatureWorks LLC and ICI (Imperial Chemical Industry), USA.

7.2 BIOPOLYMER MARKET

The global market for biodegradable polymers reached 206 million pounds at an average annual growth rate (AAGR) of 12.6% in 2010 and is expected to rise further by several folds in next decade (Aruas, Harte, Selke, 2003; Schlechter, 2005). Table 1 shows the 3–4 folds rise in utilization of biodegradable polymers for packaging and compostable packaging purpose from year 2000 to 2010. Figure 1 represents the global scenario of bioplastic packaging market by product wise in the year 2010 on percentage utilization basis.

TABLE 1 Biopolymer market from 2000 to 2010

Area of application	Production(1000 tons)			Average annual increase %
	2000	2005	2010	
Packaging	15	24	38	9.4
Compostable packaging	10	22	43	14.6
Others	3	7	13	16.6
Total	28	53	94	13.5

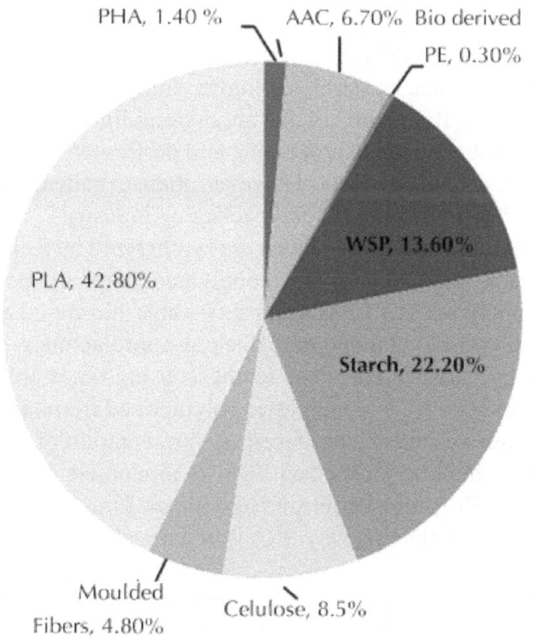

FIGURE 1 Global bio-plastic packaging market by product type (%utilization) 2010.

7.3 BIOPOLYMER BASED PACKAGING MATERIAL

Packaging materials exhibit various features in terms of mass transport properties which include: the respiration loss of food products, providing a selective barrier to gases and water vapor, increased shelf life, development of modified atmosphere in the package headspace, and minimizing the possible loss of certain food additives such as flavor, color, antioxidants, and antimicrobial agents (Tharanathan, 2003). As a matter of fact, in general, biopolymers display performance problems under this respect in comparison with synthetic polymers, which are even more pronounced in the case of polymers directly extracted from natural materials (Petersen, Nielsen, Bertelsen, Lawther, Olsen, Nilssonk, 1999).

Biopolymer find large spectrum of application in different medical, pharmaceutical, and food commodities, where these are used as collection bags for compost, agricultural foils, horticulture, food packaging, nursery products, toys, fibers, textiles, and so on. Other fields such as packaging and technical applications are also gaining importance. The biopolymer based packaging material performance depends on its capability to prevent its environmental degradation, while maintaining food quality (Arvanitoyannis, 1999). Biodegradable polymers find various applications in food-contact articles, which includes disposable cutlery, drinking cups, salad cups, plates, overwrap and lamination film, straws, stirrers, lids and cups, plates, and containers for food dispensed at delicatessen and fast-food establishments.

Biopolymer based packaging materials may be divided into three main categories based on their origin and production:

Category 1: Polymers, which are directly extracted or removed from biomass. Certain polysaccharides such as starch, cellulose, and proteins like casein and gluten constitutes this category. By nature, all this are hydrophilic and somewhat crystalline-factors and creates problems while processing and performance, especially in relation to packaging of moist food products. However, their excellent gas barrier property makes it suitable for its utilization in food packaging industry.

Category 2: Polymeric materials, which are synthesized by a classical polymerization procedure such as aliphatic aromatic copolymers, aliphatic polyesters, and polylactide aliphatic copolymer (CPLA), using renewable bio-based monomers such as poly (lactic acid) and oil-based monomers like polycaprolactones. A good example of polymer produced by classical chemical synthesis using renewable bio-based monomers is polylactic acid (PLA), a biopolyester polymerized from lactic acid monomers. The monomers themselves may be produced *via* fermentation of various carbohydrate feedstocks. The PLA may be plasticized with its monomers or, alternatively, oligomeric lactic acid. The PLA may be formed into blown film, injected molding objects, and coating all together explaining why PLA is first novel bio-based material produced at commercial scale.

Category 3: Polymers, which are produced by microorganisms or genetically modified bacteria. Till date, this group of bio-based polymers consists mainly of the polyhydroxyalkanoates but developments with bacterial cellulose and other polysaccharides are also in progress. Figure 2 represents the various sources of currently used biopolymers sources which find profound application in food industry mainly in packaging.

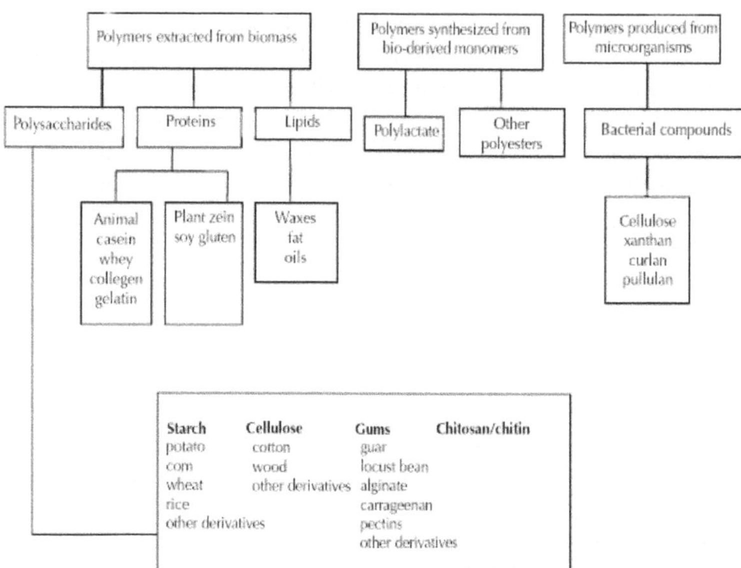

FIGURE 2 Different categories of bio-based materials (Weber et al., 2002).

7.4 POLYMERS EXTRACTED FROM CELL BIOMASS

7.4.1 POLYSACCHARIDE-BASED BIOPOLYMERS

Polysaccharide films are made from starch, alginate, cellulose, chitosan, carageenan, or pectins

Polysaccharide-derived films exhibit excellent gas permeability properties, resulting in desirable modified atmospheres that enhance the shelf life of the product without creating anaerobic conditions (Baldwin, Nisperos, Baker, 1995).

7.4.2 STARCH-BASED POLYMER

Bioplastics,-based on starch, utilize the benefits of both natural polymerization and the availability of raw material and process technology. Several bioplastic manufacturing industries such as Biotec GmbH are producing TPS® derivatives and starch esters in a continuous extrusion process. Biodegradable and compostable plastics, especially those based on renewable resources from the agricultural industry, are an essential innovation. As a packaging material, starch alone does not form films with adequate mechanical properties, unless it undergoes few pretreatments in the form of plasticization, blending with other materials, or it undergoes several modifications in terms of genetic or chemical modification. Starch can be added to commercial polymer derived from the oil based monomer with different percentage (10, 50, and 90%). Starch can be transformed also into a foamed material using water steam, replacing the polystyrene foam as packaging material. It can be molded into different forms of packaging units such as trays or disposable dishes.

It is commercially available under various trade names such as:

- Biopur (from Biotec GmbH),
- Eco-Foam (from National Starch and Chemical), and
- Envirofill (from Norel).

7.4.3 THERMOPLASTIC STARCH TPS AND TPS BLENDS

The TPS bioplastic granules termed as BIOPLAST are film extrusion and injection moulding. The BIOFLEX film has mechanical properties, such as PE films are opaque to transparent, printable, sealable, shrinkable, and can be colorized. The BIOFLEX is permeable to vapor and has good barrier properties to oxygen. The BIOFLEX can be used in the same way as conventional foils, for instance for garbage sacks, shopping bags, packing up, nets for food packaging, diapers, agricultural, or technical uses.

7.4.4 CELLULOSE AND ITS DERIVATIVES

Cellulose, an almost linear polymer of hydroglucose, is the most abundantly occurring natural polymer on earth. Films made from cellulose derivatives tend to be water soluble, resistant to fats and oils, tough, and flexible (Baldwin, Nisperos, Hagenmaier, Baker, 1997; Cutter, Sumner, 2002; Krumel, Lindsay, 1976). A number of cellulose derivatives are produced commercially, most commonly carboxy-methyl cellulose, methyl cellulose, ethyl cellulose, hydroxyethyl cellulose, hydroxypropyl cellulose, and cellulose acetate (CA). Of these derivatives, only CA is widely used in packaging of baked goods and fresh produce. Cellulosic derivatives such as hydroxy propyl methyl cellulose (HPMC) are promising raw materials for edible coatings or films associated with antimicrobial entities.

7.4.5 CHITIN/CHITOSAN

Chitosan is an edible and biodegradable polymer derived from chitin, the major organic skeletal substance in the exoskeleton of arthropods, including insects, crustaceans, and some fungi (Cutter, Sumner, 2002; Chen, Yeh, Chiang, 1996; Nisperos-Carriedo, 1994; Suyatma, Copinet, Tighzert, Coma, 2004). Chitosan films are tough, long lasting, flexible, and very difficult to tear. Some desirable properties of chitosan are that, it forms films without the addition of additives, exhibits good oxygen, and carbon dioxide permeability, as well as excellent mechanical properties (Suyatma, Copinet, Tighzert, Coma, 2004). Another interesting property of chitosan and chitin in relation to food packaging are their antimicrobial properties and their ability to absorb heavy metal ions. In addition to the formation of gas permeable films, chitosan has a dual function that is direct interference of fungal growth and activation of several defense processes. These defense mechanisms include accumulation of chitinases, synthesis of proteinase inhibitors, and lignification and induction of callous synthesis. The use of edible films and coatings to extend shelf life and improve the quality of fresh, frozen, and fabricated foods due to their eco-friendly and biodegradable nature. The biodegradable laminate consisting of chitosan–cellulose and polycaprolactone can be used in modified atmosphere packaging of fresh produce. The use of chitosan as an edible coating has several advantages such as it provides controlled moisture transfer between food and surrounding environment, controlled release of chemical agents like antimicrobial substances, reduction of oxygen partial pressure in the package that results in a decreased rate of metabolism, antioxidants, controlled rate of respiration, and high impermeability to certain vital components such as fats and oils, temperature control, and structural reinforcement of food. Due to its intrinsic ability to form semipermeable film, chitosan coating is expected to modify the internal atmosphere as well as decrease the transpirational loss and induces delayed ripening of fruits. Chitosan film is found to extend the storage life of peaches, Japanese pears and kiwifruits, cucumbers, bell peppers, strawberries, and tomatoes. Chitosan coating can be also used to minimize the postharvest handling and processing losses of fruits due to their anti-browning activity. It is reported that in case of Litchi fruit, chitosan laminates reduces the enzymatic activity of PPO (Polyphenol oxidase) and peroxidises(Zhang, Quantick, 1997). It delays changes in flavonoid, anthocyanin, and polyphenol content with increased storage time. Chitosan are not only used as packaging material in food industry, but also used as clarifying and finning agent in food processing industry. It contains a strong positive charge which acts as dehazing agent and also controls the acidity of fruit juices. Chitosan may act as an effective substitute of various clarifying agents used in fruit juice industries such as bentonite, gelatine, silica gels, tannins, and so on. It shows synergistic approach in association with pectinase enzyme in clarification of apple and grape juice, reducing haziness, and turbidity. In a similar study, Spagna et al. (Spagna, Pifferi, Rangoni, Mattivi, Nicolini, Palmonari, 1996) observed that chitosan has a good affinity for polyphenolic compounds such as catechins, proanthocyanidins, cinnamic acid, and their derivatives that can change the initial straw-yellow color of white wines into deep golden-yellow color due to their oxidative products. It has been also reported earlier, that by adding chitosan to grapefruit juice at a concentration of 0.015 g/mL, total acid content was reduced by about 52.6% due to decreasing the amount of citric acid, tartaric acid, l-malic acid, oxalic acid, and ascorbic acid by 56.6, 41.2, 38.8, 36.8, and 6.5%, respectively.

Electrostatic interaction between chitosan and milk fat globule fragment facilitated fat and protein recovery. Fernandez and Fox (Fernandez, Fox, 1997) reported the use of chitosan to remove proteins and peptides from cheese whey. Urea-PAGE (polyacrylamide gel electroplorosis) showed that chitosan gave good fractionation of water-soluble extract at pH 2, 3, and 4. At pH 5, 6, and 7, most of the nitrogen of the water-soluble extract remained soluble in 0.02% chitosan.

Chitosan and Chitosan derivatives (N-carboxymethylchitosan) are also used in preservation of cooked meat and meat products due to their anti-oxidative properties, since, these products are more susceptible to off-flavor and rancidity due to the presence of unsaturated lipids.

Effectiveness of chitosan treatment on oxidative stability of beef was studied by Darmadji and Izumimoto(Darmadji, Izumimoto, 1994) who observed that addition of chitosan at 1% resulted in a decrease of 70% in the 2-thiobarbituric acid (TBA) values of meat after 3 days of storage at 4°C. Enormous application of chitosan in food and food products is due to its functional attribute. Due to its role as dietary fiber and other functional properties, USDA has approved chitosan as a feed additive. Due to its hypocholesterolemic effect, various food products in Japan such as pasta noodles, dietary cookies, vinegar products, and chips are enriched with chitosan and its derivative.

7.4.6 GUMS

Gums are widely used for packaging of different food products such as fresh fruit and vegetables, meat, and poultry products. Alginates, carrageenan, guar, and pectins and its derivatives are some of the commonly used biopolymer for food packaging. Alginates are derived from seaweed and possess good film-forming properties that make them particularly useful in food applications (Cutter, Sumner, 2002; Nisperos-Carriedo, 1994). Carrageenan is a complex mixture of several polysaccharides. Carrageenan-based coatings have been used to prolong the shelf life of a variety of muscle foods including poultry and fish (Meyer, Winter, Weister, 1959; Pearce, Lavers, 1949; John, Gangadharan, Nampoothiri, 2008) .

Although, gums exhibit tremendous potential to be utilized in bioplastic production but it is still in nascent phase due to its high downstream processing cost.

7.4.7 LIPID-BASED BIOPOLYMERS

Lipid-based biopolymer films are made from waxes, fats, and oils. The advantages for coating foods with lipids are that it not only imparts hydrophobicity, cohesiveness, and flexibility but also make excellent moisture barriers due to the tightly packed crystalline structure of lipids that naturally restricts the passage of water vapor molecules (Cutter, Sumner, 2002; Hernandez, 1994). Edible lipid or resin coatings also may be prepared from waxes (for example, carnauba, beeswax, and paraffin), oils (vegetable, animal, and mineral), and surfactants. After having all these advantages, lipid-based films at higher storage temperatures may exhibit lower permeability to gases such as oxygen, carbon dioxide, and ethylene, leading to potentially anaerobic conditions which may present food safety issues, as well as lack structural integrity, and poor adherence to hydrophilic surfaces (Baldwin, Nisperos, Baker,1995; Hernandez, 1994; Ben, Kurth, 1995). Lipid-based films also are subjected to oxidation, cracking, flaking, retention of off-flavors, as well as bitter after tastes (Gennadios, Hanna, Kurth, 1997; Morgan, 1971; Zabik, Dawson, 1963). The primarily function of a lipid coating is to block transport of moisture due to their relative low polarity. In contrast, the hydrophobic characteristic of lipid forms thicker and more brittle films. Consequently, they must be associated with film forming agents such as proteins or cellulose derivatives (Cairncross, Becker, Ramaswamy, O'Connor, 2006). Generally, water vapor permeability decrease when the concentration of hydrophobicity phase increases. Lipid-based films are often supported on a polymer structure matrix, usually a polysaccharide, to provide mechanical strength. In food industry applications include coating of wine bottle corks with beeswax and paraffin wax which prevents the intimate contact of wine and the cork, thus controlling the release of flavor from the cork to the wine. Cocoa butter and cocoa-based films are widely used in confectionery and biscuit industries. These films are used in coating of pizza and dough products to prevent moisture from moving out and in dry and crisp food products to minimize the flavor and water loss from fat or any moist filling. Lipid-based films are mostly used as moisture barriers, especially for fresh fruits and vegetables (Cha, Chinnan, 2004). In food Industry these are used to enhance the color and appearance and thereby reducing the sticky surface. Nowadays, lacs and varnishes of natural origin are also applied to the foods as like coating agents. These lacs and varnishes are also used in pharmaceutical industry to protect the active substances. Emulsifiers and surface active agents are sometimes used as barriers to gases and/or moisture on fish, poultry, or meat products. These may be subjected to manipulations to improve their properties such as adherence between the product to be coated and the coating applied (Weber, Haugaard, Festersen, Bertelsen, 2002).

7.4.8 PROTEIN BASED BIOPOLYMERS

Protein based packaging films are made from casein, whey protein, gelatin/collagen, fibrinogen, soy protein, wheat gluten, corn zein, and egg albumen have been processed into edible films (Ben, Kurth, 1995). Protein-based films adhere well to hydrophilic surfaces, provide barriers for oxygen and carbon dioxide but do not resist water diffusion (Baldwin, Nisperos, Baker, 1995; Cutter, Sumner, 2002; Gennadios, Weller, 1990; Han, 2002). Casein and whey have been widely used in the manufacture of edible films because they possess excellent mechanical and barrier properties, solubility in water, ability to act as emulsifiers. Despite having all these advantages, problems using protein films are that they can be degraded by the enzymes present in food. Certain proteolytic enzymes associated with meat products can degrade the protein films (Gennadios, Hanna, Kurth, 1997).Additionally, application of protein films may cause health problems, especially for individuals with food allergies associated with milk, egg, peanut, soybean, or rice proteins. Formation of protein-based edible films is generally based on preparation from solutions or dispersions of the protein as the solvent evaporates. These solvents include water, ethanol, or ethanol- water mixtures. The protein needs to be denatured by heat, acids, bases, and solvents for better extended structures that are needed for film formation. Then upon extension, protein chains can associate through hydrogen, ionic, and hydrophobic and covalent bonding. Production of cohesive films is affected by the degree of chain extension, the nature, and sequence of amino acid residues. The uniform distribution of polar, hydrophobic, and/or thiol groups along the polymer chain increase the likelihood of the respective interactions. The promotion of polymer chain-to-chain interaction results in films that are stronger but less flexible and less permeable to gases, vapors, and liquids. Polymers containing groups that can associate through hydrogen or ionic bonding result in films that are excellent oxygen barriers but susceptible to moisture. Thus, protein films are expected to be good oxygen barriers at low relative humidity.(Thawien, 2012) However, protein films have higher water vapor permeability as compared to plastic films due to the hydrophilic nature of most proteins.(Tang, Kumar, Alavi, Sandeep, 2012)

Casein and whey have been widely used in the manufacture of edible films because they possess excellent mechanical and barrier properties, solubility in water, and ability to act as emulsifiers. Despite having all these advantages problems using protein films are that, they can be degraded by the enzymes present in food. Certain proteolytic enzymes associated with meat products can degrade the protein films (Gennadios, Hanna, Kurth, 1997). Additionally, application of protein films may cause health problems, especially for individuals with food allergies associated with milk, egg, peanut, soybean, or rice proteins. Due to their mechanical strength and excellent barrier properties, whey protein films especially have gained attention in recent years. Lipid-based films are mostly used as moisture barriers, especially for fresh fruits and vegetables. Protein and lipid-based polymers has good potential as antimicrobial packaging due to their mechanical and barrier properties.

Collagen is widely used in preparation of edible protein films. Collagen films have several advantages as it is biocompatible and non-toxic to most tissues and it has better structural, physical, chemical, and resistance properties, it can be converted in variety of products with simplicity and it is easily isolated and purified in large quantities (Hood, Shen, Varner, 1988). Similarly, Gelatin is used in food encapsulation preferably low moisture or oil phase food ingredients and pharmaceutical products, providing protection against oxygen and light. In addition, gelatin films have been formed as coatings on meats to reduce oxygen, moisture, and transport of oil (Gennadios, Mc Hugh, Weller, Krochta, 1994). Gelatin is also used for microencapsulation, for the manufacturing of films based on Zein proteins, the solvent process is used. Zein films prepared with this technique are brittle at room conditions and to overcome this plasticizers such as polyethylene glycol (PEG), glycerol, lactic acid, and fatty acids are often added to impart softness and flexibility to the films (Guilbert, 1986).As far as soy proteins films are considered, the strong charge and polar interactions between side chains of soy protein molecules prevents segment rotation and mobilization of molecules, which in turns increases the stiffness, yield point, and tensile strength of soy protein films (Zhang, Jane, 2001).Soy proteins posses advantage of being abundant, inexpensive, biodegradable, and nutritional, thus they could be developed as better edible and biodegradable films. Nowadays, Mung beans are being focused as a component of biopolymeric films because of their high protein content (20–30%). It is found that the films based on Mung beans have better mechanical properties (tensile strength and elongation at break) and also enhanced water vapor barrier properties.

Thus, protein-based edible films can be used for packaging of wide range of food products aiming for reduction in moisture loss and lipids, ceasing the absorption of oxygen, and also improving the mechanical properties, thereby providing better protection to the product. These could be better alternative to available non-biodegradable packaging materials. Polymers containing groups that can associate through hydrogen or ionic bonding result in films that are excellent oxygen barriers but are susceptible to moisture. Thus, protein films are expected to be good oxygen barriers at low relative humidities. Various types of protein have been used as edible films. These include gelatin, casein, whey protein, corn zein, wheat gluten, soy protein, and mung cross-linking agents are used, the edibility of those films is of concern. Zein coating have also shown an ability to reduce moisture and firmness loss and delay color change (reduce oxygen and carbon dioxide transmission) in fresh tomatoes. The actual applications of whey protein films as coatings on food systems. Whey protein films produce transparent, bland, flexible, water-based edible films with excellent oxygen, and aroma barrier properties at low relative humidity.

7.5 POLYMERS SYNTHESIZED FROM BIO-DERIVED MONOMERS

7.5.1 POLYHYDROXYALKANOATES (PHA)

The PHAs, a family of bacterial polyesters, are formed and accumulated by various bacterial species under unbalanced growth conditions. These polymers are produced in nature by bacterial fermentation of sugar and lipids. Structurally, PHAs comprise simple macromolecules composed of 3-hydroxy fatty acid monomers. The PHA has a considerably low volume of the biopolymer market, somewhere around 100,000 lbs. per year. In 2008, approximately 50 million kg of PHAs were commercially produced. The PHAs have thermo mechanical properties similar to synthetic polymers such as polypropylene (Galego, Rozsa, Sa´nchez, Fung, Va´zquez, Tomas, 2000). The PHAs polyesters are biodegradable and biocompatible and can be obtained from renewable resources (Steinbuchel, Fu chtenbusch, 1998). They have several desirable properties such as petroleum displacement and greenhouse gas minimization apart from their fully biodegradable nature (Orts, Nobes, Kawada, Nguyen, Ravanelle, 2008) as shown in Table 2.

TABLE 2 Greenhouse gas emissions from biodegradable plastics

Polymer	GHG Emission x 10[kgCO_2 eq./kg]
PCL	53
LDPE	50
HDPE	49
PVOH	42
PVOH	42

TABLE 2 *(Continued)*

TPS (Thermoplastic starch) + 60% PCL	36
TPS + 15% PVOH	17
Mater-Bi™ film grade	12
TPS	11
PLA	NA
PHA-ferment	NA

Applications of PHA as a biodegradable packaging includes: bottles, containers, sheets, films, laminates, fibers, and coatings. Over 100 monomers and copolymers can be developed from PHAs. Some of the polymers used are PHB, PHV, PHBV (Metabolix), PHBO, PHBH, and PHBD. The PHAs exhibit desirable properties such as good tensile strength, printability, flavor and odor barriers, heat sealability, grease and oil resistance, temperature stability, and are easy to dye, which boosts its application in food industry. For example, Metabolix, a US-based company, produces "Metabolix PHA", which is a blend of polyhydroxybutyrate (PHB) and poly (3-hydroxyoctanoate) that has been approved by the FDA for production of food additives and creating packages that maintain all the performance characteristics of non-degradable plastics.

The PHB is one of the commercially available PHAs which find multiple applications in Food Industry. Polymer of 3-hydroxybutyrate, have been introduced into the market in relatively large quantities as bioplastic material. The PHB is accumulated as intracellular granules by many prokaryotic organisms (including *Alcaligenes spp., Bacillus spp., Azotobacter spp., Pseudomonas spp.*) as they enter the stationary phase of growth, to be used later as an internal reserve of carbon and energy

The unique properties that make PHB a good food packaging material are:

i. 100% resistance to water
ii. 100% biodegradability
iii. Thermoplastic process ability

The PHB has similar properties to that of polypropylene (PP) and polyethylene (PE) (Lee, 1996). The PHB tends to be stiff, highly crystalline, and brittle, has a high melting point and low molecular weight. The PHB has a melting point of 175°C and glass transition temperature of 15°C. The PHB has a tensile strength of 40 MPa, which is close to that of polypropylene (Braunegg, Lefebvre, Genser, 1998). Bacteria that have been shown to efficiently produce PHB include *Alcaligenes eutrophus, Alcaligenes latus, Azotobacter vinelandii, and recombinant Escherichia coli.* The PHB biosynthesis occurs in 3 step pathways. It requires three enzymes namely Acetyl-CoA reductase, β-keto thiolase, and PHB synthase in PHB biosynthetic route (Tsuge, 2002) (Figure 3). However, industrial application of PHB has been hampered owing to its low thermal stability and excessive brittleness upon storage (Matsusaki, Abe, Doi, 2002). Due to the poor physical properties of PHB, the incorporation of a second monomer unit into PHB can significantly enhance its properties. This has led to an increased interest to produce hetero-polymers with improved qualities. The incorporation of 3-hydroxyvalerate (3HV) into the PHB has resulted in a poly-(3-hydroxybutyrate-co-3-hydroxyvalerate) [P(3HB-co-3HV)], which is more flexible and tougher than PHB, and easier to degrade when discarded into the natural environment.

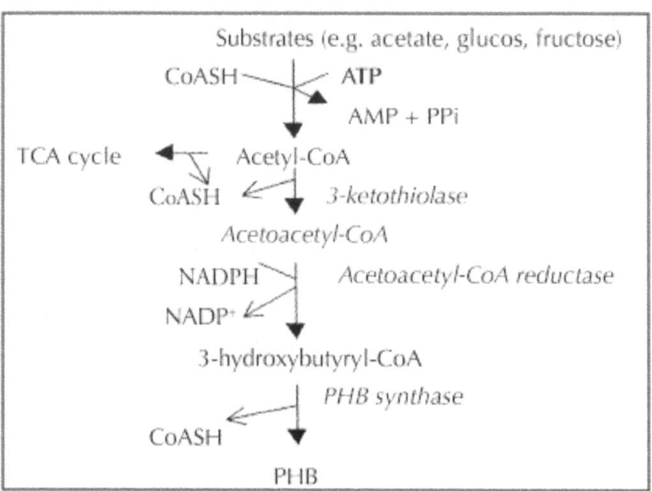

FIGURE 3 Biosynthetic pathway of PHB from acetyl-CoA (Taguchi et al.).

7.5.2 POLYLACTIC ACID (PLA)

One of the most promising biopolymer is the PLA obtained from the controlled depolymerization of the lactic acid monomer obtained from the fermentation of sugar feedstock, corn, and so on, which are renewable resources readily biodegradable (John, Nampoothiri, Pandey, 2006). The PLA is a sustainable alternative to petrochemical-derived products, since, the lactides are produced by the microbial fermentation of agricultural by-products mainly the carbohydrate rich substances (John, Nampoothiri, Pandey, 2006). The PLA is becoming a growing alternative as a green food packaging materials because it was found that in many circumstances its performance is better than synthetic plastic materials. There are several agro-industrial sources of PLA production, some of which are illustrated in Table 3. The PLA developed by Cargill Dow is one of the potentially used biopolymer in U.S. food markets. The PLA is made from starch derived directly from corn, and it biodegrades under the right composting conditions within 47 days.

Polymer architecture was used to tailor the physical, chemical, and mechanical properties of PLA so that it can replace the synthetic polymers used in fresh food packaging field like the most common oriented polystyrene (OPS) and polyethylene terephthalate (PET).

Kale et al. (Kale, Auras, Singh, 2006) studied the compostability of three commercially available biodegradable packages made of PLA, in particular water bottles, trays and, deli containers, to composting and to ambient exposure and obtained excellent results in terms of extending the shelf life of different food products. If PLA plastic can become more cost-competitive with traditional plastic, then mainstream consumers will be more likely to respond to the use of a bio-based product for contact with fresh foods such as salads, cut fruits, and deli items.

Nowadays, certain blend of polylactide and polycaprolactone are extensively used for packaging of vegetable oils. These blend exhibit comparable tensile strength, brittleness, and rigidity as packaging material derived from polypropylene or polyethylene derivative. The PLA and PCL shows improved barrier properties and prevents aroma losses but only limitation of using this blend as packaging material is its inefficient gas permeability.

TABLE 3 Starchy and cellulosic materials used for the production of lactic acid. Source: John et al. (2008)

Substrate	Microorganism	Lactic acid yield
Wheat and rice bran Lactobacillus sp.	*Lactobacillus sp*	129 g/l
Corn cob	*Rhizopus sp.MK-96–1196*	90 g/l
Pretreated wood	*Lactobacillus delbrueckii*	48–62 g/l
Cellulose	*Lactobacillus coryniformis ssp. torquens*	0.89 g/g
Barley	*Lactobacillus casei* NRRLB-441	0.87–0.98 g/g
Cassava bagasse	*L. delbrueckii* NCIM 2025, *L casei*	0.9–0.98 g/g
Wheat starch	*Lactococcus lactis ssp. lactis* ATCC 19435	0.77–1 g/g
Whole wheat	*Lactococcus lactis* and *Lactobacillus delbrueckii*	0.93–0.95 g/g
Potato starch	*Rhizopus oryzae, R. arrhizuso*	0.87–0.97 g/g
Corn, rice, wheat starches	*Lactobacillus amylovorous* ATCC 33620	< 0.70 g/g
Corn starch	*L. amylovorous* NRRL B-4542	0.935 g/g

The PLA can exist in three stereochemical forms: poly (L-lactide) (PLLA), poly (D-lactide) (PDLA), and poly (DL-lactide) (PDLLA). The properties of the PLA material are highly related to the ratio between the two forms (L or D) of the lactic acid monomer. The L-PLA is a material with a very high melting point and high crystallinity, whereas a mixture of D and L-PLA results in an amorphous polymer with a low glass transition temperature. The glass transition temperature of PLA (Tg) ranges from 50 to 80°C while the melting temperature (Tm) ranges from 130 to 180°C.

Properties that make PLA a good food packaging material are:

i. High molecular weight,
ii. Water solubility resistance,

iii. Good processability that is easy to process by thermoforming, and
iv. Biodegradable that is recyclable and compostable.

The PLA has the tensile strength, modulus, and flavor and odor barrier of polyethylene and PET or flexible PVC, the temperature stability and process ability of polystyrene, and the printability and grease-resistance of polyethylene. The PLA can be processed by several approaches which include: injection molding, sheet extrusion, blow molding, thermoforming, and film forming. Processed PLA comes in the form of films, containers and coatings for paper, and paper boards. The PLA can be further recycled by chemical conversion back to lactic acid and then re-polymerized.

Although PLA seems to be potential biodegradable polymer to be utilized in packaging of various food products, but it exhibits certain limitations in unmodified form namely, it is more brittle and degrades easily at substantial temperature rise. Further investigations related to PLA products are to be performed in order to determine the range of compatibility and its performance in real shelf-life studies. Bohlmann (Bohlmann, 2004) proposed a Life-Cycle Assessment (LCA) for the PLA polymer, where a comparison was made between PLA and conventional polypropylene (PP) based plastics in food packaging application. He found that PLA is more energy efficient than PP polymer because PLA consumes no feedstock energy and greenhouse gas emission is equivalent in both the materials. According to an independent analysis commissioned by Nature Works, PLA production consumes 65% less energy than producing conventional plastics. It also generates 68% fewer greenhouse gases and produces fewer toxins. Table 4 represents the major producers of biodegradable plastic at commercial level from different substrates.

7.6 LIFE CYCLE ASSESSMENT (LCA) OF BIOPOLYMERS

The LCA, a tool used for measuring environmental sustainability and identifying environmental performance improvement objectives. Life cycle assessment of biopolymer can be depicted by Figure 4.

TABLE 4 Classification of different commercialized biodegradable plastics

Polymer	Commercial name	Manufacturer
Cellulose acetate	Bioceta®/ Biocell	Mazzucchelli
Microbial cellulose	Biofill/ Gengiflex	bioFill Productos Biotecnologicos
PLA	Natureworks® PLA	Cargill Dow/ Natureworks
PLA-based (Blend of Ecoflex® and PLA)	Ecovio®	BASF
PLA-based (PET)	Biomax®	DuPont
Aromatic-Aliphatic copolyesters	Ecoflex®	BASF
Aromatic-Aliphatic copolyesters	Eastar Bio®	Eastman

TABLE 4 *(Continued)*

Starch/copolymer	Bioplast®	Biotec
Starch blend with PLA/PHB/V	Paragon™	Avebe
P(3HB-co-3HV)	Biopol®	Biomer
Soybean oil	Soyoyl™	Urethane Soy Systems Co
PCL	Tone®	Union Carbide

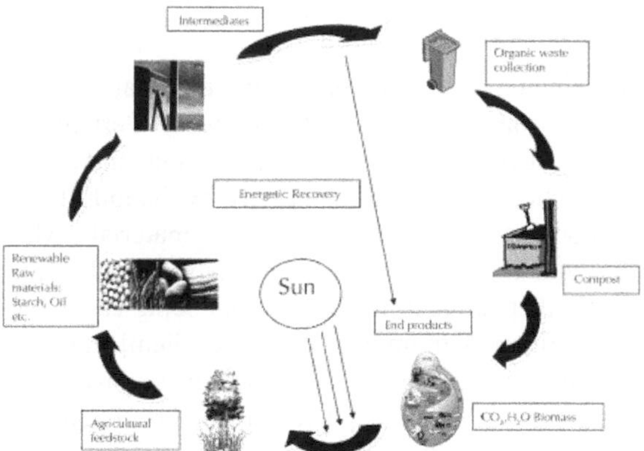

FIGURE 4 Life cycle assessment (LCA) of biopolymers.

7.7 POLYMER NANOCOMPOSITES

In modern era, consumers expect safer food of good quality, high sensory attributes, and inexpensive with increased shelf life. Nowadays, clay filled polymers termed as polymer clay nanocomposites (PCN) has been developed. The PCN exhibit more mechanical strength, increase heat resistance, improved barrier properties against moisture and volatiles as well as conserves the flavor, and taste of various drinks and beverages.

Montmorillonite (MMT) clay has been used as nanocomposite by different food packaging materials. 1–5% (w/w) nanoclay has been used in PCN development. Silicates can be used in synthesis of PCN with layer thickness of 1 nm. The thickness of PCN varies in the range of 2–1000 layers with gap between layers termed as interlayer or gallery.

Silicates clays in PCN can be arranged in the three possible ways:

i. *Non-intercalated structure*: If the polymer cannot intercalate between the silicate sheets, a non-intercalated nanocomposite is obtained.
ii. *Intercalted structure*: An interlayer spacing separates the different clay layers.

iii. *Exfoliated and laminated structure*: Different layers are formed randomly. This nanocomposite arrangement is the ideal for food packaging but this arrangement is difficult to achieve during processing or synthesis.

Polymer nanocomposites are playing significant role in food and beverage industries. Mitsubishi gas Chemical and Nanocor have jointly developed Nylon-MXD6 nanocomposites for multilayered PET bottle application. Nanocomposites can also be designed to incorporate and deliver the active components into biological system at minimum cost and with limited environmental impact. Bacteria repellent surface packages have been developed which changes color in presence of microorganism or toxins. Nanocomposites with this type of attribute have been used in wide range of minimally processed and processed food products such as meat and fish products, dairy foods, cereals, confectionary, fruit juices, beer, and carbonated drinks.

7.8 FUTURE PERSPECTIVE

Current research focusing on diversified packaging material is leading towards development of polymer nanocomposites for global industry. Certain constraints in nanocomposite development such as raw material, production, and processing cost should be minimized so that technology should be adapted by the industries and commercialized. Polymer nanaocomposites are future packaging materials, which will be used to develop the products with increased shelf life and stability and would be beneficial for the consumer. Several other research works are also being carried out to develop carbon nanotubes or other fillers with improved physicochemical properties and diversified application in the packaging industry. There is urge to cover the safety and regulatory aspects towards the use of nanocomposites as food packaging materials in future.

7.9 CONCLUSION

Biopolymers may serve as significant route for development of new and innovative food packaging material by extending the shelf life and improving the food quality, while minimizing the environmental pollution after usage. These biodegradable polymers can be used as utilized in several ways namely, films packaging for food products, loose film used for transport packaging, service packaging like carry bags, cups, plates and cutlery, biowaste bags, in agri-, and horticultural fields like bags and compostable articles. Current reports suggests that bioplastics comprises approximately 300,000.0 metric tons of the world plastics market which accounts for less than 1% of the 181.0 million metric tons per annum production of synthetic plastics.

Although biopolymer utilization for food packaging material is increasing at significant pace but, till now it covers only 5–10% of the current plastic market in Europe. Several European countries like France, Germany, England, Netherland, and Italy are vastly utilizing various biopolymers in different agricultural commodity products packaging, but other countries like Belgium, Austria, Spain, and Switzerland are utilizing it in limited applications.

Current biopolymers utilization in food packaging is not enough to quench the need of bioplastic manufacturer's satiety .There are several challenges which are yet to be overcome in order to improve world bioplastic demand. The major hurdles in biopolymer production are its complicated downstream processing operations and cost which includes recovery and purification. It is estimated that only raw materials and downstream processing cost accounts for 40% of total cost of biopolymer production. This cost can be lowered by using biotechnological intervention, which involves utilization of cheaper substrates as raw material such as waste water from food industries, molasses, whey, sugar cane bagassse and cassava bagassse, starchy materials from potato, tapioca, wheat, barley, and development of genetically modified high yielding biopolymer producing microorganisms. Novel biodegradable biopolymer/clay nanocomposite films are also developed as environment friendly material to reduce plastic waste. They too provide the improved strength and barrier properties that are desirable for food packaging. There is need to emphasize on the clay modification, dispersion, and polymer-filler interactions needed to fulfill the bridges.

KEYWORDS

- **Collagen**
- **Greenhouse gases**
- **Polylactic acid**
- **Polysaccharide**
- **Sustainable agricultural**

REFERENCES

1. Wong, P. A. L., Chua, H., Lo, W., Lawford, H. G., and Yu, P. H. Production of specific copolymers of polyhydroxyalkanoates from industrial waste. *Appl. Biochem. and Biotechnol.*, **98**, 655–662 (2002).
2. Aruas, R., Harte, B., and Selke, S. *Effect of water on the oxygen barrier properties of poly(ethylene teraphthalate) and polyactide films*, Wiley InterScience. 92–97 (2003).
3. Schlechter, M. Bcc search. http://www.bccresearch.com/report/PLS025B.html. (2005).
4. Tharanathan, R. N. Biodegradable films and composite coatings: past, present and future. *Trends Food Sci Tech.*, **14**(3), 71–78 (2003).
5. Petersen, K., Nielsen, P. V., Bertelsen, G., Lawther, M., Olsen, M. B., and Nilssonk, N. H. Potential of bio-based materials for food packaging. *Trends in Food Science & Technology*, **10**, 52–68 (1999).
6. Arvanitoyannis, I. S. Totally and partially biodegradable polymer blends based on natural synthetic macromolecules: preparation, physical properties, and potential as food packaging materials. Reviews in Macromolecular Chemistry and Physics. *Journal of Macromolecular Science*, **39**(2), 205–271 (1999).
7. Baldwin, E. A., Nisperos, N. O., and Baker, R. A. Use of edible coatings to preserve quality of lightly (and slightly) processed products. *Critical Reviews in Food Science & Nutrition.*, **35**, 509–524 (1995).
8. Baldwin, E. A., Nisperos, M. O., Hagenmaier, R. D., and Baker, R. A. Use of lipids in coatings for food products. *Food Technology*, **51**, 56–62 (1997).
9. Cutter, C. N. and Sumner, S. S. *Protein-based films and coatings*. A. Gennadios (Ed.), Application of edible coatings on muscle foods, CRC Press, Boca Raton, Florida, 467–488 (2002).
10. Krumel, K. L. and Lindsay, T. A. Nonionic cellulose ethers. *Food Technology*, **30**, 36–38 (1976).

11. Chen, M. C., Yeh, G. H. C., and Chiang, B. H. Antimicrobial and physiochemical properties of methyl-cellulose and chitosan films containing a preservative. *Journal of Food Processing and Preservation*, **20**, 379–390 (1996).
12. Nisperos-Carriedo, M. O. Edible coatings and films based on polysaccharides. J. M. Krochta, E. A. Baldwin, and M. O. Nisperos-Carriedo (Eds.), *Edible coatings and films to improve food quality*, Technomic Publishing Company, Lancaster, Pennsylvania, 305–335 (1994).
13. Suyatma, N. E., Copinet, A., Tighzert, L., and Coma, V. Mechanical and barrier properties of biodegradable films made from chitosan and poly (lactic acid) blends. *Journal of Polymers and the Environment*, **12**, 1–6 (2004).
14. Zhang, D. and Quantick, C. Effect of Chitosan Coatingon Enzymatic Browning and Decay during Post-Harvest Storage of Litchi (Litchi Chinensis Sonn.). *Fruit in Postharvest Biol.Technol.*, **12**, 195–202 (1997).
15. Spagna, G., Pifferi, P. G., Rangoni, C., Mattivi, F., Nicolini, G., and Palmonari, R. The Stabilization of White Wines by Adsorption of Phenolic Compounds on Chitin and Chitosan. *Food Res. Intern.*, **29**, 241–248 (1996).
16. Fernandez, M. and Fox, P. F. Fractionation of Cheese Nitrogen Using Chitosan. *Food Chem.*, **58**, 319–322 (1997).
17. Darmadji, P. and Izumimoto, M. Effect of Chitosan in Meat Preservation. *Meat Science*, **38**, 243–254 (1994).
18. Kester, J. J. and Fennema, O. R. Edible films and coatings: a review. *Food Technology*, 40 12, 4759, 0015–6639 (1986).
19. Meyer, R. C., Winter, A. R., and Weister, H. H. Edible protective coatings for extending the shelf life of poultry. *Food Technology*, **13**, 146–148 (1959).
20. Pearce, J. A. and Lavers, C. G. Frozen storage of poultry: effects of some processing factors on quality. *Canadian Journal of Research.*, **27**, 253–265 (1949).
21. John, R. P., Gangadharan, D., and Nampoothiri, K. M. Genome shuffling of Lactobacillus delbrueckii mutant and Bacillus amyloliquefaciens through protoplasmic fusion for L-lactic acid production from starchy wastes. *Bioresour. Technol.*, **99**, 8008–8015 (2008).
22. Hernandez, E. Edible coatings from lipids and resins. J. M. Krochta, E. A. Baldwin, and M. O. Nisperos-Carriedo (Eds.), *Edible coatings and films to improve food*, Economic Publishing Company, Lancaster, Pennsylvania, 279–303 (1994).
23. Ben, A., and Kurth, L. B. *Edible film coatings for meat cuts and primal*. Meat, the Australian Meat Industry Research Conference, September CSIRO, 10–12; 114–117 (1995).
24. Gennadios, A., Hanna, M. A., and Kurth, L. B. Application of edible coatings on meats, poultry and seafoods: a review. *Lebensmittel Wissenschaft und Technology*, **30**, 337–350 (1997).
25. Morgan, B. H. Edible packaging update. *Food Product & Development.*, **5**(6), 75–77, 108 (1971).
26. Zabik, M. E. and Dawson, L. E. The acceptability of cooked poultry protected by an edible acetylated monoglyceride coating during fresh and frozen storage. *Food Technology*, **17**, 87–91 (1963).
27. Cairncross, R. A., Becker, J. G., Ramaswamy, S., O'Connor, R., Moisture sorption, transport, and hydrolytic degradation in polylactide. *Appl. Biochem. Biotechnol.*, **31**, 774–785 (2006).
28. Cha, D. S. and Chinnan, M. S. Biopolymer-based antimicrobial packaging: A review. *Critical Review in Food Science and Nutrition*, **44**(4), 223–237 (2004).
29. Weber, C. J., Haugaard, V., Festersen, R., and Bertelsen, G. Production and applications of biobased packaging materials for the food industry. *Food Additives and Contaminants*. **19**,172–177 (2002).
30. Gennady, A. and Weller, C. L. Edible films and coatings from wheat and corn proteins. *Food Technology*, **44**, 63–69 (1990).
31. Han, J. H. Protein-based edible films and coatings carrying antimicrobial agents. A. Gennadios (Ed.), Technomic Publishing Co. Inc. Protein-based films and coatings, Pennsylvania, Lancaster, 485–500 (2002).

32. Thawien, W. Protein-Based Edible Films: Characteristics and Improvement of Properties. *In Structure and Function of Food Engineering*, Prince of Songkla University, Department of Material Product Technology, Thailand, 978-953-51-0695-1 (2012). DOI: 10.5772/48167.
33. Tang, X. Z., Kumar, P., Alavi, S., and Sandeep, K. P. Recent Advances in Biopolymers and Biopolymer-Based Nanocomposites for Food Packaging Materials. *Critical reviews in Food Science and Nutrition*, **52**(5), 426–442 (2012).
34. Hood, E. E., Shen, Q. X., and Varner, J. E. A developmentally regulated hydroxyproline-rich glycoprotein in maize pericarp cell walls. *Plant Physiology*, 138–142, ISSN: 1532–2548 (1988).
35. Gennadios, A., Mc Hugh, T. H., Weller, C. L., and Krochta, J. M. Edible coating and films based on protein. *In: Edible Coatings and Films to Improve Food Quality*, J. M. Krochta, E. A. Balwin, and M. O. Niperos Carriedo (Eds.), Technomic Publishing, 978-1-42003-198 (1994).
36. Guilbert. Technology and application of edible protective films. *Food packaging and preservation: Theory and Practice*, M. Matholouthi (Ed.) 371394. Elsevier Applied Science Publishing, 0-85334-377-2 (1986).
37. Zhang, J. and Jane, J. Mechanical and thermal properties of extruded soy protein sheets. *Polymer*, **42**, 2569–2578, 0032–3861 (2001).
38. Bourtoom, T. Factors Affecting the Properties of edible film prepared from mung bean proteins. *International Food Research Journal*, **15**, 167–180, ISSN: 2231-7546 (2008).
39. Galego, N., Rozsa, C., Sa´nchez, R., Fung, J., Va´zquez, A., and Tomas, J. S. Characterization and application of poly(b-hydroxyalkanoates)family as composite biomaterials. *Polym. Test.*, **19,** 485–492 (2000).
40. Steinbuchel, A. and Fu chtenbusch, B. Bacterial and other biological systems for polyester production. *Trends Biotechnol*, **16**, 419–427 (1998).
41. Orts, W. J, Nobes, G. A. R, Kawada, J, Nguyen, S, Yu, G, and Ravanelle, F Polyhydroxyalkanoates: biorefinery polymers with a whole range of applications. *Can J Chemistry*., **86**, 628–640 (2008).
42. Lee, S. Y. Bacterial polyhydroxyalkanoates. *Biotechnology and Bioengineering*. **49**, 1–14 (1996).
43. Braunegg, G., Lefebvre, G., and Genser, K. L. Polyhydroxyalkanoates, biopolyesters from renewable resources: Physiological and engineering aspects. *Journals of Biotechnology*, **65**, 127–161 (1998).
44. Tsuge, T. Metabolic improvements and use of inexpensive carbon sources in microbial production of polyxydroxyalkanoates. *J Biosci Bioeng*, **94**, 579–584 (2002).
45. Matsusaki, H, Abe, H, and Doi, Y. Biosynthesis and properties of poly (3-hydroxybutyrate-co-3-hydroxyalkanoates) by recombinant strains of Pseudomonas species. *Biomacromolecules*, **1**, 17–22 (2000).
46. John, R. P., Nampoothiri, K. M., and Pandey, A. Solid-state fermentation for L-lactic acid production from agro wastes using Lactobacillus delbrueckii. *Proc. Biochem*., **41**, 759–763 (2006).
47. Kale, G., Auras, R., and Singh, S. P. Degradation of commercial biodegradable packages under real composting and ambient exposure conditions. *Journal of Polymer and the Environment*., **14**, 317–334 (2006).
48. Bohlmann, G. M. Biodegradable packaging life-cycle assessment. *Environmental Progress*, **23**(4), 342–346 (2004).

CHAPTER 8

EDIBLE FILMS AND COATINGS FOR PACKAGING APPLICATIONS

RUNGSINEE SOTHORNVIT

ABSTRACT

The non-biodegradability of synthetic packaging materials rises critical environmental concerns from consumers, and the lower availability of fossil fuels makes necessary to explore alternative polymers to replace synthetic polymers for packaging. Edible films and coatings are produced from edible substances such as proteins, polysaccharides, lipids, and other natural biopolymers from fruits and vegetables, that have the potential to control the diffusion of water, oxygen, carbon dioxide, aroma and oil between the food and the environment. The objectives of this chapter are to define edible films and coatings, their functions and properties, and include recent advances related to promising packaging applications. Technologies such as compression molding, extrusion process and novel electrospinning process can be used in a commercial scale, facilitating the manufacture of edible film packaging. Several nanocomposite films have been developed to improve their properties. Additionally, by-products and low-value products from the food industry can be a promising source for edible film and coating materials with good properties. However, there is still a need to continuously research on this area since there is no exact solution for each specific problem. The different characteristics of each food product require films and coatings with different barrier and mechanical properties. Therefore, the attractive aspects of edible films and coatings, visualized in green and novel packaging, are a great driving force to be explored in both academic and industrial sectors. Moreover, their biodegradable nature makes them attractive to consumers because of the increased environmental concern.

8.1 INTRODUCTION

Synthetic packaging materials possess non-biodegradability which leads to critical environmental concerns from consumers. Moreover, the inadequacy of future fossil fuels drives efforts to explore alternative polymers to replace synthetic polymers for packaging. The requirement to extend shelf life and maintain quality of food products, while reducing packaging waste, together with an increase in consumer needs for safe and convenient foods have moved researches to look for new edible and biodegradable packaging materials from renewable resources (Krochta, 2002; Tharanathan, 2003; Embuscado and Huber, 2009; Janjarasskul and Krochta, 2010). Edible packaging is produced from edible substances such as proteins, polysaccharides, lipids and/or resins and other natural biopolymers from fruits and vegetables. Edible films and coatings are generally thin layers of biopolymer materials that have the potential to control the diffusion of water, oxygen, carbon dioxide, aroma and oil between the food and the environment. In addition, edible films and coatings can protect food from mechanical damage depending on the nature of the biopolymer film forming material (Krochta, 1997; Sothornvit and Krochta, 2000a, b, c; Sothornvit and Krochta, 2005). The attractive aspects of edible films and coatings, visualized in green and novel packaging, are a great driving force to be explored in both academic and industrial sectors. In this context, the objectives of this chapter are to define edible films and coatings, their functions and properties, including recent advances related to promising packaging applications.

8.2 DEFINITIONS, FORMATION, FUNCTIONS AND APPLICATIONS

8.2.1 DEFINITIONS

Edible films are stand-alone thin layers of biopolymer materials. Films are generally used to determine the film properties, such as water vapor and gas barrier, mechanical properties, solubility, optical and other properties. Edible films are formed with less than 0.3 mm of thickness, while at higher thickness are called sheets. The films can also be placed on or between food components as a wrapper (McHugh, 2000) and they can be potentially formed as capsules, pouches, casings, and bags (Krochta, 2002).

Edible coatings are films formed directly in the food surface by immersion of the food product in the coating solution (Krochta and De Mulder-Johnston, 1997). Therefore, coatings are part of food products to protect or enhance them in some characteristic and they can be directly eaten or removed before consumption. Moreover, edible coatings can be improved by the addition of active substances to form active food packaging, enhancing functional or specific properties of foods, such as the control of food oxidation and microbial growth (Cuq et al., 1995; Han, 2001; Jarimopas et al., 2007).

Naturally, edible films and coatings are edible and/or biodegradable due to the biopolymer and food-grade nature of the different ingredients used (example biopolymers, plasticizers, emulsifiers, acids, bases, enzymes). Moreover, formulation, fabrication method, and modification treatments are also important factors that need to be considered to define edibility of films and coatings (Guilbert et al., 1997, 2002; Krochta, 2002; Pavlath and Orts, 2009). If biopolymers react with other chemicals formulation (example chemical cross-linking or chemical grafting) or non-edible components are used in the film or coating formulations, edibility is lost. Therefore, biodegradable films and coatings for food packaging have to perform in a successful and safe manner as designed and have to be able to be degraded later by microorganisms in a composting process or at appropriate conditions of the biodegradation process.

8.3 FORMATION AND FUNCTIONS

8.3.1 FILMS AND COATINGS COMPONENTS

The main components for films and coatings are biopolymers, plasticizers, emulsifiers and other food-grade additives. The conventional structural materials are biopolymers such as proteins, polysaccharides, lipids and resins. Within these groups, by-products or wastes from the food industry have shown a potential to be used as film forming materials. Some examples are whey protein from cheese making, fish protein from surimi production, chitosan from crustacean shells and potato starch from potato processing, and so on. Recently, other natural biopolymers from fruit and vegetable purees are used as attractive film forming materials (Rojas-Grau et al., 2006; Du et al., 2008, Sothornvit and Rodsamran, 2008, 2010a and b; Azeredo et al., 2009).

Generally, most biopolymer materials form very brittle films. Thus, they usually require the addition of plasticizers to overcome this problem and improve the mechanical properties of edible films and coatings. Plasticizers are low molecular weight nonvolatile substances that reduce biopolymer chain-to-chain interaction, resulting in an increase in film flexibility (Sothornvit and Krochta, 2000a, b, and c, 2001, 2005). Moreover, edible films and coatings can carry other additives such as emulsifiers, antioxidants, antimicrobials, firming agents, nutraceuticals, volatile precursors, flavors, and colorants to enhance functional, nutritional, sensory, safety, and mechanical properties of films and coatings (Olivas and Barbosa-Canovas, 2005).

8.3.2 PROTEINS

Proteins commonly used in edible films and coatings are zein, soy protein, wheat gluten, milk protein (whey protein, casein), casein, gelatin, collagen, wheat gluten, fish myofibrillar protein, keratin, peanut protein, rice bran protein, egg white protein, sorghum protein, and sericin. The unique structure of proteins, different from polysaccharides, provides different properties to edible films and coatings (Krochta, 2002; Janjarasskul and Krochta, 2010). Modifications of secondary, tertiary, and quaternary protein structures by physical (heat, mechanical treatment, pressure, irradiation), chemical (acids, alkalis, metal ions, salts), enzymatic treatments and cross-linking affect film properties (Krochta, 2002; Han and Gennadios, 2005). For example, the tensile and barrier properties of whey protein films are improved through heat denaturation of the protein, which promotes disulfide bond formation (Perez-Gago et al., 1999).

8.3.3 POLYSACCHARIDES

Polysaccharides are abundant in nature, low cost and biodegradable. Therefore, a variety of polysaccharides have been studied as potential edible film and coating forming agents. Some polysaccharide film and coating materials are cellulose and its derivatives (carboxymethyl cellulose, methyl cellulose, hydroxypropyl cellulose, hydroxypropyl methylcellulose (HPMC), microcrystalline cellulose), starch and its derivatives (raw, modified starch, pregelatinized, dextrin, maltodextrin), pectin, seaweed extract (agar, alginate, carrageenan, furcellaron), gums (gum Arabic, guar gum, locust bean gum, xanthan gum, gellan gum) and chitosan. Their chemical nature of many repeating units with a great number of hydroxyl groups causes film formation by H-bonding, resulting in hydrophilic films. In general, polysaccharide films show good gas barrier and mechanical properties (Baldwin et al., 1995) and film properties can also be improved by physical and chemical modifications.

Hydrocolloids (proteins and polysaccharides) are common materials for edible films and coatings, since they possess excellent gas barrier, creating a gentle modified atmosphere that helps maintaining quality of foods, such as fresh produces, dry fruits, and so on. However, they exhibit poor moisture barrier due to their hydrophilic nature (Kester and Fennema, 1986; Gennadios and Weller, 1990; Park and Chinnan, 1990; Gennadios et al., 1994). To overcome this problem, lipids are added to increase film hydrophobicity.

8.3.4 LIPIDS

Unlike protein and polysaccharides, lipid and resin are not polymers so they could not form stand-alone films. Edible lipids include waxes and oils; for example, paraffin wax, beeswax (BW), candelilla wax, carnauba wax, rice bran wax, jojoba oil, triglycerides (milkfat fractions), fatty acids, fatty alcohols, and sucrose fatty acid esters. Resins include shellac and terpene (Krochta, 2002). Due to their hydrophobic nature, lipid-based films and coatings exhibit high water resistance and low surface energy (Han and Gennadios, 2005). Generally, edible lipids have been incorporated into proteins and polysaccharides as emulsion or multilayer composite films to improve moisture barrier.

8.3.5 COMPOSITE FILMS

As mentioned above, protein and polysaccharide-based edible films are excellent gas barrier and present moderately good mechanical properties at low relative humidity (RH), but they are poor moisture barrier (Janjarasskul and Krochta, 2010). Therefore, composite films by blending of a hydrocolloid with a lipid can provide film properties with desirable film structures for specific applications. Composite films can be formulated as a stable emulsion or as a bilayer. Emulsion-based films are created by the dispersion lipid globules in the hydrocolloid matrix (Perez-Gago and Krochta, 2005). This requires sometimes the use of emulsifiers to create a stable emulsion with a homogeneous distribution of the lipid in the polymer matrix. Bilayer-based films are created by the lamination of the lipid layer over the hydrocolloid layer. In practice, emulsion-based films are easier to produce than bilayer-based films, even though emulsion-based films show lower moisture barrier than bilayer-based films (Perez-Gago and Krochta, 2005; Falguera et al., 2011). Nevertheless, small lipid particle size homogeneously distributed in the polymer matrix has shown to improve the moisture barrier of emulsion-based films (Debeaufort and Voilley, 1995; McHugh and Krochta, 1994; Perez-Gago and Krochta, 2001).

8.3.6 FRUIT AND VEGETABLE PUREES

Recent works show fruit and vegetable purees as good alternative edible materials. The use of these ingredients can also help reducing postharvest losses by using overripe fruits or those fruits that present lower quality and cannot be sold as first grade quality. Fruit and vegetable purees recently explored as edible films include strawberry, apricot, apple, pear, carrot, broccoli, and mango (McHugh et al., 1996; McHugh and Olsen, 2004; Sothornvit and Rodsamran, 2008, 2010a and b). Biopolymers present in fruit and vegetable purees are mainly polysaccharides such as pectin, starch and cellulose derivatives (Kaya and Maskan, 2003) and some proteins and fats (Sothornvit and Rodsamran, 2008), which are the main components for edible films. Edible films from tropical fruits have been prepared from banana (Sothornvit and Pitak 2007) and mango (Sothornvit and Rodsamran 2008, 2010a and b, Azeredo et al. 2009). Fruit and vegetable films exhibit not only good mechanical and barrier properties, but also offer color and flavor from the pigments and volatile compounds of the fruit and vegetable that in many cases helps improving food quality.

8.3.7 PLASTICIZERS

Film formation as a coating involves two types of interaction, such as cohesion (attractive forces between film polymer molecules) and adhesion (attractive forces between film and substrate). Cohesive forces between polymer molecules cause brittleness of films, which affects integrity and film properties. To overcome this drawback, food-grade plasticizers are added to decrease intermolecular forces resulting from chain-to-chain interaction (Sothornvit and Krochta 2005). By reducing intermolecular forces and thus increasing the mobility of the polymer chains, plasticizers lower the glass transition temperature of films and improve film flexibility, elongation, and toughness (Sothornvit and Krochta 2000a, b, and c, 2001, 2005). Commonly used plasticizers in film systems are monosaccharides, disaccharides or oligosaccharides (example glucose, fructose-glucose syrups, sucrose, and honey), polyols (example glycerol, sorbitol, glyceryl derivatives, and polyethylene glycols), and some lipids and derivatives (example phospholipids, fatty acids, and surfactants). Generally, the selection of plasticizers requires considering plasticizer compatibility, efficiency, permanence, and economics (Sothornvit and Krochta 2005).

8.3.8 EMULSIFIERS AND OTHER FOOD ADDITIVES

Emulsifiers are surface active substances that modify interfacial energy at the interface of immiscible systems. Emulsifiers are often required in the film formation to disperse lipid particles in composite emulsion films or to achieve sufficient surface wettability to ensure proper surface coverage and adhesion to the coated surface (Krochta 2002). Some common emulsifiers are acetylated monoglyceride, lecithin, glycerol monopalmitate, glycerol monostearate, polysorbate 60, polysorbate 65, polysorbate 80, sodium lauryl sulfate, sodium stearoyl lactylate, sorbitan monooleate, and sorbitan monostearate (Janjarasskul and Krochta, 2010).

Edible films and coatings can be carriers of antioxidants (phenolic compounds such as butylated hydroxyanisole, propyl gallate, butylated hydroxytolene, tocopherol, citric acid, ascorbic acid, or natural compounds from herb extract) and antimicrobials (organic acids and their salts such as benzoic acid, sorbic acid, propionic acid, chitosan, plant essential oil extracts, and so on.) to enhance their functional properties as active packagings (Janjarasskul and Krochta, 2010). Moreover, they can be carriers of nutrients, flavors, and colors to improve food nutritional and sensory quality.

8.4 FILM AND COATING FORMATION

Film and coating formation occurs when biopolymer molecules interact through cohesive forces, named H-bonding, ionic bonds and covalent bonds (disulfide bonds). Factors affecting film strength are the chemical nature of the biopolymer and the rest of the components of the formulation (plasticizer type and amount and food additives), and the film forming process.

Common film and coating formation is achieved by solution casting; whereas, coatings are generally obtained by dipping, spraying, panning and enrobing the food surface. Both cases require the evaporation of the solvent, usually water or ethanol depending on the type of biopolymer, from the biopolymer solution (wet method). If ethanol is required for film and coating formation, the release of solvent to the atmosphere needs to be safe for the environment and a process of solvent recovery is needed in a commercial scale.

In the case of edible composite films and coatings containing a biopolymer and a lipid, the formulation requires to heat the lipid above its melting point, to homogenate both phases, degas, and cast on the plate or product surface, with the final evaporation of the solvent (Krochta, 2002). Some drying methods that have been studied include hot air, infrared, and microwave drying.

The time consuming of film processing and the space requirement for drying in the traditional solution casting method (wet method) is driving to the development of a faster alternative method, such as extrusion. This is a dry process, in which the water content is lower when compared to the wet process, so it is fast and facilitates the commercial production (Sothornvit et al., 2003, 2007). Generally, compression molding is the first step toward the continuous extrusion process. The compression molding utilizes high pressures and temperatures to melt and press the solid mixture, forming a sheet in a few minutes. The extrusion process utilizes dried or preconditioned solid materials which are screwed drive from a feeding zone to the compression zone with high shear rate, temperature and pressure to fluid melt the materials, with causes a reduction of the viscosity. Then the melted-low viscous mixture is expelled through the die, where contacts with ambient pressure. In this process, biopolymer structure is modified with the alignment of molecules in the direction of the flow. Moreover, the mechanical energy and high temperature used generate the denaturation, aggregation, and crosslinking of biopolymers (Hernandez-Izquierdo and Krochta, 2008). As in the casting process, plasticizers are also necessary to increase the free volume and chain mobility of the biopolymer and improve the processability (Sothornvit and Krochta, 2005). The extrusion process and compression molding can be applied for thermoplastic biopolymers such as whey protein (Sothornvit et al., 2003, 2007; Hernandez-Izquierdo and Krochta, 2008; Hernandez-Izquierdo et al., 2008), soy protein (Cunningham et al., 2000; Wu and Zhang, 2001; Zhang et al., 2001), wheat gluten (Redl et al., 1999; Pommet et al., 2005), gelatin (Park et al., 2008; Krishma et al., 2012; Andreuccetti et al., 2012), myofibrillar protein (Cuq et al., 1997), zein (Wang and Padua, 2003), starch (van Soest and Korleve, 1999; Ma et al., 2006), and banana flour (Sothornvit and Songtip, 2010, 2012). More details on commercial manufacture of edible films are totally reviewed by Rossman (2009).

Another innovative method is the electrospinning technique, which can be applied to produce nanofilms with improved properties. Electrospinning uses a high voltage to create an electrically charged jet of polymer solutions to draw very fine micro or nano scale of fibers (Reneker et al., 2007). The preparation of starch films and coatings with electrohydrodynamic atomization or electrospraying as a novel method has been described, in which the solvent evaporated very quickly because of the high surface area of the droplets generated, forming the film instantly (Pareta and Edirisinghe, 2006). This method might be an interesting method to develop edible films in a nano scale and it will be useful for food packaging applications in the future.

8.5 FILM PROPERTIES

Film forming materials, plasticizer type and amount and other additives affect film properties differently. Table 1 and 2 show the water vapor permeability (WVP) and oxygen permeability (OP) of different polysaccharide-, protein-, composite- and fruit and vegetable-based films with various plasticizers using different processes. Comparisons among them are difficult because of different film compositions, test conditions (RH, temperature), methodology, and number of replications were used in each work.

TABLE 1 Water vapor permeability (WVP), oxygen permeability (OP), and mechanical properties (tensile strength, TS; elastic modulus, EM and % elongation, %E) of various edible films from wet process (solution casting)

Film type	WVP (g mm/m^2 d kPa)	OP (cm^3 µm/ m^2 d kPa)	Mechanical Properties			Reference
			TS (MPa)	EM (MPa)	% E	
Polysaccharide						
LBG:PEG 200 = 0.58:1–2.3:1	1.51–1.83	–	–	–	–	Aydinli and Tutas, 2000
LBG:PEG 400 = 0.58:1–2.3:1	1.51–2.19	–	–	–	–	Aydinli and Tutas, 2000
LBG:PEG 600 = 0.58:1–2.3:1	1.67–2.78	–	–	–	–	Aydinli and Tutas, 2000
LBG:PEG 1000 = 0.58:1–2.3:1	1.89–2.76	–	–	–	–	Aydinli and Tutas, 2000
Gellan:Gly = 0.5:1	36	–	30	25	30	Yang and Paulson, 2000
Gellan:PEG 400 = 0.5:1	–	–	27	44	8	Yang and Paulson, 2000
HACS:Gly = 1:1–5:1	1011–1270	–	2–32	–	6–22	Ryu et al., 2002
HACS:Sor = 1:1–5:1	1011–1270	–	7–47	–	6–38	Ryu et al., 2002
Pullulan:Sor = 3.3:1	2.88	–	29.2	–	2.6	Kim et al., 2002
Pullulan:Man = 3.3:1	3.6	–	15.7	–	9.5	Kim et al., 2002

TABLE 1 *(Continued)*

Cassava starch:Gly = 1:0–1:5	34.73–65.58	–	4–26	9–737	6–46	Mali et al., 2006
Corn starch:Gly = 1:0–1:5	46.40–71.97	–	9–37	162–1188	3–28	Mali et al., 2006
Yam starch:Gly = 1:0–1:5	38.53–65.58	–	10–49	159–1003	3–25	Mali et al., 2006
CMC:Gly = 1:1.43–1:2	14.02–18.34	0.013–0.033	–	–	–	Li et al., 2008
Rice starch–chitosan:Sor = 1:2.5	4.11–7.08	–	27.5–38.1		8.1–13.0	Bourtoom and Chinnan, 2010
Rice starch:Gly = 1:4–1:6	4.01–7.49	–	1.6–10.9	21.3–532.8	2.8–59.8	Dias et al., 2010
Rice starch:Sor = 1:4–1:6	2.30–2.62	–	11.2–22.3	456.3–1052.6	2.8–3.9	Dias et al., 2010
Rice flour:Gly = 1:4–1:6	9.84–15.26	–	1.34–10.31	22.21–560.74	2.73–66.43	Dias et al., 2010, 2011
Rice flour:Sor = 1:4–1:6	3.02–3.34	–	7.23–14.99	248.57–815.99	2.24–4.32	Dias et al., 2010, 2011
galactomannan: Tween80:Gly = 1:0.2:0–1:0.2:2	5.70–11.52	–	1.70–18.55	–	3.77–38.72	Cerqueira et al., 2012
Chitosan: Tween80:Gly = 1:0.2:0–1:0.2:2	4.38–9.85	–	1.17–21.45	–	16.18–99.52	Cerqueira et al., 2012
Protein						
β-Lg:Gly = 1.70:1–3.20:1	–	20–43	4.98–16.01	150.1–705.6	11.36–76.46	Sothornvit and Krochta, 2000b, 2001
β-Lg:PG = 2.06:1–3.87:1	–	17.27	13.2–21.8	1476.4–1922.3	–	Sothornvit and Krochta, 2000b
β-Lg:Sor = 0.86:1–1.62:1	–	3–8	2.71–10.06	99.6–383.8	24.75–65.85	Sothornvit and Krochta, 2000b, 2001
β-Lg:Suc = 0.46:1–0.86:1	–	< 0.05	1.74–9.71	64.1–340.8	30.33–89.41	Sothornvit and Krochta, 2000b, 2001
β-Lg:PEG 200 = 0.78:1–1.47:1	–	110–700	1.80–6.46	67.4–255.2	41.67–77.09	Sothornvit and Krochta, 2000b, 2001
β-Lg:PEG 400 = 0.39:1–0.73:1	–	1050–2220	0.72–2.88	28.7–117.2	25.53–32.31	Sothornvit and Krochta, 2000b, 2001
WPI:Gly = 4:1–2.86:1	79.2–108	38–90	4.8–10.5	149–403	40–73	Sothornvit and Krochta, 2000a, c
Hydrolyzed WPI:Gly = 4:1–2.86:1	84–120	40–115	0.7–1.3	< 25	< 7	Sothornvit and Krochta, 2000a, c

TABLE 1 *(Continued)*

CZ:PEG+Gly = 4:1–10:1	345.6–985.0	–	19.5–22.7	–	1.3–4.3	Ryu et al., 2002
CZ:OA = 4:1–10:1	259.2–388.8	–	17.5–21.7	–	1.8–5	Ryu et al., 2002
WPI:Gly = 1:1–2:1	116–144	–	1–3.5	20–110	35–48	Shaw et al., 2002
WPI:Sor = 1:1–2:1	84–112	–	2.5–9	75–325	12–22	Shaw et al., 2002
WPI:X = 1:1–2:1	84–89	–	0.5–8.5	80–127510	2–15	Shaw et al., 2002
Soybean:Gly = 0.25:1–0.67:1	–	–	1.55–2.08	–	250–275	Cao and Chang, 2002
Soybean:TEG = 0.25:1–0.67:1	–	–	1.98–2.38	–	240–267	Cao and Chang, 2002
Soybean:Gly:TEG = 0.5:1:1–1.33:1:1	–	–	1.75–2.3	–	240–255	Cao and Chang, 2002
SPI-PLA:Gly = 1:2	3.84–10.54	–	8.5–17.4	–	82.6–349.9	Rhim et al., 2007
Composite						
Sericin-GM:Gly:BW = 1:0:0–3.33:1:0.2	168–420	–	3.7–18	10–600	3–67	Sothornvit and Chollakup, 2009
WPI:MG:Sor = 1:0:3.33–0:1:3.33	48	–	0.5–11.5	7.5–253.3	6.7–70.7	Oses et al., 2009
Nanocomposite mango puree: CNF = 1:0–1:0.36	40.08–63.84	–	4.09–8.76	31.54–44.07	19.85–322/05	Azeredo et al., 2009
Nanocomposite chtiosan-MMT:Gly = 1:0–1:4	1.00–2.16	–	9.07–32.59	–	5.37–9.60	Lavorgna et al., 2010
Agar-MMT:Gly = 1:5	99.36–194.4	–	28–37.2	–	45.1–53.6	Rhim, 2011
Rice flour-cellulose fibers:Gly = 1:4–1:6	6.55–9.62	–	13.49–20.64	492.03–1001.30	4.03–8.78	Dias et al., 2011
Rice flour-cellulose fibers:Sor = 1:4–1:6	2.71–2.81	–	16.76–22.51	728.08–1242.02	3.27–4.67	Dias et al., 2011
NaCas:OB:Gly = 1:0.3:0.25–1:0.3:1	252–364.8	–	5–30	5–80	5–1200	Matsakidou et al., 2013
Fruit and vegetable						
Apple puree	140.1	–	0.7	4.4	11.8	McHugh et al., 1996; McHugh and Olsen, 2004
Peach puree	100.4	–	1.8	5.9	23	McHugh et al., 1996; McHugh and Olsen, 2004
Carrot puree	–	–	5.3	208.9	7.3	McHugh and Olsen, 2004
Brococoli puree	–	–	7.1	421.5	4.1	McHugh and Olsen, 2004

TABLE 1 *(Continued)*

Apricot puree	103.0	–	–	–	–	McHugh et al., 1996
Pear puree	186.6	–	–	–	–	McHugh et al., 1996
Banana starch:Gly = 2:1	–	–	25	1.6	40	Romero-Bastida et al., 2005
Mango starch:Gly = 2:1	–	–	19	1.4	30	Romero-Bastida et al., 2005
Banana starch:Gly = 2:1	–	–	25	1.6	40	Romero-Bastida et al., 2005
Banana flour-pectin:Gly = 3.3:1–2:1	–	23.5–40	2.7–16	0.5–9.8	2.5–16.9	Sothornvit and Pitak, 2007
Mango puree	213.2	41.2	1.2	8.3	18.5	Sothornvit and Rodsamran, 2008

LBG = locust bean gum, PEG = polyethylene glycol, Gly = glycerol, HACS = high amylase corn starch, Sor = sorbitol, Man = manitol, CMC = carboxymethylcellulose, β-Lg = β-lactoglobulin, PG = propylene glycol, Suc = sucrose, WPI = whey protein isolate, CZ = corn zein, OA = oleic acid, X = xylitol, TEG = triethylene glycol, SPI = soy protein isolate, PLA = poly lactic acid, GM = glucomannan, BW = beeswax, CNF = cellulose nanofiber, MMT = montmorillonite, NaCas = sodium caseinate, OB = oil bodies from maize, MG = mesquite gum

TABLE 2 Water vapor permeability (WVP), oxygen permeability (OP), and mechanical properties (tensile strength, TS; elastic modulus, EM and % elongation, %E) of various edible films from dry process (compression molding and extrusion)

Film type/process	WVP (g mm/m^2 d kPa)	OP (cm^3 μm/ m^2 d kPa)	Mechanical properties			Reference
			TS (MPa)	EM (MPa)	% E	
Polysaccharide						
CS-banana/sugarcane fiber:Gly = 1:3.3/ compression molding	–	–	1.73–4.34	70.81–484.04	1.20–14.96	Guimaraes et al., 2010
Protein						
SPI:Gly = 2.5:1/ compression molding	–	–	–	2.6	74.5	Cunningham et al., 2000
Acetylated SPI / compression molding	12.07	3104	2.3	–	105	Foulk and Bunn, 2001
WPI:water = 3.33:1–2:1/ compression molding	88.8–120	–	–	–	–	Sothornvit et al., 2003, 2007
WPI:Gly = 3.33:1–2:1/ compression molding	268.8–576		4–10	60–251	43–94	Sothornvit et al., 2003, 2007
Keratin/compression molding	–	–	7.9–27.8	697–1218	1.1–4.7	Katoh et al., 2004

TABLE 2 *(Continued)*

Gelatin-yucca extract:Gly = 1:0.005–1:0.20/extrusion	2.88–3.84	–	19.5–33.8	–	40.3–157.8	Andreuccetti et al., 2012
Composite						
CS-MMT:Gly = 1:0–1:5	14.88–30	–	17–36	–	1.1–22.5	Tang et al., 2008
SPI-gelatin:Gly = 1:2.5–1:3.33/compression molding	–	–	3.45–11.31	–	135.98–163.54	Guerrero et al., 2011
Fruit and vegetable						
Banana flour:Gly = 1:3.3/compression molding	–	–	0.62–1.35	14.5–18	7–15	Sothornvit and Songtip, 2010

CS = corn starch, SPI = soy protein isolate, Gly = glycerol, WPI = whey protein isolate, MMT = montmorillonite nanoclay

Films and coatings are characterized by their moisture, oxygen, aroma, and oil barrier properties, mechanical, physical and optical properties (solubility, moisture content, color, and gloss) and other functional properties if additives, such as antimicrobials or antioxidants are added. Barrier properties are affected by the physical and chemical nature of the biopolymers. The chemical structure of the polymer backbone, the degree of crystallinity and orientation of molecular chains, including the plasticizer properties, affect the barrier properties (Salame, 1986). Like other polymer films, which are not perfect barriers, each edible film has its limit of permeability to water, oxygen, aroma, and oil (Sothornvit and Krochta, 2005). Permeability occurs when the permeant molecules (water, oxygen, aroma, and oil) run into and adsorb to the surface of the polymer film, diffuse through the polymer network, and finally desorb from the other side of the film.

Mechanical properties, such as tensile strength (TS), elastic modulus (EM), and elongation (E), reflect the toughness of films and the ability to protect from mechanical damage the coated food. TS and EM of biopolymer films and coatings decrease with increasing plasticizer content; whereas, E increases as plasticizer content increases.

Other physical and optical properties related to functionality and stability of films that affect consumer perception and acceptability of coated foods are solubility, a_w, moisture content, color, gloss, and so on.

Film and coating preparation technique also has an effect of their final properties. Dry technologies, such as compression molding and extrusion, affected edible film properties differently compared to wet technologies, such as solution casting. For example, WVP values of compression-molded WPI sheets (Table 2) were three to four times higher than those of solution cast films (Table 1), since thicker compression-molded sheets faced higher RH in WVP testing (Sothornvit et al., 2003, 2007). The use of glycerol (Gly) gave higher WVP values in compression-molded WPI sheets than using water as a plasticizer (Table 2). Mechanical properties (E and TS) of compression-molded WPI sheets were greater than those of solution cast WPI films due to a greater crosslinking of protein chains resulting from the high temperature heating and high pressure applied during the compression molding process (Sothornvit et al., 2003, 2007).

Addition of cellulose fibers to rice flour lowers the WVP of the composite films (Dias et al., 2011). In addition, composite films of blends of a hydrocolloid and a lipid improve the moisture barrier of biopolymer films. However, the lipid particle sizes, distribution and formation technique influence film barrier, and mechanical properties. For example, the higher WVP of sericin-glucomannan-beeswax (BW) composite films compared to the sericin-glucomannan film was due to the lower ability of BW to blend with sericin and glucomannan (Sothornvit and Chollakup, 2009; Sothornvit et al., 2010a). In other work, the addition of BW to mango edible films improved the film water barrier and desirable film properties could be obtained depending on the BW content used (Sothornvit, 2010).

8.5.1 MOISTURE BARRIER

In general, the main objective for using edible films and coatings is to control water transfer between the food and the environment. Water is involved in many chemical and biological reactions, such as browning, lipid oxidation, enzymatic activity, microbial growth, texture change, and so on. Therefore, it has an important effect in food quality and shelf life. Each edible film possesses a unique permeability under specific test conditions. Generally, polysaccharide- and protein-based films are water soluble and exhibit good barriers against nonpolar molecules such as oxygen, aroma, and oil, while they are poor barrier against polar molecules such as water. Efforts to improve their moisture barrier are addition of lipids, crosslinking with transgluminase, irradiation, and heat curing (Krochta, 2002).

8.5.2 OXYGEN BARRIER

Oxygen causes many food deteriorations such as oxidation of food components (lipids, vitamins, colors, and flavors) and microbial growth, leading to the loss of nutritional quality and unacceptable products for consumer consumption. Film forming materials affect films and coatings properties (Cuq et al., 1998). Generally, polysaccharide- and protein-based films are excellent oxygen barrier as shown by WPI, β-lactoglobulin films (Sothornvit and Krochta, 2000a, b, c) and other biopolymers. In fresh fruits and vegetables, the gas barrier can modify the atmosphere inside the fresh produce. Nonetheless, plasticizer content increases film oxygen permeability due to the higher free volume in the film network (Sothornvit and Krochta, 2005).

8.5.3 GREASE/OIL BARRIER

Edible films and coatings can also provide grease/oil barrier. There are few works on oil barrier of edible films (Krochta, 2002). As known, hydrophilic edible films exhibit good grease/oil barrier. Thus, the application of biopolymer coatings on paperboard can provide an alternative food application for fast-food packaging and oily food product.

8.5.4 AROMA BARRIER

Biopolymer coatings can be used to prevent the loss of volatile organic substances, such as flavor or aroma, and the migration of external off-flavors into packed food during storage and distribution (Janjarasskul and Krochta, 2010). In addition, proteins and polysaccharides can be used to encapsulate flavor and aroma compounds. The encapsulation process allows to hold organic aroma or hydrophobic compounds in the matrix of the hydrophilic polymer, which prevents aroma loss to the environment and oxidation. Studies on aroma permeability of edible films are limited compared to other barrier properties. For example, i-carrageenan and sodium alginate films showed good mechanical properties and oxygen barrier, and they can be used to encapsulate aroma compounds, such as n-hexanal and D-limonene (Hambleton et al., 2011).

8.6 FILM AND COATING APPLICATION

Edible films and coatings are widely used to improve food quality and shelf life and to partially substitute non-recyclable and synthetic polymer films. Many recent edible films and coatings have been commercialized and many more explored, as summarized in several review articles, with various biopolymers such as WPI (Krochta, 2002; Jooyandeh, 2011; Ramos et al., 2012), chitosan (Tamer and Copur, 2010; Alishahi and Aider, 2012) and galactomannan (Cerqueira et al., 2009, 2011). Most applications of edible coatings are on postharvest produces and food products. Edible coatings for fresh fruits and vegetables are reviewed by Vargas et al. (2008), Cerqueira et al. (2009), Rojas-Grau et al., 2009, Garcia et al. (2010), Perez-Gago et al. (2010), Tamer and Copur (2010), Valencia-Chamorro et al., (2011), and Campos et al. (2011). Similarly, the application of edible coatings and films for meat, poultry and seafood has been reviewed by Baker et al. (1994), Gennadios et al. (1997), and Coma (2008). Reviews of edible coatings for other food products are summarized by Atares et al. (2011) and Varela and Fiszman (2011).

Protein films and coatings have been widely used in some fruits. For example, whey protein coatings have been studied on apples (Cisneros-Zevellos and Krochta, 2003a, b), fresh-cut mangoes (Plotto et al., 2004), oranges (Sothornvit, 2005), and fresh-cut apples (Perez-Gago et al., 2005, 2006) and wheat gluten-based films and coatings on refrigerated strawberry (Tanada-Palmu and Grosso, 2005). In vegetables, some works include the use of sodium caseinate/stearic acid emulsion coatings on peeled carrots (Avena-Bustillos et al., 1994a), calcium caseinate-acetylated monoglyceride aqueous emulsions coating on zucchini (Avena-Bustillos et al., 1994b), whey protein isolate, sodium caseinate or sodium caseinate/beeswax emulsion coatings on green bell pepper (Lerdthanangkul and Krochta, 1996) and zein films on fresh broccoli (Rakotonirainy et al., 2001). The use of protein-based coatings on fresh-cut fruits and vegetables can provide nutritional value to the produce, but care should be taken to allergies or intolerances to certain proteins such as lactose or gluten for some consumers (Baldwin and Baker, 2002). Moreover, the vegetarians would not accept animal-derived protein coatings, such as gelatin and milk protein. Therefore, it is recommended to label on the marketing display to inform the type of coating used. An alternative solution to overcome these problems is the use of natural products, such as banana films (Sothornvit and Pitak, 2007) or mango films (Sothornvit and Rodsamran, 2008, 2010a, b) on fresh fruits and vegetables, as well as in other food products.

Polysaccharides mainly used in coatings are carrageenan, maltodextrin, methylcellulose, carboxymethyl cellulose, pectin, alginate, and starch. Generally, edible coatings are applied on fruits by dipping or spraying. Coating of fruits and vegetables with polysaccharides usually increases gloss and acceptability of consumers. Figure 1 shows the shiny appearance of coated guava. Some works describing the use of polysaccharide films and coatings in fruits are starch-based edible coatings on guava (Quezada-Gallo et al., 2003), hydroxypropyl methylcellulose-lipid edible composite coatings on plum (Perez-Gago et al., 2003; Navarro-Tarazaga et al., 2008), carboxymethylcellulose, chitosan and starch coatings on fresh-cut mangoes (Plotto et al., 2004), hydroxypropyl methylcellulose-lipid edible composite coatings on fresh-cut apples (Perez-Gago et al., 2005), chitosan coatings on sliced mangoes (Chien et al., 2007), and alginate coatings on fresh-cut apples (Olivas et al., 2007).

FIGURE 1 Edible coating by dipping method on guava compared to uncoated guava.

As mentioned previously, fruit and vegetable films show a potential as natural food packagings for dried foods (Figure 2) and other foods which are susceptible to oxidation at low to intermediate a_w (McHugh et al., 1996; Sothornvit and Rodsamran, 2008). They can be applied to fresh fruits, both whole and fresh-cut fruits, to reduce weight loss and retard fruit ripening, extending shelf life (Sothornvit and Rodsamran, 2008). To improve fruit and vegetable film properties for a broad food application, nanoreinforced acerola puree-alginate films and coatings were developed and applied on acerola. These coatings reduced weight loss, decay incidence and ripening rate of acerola (Azeredo et al., 2012).

FIGURE 2 Edible film from banana and its packaging application as banana film sachet used for dried food such as sugar, salt and milk powder.

Edible films and coatings functional properties can be improved by incorporation of natural extracts, nanoparticles, or modified biopolymers. For example, rice starch-based edible coating with coconut oil and tea leaf extract was found to improve the surface integrity of films and lowered the solubility and stability of the film solution. When rice starch-based solution was applied on tomatoes, the coating reduced weight loss, total soluble solid and extended the storage time (Das et al., 2013). Chitosan-based edible coatings with lime essential oil controlled the growth of pathogenic microorganisms (*Rhizopus stolonifer* and *Escherichia coli* DH5a) in fresh tomatoes (Ramos-Garcia et al., 2012). Whey protein isolate (WPI) with montmorillonite (MMT) as Cloisite 30B organo-clay composite film exhibited an opaque appearance (Sothornvit et al., 2009, 2010b). Film properties, such as surface color, optical, tensile, and water vapor barrier, varied depending on clay content. In addition, WPI / Cloisite 30B composite films showed a beneficially bacteriostatic effect against *Listeria monocytogenes* (Sothornvit et al., 2010b). Acerola puree and alginate reinforced with cellulose whiskers or MMT as nanocomposite edible film and coating reduced weight loss, decay and ripening of acerolas (Azeredo et al., 2012). Defatted mustard meal-xanthan coating retarded lipid oxidation, reduced volatile changes and improved the stability of smoked salmon during storage (Kim et al., 2012). Incorporation of antimicrobial agents (cinnamaldehyde and carvacrol) into apple and tomato edible films wrapped on baked chicken showed a protection against bacterial pathogens and spoilages with enhancing food sensory (Du et al., 2012).

In addition, edible coatings can improve the quality of fresh, frozen and processed meat, poultry and seafood products by lowering moisture loss, reducing lipid oxidation and surface discoloration, improving product appearance, functioning as carrier of food additives and reducing oil uptake during frying (Gennadios et al., 1997). Many studies have shown that edible coatings from protein, polysaccharide, and lipid-based materials help to prolong the shelf life and maintain the quality of fish (Artharn et al., 2009; Jeon et al., 2002; Sathivel, 2005; Stuchell and Krochta, 1995). Locust bean gum coatings having with different plasticizer type and amounts reduced moisture loss and extended shelf life of sausages (Dilek et al., 2011). Composite films of defatted mustard meal and xanthan retarded lipid oxidation and reduced volatile changes without imparting any negative sensory quality to smoked salmon (Kim et al., 2012). Mostly, chitosan-based edible coatings have been applied on several meat and seafood products to preserve their qualities and prolong their shelf life due to chitosan antimicrobial activity. For instance, chitosan coatings reduced lipid oxidation, chemical spoilage and growth of microorganisms of refrigerated fish fillets (Jeon et al., 2002). Moreover, chitosan film used for glazing frozen skinless pink salmon fillets had a higher thaw yield than the lactic acid-glazed and distilled water-glazed fillets (Sathivel et al., 2007). To enhance the efficiency of edible coatings and films, a few antimicrobial agents and antioxidant have been incorporated into edible coatings and films to reduce quality changes during storage of meat and fish products (Beverlya et al., 2008; Chidanandaiah et al., 2009; Haque et al., 2009; Kang et al., 2007; Song et al., 2011). For example, pectin-based edible coating with green tea powder extended the quality of irradiated pork patty (Kang et al., 2007) and chitosan coatings with acetic acid reduced microbial count of ready-to-eat roast beef (Beverlya et al., 2008). Sodium alginate-based edible coatings containing various antioxidants (vitamin C and tea polyphenol) inhibited the growth of total viable counts, reduced chemical spoilage, retarded water loss and increased the overall sensory quality of bream, extending the shelf life of the fish (Song et al., 2011). Furthermore, chitosan coatings and films containing various antimicrobial agents (sodium lactate, sodium diacetate, and potassium sorbate) controlled the growth of *Listeria moncytogenes* on cold smoked salmon (Jiang et al., 2011).

Gelatin films and coatings have been widely used to preserve the quality of meat and seafood products as well. For instance, gelatin-based films containing oregano and rosemary extracts increased the phenol content and antioxidant power of the muscle, which translated in lower lipid oxidation levels and reduced microbial growth of cold-smoked sardine (Gomez-Estaca et al., 2007). Moreover, gelatin coatings with probiotic bacteria *Lactobacillus acidophilus* and *Bifidobacterium bifidum* reduced microbial load of fish during chilled storage (Lopez de Lacey et al., 2012). Likewise, the use of gelatin film with 25% (w/w) lemongrass essential oil (LEO) to wrap sea bass slices retarded the growth of lactic acid bacteria, psychrophilic bacteria and spoilage microorganisms, including H_2S-producing bacteria and Enterobacteriaceae during 12 days of storage at 4°C. Moreover, this gelatin film containing LEO lowered the changes of color and the total volatile base nitrogen and thiobarbituric acid reactive substance values, thus it maintained the quality and extended the shelf life of sea bass slices (Ahmad et al., 2012).

Whey protein isolate (WPI) coatings with antimicrobial agents (grape seed extract, nisin, and malic acid) also decreased the growth of *L. monocytogenes* in a turkey frankfurter system (Gadang et al., 2008). Likewise, sour whey protein based edible coatings with or without carboxymethycellulose reduced oxidative degradation of cut beef steak (Haque et al., 2009).

The possibility to apply edible films and coatings in other food products has continuously being studied. For instance, there is an effort to replace ethanol-based coating, such as shellac, to finish chocolates, and other candies, due to the emission of volatile compounds by these coatings into the environment. In this sense, the use of water-based coatings, such as WPI coatings plasticized with sucrose or hydrolyzed collagen-cocoa butter plasticized with sucrose have been effective coatings, providing high gloss and reducing rancidity of nuts (Dangaran and Krochta, 2003; Fadini et al.,

2012). Similarly, a coating based on pea starch, WPI and carnauba wax helped preventing oxidative and hydrolytic rancidity of walnuts and pine nuts (Mehyar et al., 2012). Additionally, food hydrocolloids have shown a potential to reduce oil absorption in deep-fat-fried products (Albert and Mittal, 2002). For example, guar and xanthan gums at concentrations between 0.25% and 2.00% reduced oil absorption of fried chickpea products, reaching in the case of guar gum coatings an oil content reduction around 30–33% (Annapure et al., 1999). Various polysaccharides such as alginate, carboxylmethyl cellulose, and pectin reduced the oil absorption in deep-fat-fried banana chips between 2–17% depending on the coating composition (Singthong and Thongkaew, 2009). Moreover, the combination of an edible coating (guar or xanthan gum) and the highest centrifugation speed showed the greatest reduction of oil absorption (33.71%) in vacuum-fried banana chips, obtaining a healthier snack for consumers (Sothornvit, 2011). Other benefit of edible coatings is their effect maintaining nutritional quality. For example, native and modified maize and cassava starch helped maintaining carotenoid content in dehydrated pumpkin products (Lago-Vanzela et al., 2012).

Beside the wide use of edible films and coatings on food products, paper is probably the most important application of edible films and coatings. Paper is generally coated with non-degradable polymer coating materials such as polyethylene (PE), wax (Wenzel et al., 1997), polyethylene terephthalate (PET) and polybutylene terephthalate (PBT) (Overcash and Elsebaumer, 2001) to provide barrier properties. These materials are difficult to dispose and cause environmental problems; thus, introducing biopolymers as coating materials on paper and paperboard can improve its properties. For example, corn zein protein coated paper provided grease resistance and was found suitable for wrapping sandwiches or fast food (Trezza and Vergano, 1994). Coating paper with WPI also increased grease resistance and enhanced gloss without altering mechanical properties of the paper (Han and Krochta, 2001). Paperboard coated with undenatured and denatured WPI reduced oxygen permeability (Chan and Krochta, 2001). Furthermore, various polymer materials such as chitosan, WPI, whey protein concentrate and wheat gluten used to coat paper and paperboard enhanced the strength and toughness of the paper material (Gallstedt et al., 2005). Additionally, hydroxylpropyl methylcellulose (HPMC)-based coatings reduced the WVP of coated paper and a further reduction was obtained when beeswax was incorporated in the HPMC coating (Sothornvit, 2009). These results suggested that paper coated with a hydrocolloid and a lipid could be used as packaging material for agricultural produces, the food industry and it may have other applications as medical packaging (Sothornvit, 2009). A comprehensive review of biopolymer coatings on paper packaging materials is done by Khwaldia et al. (2010).

8.7 CONCLUSION

Edible films and coatings have the potential to serve as green and novel food packaging application. However, there is still a need to research on this area since there is no exact solution for each specific problem. The different characteristics of each food product require films and coatings with different barrier and mechanical properties. Edible films and coatings should be characterized as containing no toxic and non-allergic components, and to exhibit good structural stability during transportation, handling and selling. They should also provide a uniform coverage and good adhesion to the food surface, exerting adequate moisture and gas barrier for each product. Technologies such as compression molding, extrusion process and novel electrospinning process can be used in a commercial scale, facilitating the manufacture of edible film packaging. Several nanocomposite films have been developed by adding reinforcing compounds (nanofillers) to the biopolymers, improving their properties. Additionally, by-products and low-value products, such as overripe or surplus production of fruits and vegetables, from the food industry can be a promising source for edible film and coating materials with good properties. These fruit and vegetable films and coatings might be attractive to consumers due to its natural color and flavor, providing good sensory properties and safety. However, the composition of edible films and coatings must be declared on the label to consumers, especially common allergens (milk, egg, peanut, fish, shellfish, soy, wheat, tree nut, and so on). It is necessary to consider that the commercial use of films based on biopolymers has been limited depending on market price, consumer acceptance and modification of processing equipments. However, edible films and coatings are continuously studied due to their biodegradable nature, which makes them attractive to consumers and because of the increased environmental concern.

KEYWORDS

- **Edible films**
- **Lipids**
- **Polysaccharides**
- **Proteins**
- **Synthetic packaging materials**

REFERENCES

1. Ahmad, M., Benjakul, S., Sumpavapol, P., and Nirmal, N. P. *Int. Journal of Food Microbiol,* **155**, 171–178 (2012).
2. Albert, S. and Mittal, G. S. *Food Res. Int. Journal,* **35**, 445–458 (2002).
3. Alishahi, A. and Aider, M. *Food Bioprocess Tech.,* **5**, 817–830 (2012).
4. Andreuccetti, C., Carvalho, R. A., Galicia-Garcia, T., Martinez-Bustos, F., Gonzalez-Nunez, F., and Grosso, C. R. F. *Journal of Food Eng.,* **113**, 33–40 (2012).
5. Annapure, U. S., Singhal, R. S., and Kulkarni, P. R. *Fett/Lipid,* **101**, 217–221 (1999).
6. Artharn, A., Prodpran, T., and Benjakul, S. *Food Sci. Technol,* **42**, 1238–1244 (2009).
7. Atares, L., Perez-Masia, R., and Chiralt, A. *Journal of Food Eng,* **104**, 649–656 (2011).
8. Avena-Bustillos, R. J., Cisneros-Zevallos, L. A., Krochta, J. M., and Salveit, M. E., Jr. *Postharvest Biol. Technol,* **4**, 319–329(1994a).
9. Avena-Bustillos, R. J., Krochta, J. M., Salveit, M. E., Jr., Rojas-Villegas, R. J., and Sauceda-Perez, J. A. *Journal of Food Eng,* 21, 197–214 (1994b).
10. Aydinli, M. and Tutas, M. *LWT,* **33**, 63–67 (2000).
11. Azeredo, H. M. C., Mattoso, L. H. C., Wood, D., Williams, T. G., and Avena-Bustillos, R. *Journal of Food Sci,* **74**, N31–N35 (2009).
12. Azeredo, H. M. C., Miranda, K. W. E., Ribeiro, H. L., Rosa, M. F., and Nascimento, D. M. *Journal of Food Eng.,* **113**, 505–510 (2012).

13. Baker, R. A., Baldwin, E. A., and Nisperos-Carriedo, M. O. Edible coatings and films for processed foods. In *Edible Coatings and Films to Improve Food Quality*, J. M. Krochta, E. A. Baldwin, andM. O. Nisperos-Carriedo (Eds.), Technomic Publishing: Pennsylvania, USA, pp. 89–104 (1994).
14. Baldwin, E. A. and Baker, R. A. Use of proteins in edible coatings for whole and minimally processed fruits and vegetables. In *Protein-based Films and Coatings*, A. Gennadios (Ed.) CRC Press. FL, pp. 501–516 (2002).
15. Baldwin, E. A., Nísperos, M. O., and Baker, R. A. *Journal of Hortic. Science*, **30**, 35–38 (1995).
16. Beverlya, R. L., Janes, M. E., and Prinyawiwatkul, W.; No, H. K. Food Microbiol, **25**, 534–537 (2008).
17. Bourtoom, T. and Chinnan, M. S. *LWT,* **41**, 1633–1641 (2010).
18. Campos, C. A., Gerschenson, L. N., and Flores, S. K. *Food Bioprocess Tech,* **4**, 849–875 (2011).
19. Cao, Y. M. and Change, K. C. *Journal of Food Sci,* **67**, 1449–1454 (2002).
20. Cerqueira, M. A., Bourbon, A. I., Pinheiro, A. C., Martins, J. T., Souza, B. W. S., Teixeira, J. A., and Vicente, A. *Trends in Food Science Technology*, **22**, 662–671 (2011).
21. Cerqueira, M. A., Lima, A. M., Teixeira, J. A., Moreira, R. A., and Vicente, A. *Journal of Food Eng*, **94**, 372–378 (2009).
22. Cerqueira, M. A., Souza, B. W. S., Teixeira, J. A., and Vicente, A. *Food Hydrocolloids*, **27**, 175–184 (2012).
23. Chan, M. A. and Krochta, J. M. *Tappi*, **84**, 1–10 (2001).
24. Chidanandaiah, Keshri, R. C. and Sanyal, M. K. *Journal of Muscle Foods,* **20**, 275–292 (2009).
25. Chien, P. J., Sheu, F., and Yang, F. H. *Journalof Food Eng,* **78**, 225–229 (2007).
26. Cisneros-Zevallos, L. and Krochta, J. M. *Journal of Food Sci,* **68**, 176–181 (2003a).
27. Cisneros-Zevallos, L. and Krochta, J. M. *Journal of Food Sci,* **68**, 503–510 (2003b).
28. Coma, V. *Meat Sci,* **78**, 90–103 (2008).
29. Cunningham, P., Ogale, A. A., Dawson, P. L., and Acton, J. C. *Journal of Food Sci,* **65**, 668–671 (2000).
30. Cuq, B., Aymard, C., Cuq, J. L., and Guilbert, S. *Journal of Food Sci*, **60**, 1369–1374 (1995).
31. Cuq, B., Gontard, N., and Guilbert, S. *Cereal Chem,* **75**, 1–9 (1998).
32. Cuq, B., Gontard, N., and Guilbert, S. *Polymer,* **38**, 4071–4078 (1997).
33. Dangaran, K. L. and Krochta, J. M. *Manufacturing Confectioner*, **83**, 61–65 (2003).
34. Das, D. K., Dutta, H., and Mahanta, C. L. *LWT*, **50**, 272–278 (2013).
35. Debeaufort, F. and Voilley, A. *Journal of Food Sci. Technol,* **30**, 183–190 (1995).
36. Dias, A. B., Muller, C. M. O., Larotonda, F. D. S., and Laurindo, J. B. *Journal of Cereal Sci*, **51**, 213–219 (2010).
37. Dias, A. B., Muller, C. M. O., Larotonda, F. D. S., and Laurindo, J. B. *LWT,* **44**, 535–542 (2011).
38. Dilek, M., Polat, H., Kezer, F., and Korcan, E. *Journal of Food Process Pres,* **35**, 410–416 (2011).
39. Du, W. X., Olsen, C. W., Avena-Bustillos, R. J., McHugh, T. H., Levin, C. E., and Friedman, M., *Journal of Food Sci,* **73**, M378–M383 (2008).
40. Du, W. X., Avena-Bustillos, R. J., Woods, R., Breksa, A. P., McHugh, T. H., Friedman, M., Levin, C. E., and Mandrell, R. *Journal of Agric. Food Chem,* **60**, 7799–7804 (2012).
41. M. E. Embuscado andK. C. Huber (Eds.) *Edible Films and Coatings for Food Applications*, Springer Science,(2009).
42. Fadini, A. L., Rocha, F. S., Alvim, I. D., Sadahira, M. S., Queiroz, M. B., Alves, R. M. V., and Silva, L. B. *Food Hydrocolloids,* **30**, 625–631 (2012).
43. Falguera, V., Quintero, J. P., Jimenez, A., Munoz, J. A., and Ibarz, A. *Trend Food Sci. Technol,* **22**, 292–303 (2011).
44. Foulk, J. A. andBunn, J. M. *Ind. Crop Prod,* **14**, 11–22 (2001).
45. Gadang, V. P., Hettiarachchy, N. S., Johnson, M. G., and Owens, C. *Journal of Food Sci,* **73**, M389–M394 (2008).
46. Gallstedt, M., Brottman, A., and Hedenqvist, M. S. *Packaging Technol. Sci,* **18**, 161–170 (2005).
47. Garcia, M. M., Garcia, T. F., and Mate, J. I. Stewart Posthar. Rev, **6**, 1–10 (2010).

48. Gennadios, A., Hanna, M. A., andKurth, L. B. *LWT*, **30**, 337–350 (1997).
49. Gennadios, A., McHugh, T. H., Weller, C. L., and Krochta, J. M. Edible coatings and films based on proteins. In *Edible Coatings and Films to Improve Food Quality*, J. M. Krochta, E. A. Baldwin, andM. O. Nisperos- Carriedo (Eds.), Technomic, pp. 201–277 (1994).
50. Gennadios, A. and Weller, C. L. *Food Technol,* **44**, 63–69 (1990).
51. Guerrero, P., Stefani, P. M., Rusechaite, R. A., and Caba, K. de la. *Journal of Food Eng,* **105**, 65–72 (2011).
52. Gomez-Estaca, J., Montero, P., Gimenez, B., and Gomez-Guillen, M. C. *Food Chem,* **105**, 511–520 (2007).
53. Guilbert, S., Gontard, N., Morel, M. H., Chalier, P., Micard, Y., and Redl, A. Edibleand biodegradable food packaging. In *Protein-based Films and Coatings*, A. Gennadios (Ed.), CRC Press, New York, USA, pp. 69–122(2002).
54. Guilbert, S., Cuq, B., and Gontard, N. *Food Addit. Contam,* **14**, 741–751 (1997).
55. Guimaraes, J. L., Wypych, F., Saul, C. L. K., Ramos, L. P., and Satyamarayana, K. G. *Carbohyd. Polym,* **80**, 130–138 (2010).
56. Hambleton, A., Voilley, A., and Debeaufort, F. *Food Hydrocolloids,* **25**, 1128–1133 (2011).
57. Han, J. H. and Gennadios, A. Edible films and coatings: a review. In *Innovations in Food Packaging*; J. H. Han (Ed.), Elsevier Science: London, United Kingdom, pp. 239–262 (2005).
58. Han, J. H. and Krochta, J. M. *Journal of Food Sci,* **66**, 294–299 (2001).
59. Haque, Z. U., Shon, J., and Williams, J. B. *Journal Food Quality*, **32**, 381–397 (2009).
60. Hernandez-Izquierdo, V. M. and Krochta, J. M. *Journal of Food Sci,* **73**, R30–R39 (2008).
61. Hernandez-Izquierdo, V. M., Reid, D. S., McHugh, T. H., Berrios, J. D., and Krochta, J. M. *Journal Food Sci*, **73**, E169–E175 (2008).
62. Janjarasskul, T. and Krochta, J. M. *Annu. Rev. Food Sci. Technol,* **1**, 415–448 (2010).
63. Jarimopas, B., Sirisomboon, P., Sothornvit, R., and Terdwongworakul, A. The development of engineering technology to improve the quality of production of tropical fruit in developing countries. In *Focus on Food Engineering Research and Developments*, N. P. Vivian (Ed.) Nova Science Publishers, New York, USA, pp. 239–305 (2007).
64. Jeon, Y. J., Kamil, J. Y. V. A., and Shahidi, F. *Journal of Agric. Food Chem,* **50**, 5167–5178 (2002).
65. Jiang, Z., Neetoo, H., and Chen, H. *Journal of Food Science,* **76**, 22–26 (2011).
66. Jooyandeh, H. *Pakistan Journal Nut.*, **10**, 293–301 (2011).
67. Kang, H. J., Jo, C., and Kwon, J. H. *Food Control,* **18**, 430–435 (2007).
68. Katoh, K., Shibayama, M., Tanabe, T., and Yamauchi, K. *Biomaterials*, **25**, 2265–2272 (2004).
69. Kaya, S. and Maskan, A. *Journal of Food Eng,* **57**, 295–299 (2003).
70. Kester, J. J. and Fennema, O. R. *Food Technol,* **40**, 47–59 (1986).
71. Khwaldia, K., Arab-Tehrany, E., and Desobry, S. *Compr. Rev. Food Sci. F,* **9**, 82–91 (2010).
72. Kim, I. H., Yang, H. J., Noh, B. S., Chung, S. J., and Min, S. C. *Food Chem.,* **133**, 1501–1509 (2012).
73. Kim, K. W., Ko, C. J., and Park, H. J. *J. Food Sci,* **67**, 218–222 (2002).
74. Krishma, M., Nindo, C. I., andMin, S. C. *J. Food Eng,* **108**, 337–344 (2012).
75. Krochta, J. M. Film, edible. In *The Wiley Encyclopedia of Packaging Technology*. A. L. Brody, K. S. March (Eds.), John Wiley & Sons, New York, USA, pp. 397–401 (1997).
76. Krochta, J. M. and De Mulder-Johnston, C. *Food Technol,* **51**, 61–73 (1997).
77. Krochta, J. M. Protein as raw materials for films and coatings: Definitions, current status, and opportunities. In *Protein-based Films and Coatings*. A. Gennadios (Ed.), CRC Press: New York, USA, pp. 239–305 (2002).
78. Lago-Vanzela, E. S., do Nascimento, P., Fontes, E. A. F., Mauro, M. A., and Kimura, M. *LWT,* **50**, 420–425 (2013).
79. Lavorgna, M., Piscitelli, F., Mangiacapra, P., and Buonocore, G. G. *Carbohyd. Polym,* **82**, 291–298 (2010).
80. Lerdthanangkul, S. and Krochta, J. M. *Journal of Food Sci*, **61**, 176–179 (1996).
81. Li, Y., Shoemaker, C. F., Ma, J., Shen, X., andZhong, F. *Food Chem,* **109**, 616–623 (2008).

82. Lopez de Lacey, A. M., Lopez-Caballero, M. E., Gomez-Estaca, J., Gomez-Guillen, M. C., and Montero, P. *Innov. Food Sci. Emerg,* 277–282 (2012).
83. Ma, X. F., Yu, J. G., and Wan, J. *Journal of Carbohyd. Polym,* **64**, 267–273 (2006).
84. Mali, S., Grossmann, M. V. E., Garcia, M. A., Martino, M. N., and Zaritzky, N. E. *Journal of Food Eng,* **75**, 453–460 (2006).
85. Matsakidou, A., Biliaderis, C. G., and Kiasseoglou, V. *Food Hydrocolloid,* **30**, 232–240 (2013).
86. McHugh, T. H., Huxsoll, C. C., and Krochta, J. M. *Journal of Food Sci,* **61**, 88–91 (1996).
87. McHugh, T. H. *Journal of Food Sci,* **61**, 88–91 (2000).
88. McHugh, T. H. and Krochta, J. M. *J. Am. Oil Chem. Soc,* **71**, 307–311 (1994).
89. McHugh, T. H. and Olsen, C. W. *U. S.-Jpn. Cooper. Prodgram Nat. Res,* 104–108 (2004).
90. Mehyar, G. F., Al-Ismail, K., Han, J. H., and Chee, G. W. *Journal of Food Sci,* **77**, E52–E59 (2012).
91. Navarro-Tarazaga, M. L., Sothornvit, R., and Perez-Gago, M. B. *Journal of Agric. Food Chem,* **56**, 9502–9509 (2008).
92. Olivas, G. I. and Barbosa-Canovas, G. V. *Crit. Rev. Food Sci. Nut,* **45**, 657–670 (2005).
93. Olivas, G. I. and Barbosa-Canovas, G. V. *Crit. Rev. Food Sci. Nut,* **45**, 657670 (2007).
94. Overcash, D. T. and Elsebaumer, R. L. *Coated sheet method,* US Patent6, 193, 831 (2001).
95. Oses, J., Fabregat-Vazquez, M., Pedroza-Islas, R., Tomas, S. A., Cruz-Orea, A., and Mate, J. I. *Journal of Food Eng,* **92**, 56–62 (2009).
96. Pareta, R. and Edirisinghe, M. J. *Carbohyd. Polym,* **63**, 425–431 (2006).
97. Park, H. J. and Chinnan, M. S. *Properties of Edible Coatings for Fruits and Vegetables*. ASAE Paper No. 90–6510, 19 p. (1990).
98. Park, J. W., Whiteside, W. S., and Cho, S. Y. *Food Sci. Technol,* **41**, 692–700 (2008).
99. Pavlath, A. E. and Orts, W. Edible films and coatings: why, what and how? In *Edible Films and Coatings for Food Applications*, M. E. Embuscado and K. C. Huber (Eds.), Springer Science, pp. 1–23 (2009).
100. Perez-Gago, M. B., Gonzalez-Aguilar, G. A., and Olivas, G. I. *Stewart Posthar. Rev*, **6**, 1–14 (2010).
101. Perez-Gago, M. B. and Krochta, J. M. *Journal of Agric. Food Chem*, 49, 996–1002 (2001).
102. Perez-Gago, M. B. and Krochta, J. M. Emulsion and bi-layer edible films. In *Innovations in Food Packaging*; J. H. Han (Ed.); Elsevier Science: London, United Kingdom, pp. 384–402 (2005).
103. Perez-Gago, M. B., Nadaud, P., and Krochta, J. M. *Journal of Food Sci,* **64**, 1034–1037 (1999).
104. Perez-Gago, M. B., Rojas, C., anddel Rio, M. A. *Journal of Food Sci,* **68**, 879–883 (2003).
105. Perez-Gago, M. B., Scrra, M., and del Rio, M. A. *Postharvest Biol. Technol,* **39**, 84–92 (2006).
106. Perez-Gago, M. B., Serra, M., Alonso, M., Mateos, M., and del Rio, M. A. *Postharvest Biol. Technol,* **36**, 77–85 (2005).
107. Plotto, A., Goodner, K. L., and Baldwin, E. A. *Proc. Fla. State Hort. Soc,* **117**, 382–388 (2004).
108. Pommet, M., Redl, A., Guilbert, S., and Morel, M. H. *Journal of Cereal Sci,* **42**, 81–91 (2005).
109. Quezada Gallo, J. A., Diaz Amaro, M. R., Gutierrez Cabrera, D. M. B., Castaneda Alvarez, M. A., Debeaufort, F., and Voilley, A. *Acta Hort,* **599**, 589–594 (2003).
110. Rakotonirainy, A. M., Wang, Q., and Padua, G. W. *Journal of Food Sci,* **66**, 1108–1111 (2001).
111. Ramos, O. L., Fernándes, J. C., Silva, S. I., Pintado, M. E., and Malcasa, F. X. *Crit. Rev. Food Sci,* **52**, 533–552 (2012).
112. Ramos-Garcia, M., Bosquez-Molina, E., Hernandez-Romano, J., Zavala-Padilla, G., Terres-Rojas, E., Alia-Tejacal, L., Barrera-Necha, L., Hernandez-Lopez, M., and Bautista-Banos, S. *Crop Protection,* **38**, 1–6 (2012).
113. Redl, A., Morel, M. H., Bonicel, J., Vergnes, B, and Guilbert, S. *Cereal Chem*, **76**, 361–370 (1999).
114. Reneker, D. H., Yarin, A. L., Zussman, E., and Xu, H. *Adv. Appl. Mech,* **41**, 43–195 (2007).
115. Rhim, J. W. *Carbohyd. Polym,* **86**, 691–699 (2011).
116. Rhim, J. W., Lee, J. H., and Ng, P. K. W. *LWT,* **40**, 232–238 (2007).

117. Rojas-Grau, M. A., Soliva-Fortuny, R., and Martin-Belloso, O. *Trends Food Sci. Technol,* **20**, 438–447 (2009).
118. Rojas-Grau, M. A., Avena-Bustillos, R. J., Friedman, M., Henika, P. R., Martin-Belloso, O., and McHugh, T. H. *Journal of Agric. Food Chem,* **54**, 9262–9267 (2006).
119. Romero-Bastida, C. A., Bello-Perez, L. A., Garcia, M. A., Martino, M. N., Solorza-Feria, J., and Zaritzky, N. E. *Carbohyd. Polym,* **60**, 235–244 (2005).
120. Rossman, J. M. Commercial manufacture of edible films. In *Edible Films and Coatings for Food Applications*, M. E. Embuscado andK. C. Huber (Eds), Springer Science, pp. 367–390 (2009).
121. Ryu, S. Y., Rhim, J. W., Roh, H. J., and Kim, S. S. *LWT,* **35**, 680–686 (2002).
122. Salame, M. Barrier polymers. In *The Wiley Encyclopedia of Packaging Technology*, M. Bakker,(Ed.), John Wiley &Sons, New York, USA, pp. 48–54 (1986).
123. Sathivel, S. *Journal of Food Sci,* **70**, E455–E459 (2005).
124. Sathivel, S., Liu, Q., Huang, J., and Prinyawiwatkul, W. *Journal ofFood Eng,* **83**, 366–373 (2007).
125. Shaw, N. B., Monahan, F. J., O'Riordan, E. D., and O'Sullivan, M. *Journal of Food Sci,* **67**, 164–167 (2002).
126. Singthong, J. and Thongkaew, C. *LWT,* **42**, 1199–1203 (2009).
127. Song, Y., Liu, L., Shen, H., You, J., andLuo, Y. *Food Control,* **22**, 608–615 (2011).
128. Sothornvit, R. *Acta Hort,* **682**, 1731–1738 (2005).
129. Sothornvit, R. *Acta Hort,* **877**, 973–978 (2010).
130. Sothornvit, R. *Food Res. Int,* **42**, 307–311 (2009).
131. Sothornvit, R. *Journal of Food Eng,* **107**, 319–325 (2011).
132. Sothornvit, R. and Chollakup, R. *Int. Journal of Food Sci. Technol,* **44**, 1395–1400 (2009).
133. Sothornvit, R., Chollakup, R., and Suwanruji, P. *Songklanakarin Journal of Sci. Technol,* **32**, 17–22 (2010a).
134. Sothornvit, R., Hong, S. I., An, D. J., and Rhim, J. W. *LWT,* **43**, 279–284 (2010b).
135. Sothornvit, R., Krochta, J. M. *Journal of Agric. Food Chem,* **48**, 3913–3916 (2000a).
136. Sothornvit, R. and Krochta, J. M. *Journal of Agric. Food Chem,* **48**, 6298–6302 (2000b).
137. Sothornvit, R. and Krochta, J. M. *Journal of Food Sci,* **65**, 700–703 (2000c).
138. Sothornvit, R. and Krochta, J. M. *Journal of Food Eng,* **50**, 149–155 (2001).
139. Sothornvit, R. and Krochta, J. M. Plasticizers in edible films and coatings. In *Innovations in Food Packaging*, J. H. Han (Ed.) Elsevier Science: London, United Kingdom, pp. 403–433 (2005).
140. Sothornvit, R., Olsen, C. W., McHugh, T. H., and Krochta, J. M. *Journal of Food Sci,* **68**, 1985–1989 (2003).
141. Sothornvit, R., Olsen, C. W., McHugh, T. H., and Krochta, J. M. *Journal Food Eng.,* **78**, 855–860 (2007).
142. Sothornvit, R. and Pitak, N. *Food Res. Int,* **40**, 365–370 (2007).
143. Sothornvit, R., Rhim, J. W., and Hong, S. I. *Journal of Food Eng,* **91**, 468–473 (2009).
144. Sothornvit, R. and Rodsamran, P. *Acta Hort.***857**, 359–366 (2010a).
145. Sothornvit, R. and Rodsamran, P. *Int. Journal of Food Sci. Technol,* **45**, 1689–1695 (2010b).
146. Sothornvit, R. and Rodsamran, P. *Postharvest Biol. Technol,* **47**, 407–415 (2008).
147. Sothornvit, R. and Songtip, S. *Acta Hort,* **877**, 1295–1302 (2010).
148. Sothornvit, R. and Songtip, S. *Acta Hort,* **928**, 243–250 (2012).
149. Stuchell, Y. M. and Krochta, J. M. *Journal of Food Sci,* **60**, 28–31 (1995).
150. Tamer, C. E. and Copur, O. U. *Acta Hort,* **877**, 619–626 (2010).
151. Tanada-Palmu, P. S. and Grosso, C. R. F. *Postharvest Biol. Technol,* **36**, 199–208 (2005).
152. Tang, X., Alavi, S., and Herald, T. *Journal of Carbohyd. Polym,* **74**, 552–558 (2008).
153. Trezza, T. A. and Vergano, P. J. *Journal of Food Sci,* **59**, 912–915 (1994).
154. Tharanathan, R. N. *Trends Food Sci. Technol,* **14**, 71–78 (2003).
155. Valencia-Chamorro, S. A., Palou, L., Delrio, M. A., and Perez-Gago, M. B. *Crit. Rev. Food Sci,* **51**, 872–900 (2011).

156. van Soest, J. J. G. and Korleve, P. M. *Journal of Appl. Polym. Sci,* **74**, 2207–2219 (1999).
157. Varela, P. and Fiszman, S. M. *Food Hydrocolloids,* **25**, 1801–1812 (2011).
158. Vargas, M., Pastor, C, Chiralt, A., McClements, D. J., and Gonzalez-Martinez, C. *Crit. Rev. Food Sci. Nut,* **48**, 496–511 (2008).
159. Wang, Y. and Padua, G. W. *Macromol. Mater. Eng,* **288**, 886–893 (2003).
160. Wenzel, D. J., Bartholomew, W., Quick, R., Delozier, M. S., and Klass-Hoffman, M. *Recyclable and compostable coated paper stocks and related methods ofmanufacture*, US Patent 5,654,039 (1997).
161. Wu, Q. X. and Zhang, L. N. *Ind. Eng. Chem. Res,* **40**, 1879–1883 (2001).
162. Yang, L. and Paulson, A. T. *Food Res. Int,* **33**, 563–570 (2000).
163. Zhang, J., Mungara, P., and Jane, J. *Polymer,* **42**, 2569–2578 (2001).

CHAPTER 9

ENVIRONMENTAL FRIENDLY MICROBIAL POLYMERS, POLYHYDROXYALKANOATES (PHAs) FOR PACKAGING AND BIOMEDICAL APPLICATIONS

P. P. KANEKAR, S. O. KULKARNI, S. S. NILEGAONKAR, S. S. SARNAIK, P. R. KSHIRSAGAR, M. PONRAJ, and S. P. KANEKAR

ABSTRACT

Microorganisms are versatile in coping up with their environment. Under unbalanced nutritional conditions, they produce biodegradable polymer, Polyhydroxyalkanoate (PHA) which is accumulated intracellularly. This storage material is degraded to derive carbon and energy by the microorganisms that produce it. Microbial polymers, PHAs are polyesters of hydroxyacids. They are synthesized by three enzymes viz. 3-ketothiolase, Acetoacetyl-Co-A reductase and PHA synthase. Over 300 microbial species have been reported to produce PHA. *Alcaligenes eutrophus* (*Ralstonia eutrophus* or *Cupriavidus necator*) is the most studied organism and polyhydroxybutyrate (PHB) is the most extensively studied polymer. Extremophiles, the microorganisms growing at extreme conditions of pH, temperature, salinity, and so on are now recognized as the source of PHA. Moderately halophilic *Halomonas* sp. and halophilic archaea *Haloferax* sp. are reported to produce PHA. Copolymers of different hydroxyacids enhance the elasticity of the polymers and hence extremophiles are being explored for production of copolymers. A number of inexpensive substrates are tested for production of PHA to reduce cost of production. Microbial polymers are thermoplastic in nature, can be produced from renewable resources, have good oxygen barrier capacity and are biocompatible. Their properties are near to that of polypropylene and thus can be used as packaging materials. PHAs have many biomedical applications for example, in preparation of surgical sutures, swabs, for slow delivery of drugs, hormones, medicine, and so on. Microbial polymers being biodegradable in nature thus become environmental friendly polymers having various industrial, agricultural and biomedical applications.

9.1 INTRODUCTION

Plastic industry has been one of the fastest growing sectors of the world economy. This is because plastics are utilized in almost every manufacturing industry from automobiles to medicines. A majority of synthetic plastic polymers are extremely resistant to microbial attack due to their high molecular weight, high number of aromatic rings, unusual bonds or halogen substitutions (Alexander, 1981). For this reason, a large accumulation of plastic waste in the biosphere has given rise to the problem of severe environmental pollution. Over past three decades, a great deal of intensive efforts has been made upon the preparation of environmental friendly polymers, which can be easily degraded by microorganisms. As a result, many types of aliphatic polyesters including polyhydroxyalkanoates (PHAs), poly (ε-caprolactone) (PCL), and poly (L-lactides) (PLA) have been developed as biologically recyclable green polymers (Scott, 2000; Muller et al., 2001; Shimao, 2001). The biodegradable plastics, as novel materials make claims to be environmentally friendly. In order to find an alternative material, researchers have developed fully biodegradable plastics, such as polyhydroxyalkanoates (PHAs) (Verlinden et al., 2007). Until today, more than 300 microorganisms are known to synthesize and intracellularly accumulate PHA (Lee et al., 1999). The production of PHA (SCL) has been studied extensively using many microorganisms viz. *Cupriavidus necator, Bacillus cereus, Alcaligenes lactus, Azotobacter vinelandii, Methylobacterium* sp., recombinant *Escherichia coli* (Lee, 1996) and in few halophilic microorganisms including *Halomonas* sp. while the production of medium chain length PHAs have been mainly focused on *Pseudomonas* sp.

In recent years, intensive research has been carried out on bacterial production of PHA and a great effort is underway to improve this process (Khanna and Srivastava, 2005). However, the production cost of PHA is still far above the price of conventional plastic (Verlinden et al., 2007). In order to make the PHA production process economically attractive, many goals have to be addressed simultaneously. Recombinant bacterial strains are being developed to achieve high substrate conversion rate. A more efficient fermentation process (Grothe et al., 1999; Patwardhan and Srivastava, 2004) better recovery and purification processes (Jung et al., 2005) and the use of inexpensive substrates (Daniel et al., 2006) can also substantially reduce the production cost.

9.1.1 CHEMICAL STRUCTURE OF POLYHYDROXYALKANOATES

Polyhydroxyalkanoates (PHAs) are polyesters of hydroxyacids (HAs) with the general structural formula as depicted in Figure 1 Where X varies from 600 to 35,000 n = 1 R = hydrogen: Poly (3-hydroxypropionate) R = methyl: Poly (3-hydroxybutyrate) R = ethyl: Poly (3-hydroxyvalerate) R = propyl: Poly (3-hydroxyhexanoate) R = pentyl: Poly (3-hydroxyoctanoate) R = nonyl: Poly (3-hydroxydodecanoate) n = 2 R = hydrogen: Poly (4-hydroxybutyrate) n = 3 R = hydrogen: Poly (5-hydroxyvalerate) (Ojumu et al., 2004).

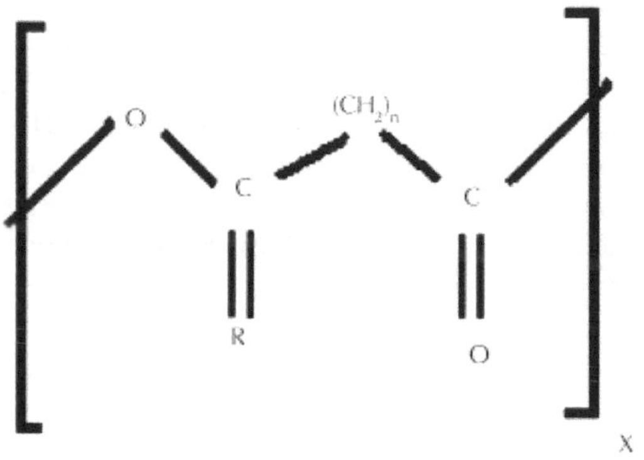

FIGURE 1 The general structure of polyhydroxyalkanoates.

Many different PHAs that have been identified to date are primarily linear, head to tail polyesters composed of 3-hydroxyfatty acid monomers. In the polymer, the carboxyl group of one onomer forms an ester bond with the hydroxyl group of neighboring monomer (Figure 2). In all PHAs that have been characterized so far, the hydroxyl substituted carbon atom is in the R configuration. At the same C3 or β position, the alkyl group which can vary from methyl to decyl is positioned (Figure 2). This alkyl side is not necessarily saturated, aromatic, unsaturated, halogenated, epoxidized, and branched monomers have been reported as well.

The variation in the alkyl substituents, the position of hydroxyl group is variable and 4-, 5-and 6- hydroxyacids have been incorporated (Figure 2). Substituents in the side chain of PHAs can be modified chemically, for instance by cross linking of unsaturated bonds. This variation in length and composition of side chains is the basis for diversity of PHA family and their vast array of potential applications (Madison and Huisman, 1999).

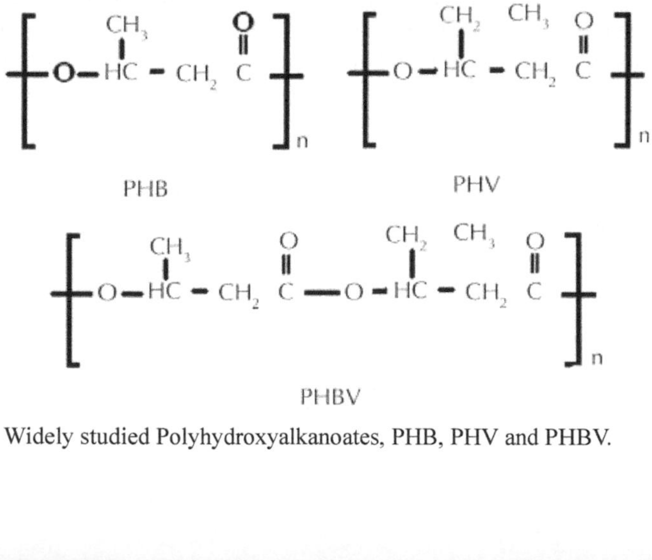

FIGURE 2 Widely studied Polyhydroxyalkanoates, PHB, PHV and PHBV.

9.1.2 DIVERSITY OF PHA

The PHAs, PHB is the most extensively characterized polymer, mainly because it was the first to be discovered. The diversity of bacterial PHAs has changed dramatically. Until 1970's, 3HB was considered as the only constituent of PHAs. In 1980's, PHAs having other monomers besides 3HB were shown to be accumulated by many bacteria with addition of certain precursors in the production medium. Figure 2 shows widely studied polyhydroxyalkanoates, which are PHB, PHV, and copolymer of PHB-co-PHV. Today more than 150 different monomers of PHAs are synthesized by different microorganisms which include the following:

1. Saturated 3 hydroxyacids (3HA) monomers, for example, 3-hydroxypropionic acid,
2. Unsaturated 3HA monomers with one or two double bonds. For example, 3-hydroxybutenoic acid,
3. 3HA with methyl group at various positions, for example, 3-hydroxy-2-methyl butyric acid,
4. Non 3HA monomers such as 4 hydroxybutyrate (4HB), 4 hydroxyvalerate (4HV), 4 hydroxyhexanoate (4HHx), 4 hydroxyoctanoate (4HO), 4hydroxydecanoate (4HD), 5 hydroxyvalerate (5HV), 5 hydroxyhexanoate (5HHx), and 5 hydroxydodecanoate (5HDD) 5. 3HA monomers with various functional groups that include carboxyl, benzyl, acetoxyl, phenoxyl, cyano and nitrophenoxyl, phenyl, cyclohexyl, epoxyl or halogen groups. PHAs can be divided into two broad categories based on the number of carbon atoms in the monomer units, viz,
5. Short Chain Length (SCL) polyhydroxyalkanoates, which consist of C_3 to C_5 atoms,
6. Medium Chain Length (MCL) polyhydroxyalkanoates consisting of C_6 to C_{14} atoms.

This grouping is due to the substrate specificity of the PHA synthase enzyme that only accepts 3-hydroxyalkanoates of a certain range of carbon length. The PHA synthases of *Cupriavidus necator* (formerly known as *Alcaligenes eutrophus*) can polymerize 3 hydroxyacids of short chain length PHAs consisting of 3–5 carbon atoms whereas PHA synthases of *Pseudomonas oleovorans* can accept 3HAs consisting of 6–14 carbon atoms. A lot of medium chain length PHAs containing various functional groups such as olefins, branched alkyls, halogens, aromatic, and cyano has been reported (Huijbert et al., 1992; Kim et al., 1992). The short chain length PHAs include poly (3-hydroxypropionate), poly (3-hydroxybutyrate), poly (4-hydroxybutyrate), poly (3-hydroxyvalerate), poly (4- hydroxyvalerate), poly (5-hydroxyvalerate), and so on. While medium chain length PHAs include 3- hydroxyoctanoate, 3-hydroxydecanoate as major monomers (Table 1).

TABLE 1 Examples of various types of hydroxyl alkanoates monomers detected in PHA isolated from microorganism

Sr. No.	Hydroalkanoate type	Hydroxyalkanoate monomer	Chemical formula
1	Saturated hydroxy alkanoates	i) 3-hydroxy propionic acid (3HP) ii) 4-hydroxy butyric acid (4HB) iii) 5-hydroxy valeric acid (5HV)	i) $COOHCH_2CH_2OH$ ii) $COOH(CH_2)_3OH$ iii) $COOH(CH_2)_4OH$
2	Unsaturated hydroxy alkanoates	i) 3-hydroxy-2-butenoic acid ii) 3-hydroxy-4-tranhexenoic acid iii) 3- hydroxyl-5-hexenoic acid	i) $COOHCH(OH)CH_3$ ii) $COOHCH_2CH(OH)CHCHCH_3$ iii) $COOHCH_2CH(OH)CH_2CHCH_2$
3	Methyl hydroxy alkanoates	i) 3-hydroxy-2 methylbutyric acid ii) 3-hydroxy-4 methylhexanoic acid	i) $COOHCH_2CH(OH)CH(CH)_3CH_3$ ii) $COOHCH_2CH(OH)CH(CH_3)$ CH_2CH_3
4	Dicarboxylic and ether hydroxy alkanoates	i) 3- hydroxyl succinic acid methyl ester ii) 3 hydroxy-9-acetoxy manioc acid	i) $COOHCH_2CH(OH)COOCH_3$ ii) $COOHCH_2(OH)(CH_2)_6OCOCH_3$
5	Aromatic hydroxy alkanoates	i)3-phenoxy-3 hydroxy-butyric acid ii)3-hydroxyl-cyclohexyl butyric acid	i) $COOHCH_2CH(OH)CHOC_6H_5$ ii) $COOHCH_2CH(OH)CH_2C_5H_{12}$
6	Dihydroxy hydroxy alkanoates	i) 3-12-dihydroxy dodecanoic acid	i) $COOHCH_2CH(OH)(CH_2)_9OH$
7	Epoxy hydroxy alkanoate	i) 3-hydroxy-4-5-epoxy decanoic acid	i) $COOHCH_2(OH)CH(O)$ $CH(CH_2)_4CH_3$
8	Cyano hydroxy alkanoates	i) 7-cyano-3-hydroxy-heptanoic acids	i) $COOHCH_2CH(OH)(CH_2)_3CN$
9	Halogenated hydroxy alkanoates	i) 3-hydroxyl-7-fluroheptanoic acid ii) 3-hydroxyl-11-bromodecanoic acid	i) $COOHCH_2(OH)(CH_2)_4F$ ii) $COOHCH(OH)(CH_2)_8Br$

9.1.3 PHA AS A STORAGE MATERIAL

The PHAs are structurally simple macromolecules synthesized by certain Gram positive and negative bacteria. They are accumulated as discrete granules to the levels as high as 90% of the cell dry weight and are generally believed to play a role as a sink of carbon and reducing equivalents. When nutrient supplies are imbalanced, it is advantageous for a microorganism to store excess nutrients intracellularly so that their fitness is not affected. PHAs exist as discrete inclusions that are typically 0.2 ± 0.5μm in diameter and possess a membrane coat of about 2 nm thick, composed of lipid and protein. The activities of PHA synthase and depolymerase enzymes are associated with this membrane protein (Madison and Huisman, 1999). These PHA granules are localized in the cell cytoplasm and can be visualized under a phase contrast light microscope due to their high refractivity (Dawes and Senior, 1973).

When the thin sections of PHA containing bacteria are observed under transmission electron microscope, the PHA inclusions appear as electron dense bodies (Joshi et al. 2007) [Figure 3]. The PHA is specifically stained by the oxazine dye Nile Blue A, exhibiting a strong orange fluorescence at an excitation wavelength of 460 nm (Ostle and Holt, 1982) [Figure 4]. Native PHA granules were isolated from cell extracts of *Bacillus megaterium* (Merrick and Doudoroff, 1964), *Azotobacter beijerinckii* (Ritchie and Dawes, 1969) and *Zooglea ramigera* (Barnard and Sanders, 1988, 1989).

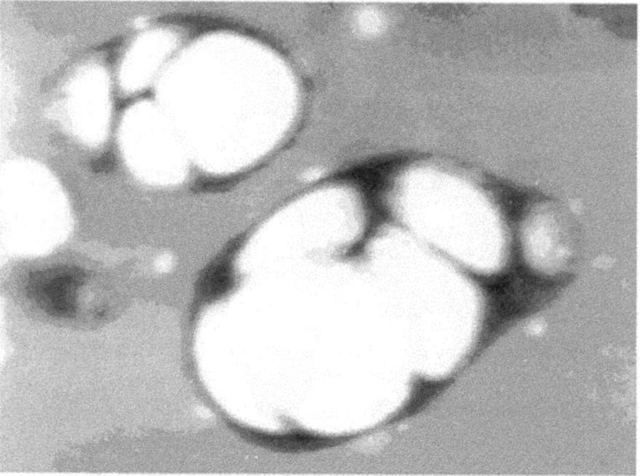

FIGURE 3 Transmission electron micrograph of *Halomonas campisalis* showing presence of PHA granules, grown in the PHA production medium of pH 9 for 24 h (Joshi et al. 2007).

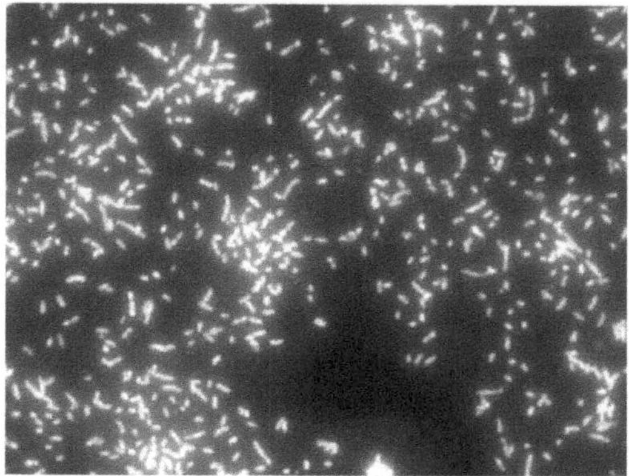

FIGURE 4 Nile blue sulphate staining of *Halomonas* cells showing bright orange fluorescence of PHA granules (Kulkarni 2010).

9.1.4 BIOSYNTHESIS OF PHA

The PHB biosynthetic pathway consists of three enzymatic reactions catalyzed by three distinct enzymes. The first reaction consists of a condensation of two acetyl coenzyme A (acetyl CoA) molecules into acetoacetyl CoA by β-ketoacetyl-CoA thiolase (encoded by *phbA*). The second reaction is the reduction of acetoacetyl- CoA to (R)-3-hydroxybutyryl CoA by an NADPH dependant acetoacetyl-CoA dehydrogenase (encoded by *phbB*). Lastly, the (R)-3-hydroxybutyryl-CoA monomers are polymerized into poly (3-hydroxybutyrate) by PHB polymerase (encoded by *phbC*) [Figure 5].

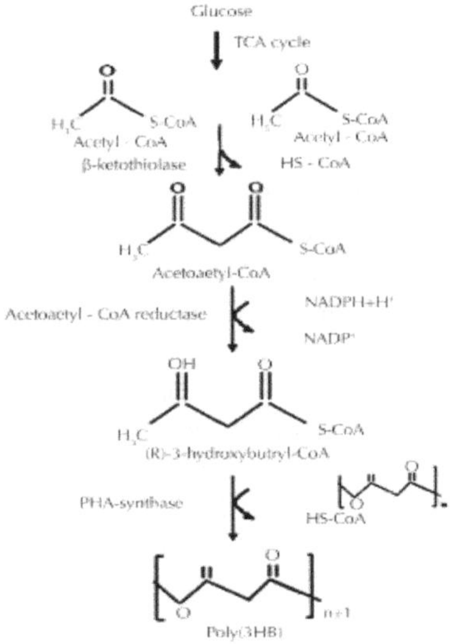

FIGURE 5 Biosynthetic pathway of PHB.

Although, PHB accumulation is a widely distributed prokaryotic phenotype, the biochemical investigations into the enzymatic mechanisms of β-ketoacetyl-CoA thiolase, acetoacetyl-CoA reductase and PHB polymerase have focused only on two of the natural producers, *Zooglea ramigera* and *Cupriavidus necator*.

Four different pathways have been elucidated so far for the biosynthesis of PHA (Doi and Abe, 1990; Steinbuchel, 1991; Byrom, 1994). In the bacterium *C. necator*, PHB synthesis is thoroughly studied. *C. necator* can accumulate over 90% PHA on dry cell weight basis when the media contains excess carbon source such as glucose, but limited in one essential nutrient such as nitrogen or phosphorus. Varieties of PHAs with different C_3–C_5 monomers have been produced by *C. necator*, but the nature and proportions of these monomers are influenced by the type and relative quantity of the carbon source supplied to the growth medium (Doi et al., 1988; Steinbuchel 1991; Steinbuchel et al., 1993). For example, it has been reported that addition of propionic acid or valeric acid in the growth media containing glucose leads to the production of copolymer composed of 3-hydroxybutyric acid and 3-hydroxyvaleric acid. If valeric acid is used as a sole carbon source, the proportion of 3HV monomer in the PHA produced by *C. necator* has reached up to 90% (Ojumu et al., 2004).

In *Rhodopseudomonas rubrum*, the biosynthetic pathway of PHA differs after second step, where acetoacetyl–CoA formed by β-ketothiolase is reduced by a NADH dependant reductase to L-(+)-3-hydroxybutyryl-CoA which is then converted to D-(-) -3-hydroxybutyryl –CoA by two enoyl-CoA hydratases. A third type of PHA biosynthetic pathway is found in most *Pseudomonas* sp. belonging to the rRNA homology group I. *Pseudomonas oleovorans* and other *Pseudomonas* species accumulate PHA consisting of 3-hydroxyalkanoic acid of medium chain length if the cells are cultivated on alkanes, alcohols or alkanoic acids (Brandl et al., 1988; Lagaveen et al., 1988). The forth type of biosynthetic pathway is present in almost all *Pseudomonas* species belonging to rRNA homology group II. This pathway involves the synthesis of copolyesters consisting of medium chain length 3-hydroxyacids from acetyl –CoA. This pathway has not been studied in detail. Some PHA producers use alternative pathways for PHA production. In the absence of thiolase and reductase enzyme, *Aeromonas cavie* employs an enoyl-CoA hydratase for the formation of the (R)-3-hydroxymonomer from either crotonyl CoA or hexenoyl-CoA. It produces random copolymer of 3-hydroxybutyrate and 3-hydroxyhexanoate (3HH) when growing on olive oil as a sole carbon source.

Few microorganisms like *Rhodococcus rubber* and *Nocardia corrallina* produce a copolymer, PHB-co-PHV even in the absence of typical HV precursors such as propionate or valerate in the feed. These organisms produce PHA from simple sugars using a methylmalonyl CoA pathway in which 3HV monomer is derived from acetyl-CoA and propionyl-CoA where propionyl CoA is the product of methylmalonyl CoA pathway. In this pathway, succinyl–CoA is converted to methylmalonyl-CoA, which is decarboxylated to propionyl-CoA.

9.1.5 PRODUCTION OF PHA USING DIFFERENT MICROORGANISMS

Among the candidates for biodegradable plastics, PHAs have been drawing much attention because of their similar material properties to conventional plastics and complete biodegradability. For past two decades, the whole research was focused on the preparation of biodegradable polymer. There exist a number of reports on production of PHAs (Anderson and Dawes, 1990l; Mergaert et al., 1992; Lee et al., 1996; Lee, 1999; Madison and Huisman, 1999; Ojumu et al., 2004; Khanna and Srivastava, 2005; Verlinden et al., 2007) (Table 2). The fact that PHA can be produced from renewable resources, it does not lead to the depletion of finite resources and its good processibility make PHA suitable for applications in several areas as a partial substitute for non biodegradable synthetic polymers. Over 250 different bacteria including Gram–negative and Gram-positive species phylogenetically representing both eubacteria and archaebacteria have been reported to accumulate various PHAs (Steinbuchel, 1991). Apart from PHA accumulation by microorganisms, transgenic plants can also accumulate these polymers in considerable amounts.

TABLE 2 Production of PHA by different microorganisms

Microorganism	Substrate	PHA (% w/w)	Reference
Azotobacter beijernckii	Glucose	35–75	Senior, 1972
Azotobacter vinelandii UWD	Valeric acid	29–94	Page et al., 1992, 1995
Rhodococcus sp. NCIMB 40126	Acetic acid	29	Haywood et al., 1991
	Lactose	25	
	Glucose	21	
Methylobacterium extorquens	Methanol	60–70	Bourque et al., 1995
Methylobacterium sp. ZP 24	Lactose	40–60	Yellore and Desai, 1998
Rhodococcus sp. ATCC 19070	Acetic acid	4	Haywood et al.1991
	Lactose	2	
	Glucose	14	
Rhodococcus rubber NCIMB 40126	Glucose	16.2	Anderson and Dawes, 1990
	Valeric acid	26.2	
	Glucose + Valeric acid	27.7	
Corneybacterium hydrocarboxydans ATCC 21767	Acetic acid	21	Haywood et al., 1991
	Lactose	2	
	Glucose	8	
Alcaligenes latus	Sucrose	50	Yamane et al., 1996
R. eutropha strain R3	Fructose	47	Steinbuchel and Pieper, 1992
	Galactose	35.5	
	Aactic acid	29.5	
	Sucrose	21.5	
	Lactose	43.2	
	Fructose	45	
	Fructose	47	
	Galactose	33	
	Galactose	38	
R.eutropha	Digested sludege supernatent	34	Lee and Yu, 1997
R. eutropha H16	Acetic acid	53	Doi et al., 1986
	Acetic acid	51	
	Propionic acid	35	

TABLE 2 *(Continued)*

R. eutropha H16 (ATCC 17699)	Butyrate	44–75	Shimizu et al., 1994
	Valeric acid	38	
Bacills cereus SPV	Glucose	33	Valappil et al., 2007a, 2008

Exploring many more microorganisms for production of PHA continues. Pal et al. (2009) reported production of PHB by *Bacillus thuringensis.* Patel et al. (2002) have described production of PHA by mixed culture nitrogen fixing bacteria. Reddy et al. (2009) have reported a novel *Bacillus* sp. accumulating PHB from a single carbon source. Santhanam and Sasidharan (2010) studied production of PHA by *Alkaligenes* sp. and *Pseudomonas oleovorans* using different carbon sources. Guo et al. (2011) have reported simultaneous production and characterization of PHA and alginate oligosaccharides by *Pseudomonas mendocina* NK-01.

9.1.6 OCCURRENCE OF PHA IN PHOTOTROPHIC BACTERIA

Oxygenic phototrophic bacteria (previously referred to as *Cyanobacteria*) like *Chlorogloea fritschii, Nostoc* sp., *Microcystis aeruginosa, Oscillatoria* sp., *Spirullina platensis, Schizothrix* sp., and so on are capable of accumulating PHB granules. The PHB content in this group of organisms is usually less than 1% and is not more than 2.5–9% of their cellular dry weight. Another group of photosynthetic microorganisms that can accumulate PHAs are the anoxygenic phototrophic bacteria which include purple sulphur bacteria and purple non sulphur bacteria. Purple non sulfur bacteria are certainly better producers of PHA compared to purple sulfur and green sulphur bacteria.

Purple non sulphur bacteria include *Rhodobacter capsulatus, Rhodobacter spheroides, Rhodococcus flavum, Rhodococcus rubrum, Rhodococcus centenum*, and so on which generally produce copolymers of PHB-co-PHV using acetate or butyrate as carbon source. Considerable variation occurs in PHA accumulation ranging from 30 to 70% on their dry cell weight (Sasikala and Ramana, 1996). Recently Fradinoho et al. (2013) have described production of PHA by mixed photosynthetic consortium of bacteria and algae.

9.1.7 OCCURRENCE OF PHA IN CHEMOTROPHIC BACTERIA

A number of chemotrophs viz. *Cupriavidus necator*, *Alcaligenes lactus, Azotobacter* sp., few species of *Bacillus* and few species of *Pseudomonas, Halomonas,* and *Methylotrophs,* and so on. Accumulate considerable amount of PHAs intracellularly. Among more than 300 different microorganisms that are known to synthesize PHAs, only a few bacteria are employed for large scale production of PHAs namely *C. necator, A. lactus, A. vinelandii, Methylotrophs, P. oleovorans,* and recombinant *E. coli*. Table 3 describes the polyhydroxyalkanoate content of various microorganisms suitable for biotechnological production of PHA.

TABLE 3 Polyhydroxyalkanoate (PHA) content of various microorganisms

Organism	Carbon source used	PHA composition	Cell density (g/L)	PHA content (% PHA on dry cell weight basis)	References
Cupriavidus necator	Glucose	PHB	164	74	Kim, 1994
Azotobacter vinelandii	Glucose	PHB	40	79.8	Page and Cornish, 1993
Alcaligenes lactus	Sucrose	PHB	112	88	Wang and Lee, 1997
Pseudomonas strain K	Methanol	PHB	233	64	Suzuki et al., 1986
Haloferax mediterranei	Starch	PHB	9.7	67	Lillo and Rodriguez-Valera, 1990
Halomonas boliviensis	Sucrose	PHB	14	54	Quillaguaman et al., 2007
Recombinant E.coli	LB + Glucose	PHB	117	76	Kim et al., 1992
Halomonas campisalis	Maltose	PHB-co-PHV	1–2.2	45–81%	Kulkarni et al. 2010

9.1.8 PRODUCTION OF COPOLYMERS OF PHA

Biosynthesis and characterization of various copolymers including copolymers of HB with 3-hydroxyvalerate (HV), 3-hydroxypropionate (HP), 3-hydroxyhexanoate (3HH) (Doi et al., 1995), and 4- hydroxybutyrate (4HB) (Hiramitsu et al., 1993), have been reported. Copolymer (PHB-co-PHV) formation with addition of certain precursors like propionic acid, valeric acid, and so on was studied by many researchers and in many organisms including *Cupriavidus necator*, *Pseudomonas* sp., and so on (Madison and Huisman, 1999). The phenomenon of copolymer formation without addition of precursors has been observed in a few organisms viz. *Rhodobacter rubber, Nocardia corallina, C. necator*- SH-69, *Agrobacterium* sp. strain SH1 and GW-014 (Madison and Huisman, 1999). Thus, it was thought worthwhile to search for microorganisms capable of producing copolymer from a simple carbon source without addition of any precursors. Many researchers have reported production of copolymer. Dai et al. (2007) studied production of targeted PHA copolymers by glycogen accumulating organisms using acetate as a carbon source. Amirul et al. (2008) have reported biosynthesis of copolymer poly (3HB-co-4HB) by *Cupriavidus* sp. USMAA1020 isolated from Lake Kulin, Malaysia. Biosynthesis and characterization of copolymer poly (3HB-co-3HV) by *Pseudomonas oleovorans* has been described by Allen et al. (2010).

PRODUCTION OF PHA USING DIFFERENT PRECURSORS

The physical and mechanical properties of PHB can be improved by the production of copolymers like PHB-co-PHV. The properties of these copolymers can change considerably as a function of the monomer composition and distribution. Thus, it was thought desirable to incorporate different types of precursors like propionic acid, acetic acid, valeric acid, 4- hydroxybutyric acid, and so on. Into the production medium in order to produce the copolymer with specific requirements for practical application. PHA copolymers can improve the crystallinity, melting temperature, stiffness, and toughness of polymers. Random copolymers containing 3HB as a constituent along with other hydroxyacids of chain length ranging from 3–14 carbon atoms have been produced by variety of microorganisms viz. *C. necator*, *A. lactus*, *Pseudomonas* sp., *A. cavie*, *Commamonas acidovorans*, and so on as reviewed by Khanna and Srivastava (2005). Copolymer (PHB-co-PHV) formation with addition of certain precursors like propionic acid and valeric acid was studied by many researchers and in many organisms including *Cupriavidus necator*, *Pseudomonas* sp., and so on. These precursors are structurally related to the constituents that are to be incorporated into PHAs and thus fed as a sole source of carbon or as a co- substrate. Synthesis of PHB-co-PHV occurs in mostly PHB accumulating microorganisms only when substrates like propionic acid or valeric acid are fed alone or along with the glucose (Steinbuchel and Fuchtenbusch, 1998). There are a very few studies reported for the use of precursors for copolymer production using halophilic microorganisms. Quillaguaman et al. (2008) reported the use of butyric acid and sodium acetate as carbon sources for production of PHA. Yang et al. (2010) optimized growth media components for production of PHA by *R. eutropha* using organic acids.

9.1.9 PRODUCTION OF PHA BY EXTREMOPHILIC BACTERIA

The most concentrated and wide spread occurrence of microorganisms is generally observed in "moderate" environments. It has also been known that there are "extreme" environments on earth which are thought to prevent the existence of life (Horikoshi, 1991). In these habitats, environmental conditions such as pH, temperature and salinity are extremely high or low. Extreme environments are populated by a group of microorganisms that are specifically adapted to these particular conditions and these types of extreme microorganisms are usually referred to as alkaliphiles, halophiles, thermophiles, and acidophiles reflecting a particular type of extreme environment which they inhabit.

ALKALIPHILES

Microorganisms occurring in alkaline environment can be classified into two main groups viz. alkaliphiles and alkalitolerants. The term alkaliphiles is generally restricted to those microorganisms that actually require alkaline media for growth. The optimum growth rate of these microorganisms is observed in at least two pH units above neutrality. They grow optimally at pH 9.0 and above but do not grow or grow slowly at near neutral pH (Horikoshi, 1999). Alkalitolerant microorganisms are capable of growing at pH values more than 9 or 10, but grow optimally at pH near neutrality.

Soda lakes or soda deserts represent the most stable, naturally occurring alkaline environments found worldwide. The best studied soda lakes are those of the East African Rift valley, where the detailed limnological and microbiological investigations have been carried out over many years (Grant et al., 1990; Jones et al., 1994). Microbiological studies of central Asian soda lakes have also been well documented (Zhilina and Zavarzin, 1994). Recently Joshi et al. (2008) have studied alkaliphilic bacteria from alkaline soda lake of Lonar, India. Kulkarni et al. (2010) have described production of PHA by moderately haloalkaliphilic bacteria from Lonar lake.

HALOPHILES

Halophiles are the salt loving organisms that inhabit hypersaline environments. Among the halophiles are a variety of heterotrophic, photosynthetic and lithotrophic, bacteria, and archaea. Examples of well adapted and well distributed extremely halophilic microorganisms include archaeal *Halobacterium* sp., cyanobacteria and blue green algae *Dunaliella salina.* They are grouped as slight halophiles which grow optimally at 0.2–0.8 mol/L (2–5%) NaCl, moderate halophiles which grow optimally at 0.85–3.4 mol/L (5–20%) NaCl and extremely halophiles that grow optimally above 3.4–5.1 mol/L (20–30%) NaCl. Halotolerant organisms can grow in high salinity as well as in the absence of a high concentration of salt. Many halophiles and halotolerant microorganisms can grow over a wide range of salt concentrations with requirement or tolerance for salts, sometimes depending upon the environmental and nutritional factors (Das Sarma and Arora, 2001). Moderately halophilic bacteria inhabit a wide range of habitats such as saline lakes, saltern ponds, deserts, hypersaline soils, salted foods, and so on. Moderate haloalkaliphilic microorganisms can grow at NaCl concentration ranging from 0.2–4.5 M with optimum growth occurring at 1.5 M and pH values ranging from 6 to 12 with 9.5 being optimum pH.

During the last decade, extensive studies of hypersaline environments in different geographical locations have led to the isolation and characterization of large number of moderately halophilic species (Ventosa et al., 1998, Oren, 2002). Many Gram negative, halotolerant or halophilic species are currently included in the family *Halomonadaceae*, which belong to the γ- subclass of the Proteobacteria. Among the genera *Halomonas*, which covers the major number of species (>20) with heterogeneous features has recently been distinguished into two phylogenetic groups based on the 16S and 23S rDNA sequences of the different species (Arahal et al., 2002). Several *Halomonas* species viz. *Halomonas elongata*, isolated from a solar saltern, *H. halodinitrificans*, isolated from meat curing brines, *H. eurihalina*, *H. salina,* and *H. halophila*, isolated from saline soil, *H. halodurans*, isolated from estuarine water, *H. subglaciescola* from beneath the ice of the organic lake in Antarctica, *H. boliviensis* isolated from soil around a Bolivian hypersaline lake (Quillaguaman et al., 2004) and *H. campisalis* from saline and alkaline lakes located in Grant Country, Washington state have been isolated and many of them have been reported earlier for their biotechnological potential like exopolysaccharide production, nitrate degradation, degradation of aromatic compounds, and so on. (Das Sarma and Arora, 2001).

PRODUCTION OF PHA BY HALOPHILES

A few species of *Halomonas*, viz. *H. maura*, isolated from hypersaline soil and *H. ventosa* isolated from saline soils in Jaen (South Eastern Spain) have been reported to accumulate PHB granules intracellularly (Bouchotroch et al., 2001; Martinez-Canovas et al., 2004). Quillaguaman et al., (2005) for the first time reported production of PHB by moderately halophile *Halomonas boliviensis* LC1 isolated from the shores of a hypersaline lake located in the Andean region of Bolivia. Another species, *Halomonas profoundus*, sp. nov, isolated from a deep sea hydrothermal vent shrimp was reported to produce PHB and PHB-co-PHV using carbon source like glucose and glucose plus valeric acid respectively (Simon-Colin et al., 2008). Mothes et al. (2008) reported a moderately halophilic *H. elongata*, which was found to synthesize PHA and ectoine simultaneously. Recently, Biswas et al. (2009) studied the production of PHB in another moderately halophile *H. marina* HMA103, isolated from a solar saltern in Orissa, India. Quillaguaman et al. (2010) have reviewed synthesis and production of PHA by halophiles. Salgaonkar et al. (2013) studied characterization accumulated by moderately halophilic salt pan isolate *Bacillus megaterium.*

HALOARCHAEA

Haloarchaea, the halophilic aerobic archaea inhabit the hypersaline environments distributed throughout the world. Though the oceans are the largest saline body of water, hypersaline environments are generally defined as those containing salt concentrations in excess of seawater (more than 3.5 g% NaCl) (Das Sarma et al. 2001; Oren et al. 1994; Tindall et al. 1991). These organisms have been routinely isolated from marine salterns and hypersaline lakes with 3.5–4.5 M (20–30 g% NaCl). Montalvo-Rodriguez et al. (1998), Stan-Lotter et al. (2002), Upasani and Desai (1990) and Elevi et al. (2004) reported seven extremely halophilic strains from the Ayvalik saltern in north-eastern part of Turkey. Extremely halophilic Archaea were characterized from saline environment in different parts of Turkey by Ozcan et al. (2006). Lizama et al. (2001) studied taxonomy of extremely halophilic Archaea from "Salar de Atacama", Chile. Two novel species of genus *Halorubrum* were isolated from two salt lakes in Xin-Jiang, China (Heng-Lin Cui et al. 2006). The family Halobacteriaceae currently contains 96 species classified in 27 genera (Oren et al. 2009). Recently, Kanekar et al. (2012) have compiled information on halophiles including their potential in production of PHA.

There are few reports indicating the presence of such extreme halophiles from sea waters (Rodriguez-Valera et al. 1979; Purdy et al. 2004; Fukushima et al. 2007; Braganca et al. 2009). Evidence for the widespread occurrence of unusual archaea in coastal surface waters has been reported (Delong et al. 1992). Although these haloarchaeal isolates remain viable in sea water, they still require high salt to grow.

Manikandan et al. (2009) have reported culture-based and culture independent approaches to study the diversity of microorganisms in solar salterns of Tamilnadu, India and has obtained pure cultures of genera *Haloferax, Halorubrum, Haloarcula, Halobacterium,* and *Halogeometricum. Natrinema thermotolerance* has been isolated by Dave et al. (2006) while studying microbial diversity at marine salterns near Bhavnagar, Gujarat, India. Braganca et al. (2009) have isolated haloarchaea from low salinity coastal sediment and waters of Goa. *Natrialba* sp. was isolated from solar saltern in Egypt (Asker et al., 2002). Fukushima et al. (2007) have demonstrated that haloarchaea are really thriving in the soil of Japanese style salt field.

PRODUCTION OF PHA BY HALOARCHAEA

Literature survey of haloarchaea reveals that species from genus *Haloferax* have been reported to produce PHA (Fernandez-Castillo et al. 1986; Huang et al. 2006). Extremely halophilic Archaebacteria, *Haloferax mediterranei*, and *Halobacterium volcanii* have been shown to accumulate large amount of PHB under nitrogen limiting condition and in the presence of excess carbon source (Fernandez-Castillo et al., 1986). Lillo and Rodriguez Valera, (1990) reported that the *H. mediterranei* grows optimally with 25% (w/v) salts and glucose or starch or extruded rice bran as sole carbon source in the medium and accumulates 60–65% PHB intracellularly. This organism also accumulates a copolymer of PHB-co-PHV using a enzymatic extruded starch without addition of any precursor (Chen et al., 2006).

9.1.10 FERMENTATION STRATEGIES

Microorganisms that are used for production of PHAs in fermenter can be divided into 2 groups based on the culture conditions required for PHA synthesis. The first group of bacteria requires the limitation of essential nutrients such as N, P, S, O, Mg or K for efficient synthesis of PHA from an excess carbon source. The second group of bacteria does not require any kind of nutrient limitation for PHA synthesis and can accumulate PHA during growth.

C. necator, Pseudomonas, Halomonas boliviensis, Bacillus sp., and many other bacteria belong to the first group while bacteria such as *Alcaligenes lactus,* a mutant strain of *Azotobacter vinelandii* and recombinant *E. coli* belong to the second group. Either fed batch or continuous fermentation technique can be used for the production of PHA with high productivity. For fed batch culture of bacteria belonging to first group, a two step cultivation method is most commonly employed where cells are first grown to a desired concentration without nutrient limitation, after which an essential nutrient is limited to allow efficient PHA synthesis. For the fed batch culture of bacteria belonging to a second group, the development of nutrient feeding strategy is the success of fermentation. Complex nitrogen sources like yeast extract, fish peptone, and so on. can be supplemented to the cells to enhance cell growth as well as PHA accumulation since PHA synthesis is not dependent on nutrient limitation in these bacteria. Thus selection of microorganisms for the production of PHA is based on several factors like ability of microorganisms to utilize inexpensive carbon sources, growth rate, PHA synthesis rate, and maximum extent of PHA accumulation.

There are a few reports on PHA fermentation. Kasemsap and Wantawin (2007) have described batch production of PHA by low phosphate content activated sludge. Atlic et al. (2011) studied continuous production of PHB by *C. necator* in a multistage bioreactor cascade. Follonier et al. (2012) used the strategy of putting cells under pressure to enhance the productivity of *P. putida* for medium chain length (mcl) PHA. Kshirsagar et al. (2012) have carried out scale up production of PHA using *Halomonas campisalis*. Berezina (2013) has proposed novel approach for productivity enhancement of PHA production by *C. necator* DSM 545 using supplementation of different nutrient media with sodium glutamate.

9.1.11 PRODUCTION OF PHA FROM INEXPENSIVE SUBSTRATES

Despite a basic attractiveness of PHA as a substitute for petroleum–derived polymers, the major hurdle facing commercial production and application of PHA in consumer products is the high cost of bacterial fermentation, making PHA 5–10 times more expensive than petroleum –derived polymers such as polypropylene, which costs approximately US $ 0.25–0.5/kg. The most significant factor for increasing the cost of PHA is the cost of substrates mainly carbon source. In order to reduce this cost, several inexpensive carbon sources and strains utilizing those cheap carbon substrates have been developed. There are several reports on production of PHA using cheap carbon sources. *C. necator, Pseudomonas* sp. have been described to utilize residual oils as a carbon source for production of PHA (Kocer et al., 2003). Wong et al. (2005) studied accumulation of PHB by *A. lactus* and *Staphylococcus epidermis* grown on several types of food wastes. Dionisi et al. (2005) reported PHA production from olive oil mill effluents. Quillaguaman et al. (2005) were able to carry out biosynthesis of PHA with *Halomonas boliviensis* using starch hydrolysate as carbon source.

Fernandez et al. (2005) reported the ability of *Pseudomonas aeruginosa* to feed on fatty acids and frying oil, with a maximum production of 66% (w/w) PHA. Alias and Tan (2005) were able to obtain PHAs [54.4% (w/w)] from palm-oil-utilizing bacteria. Chen et al. (2006) and Koller et al. (2007) studied production of a copolymer, PHB-co-PHV (10% HV) by an extremely halophilic archeon, *Haloferax mediterranei* using enzymatic extruded starch and hydrolyzed whey as a carbon source on dry cell weight basis. Ramadas et al. (2009) have reported the use of hydrolysates of agroindustrial residues as substrates for production of PHB using *Bacillus sphaericus*.

Though there are several reports on production of PHAs from cheap carbon substrates by wild type producers, the polymer concentration and the content obtained were relatively low than those obtained using purified carbon sources. Therefore there is a need for the development of more efficient fermentation strategies for production of PHA from cheap carbon sources using different microbial strains.

One of the problems preventing the commercial application of PHB is its high cost of production. From the economical point of view, the cost of the substrates, mainly the carbon source contributes to 50% of the overall cost of production. To reduce the cost of substrate, search for the different cheap substrates and the microbial strains which could utilize those cheap substrates is being done globally. Aside from the importance to the survival and general welfare of mankind, agriculture and its related industries produce large quantities of feedstocks and co-products that can be used as inexpensive substrates for fermentation processes. Successful adoption of these materials into commercial processes could further lead to the realization of a biorefinery industry based on agriculturally derived feedstocks. (Daniel et al., 2006). Sasikala and Ramana (1996) have reviewed production of PHA using inexpensive carbon sources. In the light of high cost of production of PHA using carbohydrates as the substrate, many researchers across the globe have explored different inexpensive substrates for production of PHA. The emphasis appears to be on use of plant or agricultural or industrial waste based substrates (Table 4).

TABLE 4 Production of PHA using plant or agricultural or industrial waste based substrates

Sr. No	Inexpensive substrate used	Name of the organism producing PHA	Reference
1	Mahua (Madhuca sp.) flowers	*Cupriavidus* sp. USMAA1020	Anil Kumar et al., 2007
2	Glycerine and levulinic acid	*Pseudomonas oleovorans* NRRL B-14682	Ashby et al. 2012
3	Plant oil medium	*Ralstonia eutropha*	Budde et al., 2011
4	C_1 (Methane, methanol, CO_2) carbon sources	Hydrogen-oxidizing bacteria and methanotrophs	Darani et al. 2013
5	Biodiesel industry	*Cupriavidus necator*	Garcia et al. 2013
6	Fatty acids (C_8-C_{18})	*Pseudomonas putida* Bet001	Gumel et al. 2012
7	Waste vegetable oil	*Psuedomonas* sp. Strain DR 2	Hwan et al., 2008
8	Hydrolyzed starch	Stain SP-Y1	Jayaseelan et al., 2013
9	Whey	Pseudomonas hydrogenovora	Koller et al. 2008
10	Cassava starch by-product	Bacillus megaterium	Krueger et al., 2012
11	Palm oil	*Alcaligenes* sp. AK201, *Burkholderia* sp. FLP1, *P. aeruginosa* IFO3924, *C. necator* PHB–4 (phaCAc)	Kumar at al 2011
12	Mixture of plant oils	*Cupriavidus necator* H16	Lee et al., 2008
13	Rape seed oil	*Cupriavidus necator*	Obruca et al. 2010
14	Soybean oil	*Ralstonia eutropha*	Park and Kim, 2011
15	Whey from dairy industry	*Hydrogenophaga pseudoflava* DSM1034	Povolo et al. 2013
16	Fruit watse	Strain SPY-1	Preethi et al., 2012
17	Palm oil	*Ralstonia eutropha*	Riedel et al. 2012
18	Agro-industrial residues and starch	*Bacillus* Sp. CFR-67	Shamala et al. 2012
19	whey, vegetable oils (palm, mustard, soybean and coconut) rice and wheat bran, and mustard and palm oil cakes	*Pseudomonas aeruginosa* MTCC 7925	Singh and Mallick 2009

TABLE 4 *(Continued)*

20	Glycerol and fatty acid methyl esters-FAME (Stemming from biodiesel production)	*Cupriavidus necator* DSM 545	Špoljaric et al. 2013
21	Crude palm kernel oil	*Cupriavidus necator*	Wong et al., 2012

A few halophilic microorganisms have been reported to produce PHA from inexpensive carbon sources for example, *Halomonas boliviensis* LC1 (Quillaguaman et al. 2005; Van-Thuoc et al., 2008) is reported to produce 58% (w/w) PHB by utilizing starch hydrolysate as a carbon source in flask culture condition. Using wheat bran hydrolysate and potato waste, the culture produced 35 and 43% PHB on dry cell weight basis, respectively. Chen et al. (2006) and Koller et al. (2007) studied production of a copolymer, PHB-co-PHV (10% HV) by an extremely halophilic archeon, *Haloferax mediterranei* using enzymatic extruded starch and hydrolyzed whey as a carbon source. Kanekar et al. (2007) have described production of PHA using orange peel powder. Kulkarni (2010) has reported use of sugarcane bagasse extract, orange peel and potato peel powder as inexpensive substrates for production of PHA using *Halomonas campisalis*. Chee et al. (2010) have reviewed conversion of renewable resources into PHA by different bacteria.

9.1.12 INDUSTRIAL SCALE PRODUCTION OF PHA AND PRODUCT DEVELOPMENT

The commercial production of PHA was initially developed by W. R. Grace in 1960's and later developed by Imperial Chemical Industries, (ICI) Ltd., in United Kingdom during 1970's to 1980's (Madison and Huisman, 1999). The development of PHB was started as a response to increase in the oil prices. ICI produced PHB and PHB-co-PHV using *C. necator* H16 and started marketing it under the trade name of BIOPOL at a price of US $ 16/kg. In 1982, the production of BIOPOL was 300 tons per year and they had plans to raise the production to 5000 tons per year. ICI has produced a range of thermoplastic polymers which can be processed with conventional techniques to make bottles, moldings, fibers, and films.

High grade polymer is being used for biomedical application including woven patches. ICI's biodegradable plastics has found applications as a blow moulded bottle, injection moulded caps for hair care products, disposable razors with BIOPOL handle, rubbish bags, disposable nappies, paper plates, and cups coated with thin plastic films made up of BIOPOL, and so on. In 1990's, Metabolix and Monsanto have been the driving forces for the commercial production of PHA polymers in United States.

Even though the price of PHA is very high, there are several companies producing PHAs worldwide to meet the demand of the market and these include Kaneka in Japan, P & G chemicals, BP and Metabolix in US and Imperial Chemical Industries Ltd. in UK (DelMarco, 2005). There are few products of PHA in the market such as

Biopol, Mirel and Nodax made in US, Biomer made in Germany, Biocycle in Brazil, DegraPol in Italy, Tianan PHBV, PHB in China (Tian et al., 2009).

In a typical large scale fermentation of PHA using *C. necator* under phosphate limiting condition, the final cell mass of 164 g/L, PHB content of 76% on dry cell weight basis while PHB concentration of 121 g/L was obtained after 120 h of incubation. At present, bacterial fermentation of *C. necator* seems to be the most cost effective process and even when the production of PHA is switched over to other microorganisms or in agricultural crops, the PHA biosynthesis genes from *C. necator* are inserted in the crops and other microorganisms to produce recombinant strains. *A. lactus* has also been considered as a good candidate for production of PHB since it grows quickly and accumulates large amount of PHB (~ 80% of dry cell weight) during growth using inexpensive carbon sources like molasses, sugar syrup, and so on. The highest PHB productivity of 4.94 g/L/h was reported using *A. lactus* (Wang and Lee, 1997).

Use of *Methylobacterium extorquens* for PHA production was considered by Imperial Chemical Industries (ICI), UK for cost effective production of PHA. The culture uses methanol as a carbon source for growth as well as for PHB production. It was demonstrated that a very high concentration of PHB that is 149 g/L could be obtained by fully automatic fed batch culture of *M. extorquens* using methanol as a carbon source after 120 h of incubation resulting in the PHB productivity of 0.88 g/L/h (Bourque et al., 1992). Use of activated sludge for commercial production of PHA has been explored by Basak et al. (2011).

The production of PHA by fermentation processes has been intensively studied over the last 30 years. During production of PHA in the fermenter, mixing of gaseous and liquid phase and biomass is important which can be achieved by means of aeration and agitation. For aerobic fermentation, oxygen transfer is a key variable and is a function of aeration and agitation (Castilho et al., 2009). Therefore, it is necessary to establish optimum combination of aeration and agitation for maximum production of PHA.

9.1.13 EXTRACTION OF PHA

After the production of PHA within cells, the biomass containing PHA are separated from fermentation broth by centrifugation. The harvested cells are then lysed for recovery of PHA. A number of different methods are available in literature for recovery of PHAs. Organic solvents such as chloroform, methylene chloride, dichloroethane, acetone or propylene carbonate can be used to dissolve PHA. However the necessity of large quantities of solvent makes the process economically and environmentally unattractive. For quantification of PHA, the extracted PHA was converted to crotonic acid by reacting it with concentrated sulphuric acid and estimated spectrophotometrically at 235 nm. Detection of PHA with chloroform can also be done by IR spectroscopy. But the above mentioned methods are time consuming and not accurate for low PHA concentrations. Thus the use of gas chromatography (GC) for the identification of PHA components was proposed. The other methods for quantification of PHA include HPLC, ionic chromatography, and enzymatic determination. The determination of PHA inside intact cells by two dimensional fluorescence spectroscopy and flow cytometry has also been proposed recently. Ease of recovery of PHA is an important parameter in the economical production of PHA. Thus there is a need for the development of methods for extraction of PHA so that the overall process could be made much simpler and cheaper.

As an alternative to solvent extraction, aqueous enzymatic digestion methods have been developed by ICI Ltd. This process is very efficient when a less pure product is suitable for some applications of PHB. However, this process often requires additional digestion or solvent extraction steps for increasing product purity, rendering the recovery costs higher (Hahn et al., 1994).

Berger et al. (1989) and Ramsay et al. (1990) described a differential digestion method employing sodium hypochlorite. Although this method is effective in the digestion of non PHA cellular materials (NPCMs), at the same time it causes severe degradation of PHB rendering it unsuitable for many applications.

Hahn et al. (1994) reported the use of dispersion of sodium hypochlorite and chloroform for the efficient recovery of PHB from *C. necator*. Sodium hypochlorite isolates PHB from the cell in the aqueous phase, then released PHB immediately migrates to the chloroform phase where the chloroform at least partially protects the PHB molecules from further degradation by the action of hypochlorite.

Recently, supercritical fluid disruption (Hejazi et al., 2003; Khosravi-Darani et al., 2004) dissolved air flotation (Van Hee et al., 2006) and the selective dissolution of cell mass (Yu and Chen, 2006) for the recovery of PHAs were studied. A new cultivation method allows spontaneous release of up to 80% of intracellular PHB from *E. coli* (Jung et al., 2005). All these methods are promising alternatives to the solvent extraction. Kshirsagar et al. (2013) have described kinetics and modeling of solvent extraction of PHA from *H. campisalis.* Rathi et al. (2012) have reported simplified polymer recovery from a novel moderately halophilic bacterium.

For biomedical applications, the solvent extraction should be preferred because the resulting PHAs have a high purity (Chen and Wu, 2005).

9.1.14 CHARACTERIZATION OF PHA

A homopolymer of (R)-3-hydroxybutyrate (PHB) is the most common type of PHA that bacteria accumulate in nature and has been studied and characterized extensively by many researchers. Poly (β-hydroxybutyrate) (PHB) has a number of interesting characteristics and can be used in various ways similar to many conventional synthetic plastics now in use. The properties of polyhydroxyalkanoates viz. PHB, PHV and comparison with synthetic polymer polypropylene (PP) have been described by Sasikala and Ramana (1996).

The characterization of PHA polymers has been carried out using various methods viz. Nuclear Magnetic Resonance (NMR) spectroscopy, Gel Permeation Chromatography (GPC), Gas Chromatography (GC), and Differential Scanning Calorimetry (DSC). After determining the basic properties of PHA polymer, the structural and mechanical properties of the PHA films are studied using techniques like Fourier Transform Infrared (FTIR) spectroscopy, optical microscopy, tensile testing, Dynamic Mechanical Thermal Analysis (DMTA), and Dielectric Relaxation Spectroscopy (DRS) (Kulkarni et al., 2011).

9.1.15 PHYSICAL PROPERTIES OF PHA

Marchessault and colleagues showed that PHB is a compact, right handed helix with a two folds screw axis and a fiber repeat of 0.596 nm (Cornibert and Marchessault, 1972). It is optically active with a chiral center of a monomer unit always in the R absolute configuration. [(D-(-) in the traditional nomenclature]. The similarity of PHB structure is with that of the polypropylene, which also has a compact helical configuration and a melting temperature of about 180°C. PHB and polypropylene display similar degrees of crystallinity and glass transition temperature, although their chemical properties are completely different (Anderson and Dawes, 1990). The material properties of PHAs are similar to the conventional plastics such as polypropylene, hence majority of their applications are aimed as a packaging and coating material to replace the synthetic plastic. PHAs exhibit some interesting properties as follows:

1. The PHA is thermoplastic in nature and can be produced using many renewable resources.
2. It is water insoluble and this differentiates PHA from most other currently available biodegradable plastics, which are either water soluble or moisture sensitive.
3. It shows good oxygen barrier capacity and UV resistance, but has poor resistance to acids and bases.
4. Most important property of PHA is its biocompatible nature and hence it's suitability for biomedical application.
5. The molecular mass of PHA is in the range of 105–106 Da.
6. PHAs are partially crystalline polymers with a degree of crystallinity in the range of 60–80%. PHB homopolymer is highly crystalline, stiff but brittle material, when spun into fiber it behaves as a hard elastic material. It becomes more brittle over a period of several days upon storage under ambient conditions and this effect is called as "ageing effect" of PHB (De. Koning et al., 1994). Much work has been carried out to improve the physical properties of PHB. The formation of copolymer with incorporation of other hydroxyacid units was found to be a valuable approach to improve the polymer's crystallinity, melting temperature, stiffness and toughness.
7. Copolymers of PHB that is PHB-co-PHV are less stiff and brittle than PHB, while retaining most of the other mechanical properties of PHB. As the fraction of hydroxyvalerate unit (HV) increases, the polymer becomes tougher and flexible. The elongation to break also increases. Furthermore, the decrease in the melting temperature with increasing 3HV fraction without affecting the degradation temperature allows thermal processing of copolymer without thermal degradation. The thermal degradation temperature of PHB is 246.3°C, which is close to its melting temperature. This limits its processing by injection molding. The thermal degradation of PHB-co-PHV starts at 260.4°C. This indicates that the presence of valerate in the PHB chain increases its thermal stability.
8. The medium chain length PHAs are elastomers with much lower melting temperature and crystallinity as compared to PHB. This means that family of PHAs exhibit a wide variety of mechanical properties from hard, crystalline to

elastic depending upon monomer composition, which broadens its application area. For example, medium chain length PHAs are semi crystalline elastomers with low melting point, low tensile strength, and high elongation to break and can be used as a biodegradable rubber after cross linking.

9. The melting temperature of PHB is 175°C with a glass transition temperature in the range of 0–20°C.
10. The material properties of PHB including Young's modulus (3.5 Gpa) and tensile strength (40 Mpa) are similar to that of polypropylene, but the elongation to break of PHB (6%) is significantly lower than polypropylene (400%).
11. Density of PHB is higher than polypropylene and thus PHB sinks in water, while polypropylene floats. Sinking of PHB in water facilitates its anaerobic biodegradation in sediments.
12. PHB is non toxic. It is found in variety of plant and animal tissues. In human plasma, PHB can be found associated with very low density lipoprotein and low density lipoprotein. In addition, a significant portion of PHB is found associated with serum albumin (Madison and Huisman, 1999). Comparative account of properties of PHA and synthetic plastics as presented in Table 5.

TABLE 5 Properties of PHAs compared with synthetic polymers like polypropylene, poly (ethylene terephthalate) and Nylon 6.6 (Sasikala and Ramana, 1996)

Property	PHB	PHV	PHBV (4–29% HV)	PP	PETP	Nylon 6.6
Melting temperature, T_m (°C)	175	107	157–102	176	267	265
Crystallinity (%)	80	80	69–39	70	30–50	40–60
Average molecular weight, M_w (x10^5)	1–8	2	6	2–7	–	–
Glass transition temperature, T_g (°C)	5–15	–16	2–8	–10	69	50
Density, ρ (gcm^{-1})	1.25	1.2	1.2	0.905	1.385	1.14
Water uptake, (wt %)	0.2	–	–	0.0	0.4	4.5
UV resistance	Good	Good	Good	Poor	–	–
Solvent resistance	Poor	Poor	Poor	Good	–	–
Biodegradability	+	+	+	–	–	–
Oxygen permeability, $cm^3m^2atm^{-1}day^{-1}$	45	–	–	1700	70	–

9.1.16 BIODEGRADABILITY OF PHA

Besides the typical polymeric properties of PHA, an important characteristic of PHA is their biodegradability. Microorganisms in nature are able to degrade PHAs using PHA hydrolases and PHA depolymerases. Microorganisms colonize on the surface of the polymer and secrete enzymes which degrade PHB or PHB-co-PHV into HB and HB and HV units respectively. These units are then used up by the cell as a carbon source for the biomass growth. The rate of polymer biodegradation depends on the variety of factors including surface area, microbial activity of the disposal environment, pH, temperature, moisture. and pressure, and so on. The end products of PHA degradation in aerobic environment are CO_2 and water while methane isproduced in anaerobic condition (Ojumu et al., 2004). The effect of different environments on the rate of biodegradation of PHB and PHB-co-PHV has been studied by several researchers (Doi et al., 1992; Mergaert et al., 1992, 1993, 1994; Lee, 1996). PHA degrading microorganisms have been isolated from various environments for example, soil, river water, compost, sludge, and so on. Various microorganisms excrete extracellular PHA depolymerases that hydrolyze PHA into water soluble oligomers and monomers and subsequently utilize those resulting products as nutrients within cells. The properties of an extracellular PHB depolymerase from *Alcaligenes faecalis* have been extensively investigated. Biodegradability of PHA can be tested by ASTM method and soil burial method (Kulkarni et al., 2011).

9.1.17 APPLICATIONS OF PHA

PHAs have wide range of applications owing to their novel features ~~(Fig. 8.6)~~. They are natural thermoplastic polyesters and hence majority of their applications are as replacements for petrochemical polymers currently in use for packaging and coating application. Initial efforts focused on molding applications in particular for consumer packaging items such as bottles, cosmetic containers, pens, diper back sheets, and so on. The films prepared by using PHA can be used to make laminates with other polymers like polyvinyl alcohol. PHAs also have been processed into fibers which then were used to construct materials such as non woven fabrics. They have also been described as hot melt adhesives. PHA promises to be a new source of molecules. It can be hydrolyzed chemically and the monomers can be converted to commercially important molecules such as β-hydroxyacids, 2-alkenoic acids, β-hydroxyalkanols, and so on. (Madison and Huisman, 1999). Composites of bioplastics are already used in electronic products like mobile phones (NEC Corporation and UNITIKA Ltd., 2006). Potential agricultural applications include encapsulation of seeds, encapsulation of fertilizers for slow release, biodegradable plastic films for crop protection and biodegradable containers, and so on. Kanekar et al. (2012) have demonstrated application of a co-polymer PHB-co-PHV produced by moderately haloalkaliphilic *Halomonas campisalis* in packaging food commodity items like milk, vegetable oil, wheat semolina, and so on.

Singh and Mallick (2009) exploited inexpensive substrates for production of a novel SCL-LCL-PHA copolymer by *P. aeruginosa* MTCC7925, having good tensile strength indicating its potential application in packaging. Sankhla et al. (2010) has reported high elasticity of PHBV copolymer produced by *Brevibacillus invocatus* suggesting potential application in packaging. Salim et al. (2011) studied biosynthesis of PHBV by *Cupriavidus* sp. and characterization of its blend with oil palm empty fruit bunch fibers. The tensile strength of the polymer blend indicated its possible application in packaging. Kumar et al. (2011) have reviewed synthesis of PHA using palm oil and their application. Enhancement in mechanical properties of PHA with high content of HV monomer has been reported by Wong et al. (2012) and Yang et al (2012).

The PHAs have numerous biomedical applications. The main advantage in the biomedical field is that a PHA can be inserted into a human body and does not need to be removed again. The PHA has an ideal biocompatibility as it is a product of cell metabolism and also 3-hydroxybutyric acid is normally present in the blood at concentration between 0.3 and 1.3 mMol/L (Zinn et al., 2001). In the pure form or as a composite, PHAs are used as sutures, swabs, repair patches, orthopedic pins, adhesion barriers, stents, nerve guides, bone marrow scaffolds, and so on. (Verlinden et al., 2007). The PHAs are used particularly as the osteosynthetic materials in the stimulation of bone growth, in bone plates, surgical sutures, and blood vessel replacements. PHAs are used for the slow delivery of drugs, hormones, medicines, and so on. Vigneswari et al. (2009) tested biocompatibility of PHA supporting growth and proliferation of fibroblast cells indicating its biomedical applications. Teeka et al. (2012) have reported endotoxin free PHA from *Nobosphingobium* sp. having potential biomedical application.

9.2 FUTURE PROSPECTS

In order to make the production of PHA feasible for industrial application, obtaining high yield is essential. Several factors are known to influence PHA biosynthesis for example,. bacterial strain used carbon and nitrogen source provided and cultural conditions, and so on. Thus optimization studies conducted by either conventional methods or by use of any statistical tool are necessary to improve yield of the PHA.

One of the advantages of using haloarchaea for the production of PHA is the reduced cost of extraction. Haloarchaea could be lysed by simply suspending in distilled water. PHA can be extracted in distilled water and can be recovered by evaporation. The PHA pellet can be redissolved in a small volume of chloroform and a PHA film can be casted. Thus, the cost of solvent extraction would be significantly reduced on an industrial scale. Further, chances of microbial contamination during fermentation of PHA could be minimized since the haloarchaea outgrow contaminating bacteria in the presence of high salinity. Moreover, the organism can grow on simple carbon source like glucose and low amount of yeast extract (0.1 g%) to produce PHA which also reduces the cost of production (Asker and Ohta 2002).

9.3 CONCLUSION

Microbial polymers are versatile being biodegradable in nature. Many microorganisms have been investigated for production of PHA. They could be exploited for commercial production. Inexpensive substrates have been tested to reduce cost of their production. Extremophiles, especially the halophiles are recognized as a source of PHA and hence need to be explored for production. Microbial PHA with enhanced mechanical properties can be used for packaging applications in food industry and agriculture. They being biocompatible and biodegradable in nature can be used for biomedical applications.

KEYWORDS

- **Microbial polymers**
- **Polyhydroxyalkanoates (PHAs)**
- **Extremophiles**
- **Halophiles**
- **Food packaging**
- **Slow drug delivery**

ACKNOWLEDGEMENT

The authors are thankful to Department of biotechnology, Govt. of India for financial support to carry out research on biodegradable polymer, PHA from haloalkaliphilic bacteria. The authors are also grateful to Director, MACS-Agharkar Research Institute, Pune for extending infrastructural facilities to work on PHA. S. P. Kanekar is thankful to Council for Scientific and Industrial Research (CSIR), Govt. of India for Senior Research Fellowship to carry out work on production of PHA by haloarchaea. Thanks are due to Dr. Mani and Dr. Gore, Armed Forces Medical College (AFMC), Pune for using the Transmission Electron Microscope facility for having TEM of PHA granules in the cells of *Halomonas campisalis*. The authors thank Dr. Sasikala and Dr. Ramana for information provided in Table 5 of the manuscript.

REFERENCES

1. Alexander, M. *Science*, **211**, 132–138 (1981).
2. Alias, Z. and Tan, I. K. P. *Bioresour. Technol*, **96**, 1229–1234 (2005).
3. Allen, A. A., Anderson, W. A., Ayorinde, F. O., and Erilo, B. E. *J. Ind. Microbiol. Biotechnol*, **37**, 849–856 (2010).
4. Amirul, A. A., Yahya, A. R. M., Sudesh, K., Azizan, M. N. M., and Majid, M. I. A. *Bioresour. Technol*, **99**, 4903–4909 (2008).
5. Anil Kumar, P. K., Shamala, T. R., Kshama, L., Prakash, M. H., Joshi, G. J., Chandrashekar, A., Latha Kumari K. S., and Divyashree M. S. *J. Appl. Microbiol.*, **103**, 204–209 (2007).
6. Anderson, A. J. and Dawes, E. A. *Microbiol. Rev.*, **54**, 450–472 (1990).
7. Arahal, D. R., Ludwig, W., Schleifer, K. H., and Ventosa, A. *Int. J. Syst. Evol. Microbiol.*, **52**, 241–249 (2002).
8. Ashby, R. D., Solaiman, D. K. Y., Strahan, G. D., Zhu, C., Tappel, R. C., and Nomura, C. T. *Bioresour. Technol*, **118**, 272–280 (2012).
9. Atlić, A., Koller, M., Scherzer, D., Kutschera, C., Grillo-Fernandes, E., Horvat, P., Chiellini, E., and Braunegg, G. *Appl Microbiol Biotechnol*, **91**, 295–304 (2011).
10. Barnard, G. N. and Sanders, J. K. M. *FEBS lett*, **231**, 16–18 (1988).
11. Barnard, G. N. and Sanders, J. K. M. *J. Biol. Chem.*, **264**, 3286–3291 (1989).
12. Basak, B., Ince, O., Artan, N., Yagci, N., and Ince, B. K. *Bioprocess Biosyst. Eng*, **34**, 1007–1016 (2011).
13. Berekaa, M. M., **4**(9), 518–527 (2012).
14. Berezina, N. *New Biotechnol*, **30**(2), 192–195 (2013).
15. Berger, E., Ramsay, B. A., and Ramsay, J. A. *Biotechnol. Tech.*, **3**, 227–232 (1989).
16. Biswas, A., Patra, A., and Paul, A. K. *Acta Microbiol. Immunol. Hung.*, **56**, 125–143 (2009).

17. Bouchotroch, S., Quesada, E., Delmoral, A., Llamas, I., Bejar, V. *Int. J. Syst. Evol. Microbiol*, **51**, 1625–1632 (2001).
18. Bourque, D., Ouellette, B., Andre, G., and Groleau, D. *Appl. Microbiol. Biotechnol*, **37**, 7–12 (1992).
19. Brandl, H., Gross, R. A., Lenz, R. W., and Fuller, R. C. *Appl. Environ. Microbiol.*, **54**, 1977–1982 (1988).
20. Braganca, J. M. and Furtado, J. *Curr. Sci.*, **96**(9), 1182–1184 (2009).
21. Budde, C. F., Riedel, S. L., Hübner, F., Risch, S., Popović, M. K., Rha, C., and Sinskey, A. J. *Appl. Microbiol. Biotechnol.*, **89**, 1611–1619 (2011).
22. Byrom, D. *Polyhydroxyalkanoates*, In Plastic from Microbes, Microbial Synthesis of Polymers and Polymer Precursors, D. P. Mobley (Ed.), Hanser Munich, pp. 5–33 (1994).
23. Castilho, L. R., Mitchell, D. A., and Freire, M. G. D. *Biotechnol. Tech*, **100**, 5996–6009 (2009).
24. Chee, J. Y., Yogal, S. S., Lau N. S., Ling S. C., Abed, R. M. M., and Sudesh, K. Bacterially Produced Polyhydroxyalkanoate (PHA): Converting Renewable Resources into Bioplastics. In Current research, technology and education topics in applied microbiology and microbial biotechnology. A. Mendez-vilas (Ed.), FORMATEX, pp. 1395–1404 (2010).
25. Chen, C. W., Don, T. R., and Yen, H. F. *Process Biochem.* **41**, 2289–2296 (2006).
26. Chen, G. Q. and Wu, Q. *Biomaterials*, **26**, 6565–6578 (2005).
27. Cornibert, J. and Machessault, R. H. *J. Mol. Biol.*, **71**, 735–756 (1972).
28. Dai, Y., Yuan, Z., Jack, K., and Keller, J. *J. Biotechnol.*, **129**, 489–497 (2007).
29. Daniel, K., Solaiman, Y., Richard, D., Ashby, T., Foglia, A., and William, N. *Appl. Microbiol. Biotechnol.*, **71**, 783–789 (2006).
30. Das, S. S. and Arora, P. *Halophiles, Encyclopedia of Life Sciences*, Nature Publishing Group, **8**, 458–466 (2001).
31. Dave, S. R. and Desai, H. B. *Curr. Sci.*, **90**(4):497–500 (2006).
32. Dawes, E. A. and Senior, P. J. *The role and regulation of energy reserve polymers in microorganisms*. Adv. Microb. Physiol. **10**, 135–266 (1973).
33. De Koning, G. J. M., Scheeren, A. H. C., Lemstra, P. J., Peeters, M., and Reynaers, H. *Polymer*, **35**, 4598–4605 (1994).
34. Del Marco, S. *Advances in Polyhydroxyalkanoate Production in Bacteria for Biodegradable Plastics*. MMG445, Basic Biotechnology e-Journal, www.msu.edu/course/mmg/445, pp. 1–4 (2005).
35. Delong, E. F. *Archaea in Coastal Marine Environments*. Proceedings of the National Academy of Sciences of USA, **89** (12), 5685–5689 (1992).
36. Devi, B. A. and Nachiyar, C. V., *IEEE*, 212–215 (2011).
37. Dionisi, D., Carucci, G., Papini, M. P., Riccardi, C., Majone, M., and Carrasco, F. *Water Res*, **39**, 2076–2084 (2005).
38. Doi, Y. and Abe, C. *Macromolecules*, **23**, 3705–3707 (1990).
39. Doi, Y., Kawaguchi, Y., Koyama, N., Nakamura, S., Hiramitstu, M., Toshida, Y., and Kimura, U. FEMS Microbiol. Rev, **103**, 103–108 (1992).
40. Doi, Y., Kitamura, S., and Abe, H. *Macromolecules*, **28**, 4822–4828 (1995).
41. Doi, Y., Tamaki, A., Kunioka, M., and Soga, K. *Appl. Microbiol. Biotechnol*, **28**, 330–334 (1988).
42. D'Souza, S. E., Altekar, W., and D'Souza, S. F. *Arch. Microbiol.*, **168**, 68–71 (1997).
43. Elevi, R., Assa, P., Birbir, M., and Ogan, A., *Oren. World J. Microbiol. Biotechnol.*, **20**, 719–725 (2004).
44. Fernandez, D., Rodriguez, E., Bassas, M., Vinas, M., Solanas, A. M., Llorens, J., marques, A. M., and Manresa, A. *Biochem. Eng. J*, **26**, 159–167 (2005).
45. Follonier, S., Henes, B., Panke S., and Zinn, M. *Biotechnol. Bioeng*, **109** (2), 452–461 (2011).
46. Fernandez-Castillo, R., Rodriguez-Valera, F., González-Ramos, J., and Ruiz-Berraquero, F. *Appl. Environ. Microbiol*, **51**(1), 214–216 (1986).
47. Fradinho, J. C., Domingos, J. M. B., Carvalho, G., Oehmen, A., and Reis, M. A. M. *Bioresour. Technol*, **132**, 146–153 (2013).

48. Fukushima, T., Usami R., and Kamekura, M. *Saline syst*, **3**, p. 2 (2007).
49. García, I. L., López, J. A., Dorado, M. P., et al. *Bioresour. Technol.*, **130**, 16–22 (2013).
50. Grant, W. D., Jones, B. E., and Mwatha, W. E. *FEMS Microbiol. Rev.*, **75**, 255–270 (1990).
51. Grothe, E., Moo-Young, M., and Chisti, Y. *Enzyme Microb. Technol.*, **25**, 132–141 (1999).
52. Gumel A. M., Annuar, M. S. M., and Heidelberg, T. *PLOS one*, **7** (9), 1–8 (2012).
53. Guo, W., Song, C., Kong, M., Geng, W., Wang, Y., and Wang, S. *Appl. Microbiol. Biotechnol.*, **92**, 791–801 (2011).
54. Hahn, S. K., Chang, Y. K., Kim, B. S., and Chang, H. N. *Biotechnol.* Bioeng, **44**, 256–261 (1994).
55. Haywood, G. W., Aderson, A. J., William, D. R., Dawes, E. A., and Ewing, D. F. *Int. J. Biol. Macromol*, **13**, 83–88 (1991).
56. Hejazi, P., Vasheghani-Farahani, E., and Yamini, Y. *Prog. Biotechnol*, **19**, 1519–1523 (2003).
57. Heng-Lin, C., Tohty, D., Zhou, P., and Liu, S. *Int. J. Syst. Evol. Microbiol*, **56**, 631–1634 (2006).
58. Hiramitsu, M., Koyama, N., and Doi, Y. *Biotechnol. Lett*, **15**, 461–464 (1993).
59. Hiroe A., Ushimaru, K., and Tsuge, T. J. *Biosci. Bioeng*. In Press, pp. 1–6 (2012).
60. Horikoshi, K. *General View of Alkaliphiles and Thermophiles*. In Superbugs, Microorganisms in Extreme Environments. K. Horikoshi. and W. D. Grant (Eds.), Springer Verlag Berlin, pp. 3–13 (1991).
61. Horikoshi, K. Alkaliphiles: Some applications of their products for biotechnology. *Microbiol Mol. Biol. Rev.*, **63** (4), 735–750 (1999).
62. Huang, T. Y., Duan, K. J., Huang, S. Y., and Chen, C. W. *J. Ind. Microbiol. Biotechnol*, **33**(**8**), 701–706 (2006).
63. Huijberts, G. N. M., Eggink, G., De Waard, P., Huisman, G. W., and Witholt, B. *Appl. Environ. Microbiol*, **58**, 536–544 (1992).
64. Hwan, S. J., Jeon, C. O., Choi, M. H., Yoon, S. C., and Park, W. J. *Microbiol. Biotechnol*, **18**(8), 1408–1415 (2008).
65. Ibrahim, M. H. A. and Steinbuchel, A. *Appl. Environ. Microbiol*, **75**(19), 6222–6231 (2009).
66. Ienczak, J. L., Schmidell, W., and de Aragaˇo, G. M. F. *J. Ind. Microbiol. Biotechnol.*, (2013).
67. Jeyaseelan, A., Pandiyan, S., and Ravi, P. J. *Microbiol. Biotechnol. food Sci.*, **2**(3), 970–982 (2012–2013).
68. Jones, B. E., Grant, W. D., Collins, N. C., and Mwatha, W. E. *Alkaliphiles: Diversity and Identification*. In Bacterial Diversity and Systematics, F. G. Priest (Ed.), Plenum Press New York, pp. 195–230 (1994).
69. Joshi, A. A., Kanekar, P. P., Kelkar, A. S., Sarnaik, S. S., Shouche, Y., and Wani, A. *J. Basic Microbiol*, **47**, 213–221 (2007).
70. Joshi, A. A., Kanekar, P. P., Kelkar, A. S., Shouche, Y. S., Wani, A., Borgave, S. B., and Sarnaik, S. S. *Microb. Ecol*, **55**, 163–172 (2008).
71. Jung, I. L., Phyo, K. H., Kim, K. C., Park, H. K., and Kim, I. G. *Res. Microbiol.*, **156**, 865–873 (2005).
72. Kanekar, P. P., Nilegaonkar, S. S., Sarnaik, S. S., Ponraj, M., and Jog, J. P. *A process for production of a co-polymer PHB–Co–PHV* by *Bacillus cereus*. Indian Patent No. 244785, (2007).
73. Kanekar P. P., Kanekar, S. P., Kelkar, A. S., and Dhakephalkar, P. K. *Halophiles – Taxonomy, Diversity, Physiology and Applications*. In Microorganisms in Environmental Management, T. Satyanarayana, N. J. Bhavdish (Eds.), Springer Science + Business media B. V., pp. 1–34 (2012).
74. Kanekar, P. P., Kulkarni, S. O., Nilegaonkar, S. S., Sarnaik, S. S., Kshirsagar, P. R., and Jog J. P. *Microbial Biodegradable Polymer Having Potential Application in Packaging*. Proceedings of Indo-US International Conference on Polymers for Packaging Applications (ICPPA), (2012).
75. Kasemsap, C. and Wantawin C. *Bioresour. Technol.*, **98**, 1020–1027 (2007)
76. Keenan T. M., Nakas, J. P., and Tanenbaum, S. W. *J. Ind. Microbiol. Biotechnol.*, **33**, 616–626 (2006).
77. Khosravi-Darani, K., Mokhtari Z. B., Amai T., and Tanaka K. *Appl. Microbiol. Biotechnol.*, **97**, 1407–1424 (2013).

78. Khosravi-Darani, K., Vasheghani-Farahani, E., Shojaosadati, S. A., and Yamini, Y. *Biotechnol. Prog.*, **20**, 1757–1765 (2004).
79. Khanna, S. and Srivastava, A. *Process Biochem.*, **40**, 607–619 (2005).
80. Kim, B. S., Lee, S. Y., and Chang, H. N. *Biotechnol. Lett*, **14**, 811–816 (1992).
81. Kim, B. S., Lee, S. C., Lee, S. Y., Chang, H. N., Chang, Y. K., and Woo, S. I. *Biotechnol Bioeng*, **43**, 892–898 (1994).
82. Kim, B. S., Lee, S. Y., and Chang, H. N. *Biotechnol. Lett*, **14**, 811–816 (1992).
83. Kocer, H., Borcakli, M., and Demirel, S. *Turk J. Chem.*, **27**, 365–373 (2003).
84. Koller, M., Bona, R., Chiellini, E., Fernandes, E. G., Horvat, P., Kutschera, C., Hesse. P., and Braunegg, G. *Bioresour. Technol.*, **99**, 4854–4863 (2008).
85. Koller, M., Hesse, P., Bona, R., Kutschera, C., Atlic, A., and Braunegg, G. *Macromol. Biosci.*, 7, 218–226 (2007).
86. Kshirsagar, P. R., Kulkarni, S. O., Nilegaonkar, S. S., Niveditha, M., and Kanekar, P. P. *Sep. Purif. Technol*, **103**, 151–160 (2013).
87. Kshirsagar, P. R., Suttar, R., Nilegaonkar S. S., Kulkarni, S. O., and Kanekar P. P. *J. Biochem. Technol.*, (Accepted) (2012).
88. Krueger, C. L., Radetski, C. M., Bendia, A. G., Oliveira, I. M., Marcus, A., Castro-Silva, M. A., Rambo, C. R., Antonio, R. V., and Lima, A. O. S. *Electron. J. Biotechnol.*, **15**(3), 1–13 (2012).
89. Kulkarni, S. O., Ph. D. Thesis: Production of Biodegradable Polymer by Extremophilic Bacteria Isolated from Lonar Lake, University of Pune, Pune, India, (2010).
90. Kulkarni, S. O., Kanekar, P. P., Nilegaonkar, S. S., Sarnaik, S. S., and Jog, J. P. *Bioresour. Technol.*, **101**(24), 9765–9771 (2010).
91. Kulkarni, S. O., Kanekar, P. P., Jog, J. P., Patil, P. A., Nilegaonkar, S. S., Sarnaik, S. S., and Kshirsagar, P. R. *Bioresour. Technol.*, **102**(11), 6625–6628 (2011).
92. Lageveen, R. G., Huisman, G. W., Preusting, H., Ketelaar, P., Eggink, G., and Witholt, B. *Appl. Environ. Microbiol.*, **54**, 2924–2932 (1988).
93. Lee, S. Y. *Trends Biotechnol*, **14**, 431–438 (1996).
94. Lee, S. and Yu, J., *Resour., Conserv. Recycl*, **19**, 151–164 (1997).
95. Lee, W. H., Looa, C. Y., Nomura C. T., and Sudesh K. *Bioresour. Technol.*, **99**, 6844–6851 (2008).
96. Lee, S. Y., Choi, J., and Wong, H. H. *Int. J. Biol. Macromol*, **25**, 31–36 (1999).
97. Lillo, J. G. and Rodriguez-Valera, F. *Appl. Environ. Microbiol*, **56**, 2517–2521 (1990).
98. Lizama, C., Sanchez, M. M., Prado, B., Ramos-Cormenzana, A., Weckesser, J., and Campos, V. *Appl. Microbiol*, **24**, 464–474 (2001).
99. Madison, L. L. and Huisman, G. W. *Microbiol. Mol. Biol. Rev.*, **63**, 21–53 (1999).
100. Manikandan, M., Kannan, V., and Pasic, L., *World J. Microbiol. Biotechnol*, **25**, 1007–1017 (2009).
101. Martinez – Canovas, M., Quesada, E., Llamas, I., and Bejar, V. *Int. J. Syst. Evol. Microbiol.*, **54**, 733–737 (2004).
102. Mergaert, J., Anderson, C., Wouters, A., and Swings, J. *J. Environ. Polym. Degrad*, **2**, 177–183 (1994).
103. Mergaert, J., Anderson, C., Wouters, A., Swings, J., and Kersters, K. *FEMS Microbiol Rev.*, **103**, 317–322 (1992a).
104. Mergaert, J., Webb, A., Anderson, C., Wouters, A., and Swings, J. *Appl. Environ. Microbiol.*, **59**, 3233–3238 (1993).
105. Montalvo-Rodríguez R., Vreeland R. H., Oren, A., Kessel, M., Betancourt, C., and Lopez-Garriga, J. *Int J. Syst. Bacteriol*, **48**, 1305–1312 (1998).
106. Mothes, G., Schubert, T., Harms, H., and Maskow, T. *Eng. Life Sci*, **8**, 658–662 (2008).
107. Muller, R. J., Kleeberg, I., and Deckwer, W. D. *J. Biotechnol*, **86**, 87–95 (2001).
108. Obruca, S., Marova, I., Snajdar, O., Mravcova, L., and Svoboda, Z. *Biotechnol. Lett*, **32**, 1925–1932 (2010).
109. Ojumu, T. V., Yu, J., and Solomon, B. O. *Afr. J. Biotechnol*, **3**, 18–24 (2004).
110. Oren, A. *FEMS Microbiol Rev*, **13**, 415–39 (1994).

111. Oren, A. *Indian J. Microbiol. Biotechnol*, **28**, 56-63 (2002).
112. Oren, A., Arahal, D. R., and Antonio, V. *Int. J. Syst. Evol. Microbiol.*, **59**, 637–642 (2009).
113. Ostle, A. G. and Holt, J. G. Nile blue A as a fluorescent stain for poly-β - hydroxybutyrate. *Applied and Environmental Microbiology*, **44**, 238–241 (1982).
114. Ozcan, B., Cokmus, C., Coleri, A., and Caliskan, M. *Microbiol*, **75**(b), 739–746 (2006).
115. Page, W. J. and Cornish A. *Appl. Environ. Microbiol.*, **59**, 4236–4244 (1993).
116. Pal, A., Prabhu A., Kumar, A. A., Rajgopal, B., Dadhe, K., Ponnamma, V., and Shivkumar, S. *Polish J. Microbiol.*, **58**(2), 149–154 (2009).
117. Park, D. H., and eom Soo Kim, B. S. *New Biotechnol.*, **28**(6), 719–724 (2011).
118. Patel, M., Gapes, D. J., Newman, R. H., and Dare, P. H. *Appl. Microbiol. Biotechnol.*, **82**, 545–555 (2009).
119. Patwardhan, P. R., and Srivastava, A. K. *Biochem. Eng. J.*, **20**, 21–28 (2004).
120. Povolo, S., Romanelli, M. G., Basaglia, M., Ilieva, I., Andrea Corti, A., Morelli, A., Chiellini E., and Casella S. *New Biotechnol.*, 00 (00), 1–6 (2012).
121. Preethi, R., Sasikala, P., and Aravind, J. *Res. Biotechnol.*, **3**(1), 61–69 (2012).
122. Purdy, K. J., Cresswell, M. T. D., Nedwell, D. B., McGenity, T. J., Grant. W. D., Timmis, K. N., and Embley, T. M. *Environ. Microbiol.*, **6**(6), 591–595 (2004).
123. Quillaguaman, J., Doan-Van, T., Guzman, H., Guzman, M. J., Everest, A., and Hatti-Kaul, R., *Appl. Microbiol. Biotechnol.*, **78**, 227–232 (2008).
124. Quillaguamán, J., Guzmán H., Van-Thuoc D., and Hatti-Kaul, R. *Appl Microbiol Biotechnol.*, **85**, 1687–1696 (2010).
125. Quillaguaman, J., Hasher, S., Bento, F., Mattiason, B., and Hatti-Kaul, R. *J. Appl. Microbiol*, **99**, 151–157 (2005).
126. Quillaguaman, J., Atti-Kaul, R., Attiasson, B., Alvarez, T. M., and Delgado, O. *Int. J. Syst. Evol. Microbiol.*, **54**, 721–725 (2004).
127. Quillaguaman, J., Munoz, M., and Mattiason, B. *Appl. Microbiol. Biotechnol.* **74**, 981–986 (2007).
128. Rathi D. N., Amir, H. G., Abed, R. M. M., Kosugi, A., Arai T., Sulaiman, O., Hashim, R., and Sudesh, K. *J. Appl. Microbiol.*, **114**, 384–395 (2012).
129. Ramdas, N. V., Singh, S. K., Soccol, C. R., and Pandey, A. *Braz. Arch. Biol. Technol.* **52**, 17–23 (2009).
130. Ramsay, J. A., Berger, E., Ramsay, B. A., et al. *Biotechnol. Tech.*, **8**, 589–594 (1990).
131. Reddy, S. V., Thirumala, M., and Mahmood, S. K. *J. Ind. Microbiol. Biotechnol.*, **36**, 837–843 (2009).
132. Riedel, S. L., Bader, J., Brigham, C. J., Budde, C. F., Yusof, Z. A. M., Rha, C. K., and Sinskey, A. *J. Biotechnol. Bioeng*, **109** (1), 74–83 (2011).
133. Rodriguez-Valera, F., Ruiz-Berraquero, F., and Ramos-Cormenzana, A. *Appl. Environ. Microbiol*, **38**(1), 164–166 (1979).
134. Salgaonkar, B. B., Mani, K., and Braganca, J. M. *J. Appl. Microbiol.*, pp.1–10 (2013).
135. Salim, Y. S., Abdullah, A. A., Sipaut, C. S., Nasri, M., Ibrahim, M. N. M. *Bioresour. Technol.*, **102**, 3626–3628 (2011).
136. Sankhla, I. S., Bhati R., Singh, A. K., and Mallick, N. *Bioresour. Technol.*, **101**, 1947–1953 (2010).
137. Sasikala, C. and Ramana C. *Adv. Appl. Microbiol*, **42**, 97–218 (1996).
138. Santhanam, A. and Sasidharan, S. *African J. Biotechnol.*, **9**(21), 3144–3150 (2010).
139. Scott, G. *Polym. Degrad. Stab.* **68**, 1–7 (2000).
140. Shamala, T. R., Vijayendra, S. V. N., and Joshi, G. J. *Brazilian J. Microbiol.*, 1094–1102 (2012).
141. Shimizu, H., Tamura, S., Ishihara, Y., Shioya, S., and Suga, K. *Control of Molecular Weight Distribution and Mole Fraction in Poly(-D)(-)-(3-hydroxyalkanoates) (PHA)* Production by *Ralstonia eutropha*. In Biodegradable Plastics and Polymers, Y. Doi and K. Fukuda (Eds), Elsevier B. V.: New York, pp. 365–372 (1994).
142. Shimao, M. *Curr. Opin. Biotechnol.*, **12**, 242–247 (2001).

143. Simon-Colin, C., Raguenes, G., Cozien, J., and Guezennec, J. G. *J. Appl. Microbiol.,* **104**, 1425–1432 (2008).
144. Singh, A. K. and Mallick, N. *J. Ind. Microbiol. Biotechnol.*, **36**, 347–354 (2009).
145. Špoljaric I. V., Lopar, M., Koller, M., et al. *Bioresour. Technol.*, **133**, 482–494 (2013).
146. Stan-Lotter, H., Pfaffenhuemer, M., Legat, A., Busse, H., Radax, C., and Gruber, C. *Int. J. Syst. Evol. Microbiol.*, **52**, 1807–1814 (2002).
147. Steinbuchel, A. *Polyhydroxyalkanoic acids*, In: Biomaterials: novel materials from biological sources. D. Byrom (Ed.), Stockton: New York, 124–213 (1991).
148. Steinbuchel, A. and Pieper, U. *Appl. Microbiol. Biotehnol.*, **7**, 1–6 (1992).
149. Steinbuchel, A., Debzi, E. M., Marchessault, R. H., and Timm, A. *Appl. Microbiol. Biotechnol.* 1993 39, 443–449 (1993).
150. Steinbuchel, A. and Fuchtenbusch, B. *Trends Biotechnol.*, **16**, 419–427 (1998).
151. Sudesh, K., Bhubalan K., Chuah, J. A., Kek, Y. K., Kamilah, H., Sridewi N., and Lee, Y. F. *Appl. Microbiol. Biotechnol.*, **89**, 1373–1386 (2011).
152. Suzuki, T., Yamane, T., and Shimizu, S. *Appl. Microbiol. Biotechnol.* **24**, 370–374 (1986).
153. Teeka, J., Imai, T., Reungsang, A., Cheng, X., Yuliani, E., Thiantanankul, J., Poomipuk, N., Yamaguchi, J., Jeenanong, A., Higuchi, T., Yamamoto, K., and Sekine, M. *J. Ind. Microbiol. Biotechnol.*, **39**, 749–758 (2012).
154. Tian, P. Y., Shang, L., Ren, H., Mi, Y. L., Fan, D. D., and Jiang, M. *Afr. J. Biotechnol.*, **8**, 709–714 (2009).
155. Tindall, B. J. In The Prokaryotes (A. J. Balows (Ed), Springer-Verlag: New York, **1**, 754–808 (1991).
156. Upasani, V., and Desai, S. *Arch. Microbiol.,* **154**:589–593 (1990).
157. Valappil, S. P., Peiris, D., Langley, G. J., Herniman, J. M., Boccaccini, A. R., Bucke, C., and Roy, I. *J. Biotechnol.*, **127**, 475–487 (2007).
158. Valappil, S. P., Rai, R., Bucke, C., Roy, I. *J. Appl. Microbiol.*, **104**, 1624–1635 (2008).
159. Van, H. P., Elumbaring, A. C., Van der Lans, M. R., Van der Wielen, L. A. M. *J. Colloid Interface Sci.*, **297**, 595–606 (2006).
160. Van-Thuoc, D., Quillaguaman, J., Mamo, G., and Mattiasson, B. *J. Appl. Microbiol.*, **104**, 420–428 (2008).
161. Ventosa, A., Nieto, J., and Oren, A. *Mol. Biol. Rev.*, **62**, 504–544 (1998).
162. Verlinden, R. A. J., Hill, D. J., Kenward, M. A., Williams, C. D., and Radecka, I. *J. Appl. Microbiol.*, **102**, 1437–1449 (2007).
163. Vigneswari, S., Vijaya, S., Majid M. I. A., Sudesh, K., Sipaut, C. S., Azizan M. N. M., and Amir A. A. *J. Ind. Microbiol. Biotechnol.*, **36**, 547–556 (2009).
164. Wang, F. and Lee, S. Y. *Appl. Environ. Microbiol.*, **63**, 3703–3706 (1997).
165. Wong, P. A. L., Cheung, M. K., Lo, W. L., Chua, H., and Yu, P. H. F. *Material Res. Inovations*, **9**, 4–5 (2005).
166. Wonga, Y. M., Brigham C. J., Rha, C. K., Sinskey, A. J., and Sudesh, K. *Bioresour. Technol.*, **121**, 320–327 (2012).
167. Yamane, T., Chen, X. F., and Ueda, S. *Appl. Environ. Microbiol.*, **62**(2), 380–384 (1996).
168. Yellore, V. and Desai, A. *Lett. Appl Microbiol*, **26**, 391–394 (1998).
169. Yu, J. and Chen, L. X. L. *Biotechnol. Prog.*, **22**, 547–553 (2006).
170. Zhilina, T. N. and Zavarzin, G. A. *Curr. Microbiol.*, **29**, 109–112 (1994).
171. Zinn, M., Withholt, B., and Egli, T. *Adv. Drug Delivery Rev.*, **53**, 5–21 (2001).

Part 3: Bio-Nanocomposites in Packaging Applications

CHAPTER 10

BIO-NANOCOMPOSITES AND THEIR POTENTIAL APPLICATIONS IN FOOD PACKAGING

C. ANANDHARAMAKRISHNAN and USHA KIRAN KOLLI

ABSTRACT

Polymeric materials are widely used for food packaging applications and they are practically non-degradable, representing a serious global environmental problem. Besides, consumers are looking forward for high quality food products. The use of biopolymer based packaging materials can solve the waste disposal problem to a certain extent and acts as carrier to functional ingredients that ensures safety and quality of food. However, the use of biopolymers has been limited because of their poor mechanical and barrier properties. The problems with biopolymers are taken care by entirely new emerging group of materials "Bio-nanocomposites". These bio-nanocomposites are developed under the aegis of nanotechnology, has potential applications in the field of food packaging. They are formed by reinforcing biopolymers with small quantities of nanoscale inclusions that have high aspect ratios and able to improve the mechanical and barrier properties of the biopolymers. Additionally, bio-nanocomposites based packaging materials have great potential for enhancing food quality, safety, and stability. This chapter describes different types of bio-nanocomposites and their application in food packaging. This chapter also explores the antimicrobial effectiveness of biopolymer with metallic nanocomposites based on zinc oxide (Zno), titanium oxide (TiO), and silver (Ag). In addition, modeling of nanocomposite films for investigating the gas barrier properties, challenges involved in the nanocomposites application in food packaging, and safety issues are discussed.

10.1 INTRODUCTION

10.1.1 FOOD PACKAGING

Food products and beverages are usually packed to ensure safety, quality, and wholesomeness from the time of production to its consumption. Packaging material that is intended for food packaging should withstand harsh conditions during transport, distribution, marketing, and storage. Primarily package has to protect the food from chemical, biological, and physical agents. Besides protecting food, packaging material should offer traceability, convenience of handling, and tamper indication. But, the ultimate aim of food packaging is to contain food in a cost effective way that suits to the requirement of industries, consumer preferences more importantly with minimum environmental pollution (Marsh and Bugusu, 2007).

10.1.2 ROLE OF BIOPOLYMERS

Though there are so many packaging materials are available to satisfy the required objectives, use of plastics has its own importance becausethey are chemically resistant, inexpensive, lightweight, heat sealable, easy to print, and moldable in to any shape. However, plastics are failed in protecting environmental interests due to its poor biodegradability besides its wide range of physical, optical properties, and functional advantages (Marsh and Bugusu, 2007). Hence, over the last decade, most of the research works focused on the development of biodegradable polymers to contain food as well as environment. A variety of biodegradable packaging biopolymers such as polysaccharides, proteins, lipids, and their composites, derived from plant and animal resources have been investigated for the various packaging applications (Cuq et al., 1998; Rhim and Ng, 2007). Biopolymers have been considered as effective alternatives for nonbiodegradable petroleum based polymer films, since they are abundant, renewable, inexpensive, biodegradable, and environmental friendly (Rhim and Ng, 2007).

Biopolymers are natural polymers obtained from agricultural products or animals. Various biopolymers such as chitosan, polyethylene glycol (PEG), ethyl cellulose, methyl cellulose, soy protein, whey protein, cellulose and so onare widely used for preparation of biopolymer films. They serve as a barrier against moisture, oxygen, aroma, flavor as well as oil (Rhim and Ng, 2007). Biopolymers are biodegradable in nature and it has some additional beneficial properties such as improving food quality and extending the shelf-life of foods due to its antimicrobial properties. In addition, biopolymers are excellent carriers of food additives like antioxidants, antifungal agents, antimicrobials, colors, and other nutrients (Rhim and Ng, 2007). Proteins can be used to form networks and thus exhibits plasticity and film-forming nature. The film forming ability of several protein substances has been utilized in industrial application for a long time (Cuq et al., 1998). Among all the proteins, soy protein isolate(SPI) (a commercial form of soy protein containing more than 90% protein) and whey protein isolate (WPI) (a by-products from cheese making) are the potential source for biobased packaging materials due to their film forming properties (Kurose et al., 2007; Foegeding et al., 2002). Starch is also tested and used as alternative to synthetic polymers as it is abundantly available, renewable, less cost, and biodegradable (Fama et al, 2012). Chitosan, a polysaccharide biopolymer derived from naturally occurring chitin, displays unique poly cationic, chelating, and film-forming properties along with antimicrobial properties. Moreover, chitosan is biocompatible, non-toxic, biodegradable, and heat resistance. Blended use of more than two natural biopolymers can also provide a good opportunity for improving the property of packaging materials. However, the problems associated with biopolymers that limit the application in food packaging are low barrier properties, low heat distortion temperature, and poor resistance to processing operations. The emergence of nanotechnology provided a solution for the problem with an entirely new class of materials called bio-nanocomposites, which are made by incorporating nanomaterials (NMs) into the biopolymer matrix significantly improves its mechanical properties, barrier properties, and heat resistant properties.

10.2 EMERGENCE OF BIO-NANOCOMPOSITES

10.2.1 NANOTECHNOLOGY

Before proceeding toward the development of bio-nanocomposites for food packaging, it is important first to define the term "nanotechnology". Nanotechnology is the manipulation, arranging or self-assembly of individual atoms, molecules, or molecular clusters smaller than 100 nm into structures to create materials and devices with new or vastly different properties. Lagaron et al (2005) defined nanotechnology in a precise way that it is the creation and utilization of structures with at least one dimension in the nanometer length scale that creates novel properties and phenomena otherwise not displayed by either isolated molecules or bulk materials. The word nanotechnology is generally tagged, when referring to materials with the size of 0.1–100 nm. However, there is no scientific reason in support of this specific upper limit (Cushen et al., 2012). In addition, nanotechnology is a multidisciplinary science,in which the concepts of physics, chemistry, and biology are to be applied.

10.2.2 ROLE OF NANOTECHNOLOGY IN FOOD PACKAGING

Application of nanotechnology in food packaging sector is increasing day by day and opens new possibilities with great benefits. The market for food packaging containing NM is around $860 million in sales world-wide in 2006 and has been predicted to reach $30billion in the next 10 years (Coles et al., 2003).

Nanoclays especially montmorillonite (a volcanic material consisting of nanometer thick platelets as a common source for producing nanoclays) have been recognized as a widely used NM in food packaging. Its availability, low cost, significant enhancement in properties, simple processability, and biodegradability makes it as most sort after NM used in manufacture of nanocomposites for food processing (Quarmley and Rossi, 2001,Azeredo, 2009).

Similar to plastic polymer laminates, nanolaminates consists of 2 or more layers of NMs bonded either physically or chemically are made into an extremely thin food grade film. Nanolaminates allow themselves in preparation of edible films, which are used in packaging a variety of foods such as fruits, vegetables, meats, chocolate, candies, bakery products, and French fries. These edible nanolaminates are manufactured from polysaccharides, proteins,and lipids, offer protection against moisture, lipids, and gases, help in improving the textural properties of foods and serves as carriers of colors, flavors, antioxidants, nutrients, antimicrobials, and anti-browning agents (Morillon et al., 2002; Cagri et al., 2004; Cha and Chinnan, 2004; Rhim, 2004; Phan et al., 2008; Ponce et al., 2008). By nature, nanolaminates made out of polysaccharides, proteins are good barriers of oxygen and carbon dioxide but poor barriers against moisture. While the lipid based nanolaminates are good moisture barriers but poor in mechanical and gas barrier properties (Park, 1999). Because, nanolaminates based on polysaccharides, proteins, or lipids individually cannot provide all desired properties researchers are finding ways for alternatives to improve them with suitable additives.

Proper monitoring of food material during storage is necessary as food preservation is equally important to food processing. Sensors made using nanoparticles (NPs) as reactive particles can be used in food packaging to monitor food material inside the package. These nanosensors are able to respond to environmental changes inside the package like temperature, relative humidity, oxygen exposure, and also product degradation or microbial contamination in real time with high sensitivity (Bouwmeester et al., 2009). Nanosensors helps in reducing the time of detection so these could be directly used in packaging material and also serve as electronic tongue or nose by detecting chemicals released during food spoilage (Lange, 2002, Garcia, 2006, Bhattacharyaand Biomems, 2007).

Active food packaging is an innovative solution to meet the continuous changes in current consumer demands and market trends. Antimicrobial packaging is one of the active food packaging technology offers several advantages compared with the direct addition of preservatives into food products. The incorporation of antimicrobial agents into polymeric films allows industry to combine the preservative functions of antimicrobials with the protective functions of the pre-existing packaging concepts (Appendini and Hotchkiss, 2002). AgNPs have been shown to be effective antimicrobials (Tankhiwale and Bajpai, 2009), even more effective than larger Ag particles, due to larger surface area available for interaction with microbial cells and also since, they are able to attach more copies of microbial molecules and cells (Luo and Stutzenberger, 2008; Kvitek et al., 2008).

10.2.3 NANOCOMPOSITES

Nanocomposite is a polymer film, in which a few weight percentages of mineral fillers such as clay, silica, and talc, in nanoscale range are randomly and homogeneously dispersed on a molecular level in order to improve the mechanical and physical properties (Rhim, 2007). The incorporation of these reinforcements significantly increases the mechanical and barrier properties and produce films superior to the pure polymers or conventional composites. The first nanocomposite (nylon-clay) was invented by Toyota Central R&D Labs, Inc. in 1985, opened new doors in the field of polymer science (Okada and Usuki, 2006). Initially, it has been used in automobile industry and at later stage the concept of nanocomposites was extended to other areas like food packaging and electrical industries (Okada and Usuki, 2006). When these nanocomposites used in food packaging, they are able to withstand the stress of thermal food processing, transportation, and storage (Sinha Ray and Okamoto 2003; Thostenson et al., 2005). If the inorganic polymer matrix in nanocomposite replaced with bio-based polymer that is called as bio-nanocomposite and it is readily degradable in the environment without accumulation in the landfills.

10.2.4 NANO-REINFORCEMENTS

Nano-reinforcements are either inorganic or organic fillers with certain geometries such as fibers, flakes, spheres, and particulates, which have at least one dimension in the nanometric range (NPs). Based on how many dimensions are in the nanometric range, these are divided into three categories such as isodimensionalNPs, nanotubes, and polymer-layered crystals (Figure1). Spherical silica NPs or semiconductor nanoclusters, also known as IsodimensionalNPs, which have three dimensions in nanometric range. Nanotubes or whiskers are elongated structures, which have two nanometric dimensions, while the third is larger. Third category is having only one dimension in the nanometer range, known as polymer-layered crystal nanocomposites (Alexandre andDubois, 2000). In addition, newer forms of carbon reinforcements as buckyballs, nanofibers, and nanotubes having extraordinarily high aspect ratio could hasten the commercialization of nanocomposites despite their higher cost.

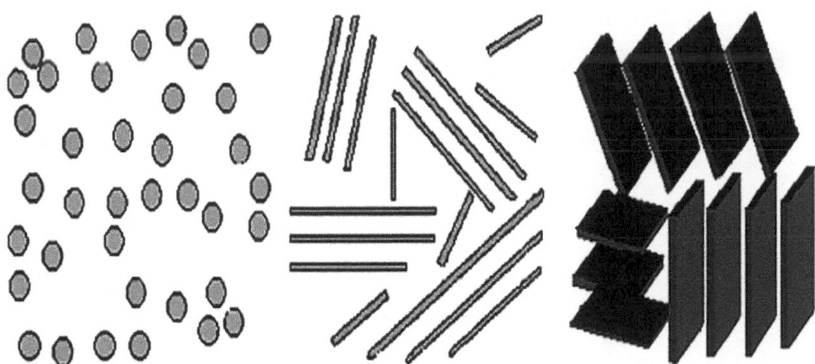

FIGURE 1 Nanomaterials: (i) in 3-dimensions as spheres, (ii) in 2-Dimensions as rods or fibres, and (iii) in 1-dimension as plates (Bradley et al., 2011).

Graphene sheets, with single layers of graphite, as nano-reinforcements enhance electrical conductivity, heatresistance, barrier properties, and mechanical properties of polymers are promising in food packaging applications. Use of graphene sheets is cost effective comparatively carbon nanotubes but delamination is problem (Goettler et al., 2007). However, the most widely used nanoreinforcements in nanocomposites include layered silicate mineral materials such as montmorillonite (MMT), hectorite, saponite, and kaolinite (Table1).

TABLE 1 Major groups of clay minerals (Faheemuddin, 2008)

Solution	Group Name	Member Minerals	General Formula	Remarks
1	Kaolinite	Kaolinite, Dickite, Nacrite	$Al_2Si_2O_5\ (OH)_4$	Members are polymorphs (composed of the same formula and different structure)
2	Montmorill-onite or smectite	Montmorillonite, pyrophyllite, talc, vermiculite, sauconite, saponite, nontronite	$(Ca,Na,H)(Al,Mg,Fe,Zn)_2\ (Si,Al)_4O_{10}(OH)_2\text{-}XH_2O$	X indicates varying level of water in mineral type

TABLE 1 *(Continued)*

3	Illite	Illite	$(K,H)Al_2 (Si,Al)_4 O_{10}(OH)_2$-$XH_2O$	X indicates varying level of water in mineral type
4	Chlorite	(i) Amesite, (ii) Chamosite, (iii) Cookeite, (iv) Nimite etc.	(i)$(Mg,Fe)_4Al_4Si_2O_{10}(OH)_8$ (ii) $(Fe,Mg)_3 Fe_3 AlSi_3O_{10}(OH)_8$ (iii) $LiAl_5 Si_3O_{10} (OH)_8$ (iv) $(Ni,Mg,Fe,Al)_6AlSi_3O_{10}(OH)_8$	Each member mineral has separate formula, this group has relatively larger member minerals and is sometimes considered as a separate group, not as part of clays

10.2.5 STRUCTURE AND DIFFERENT FORMATIONS OF NANOCOMPOSITES

The MMT, a hydrated alumina-silicate layered clay consisting of an edge-shared octahedral sheet of aluminum hydroxide between two silica tetrahedral layers (Figure 2) with dimensions, 1 nm thick and 100–500 nm in diameter, result in platelets of high aspect ratio (Weiss et al., 2006; Uyama et al., 2003).

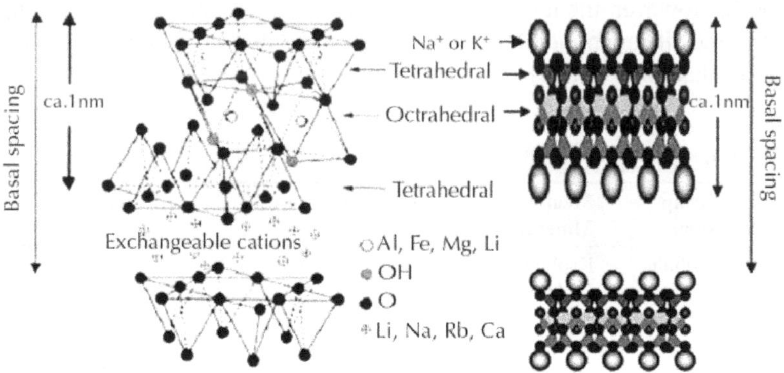

FIGURE 2 Structure of 2:1 layered silicates (Sorrentino et al., 2007).

The total clay structure is formed by hundreds of layered platelets stacked into particles or tactoids 8–10 μm in diameter. The imbalance of the surface negative charges is compensated by exchangeable cations typically Na^+ and Ca^{2+} and forms two different MMTs:Sodium MMT(Na^+MMT) and calcium MMT(Ca^{2+}MMT), respectively (Azeredo, 2009). The MMT has a property that it undergoes intercalation and swelling in the presence of water and organic cations.

There are 3 types of polymer–clay formations (Figure 3), namely:

(a) Tactoidor phase separated,
(b) Intercalated, and
(c) Exfoliated(Carrado 2003,Sinha Ray and Okamoto 2003, Okada and Usuki 2006).

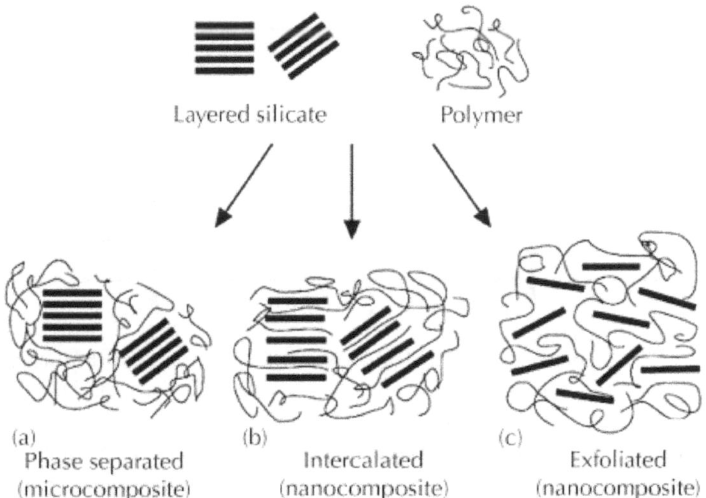

FIGURE 3 Types of composite derived from interaction between clays and polymers:(a) phase-separated microcomposite, (b) intercalatednanocomposite,and (c)exfoliated nanocomposite (Alexandre and Dubois, 2000).

Micro composite or tactoid structures will be formed, when interlayer spaces of the clay gallery remain unexpandeddue to poor affinity of filler material with the polymer. This formation is not desirable in nanocomposites (Alexandre and Dubois, 2000). The other formation intercalated structures are derived at moderate expansion of the clay interlayer due to moderate affinity between polymer and clay. Polymer chains penetrate the basal spacing of clay result insight expansion of spaces between the layers. But, the shape of the layered stack remains unchanged. In the case of exfoliated structures due to a high affinity between polymer and clay (Figure 3), the clay clusters lose their layered identity and are separated into single sheets within the continuous polymer phase and distributed homogeneously throughout the polymer phase. This is the most desirable formation to take advantage of nanoclays high surface area to its fullest extent (Arora and Padua, 2010).

10.2.6 PERMEABILITY MECHANISM OF BIO-NANOCOMPOSITES

Bio-nanocomposite film showed improved barrier properties by dispersing NPs in the biopolymer matrix, which in turn provide a tortuous path for gas molecules to pass through.The Figure 4 clearly depicts the difference between the gas permeability in pure virgin polymer films and through incorporated nanocomposite polymer matrix. In the case of nanocomposite polymer matrix (Figure 3(b)), gas molecules diffuse around the NPs coated region (tortuous pathway) and thus resulted in longer mean path for gas diffusion through the film. This increase in effective path length for water vapor diffusion improves the film barrier properties (Rhim and Ng, 2007). This allows the manufacturer to attain larger effective film thicknesses, while using smaller amounts of polymer (Duncan, 2011).

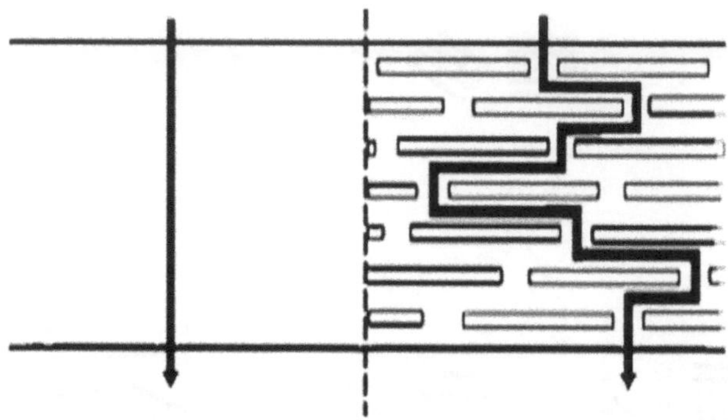

FIGURE 4 (a) Penetration of gas molecules in a pure polymer matrix and (b) Tortuous pathway created by incorporation of nanocomposite on polymer matrix (Duncan, 2011).

The NPs in the biopolymer film have much higher aspect ratios due to their high surface area to volume ratios. Moreover, NP interfacial volume element in a polymer nanocompositefilmis significantly greater than that of a polymer micro-composite created from the same materials. The barrier properties are further enhanced if the filler is less permeable, well dispersed in the matrix and with a high aspect ratio (Lagaron et al., 2004).

10.3 BIO-NANOCOMPOSITES APPLICATIONS IN FOODPACKAGING

Recently, a new class of materials called bio-nanocomposites has proven to be a promising option in improving mechanical and barrier properties of biopolymers and also to give a best solution to food industry concerns about environmental pollution. The bio-nanocompositesconsist of a biopolymer matrix reinforced with nanoscale inclusions having at least one dimension in the nanometer range (1–100 nm) and exhibit much improved properties due to high aspect ratio and high surface area of NPs (Zhao et al., 2008). Various biopolymers used in the production of bio-nanocompositesinclude plant-derived materials (Table 2) such as plant proteins, starch, cellulose, other polysaccharides, and animal products such as proteins and polymers synthesized chemically from naturally derived monomers such as PLA (Arora and Padua, 2010).

TABLE 2 Different loading levels of bio-nanocompositesand its effects

Biopolymer	Nanofiler	% Loading	Improvements	Authors
Agar	Montmorillonite (MMT)	10%	TS increased 22% WVP lowered 50%	Rhim, 2011

TABLE 2 *(Continued)*

Starch	Montmorillonite (MMT)	6%	TS increased 1.9 times, WVP lowered 2.1 times	Maksimov et al., 2009
Starch	Carbon nanotubes	0.055%	WVP lowered 43%	Fama et al., 2012
Starch	Na + MMT	5%	TS increased 35%	Chung et al., 2010
Cellulose	Tourmaline	4–8%	Good TS 92–107 Mpa, Antibacterial properties	Ruan et al., 2003
Pectin	Montmorillonite (MMT)	10%	WVP lowered 35%	Castello et al., 2010
Carrageenan	Mica	1 or 5%	WVP lowered 86 and 83% reduce UV-light transmission	Sanchez-Garcia et al., 2010
Soy Protein	Montmorillonite (MMT)	10%	TS increased 6times WVP lowered 43%	Kumar et al., 2010
Wheat protein	Montmorillonite (MMT)	5%	TS increased 6 times	Angellier-Coussy et al., 2008
Bovine gelatin	Montmorillonite (MMT)	5%	WVP lowered 3 times	Martucci & Ruseckaite, 2010
Fish gelatin	Montmorillonite (MMT)	5%	TS increased 33% WVP lowered 71%	Bae et al., 2009
Whey protein	Montmorillonite (MMT)	5%	WVP lowered 28%, Antibacterial properties	Sothornvit et al., 2009
PLA	Kaolinite	5%	OP lowered 50%	Cabedo et al., 2006
PHBV	Montmorillonite (MMT)	3%	Improved thermal, mechanical properties	Chen et al., 2002
Chitosan	Montmorillonite (MMT) and Rosemary Essential oil	1 or 3%	WVP lowered 40 and 45% Antimicrobial activity Antioxidant property	Abdollahi et al., 2012

10.3.1 STARCH BASED BIO-NANOCOMPOSITES

In recent days, starch has received great attention of researchers in the food packaging sector due to its abundant availability, biodegradability, and the low cost (Avella et al., 2005). Starch is a semi crystalline storage polysaccharide in most of the plants made up of repeating units of linear amylose (1,4-α-D-glucopyranosyl units) and branched amylopectin (α-1,4-linked backbone and α-1,6-D-glucopyranosylunits linked branches) (Dean et al, 2008). The films made out of starch alone are brittle and low moisture barriers. Therefore, it needs to be blended with other biodegradable materials to improve the properties. Polyvinyl alcohol (PVOH) one of such material produced through hydrolysis of polyvinyl acetate in producing nanocomposites (Ali et al., 2011).The composite produced by these two polymers are highly compatible, biodegradable, and also showed improved mechanical properties compared to the films made alone by starch. However, compared to starch, the PVOH is expensive and also poor moisture barrier. Ali et al. (2011) incorporated MMTNM into composite and reported that significant increase in percent elongation and tensile properties compared to starch/MMT nanocomposite. The results exhibited that the starch content can be increased up to 50% without any change in mechanical properties of the film, which also ensures low production cost.

Recently, Fama et al. (2012) reported nanocomposites based on starch reinforced with CNTs wrapped with a starch-iodine complex allows in getting starch based nanocomposites with very small amounts, which otherwise costs more. The resultant nanocomposites exhibited important improvements in storage modulus by 100% and water vapor permeability (WVP) properties lowered by 43% respect to the starch matrix. However, the so formed packaging film reinforced with CNTs cannot be used directly as food contact materials but could serve as secondary packaging application.

10.3.2 CELLULOSE BASED BIO-NANOCOMPOSITES

Cellulose is the most abundant naturally occurring low cost resource with unbranched, linear chains of D-glucose molecules linked by 1,4-D-glucosidic bonds. Because of its attractive properties like biodegradability, renewability, biocompatibility, non-toxicity, and low cost (Klemm,et al., 2005; Simkovic, 2008), it has a great potential of being used as a biopolymer, which can be used in food packaging. However, the processing of cellulose is difficult due to its crystalline structure and high molecular weight. It is neither moldable nor soluble in water or any other common solvents due to its aggressive hydrogen bonds. Cellulose regeneration by using suitable solvents is a recommended solution for the above mentioned problem. N-methylmorpholine-N-oxide is a solvent one among them, which is environment friendly system that can convert cellulose into its derivatives (Rhim and Ng, 2007). Recently, room temperature ionic liquids have been employed for cellulose regeneration due to its attractive properties such as good chemical and thermal stability, low flammability, low melting point, ease of recycling, and eco-friendly nature (Mahmoudian et al, 2012). The derivatives of cellulose after regeneration includes methylcellulose (MC), carboxymethyl cellulose (CMC), hydroxypropyl cellulose (HPC), and hydroxypropyl methylcellulose (HPMC) and cellulose esters such as cellulose acetate (CA), cellulose acetate propionate (CAP), and cellulose acetatebutyrate (CAB). Among them,CA is used particularly for preparation of nanocomposites because of its functional properties like biodegradable polymer and has excellent optical clarity and high toughness. Ruan et al. (2003) prepared cellulose/tourmaline nanocrystal composite films in solvent casting method by usingTourmaline, a mixed stone, is a naturally complex group of hydrous silicate minerals containing Li, Al, B, and Si and various quantities of alkalis (K and Na) and metals (Fe, Mg, and Mn). Its structural formula is $Na(Li, Al)_3Al_6(BO_3)_3Si_6O_{18}(OHF)_4$.

Preparation of cellulose/tourmaline nanocomposite was by dispersing tourmaline nanocrystalsNaOH/thiourea aqueous solutions and then blended with cellulose aqueous solutions. These aqueous solutions are coagulated with $CaCl_2$. Characterization of the filmrevealed tourmaline nanocrystals structure is remain unaltered in the cellulose matrix and tourmaline contentof 4–8 wt% composite films possessed good tensile strength (92–107 MPa) and exhibited antibacterial properties against *Staphylococcus aureus* (Ruan et al., 2003).

In recent times, some extracellular microbial polysaccharides, namely bacterial cellulose (BC), and pullulan (P), are attracting researchers with their unique and peculiar features (Klemm et al., 2005). Trovatti et al. (2012) prepared highly translucent bio-nanocompositeswith improved properties based on two microbial polysaccharides, BC, and pullulan, reported interesting results. Microorganisms belong to the *Gluconacetobacter* genus produce BC, in the form of a high purity swollen membrane with excellent physical and mechanical properties, arising from its tridimensional and branched nano and microfibrillar structure. Pullulan, a water soluble linear polysaccharide is produced aerobically by certain strains of the polymorphic fungus *Aureobasidiumpullulans*. This homopolysaccharide is made up of maltotriose units linked with α-(1–6) linkages produces flexible polymeric chains with enhanced solubility. Pullulan forms films that are clear, highly oxygen-impermeable, non-toxic, edible, biodegradable, biocompatible to human and environment, and present good mechanical properties, allowing their use in packaging of dried foods (Krochta and DeMulder, 1997). Trovatti et al. (2012) prepared highly translucent bio-nanocompositeswith improved properties based on two microbial polysaccharides, BC, and pullulan, reported interesting results. The BC and pullulan used as raw materials for the development of sustainable composite materials because of their inherent properties, fibrillar morphology (reinforce element), and good film forming ability (matrix), respectively. All bio-nanocomposites incorporated BC showed considerable improvement in thermal stability and mechanical properties, it is evident that significant increase in the degradation temperature (up to 40°C) and on both Young's modulus and tensile strength increments of up to 100and 50%, for films without glycerol and up to 8,000 and 7,000% for those plasticized with glycerol. It is concluded that these novel sustainable nanocomposite films, which are completely biodegradable find place in packaging of dry foods, transparent organic electronics, and also in biomedical applications.

Mahmoudian et al. (2012) developed regenerated cellulose/MMTnanocomposite films by regenerating cellulose using ionic liquid 1-butyl-3-methylimidazolium chloride (BMIMCI). This is the first work, where ionic liquids are used for regeneration of cellulose. Properties of ionic liquids such as good chemical and thermal stability, low flammability, low melting point, ease of recycling, and eco-friendly nature facilitated to produce an environmentally friendly regenerated cellulose/MMTnanocomposites with improved tensile strength, moisture, and gas barrier properties compared to regenerated cellulose alone.

10.3.4 PROTEIN-BASED BIO-NANOCOMPOSITES

SOY PROTEIN

The brilliant film forming properties allowed soy proteins to use as a potential biopolymer source. But, the poor mechanical and barrier properties limit its use as a packaging material. In addition, soy based films are brittle and needs plasticizers to be added in order to get the required flexibility, which leads to decrease in tensile strength (Wang et al., 1996). Though their oxygen permeability values are more or less similar to plastics, their WVP is higher compared to normal plastic polymers due to their hydrophilicity (Brandenburg et al., 1993; Kumar et al., 2010). Therefore, it is necessary to come out with techniques to improve mechanical and water vapor barrier properties of soy protein-based packaging materials. One of such technique is reinforcing the soy protein polymer matrix with MMT. Kumar et al. (2010) prepared bio-nanocomposite films based on SPI and MMT through melt extrusion with use of glycerol as plasticizer. Effects of the pH, MMT content, and extrusion processing parameters (screw speed and barrel temperature distribution) on the structure and properties of SPI-MMT bio-nanocomposite films were analysed. Addition of MMT at 10% and 15% to the polymer matrix significantly increases the tensile strength to a value that can be comparable to the mostly used packaging materials for food packaging. However, their very less percent elongation values limit the application of these SPI-MMT films as food packaging materials. The inherent hydrophilic nature of protein-based films increases the WVP, which is a major drawback in using protein based films as a packaging material (Kumar et al., 2010). Though the addition of MMT increased from 0–15% significantly reduces WVP by as much as 42.9%, the WVP values for SPI-MMT films are still much higher as compared to those for plastics such as low density polyethylene (LDPE), polypropylene (PP), and polyvinylidenechloride (PVDC). This might limit the application of these bio-nanocomposite films to packaging of high moisture foods like fresh fruits and vegetables.

WHEAT PROTEIN

Gluten is a mixture of two main proteins, gliadins, and glutenins and it is a byproduct of the wheat starch industry, commercially available at low cost. It shows unique viscoelastic properties and low water solubility (Mauricio-Iglesias et al., 2010). In addition, wheat gluten (WG) films displays gas barrier properties particularly against oxygen and carbon dioxide, which make the film suitable for packaging of fruits and vegetables even at high relative humidity (Mujica Paz et al., 1997; Barron et al., 2002). However, their low mechanical resistance and high water sensitivity narrow down their application in food packaging especially for dry and intermediate water activity foods. With the aim of widen the applications of WG films, Angellier-Coussy et al. (2008) added MMT to modify its mechanical properties and water sensitivity. They prepared WG/MMTnanocomposites by a thermo mechanical process consisting of mixing the components in a two-blade, counter-rotating device, and followed by thermoforming the obtained dough. The introduction of MMT into WG significantly increased the resistance at filler content of 5 wt% (Shewry et al., 1997).The formation of covalent bonds within the films led to a decrease in the hydrophilicity of the films (Schofield et al., 1983). Besides all the above, the use of WG-based nanocomposite materials in packaging are restricted because of color changes induced by the use of high thermoforming temperatures and/or high filler contents.

ANIMAL PROTEIN (GELATIN)

Gelatin is a complex polypeptide widely used in the food, pharmaceutical, photographic, and cosmetic manufacturing. Gelatin was one of the primitive materials used for the formation of biopolymer films due to the abundance of raw material, low production cost, global availability, and excellent film forming properties (Vanin et al., 2005). Hence, inexpensive bovine and fish gelatin containing a large number of suspended functional groups can facilitate chemical cross-linking and derivatization and used as raw material for development of food packaging films (Dean and Yu 2005; Jongjareonrak et al., 2006). However, the inherent sensitivity to moisture and susceptible to lose their dimensional stability easily at room temperature, poor mechanical, and barrier properties limit their application as a food packaging material (Krochta and DeMulder, 1997; Gomez-Guillen et al., 2009). Further, fish gelatin is having relatively low gelling and melting temperatures and poor mechanical properties, when compared to mammalian gelatin due to lower amounts of the proline and hydroxyproline (Bower et al., 2006; Yi et al., 2006). Bae et al. (2009) combined fish gelatin with layered Na-MMT(nanoclay) and they showed significant increase in mechanical properties and the barrier properties against oxygen and water vapor. Martucci and Ruseckaite (2010) obtained biodegradable film by mixing of bovine gelatin solutions with ultrasonically pre-treated clay suspensions under controlled conditions. Characterization of film revealed that produced nanocomposites maintained the good optical transparency with MMT loading, indicating that filler was mostly distributed at the nanoscale. The water vapor barrier of film improved threetimes and comparable to synthetic polymers CA but lesser than those of high-density polyethylene (HDPE), polyvinyl chloride (PVC), and low-density polyethylene(LDPE).

WHEY PROTEIN

The WPI, which is a by-product in the cheese industry, has received much attention for use as a potential edible or biodegradable food packaging material. It has been shown to produce transparent films besides excellent oxygen, aroma, barrier properties, and mechanical properties (Sothornvit and Krochta, 2000 and 2005). Whey protein by nature, form brittle films, and makes unsuitable for packaging application.To overcome this problem, plasticizer (a low-molecular-weight nonvolatile substance) can be added into the film to reduce protein chain-to-chain interaction. It results in increase the mobility of polymer chains and more flexible films (Banker, 1966). But, plasticizers also increase the film permeability (Gontard et al., 1993), which is undesirable for food quality. Sothornvit et al. (2009) prepared WPI-based composite films with three different types of nano-clays such as Cloisite Na^+, Cloisite 20A, and Cloisite 30B, at 5% (w/w) using a solution casting method, to demonstrate the effect of incorporation of clay in the film in order to increase the barrier properties by using glycerol as a plasticizer. Results exhibited that the film tensile strength (TS) is reduced with addition of Cloisite 20A clay because of the incomplete dispersion of the nano-clay (Cloisite 20A) into the polymer matrix, which is caused by the incompatibility of hydrophobic nano-clay with hydrophilic biopolymer. Tensile strength of WPI/Cloisite Na^+ and WPI/

Cloisite 30B composite films are not much affected compared to control WPI film. Water vapor barrier properties were increased by the addition of clay. Interestingly, the results reported that nano-clays especially Cloisite 30B incorporated into polymer matrix exert antimicrobial properties distinctively on Gram positive bacteria. This is due to the fact of quaternary ammonium group in the silicate layer of Cloisite 30B, which disrupts bacterial cell membranes and causes cell lysis (Rhimet al., 2006). This property will benefit food packaging applications for a variety of foods such as meat, fish, poultry, cereals, cheese, fruits, and vegetables to extend the shelf-life, improve quality, and enhance safety of food (Labuza and Breene, 1988; Cha and Chinnan, 2004; Cagri et al., 2004; Han, 2005; Sothornvit et al., 2010).

10.3.4 PECTIN BASED BIO-NANOCOMPOSITES

Pectins are natural polymers, which are white, amorphous, and complex carbohydrates that occur in ripe fruits and certain vegetables. Pectin rich fruits are the peach, apple, currant, and plum (Coffin and Fishman, 1994; Suvorova et al., 2003). They are based on chains of linear regions of 1,4-α-D-galacturonosyl units and their methyl esters, interrupted in places by 1,2-α-L-rhamnopyranosyl units (Mitchell et al., 1998). Castello et al. (2010) developed nanocomposites using pectin obtained from food processing industry wastes (citrus fruit wastes) and carrageenan from seaweeds. Incorporation of modified MMTat 10% reduced the WVP by 35%. Biopolymers obtained from pectin and carrageenan are cohesive and transparent and also exhibits good oxygen and carbon dioxide barrier properties but they have poor vapor barrier properties, limiting their application in food packaging (Castello et al., 2010).

10.3.5 POLYLACTIC ACID (PLA) BASED BIO-NANOCOMPOSITES

The PLA is a polymer of lactic acid produced either by chemical synthesis from bio-derived (homo-fermentation and hetero-fermentation) monomer. The microorganisms and process conditions includes species of the *Lactobacillus* genus such as *Lactobacillus delbrueckii, L. amylophilus, L. bulgaricus, L. leichmanii,*and a pH range of 5.4–6.4, a temperature range of 38–42°C, and a low oxygen concentration (Jamshidian et al., 2010). The PLA has caught attention as it found potential applications of food packaging because of its sustainable, biocompatible, and biodegradable nature with good mechanical and optical properties (Balakrishnan et al., 2012). Moreover, It has been classified as generally recognized as safe (GRAS) by the United State Food and Drug Administration (FDA) and it is safe for all food packaging applications (Conn et al., 1995; FDA, 2002). The PLA is commercialized and some of the companies such as Cargill Dow, McDonalds Mitsui, and Ecocard are using as food packaging material (Jamshidian et al., 2010). The PLA is suitable packaging material for packaging fresh food like fruits, vegetables,salads, and foods, which are not sensitive to oxygen due to its poor gas barrier properties. Addition of MMT-layered silicate to PLA may result in a nanocomposite with good barrier properties that is suitable for film packaging material. Cabedo et al. (2006) prepared a PLA nanocomposite by incorporating chemically modified kaolinite into the PLA matrix. Kaolinite showed a good interaction with PLA and thus resulted in 50% increase in oxygen barrier properties. The negative effect of plasticizer on gas barrier properties of the film is overcome by formation of kaolinite nanocomposite. Kim et al. (2009) studied the effect on transparency of the PLA nanocomposites made by using BC as nano reinforcements. The results revealed that the transparency was not affected by addition of BC but the Young's modulus and tensile strength of the nanocomosites were increased by 146% and 203%, respectively compared to virgin PLA polymer.

10.3.6 CHITOSAN BASED BIO-NANOCOMPOSITES

Chitosan, a deacetylated derivative of chitin, found in shellfish is the secondmost abundant polysaccharide in the nature after cellulose. It has been found to be non-toxic, biodegradable, biofunctional, and biocompatible. Unlike other polysaccharides, it has amino group, NH_2, which can be readily protonated in to NH_3^+ in an acid environment, exerts antimicrobial activity (Shahidi et al., 1999; Xu et al., 2006; Rinaudo, 2006; Srinivasa and Tharanathan, 2007). However, poor mechanical and gas barrier properties, hydrophilic character, and weak water resistance limit its application particularly in the presence of water and humidity (Wang et al., 2005; Xuet al., 2006). Abdollahi et al. (2012) prepared chitosan nanocomposites incorporated with MMT-nanoclay and rosemary essential oil (REO) to improve its physical and mechanical properties as well as antimicrobial and antioxidant behavior. The REO is derived from rosemary plant (*Rosmarinusofficinalis L.*) and it is drawing interest because of its effective natural antioxidant properties (Waszkowiak, 2008) for which most of the food industries are looking for (Du Plooyet al., 2009; Sanchez-Gonzalez et al., 2010).Results of XRD pattern showed MMT exfoliation and FTIR spectra demonstrated good interaction between chitosan and MMT. Incorporation of 1 or 3 w% MMT into chitosan decreased WVP about 40 and 45%, respectively. The WVP decreased further in the presence of REO. The antimicrobial properties of REO incorporated chitosan/MMT nanocomposites make them suitable for food preservation. The results also confirmed the compatibility of REO with chitosan/MMT nanocomposite to produce an active bio-nanocomposite for food packaging (Abdollahi et al., 2012).

10.3.7 ANTIMICROBIAL BIO-NANOCOMPOSITES

Active packaging is important technique to be considered to maintain the quality, wholesomeness and safety of foods. Antimicrobial packaging is one such active packaging, which has receiving attention in packaging of foods includes meat, fish, poultry, bread, cheese, fruits, and vegetables. Here, mode of application of antimicrobial agents is a key parameter to realize the benefits of antimicrobial packaging to fullest extent (Rhim and Ng, 2007). The ideal way would be development of biopolymer-based antimicrobial nanocomposite films impregnating with food grade antimicrobials into nanocomposite matrix. Chitosan is considered to be ideal antimicrobial film due to its compatibility,inherent antimicrobial activity, and film forming capacity (Rhim and Ng, 2007). The drawback in use of chitosan film includes moisture sensitivity can be overcome by incorporating nano-reinforcements into the polymer matrix. Various antimicrobial agents includes organic systemssuch as sorbate, propionate, and benzoate, bacteriocins, lysozyme and inorganic systems like Ag, copper, TiO, and Zno have been identified for the use in antimicrobial bio-nanocomposites.However, noble metals and metal oxidessuch as Ag, titanium dioxide (TiO_2), ZnO have got researchers interest due to their stability at high temperatures and antimicrobial activity (Simoncic and Tomsic, 2010).

SILVER NANOPARTICLES

Among inorganic antimicrobial systems ionic Ag has great role to play because of its great antimicrobial capacity against broad range of Gram-positiveand Gram-negative-microorganisms. Elemental or metallic Ag is a malleable and ductile transition metal with a white metallic luster appearance and has high optical reflectivity compared to other metals (Edwards and Petersen., 1936). Nanometer-sized Ag particles have been known for long time but had paid little attention (Loket al., 2006). Later, the recent advances in research on metal NPs appear to revive the use of AgNPs for antimicrobial applications. It has been revealed that Ag NPs prepared with a variety of synthetic methods have efficient antimicrobial activity (Loket al., 2006). Nanosilver have chemical and biological properties that are appealing to the consumer products, food technology, textiles/fabrics, and medical industries. Nanosilver was an effective killing agent against broad spectrum of Gram-negative and Gram-positive bacteria (Wijnhovenet al., 2009), including antibiotic-resistant strains (Wright et al., 1998). Gram-negative bacteria include genera such as *Acinetobacter*, *Escherichia*, *Klebsiella*, *Pseudomonas*, *Salmonella*, *Shigella*, and *Vibrio*. Gram-positive bacteria include several well-known genera such as *Bacillus*, *Clostridium*, *Enterococcus*, *Listeria*, *Staphylococcus*, and *Streptococcus*. Antibiotic-resistant bacteria include strains such as methicillin-resistant and vancomycin-resistant *Staphylococcus aureus*, *Enterococcus faecium*, and extended spectrum β-lactamase producing *Klebsiella*. In addition, Ag NPs are toxic to fungi such as *Aspergillusniger*, *Candida albicans*, and *Trichophytonmentagrophytes* and yeast isolated from bovine mastitis. Likewise, they were also resistant towards algae(for example,*Chlamydomonasreinhardtii*) and phytoplankton (for example, *Thalassiosiraweissflogii*) (Percival et al., 2007). Though the antimicrobial effect of Ag ions has been studied extensively, the effects of nanosilver on bacteria and the bactericidal mechanism are only partially understood. Various studies made known that AgNPs anchor to and penetrate the cell wall of Gram-negative bacteria (Moroneset al., 2005; Sondi and Salopek-Sondi, 2004). It is practical to suggest that the resultant structural change in the cell membrane could cause an increase in cell permeability, leading to an uncontrolled transport through the cytoplasmic membrane and ultimately cell death. Moreover, it has been proposed that the antibacterial mechanism of AgNPs is related to the formation of free radicals and subsequent free radical-induced membrane damage (Danilczuket al., 2006; Kim et al., 2007). Because of the membrane damage caused by the NPs, the cells cannot efficiently extrude the Ag ions and limit their effect based on the greater tendency of Ag ions to efficiently interact with thiol groups of essential enzymes and phosphorus-containing bases (Moroneset al., 2005). It is probable that further damage could be caused by interactions with DNA. This interaction may prevent cell division and DNA replication from occurring and ultimately leading to cell death.

Figure 5shows nano-scaled Ag interaction with bacterial cells. Nano-scaled Ag may:

1. Release Ag ions and generate ROS,
2. Interact with membrane proteins affecting their correct function,
3. Accumulate in the cell membrane affecting membrane permeability, and
4. Enter into the cell, where it can generate ROS, release Ag ions, and affect DNA.

Generated ROS may also affect DNA, cell membrane, and membrane proteins, and Ag ion release will likely affect DNA and membrane proteins (Marambio-Jonesand Hoek, 2010).

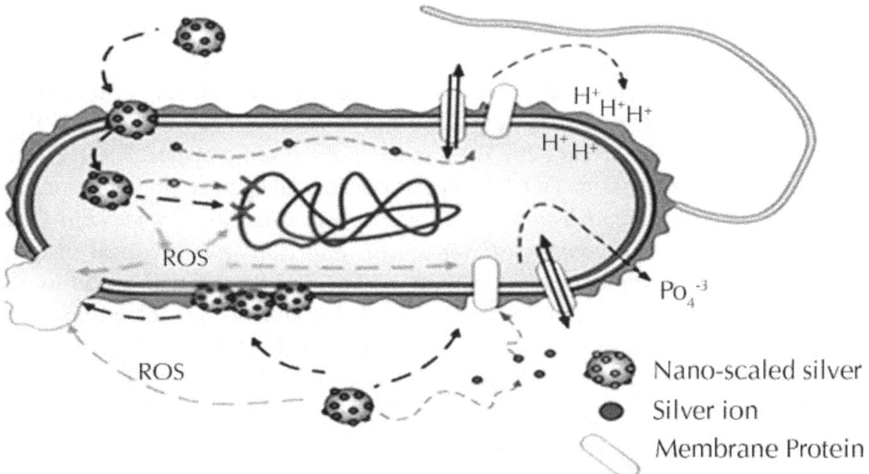

FIGURE 5 Diagram showing nanoscaledAg interaction with bacterial cells (Marambio-Jones and Hoek, 2010).

Other studies have suggested that AgNPs may alter the phosphotyrosine profile of putative bacterial peptides that could influence cellular signaling and consequently, inhibit the growth of bacteria (Shrivastavaet al., 2007). Ag^+ inhibits phosphate uptake and exchange in *Escherichia coli* and causes efflux of accumulated phosphate with mannitol, succinate, glutamine, and proline (Schreurs and Rosenberg, 1982). Moreover, Dibrov et al (2002)confirmed that Ag^+ carried the ability to collapse the proton motive force of *Vibriocholerae*. Considering the well-documented crucial importance of the transmembrane proton gradient in overall microbial metabolism, it seems inevitable that the exclusion of proton motive force be supposed to result in cell death (Dibrov et al., 2002). In addition, Ag^+ can lead to enzyme inactivation *via* formatting Ag complexes with electron donors containing sulfur, oxygen, or nitrogen (thiols, carboxylates, phosphates, hydroxyl, amines, imidazoles, and indoles)(Ahearn et al., 1995). Ag^+ may displace native metal cations from their normal binding sites in enzymes (Ghandouret al., 1988). Furthermore, Ag ions reduce oxidation of glucose, glycerol, fumarate,and succinate in *E. coli* (Ahearn et al., 1995). WhenAgNPs enter the bacterial cell, they form a low molecular weight area inside the bacteria. Therefore, the bacteria conglomerate to protect the DNA from the Ag NPs. Consequently, the NPs preferably attack the respiratory chain, cell division finally leading to cell death (Rai et al., 2009).

TITANIUM DIOXIDE NANOPARTICLES

The TiO_2 evaluated as a food additive (Directive 94/36/EC, 1994) and it is used in antimicrobial active packaging systems. Unlike AgNPs, TiO_2 based antimicrobials are active only in the presence of UV light. The photo catalytic activity is due to its crystal structure, which in turn related to the band gap. Irradiation of TiO_2at higher energies more than the band gap produces electron hole pairs, induce redox reactions simultaneously. Negative electrons induce generation of O^{2-}, while positive electric holes generate hydroxyl radicals. Reactive oxygen species oxidize organic molecules and kill bacteria and viruses. The TiO_2 is one of the most promising applications as antimicrobial especially againstcommon food-borne pathogens including *Salmonella choleraesuis* subsp., *Vibrio parahaemolyticus*, and *L. monocytogenes*under UV illumination but not in the dark (Kim, 2003). Cerrade et al. (2008) prepared EVOH films incorporated ultrasonically with TiO_2 NPs observed biocidal against nine microorganisms (*Bacillus* sp., *B. Stereothermophilus, Staphylococcusaureus, Escherichia Coli, E.Caratovora, Z.rouxii, Pseudomonasfluourescens, and P.Jadinii*) involved in food poisoning and spoilage. Photocatalytic activity of TiO_2 has been found useful to decontaminate water, wash water and also for cleaning minimally processed products (Chaleshtori et al., 2008). Chawengkijwanich et al. (2008) showed TiO_2 coated PP film effective against *E.Coli* growth on fresh cut lettuce. The TiO_2NPs apart from antimicrobial properties exhibit protection of food content from oxidation of UV irradiation without losing its optical clarity because of high photostability and short wavelength light absorbing capacity. There are evidences that TiO_2 NPs are combined with AgNPs to create films with high antimicrobial activity (Li et al., 2009; Cheng et al., 2006; Wu et al., 2010).

ZINC OXIDE NANOPARTICLES

The ZnoNPs have shown antimicrobial properties and potential applications in food preservation. Moreover, the FDA listed ZnO as GRAS and can be used as food additive. The ZnONPs have been tested against some Gram-positive bacteria such as *Bacillus subtilis* and *Staphylococcus aureus* and Gram negative bacteria such as *Pseudomonas aeruginosa, Campylobacter jejuni,and Escherichia coli,*which have presented susceptibility to these NPs (Espitia et al., 2012).

The antimicrobial activity of these NPs is due to several reasons butthe mechanism, in which release of Zn^{2+} antimicrobial ions has been considered as reasonable hypothesis, while the exact mechanism of action of ZnONPs is yet to be known. Some of the assumptions include interaction of NPs with microorganisms, subsequently damaging the integrity of bacterial cell and formation of ROS by effect of light radiation. Vicentini et al. (2010) prepared nanocomposites by incorporating ZnONPs (25–30 nm) into the blending films of chitosan and polyvinyl alcohol (PVA), showed antimicrobial S. aureus

10.3.8 COMPARATIVE STUDY OF DIFFERENT FILLERS AND ITS EFFECTS ON THE FOOD PACKAGING

NPs are used in the preparation of bio-nanocompositesmainly for two reasons, to create tortuous path in order to improve the barrier properties (as filler) and to induce various desired effects (as an additive) such as immobilization of enzymes, antimicrobial activity, biosensing, and so on. Various types of NPs, including clay and silicates, Cellulose based nano-reinforcements, CNTs, Silica, Starch nanocrystals, chitin and chitosan NPs, and metal oxides, are presently used to enhance the biopolymer performance.

Clay and silicates are popular among the packaging industry due to their abundant availability, low cost, significant enhancements, and simple processability. Research publications on Clay nanocomposites intended for food packaging started only in late 1990s after the advent of its development in 1980 (Ray et al., 2006). Clays have been reported the greater improvement in the barrier properties as well as mechanical strength of the biopolymers (AdameandBeall, 2009; Weiss et al., 2006). In addition, it also been reported the increase in glass transition temperature and thermal degradation temperatures of biopolymer (Azeredo, 2009). The barrier, mechanical properties of packaging film depends on the degree of exfoliation and the compatibility of clay with the biopolymer. The compatibility can be increased by the process of organophilization, which reduces the energy of the clay (Paiva et al., 2008). The nanocomposites that are impregnated by clay material can be used as an oxygen barrier layer in the extrusion manufacturing of bottles for fruit juices, dairy foods, beer, and carbonated drinks, or as nanocomposite layers in multilayer films to enhance the shelf life of a variety of foods such as processed meats, cheese, confectionery, cereals, and boil-in-bag foods (Azeredo, 2009).

Cellulose is an abundantly available strong natural polymer, which can be used as a NM in the preparation of bio-nanocomposites. Besides, cellulose nanofibers are low cost, light weight, and environmentally safe materials (Azeredo, 2009). These attractive properties make cellulose nanofibers a novel class of NMs in the manufacture of nanocomposites. Unlike clay NMs, cellulose based nano-reinforcements exhibits good compatibility with hydrophilic polymer matrix because of its inherent hydrophilic nature (BondesonandOksman, 2007). This hydrophilic nature associated with cellulose nanofibrils limit its application due to high water absorption capacity, which is not desirable in many food packaging applications. Like the other reinforcements cellulose based nano-reinforcements exhibits mechanical and barrier properties. Cellulose fibrils have been also reported to improve thermal properties of polymers. The thermal stability of polymers in nanocomposites with cellulose whiskers was showed to be improved, when compared to those of the corresponding bulk polymers (Azeredo, 2009).

The CNTs are two types, one-atom thick single-wall nanotube (SWNT) and number of concentric tubes called multiwalled nanotubes (MWNT), having excellent high aspect ratios. Even at low concentrations (0.1 Wt%) CNTs reported greater improvement in thermal stability as well as tensile strength (Azeredo, 2009). According to, researches from Natick CNT nanocomposite with PLA showed a 200% more water vapor transmission rate and toughness than pure PLA (Brody, 2006).

Silica NPs have been reported to improve mechanical and barrier properties of several polymer matrices. Addition of silica NPs to starch polymer matrix not only increases the tensile properties but also increases elongation (Azeredo, 2009). Moreover, silica NPs addition decreased water absorption by starch (Xiong et al., 2008).

Starch nanocrystals are another type of NM, which reported that the addition improves tensile strength and modulus of pullulan films but decreased their elongation (Kristo and Biliaderis, 2007). However, the WVP of pullulan films was decreased by addition of 20% or more starch NPs.

Chitin whiskers incorporated to SPI thermoplastics reported that the whiskers significantly improved not only the tensile properties (tensile strength and elastic modulus) of the matrix but also its water resistance (Lu et al., 2004).

10.4 MODELLING OF BIO-NANOCOMPOSITES

Modeling helps in efficiently predicting the properties of nanocomposites and comparing them to the measured properties. Finite Element Modeling is a technique and it is being used in better understanding the gas flow properties through nanocomposite films. Recently, David and Gupta (2007) proposed a Finite element model for predicting diffusion through nanocomposite films and also to determine optimum loading level of nanocomposite materials in packaging films. Steady-state Fick's second law (Equation1) was used to solve equation for diffusion.

$$\frac{\partial c}{\partial t} = \nabla\left(D\nabla c\right) \tag{1}$$

The model determines the permeability levels for nanocomposite material at various loading levels of clay nanocomposites, which gives an understanding of barrier properties of film. Nielsen (1967) model (Equation 2) and Cussler (1988) model (Equation 3) were also compared for the prediction of barrier properties of biodegradable protein based nanocomposite material.

$$\frac{D_0}{D} = 1 + \left(\frac{h}{2t}\right)\phi \tag{2}$$

$$\frac{D_0}{D} = 1 + \left(\frac{(h/t)^2\phi^2}{4(1-\phi)}\right) \tag{3}$$

Where, D_0 is the diffusivity of the molecule in the neat polymer and D is the diffusivity through the nanocomposite, φis volume fraction of nanofiller, t and h are thickness and height of each platelet respectively (David and Gupta, 2007).

Geometry of nanocomposite films (Figure 6) was created (based on TEM analysis) according to size, shape and volume fraction of nanofillers to solve its barrier properties. They observed the ratio of diffusivity of nanocomposite to the diffusivity of virgin polymer decreases with increasing layer spacing by 1 nm at different loading levels from 1–20%. Diffusivity decreases at faster rate during 1–10% loading levels, beyond which only a slight variation was observed.

Experimentally, it is very difficult and time consuming to determine the particular loading level for enhancing material barrier properties. Hence, a finite element model can be effectively used to design nanocomposites for enhanced barrier applications.

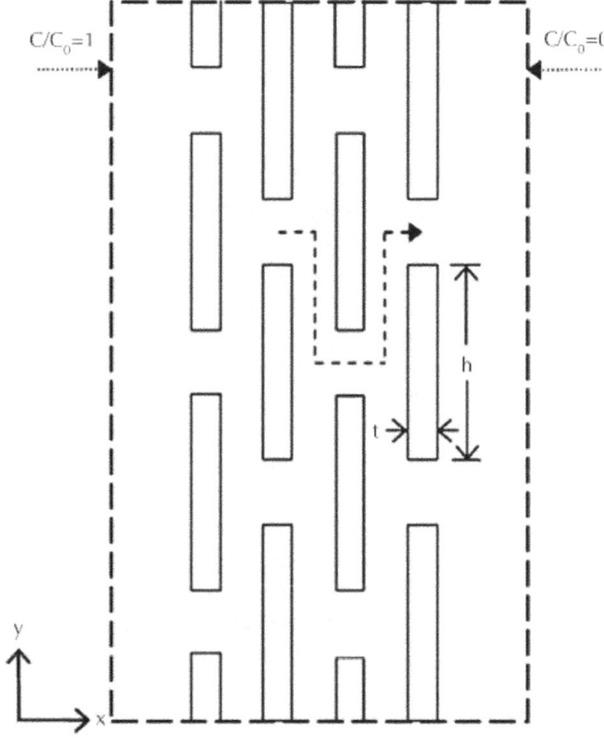

FIGURE 6 Idealized morphology for diffusion through polymer nanocomposite containing dispersed clay platelets (David and Gupta, 2007).

Later, Kumar et al. (2011) used mathematical modeling technique to determine properties of bio-nanocomposites based on starch, proteins, and cellulosic polymers. They also discussed experimental techniques to determine the mechanical, barrier, thermal, and rheological properties of bio-nanocomposites,which are significant in assessing the performance for the application as food packaging materials.

A three-phase model was developed by Luo and Daniel (2003) to explain the properties of nanocomposites of clay nanolayers exfoliated in epoxy matrix assuming near uniform dispersion and random orientation.The results and model predictions exhibited good agreement with experimental results. Minelli (2009) studied the effect of filler loading on the barrier properties of nanocomposites with help of CFD algorithm based on finite volumes method in 2D ordered structures. The results obtained are compared with previously accepted empirical models. The results are agreed with other methods and it can be used to obtain relevant information on gas permeability in real nanocomposite systems. Bharadwaj (2001) proposed a model to predict the permeability based on tortuosity. However, author considered nanocomposites in a rectangular shape and it distributed in a uniform manner in the film and this is an approximated method.

10.5 SAFETY CONCERNS AND REGULATORY ISSUES

It is undisputed that the application of nanotechnology in packaging is growing rapidly and nanocomposites are going to occupy a larger portion of food packaging in near future. But, the risk involved in terms of food safety, nanotoxicity, and regulations. Unless, these concerns are answered properly, it is quite difficult to reap the benefits of nanotechnology in food packaging to its fullest extent. It is more important to address these issues at the initial stage otherwise it follows the fate of GM foods, in which consumers are hesitant to buy them even though clear cut benefits had been communicated (Cox et al., 2004). The public acceptance study conducted by Siegrist et al. (2007) revealed that nanotechnology packaging (nano outside) is perceived as more acceptable than nanotechnology foods (nano inside). This clearly highlights the advantage of use of nanotechnology in food packaging over food processing. As discussed earlier biodegradable clay nanocomposite films were obtained by homogeneously dispersing functionalized layered MMT NPs in different biodegradable polymers *via* different processing techniques. The primary concern of consumer exposure to NPs from food packaging is likely to be through potential migration of NPs into foods. Two types of migration limits have been established in the area of plastic materials. An overall migration limit (OML) of 60 mg (of substances)/kg (of foodstuff or food simulants) that applies to all substances that can migrate from the food contact material to the foodstuff and a specific migration limit (SML), which defined to individual authorized substances and is fixed on the basis of the toxicological evaluation of the substance (Avella et al., 2005). However, it is not declared that the consumption of foods containing NMs migrated from FCMs pose a significant health risk (Cushen et al., 2012).

As per the scientific report from the European Centre for Ecotoxicology and Toxicology of Chemicals, there are three main routes for human exposure to NPs: Inhalation, dermal contact, and oral ingestion (Paul et al., 2006).Among them oral uptake from food with any migrations from the nanocomposites, is the most significant exposure source for human beings. It creates cardiovascular effects, which include heart rate changes, prothrombosis and acute myocardial infarction. The NPs due to its small size enter in to blood circulation through lungs during inhalation. Later, these NPs will be transported to other systems such as the liver, spleen, kidney, brain, heart, and causes toxicity based on their chemical composition. Exposure through dermal contact is comparatively less significant (Wei et al., 2011). Avella et al. (2005) studied the migration of the NPs to vegetable samples, including lettuce, and spinach packed in bags made out of starch based clay nanocomposites prepared *via* polymer melt processing technique. The results obtained demonstrated an insignificant trend in the levels of Fe and Mg in packaged vegetables but a consistent increase in the amount of silicon, which is the main component of the MMT NPs.The SiO_2particles of size 50 nm and 70 nm particles taken up into cell nucleus will cause aberrant protein formation, inhibited cell growth and on-set of pathology similar to neurodegenerative disorders (Wei et al., 2011).

Even, eco-friendly bio-nanocomposites reported the risk of environmental contamination due to release of NMs into environment upon degradation of polymers. The risks posed by NMs on the environment are determined by eco-toxicity tests and results to be displayed in an understandable manner on packaging to enable the consumer to make an informed choice. Priority should be given to NMs that are already in use in this way (Cushen et al., 2012).

While the risks arising from exposure to different types of NPs are not yet clearly understood, the allowance of nanotechnology-based food products, to come to the food markets in the absence of clear definition, public debate, food safety assessments, and may deny the potential benefits of nanotechnologies to food industry. In this scenario proper regulations are necessary to direct the research to improve the protection of human health and the environment. The Institute of Food Science and Technology has reminded the deficiencies in current regulations concerning the impact of nanotechnology on food and packaging (IFST, 2006).

The Regulation (EC) No 1935/2004 on Materials and Articles Intended for Food Contact regulates food packaging including new materials that can be used for active maintenance or improvement of the conditions around the food that is packed (Cushen et al., 2012). NMs will also fall in the category of new materials and can be put into the same regulation fold as they are intended to increase the performance of packaging material. Only some authorized compounds are permitted to be migrated from the FCM to food, provided they are permitted in food legislation.In addition, labeling of such systems is also to be in accordance with the Food Additive Directive (89/109/EEC) (Cushenet al., 2012). Many centers, organizations, and commissions are established around the globe aimed at development and control of nanotechnologies. These organizations or research centers play a vital role in performing or supporting nanotechnology researches, including the basic researches on nanotechnology, the applications of nanotechnology, safety assessment of NMs, and the development of regulatory control. Some major initiatives, centers, institutes, or government organizations are listed in Table 3.The safety assessment and regulations of nanocomposites, NMs are not intended for restraining its application in the food industry. But, it is obvious for us to properly assess the safety and control use of NMs, before it extensively infuses people's daily life. This will give the NMs a reasonable and justified evaluation, which undoubtedly can be considered as a warranty for the public to safely enjoy high technology products in the food industry.

TABLE 3 Examples of major initiatives, centers, institutes, or government organizations supporting the development of nanotechnology (Chau et al., 2007)

Country or District	Initiatives, centers, institutes or government organizations	**Links**
USA	National Nanotech Initiative (NNI)	http://www.nano.gov
	Food and Drug Administration (FDA)	http://www.fda.gov/nanotechnology
	Environmental Protection Administration (EPA)	http://es.epa.gov/ncer/nano
	Center for Nanotechnology	http://www.ipt.arc.nasa.gov/index.html
	National Science Foundation	http://www.nsf.gov/crssprgm/nano
	Project on Emerging Nanotechnologies	http://www.nanotechproject.org
European Union	Community Research & Development Information Service (CORDIS)	http://cordis.europa.eu/nanotechnology
	European Nanotechnology Gateway	http://www.nanoforum.org
	European Nanobusiness Association (ENA)	http://www.nanoeurope.org
UK	Institute of Nanotechnology (IoN)	http://www.nano.org.uk
	The Royal Society and Royal Academy of Engineering	http://www.royalsoc.ac.uk/page.asp?id-1212
	Health and Safety Executive (HSE)	http://www.hse.gov.uk/horizons/nanotech/index.htm
	Institute of Food Science and Technology (IFST)	http://www.ifst.org
	Department for Environment, Food and Rural Affairs (Defra)	http://www.defra.gov.uk/environment/nanotech/index.htm
France	Ministry for Research and New Technology	http://www.nanomicro.recherche.gouv.fr/uk_index.html
	French Research Network in Micro and NanoTechnologies (RMNT)	http://www.rmnt.org/EN/index.html
	Centre National de la Recherche Scientifique (National Center for Scientific Research) (CNRS)	http://www.cnrs.fr/index.html
Germany	German Federal Institute for Risk Assessment (BfR)	http://www.bfr.bund.de/cd/template/index_en
	Federal Ministry of Education and Research (BMBF)	http://www.bmbf.de/en/nanotechnologie.php

TABLE 3 *(Continued)*

Italy	National Institute for the Physics of Matter (INFM)	http://www.infm.it/
	Association italiana per la ricerca industriale (AIRI)	http://www.airi.it/2005/index.php
Canada	National Institute for Nanotechnology	http://nint-innt.nrc-cnrc.gc.ca/home/index_e.html
Japan	Nanotechnology Researchers Network Center (Nanonet)	http://www.nanonet.go.jp/english
	National Food Research Institute (NFRI)	http://www.nfri.affrc.go.jp/english/our-roles/index.html
	RIKEN	http://www.rikenresearch.riken.jp
	Japan Society for the Promotion of Science (JSPS)	http://www.jsps.go.jp/english/index.html
	Ministry of Education, Culture, Sports, Science and Technology (MEXT)	http://www.mext.go.jp/english
	National Institute of Health Sciences (NIHS)	http://www.nihs.go.jp/index.html
	National Institute for Environmental Studies (NIES)	http://www.nies.go.jp/index.html
	National Institute of Advanced Industrial Science and Technology (AIST)	http://www.aist.go.jp/index_en.html
	National Institute for Materials Science (NIMS)	http://www.nims.go.jp/eng/index.html
	Ministry of Economy, Trade and Industries (METI)	http://www.meti.go.jp/english
China	Chinese Academy of Sciences	http://www.cas.ac.cn/Index/0C/Index.htm
	National Center for Nanoscience and Technology (NCNST)	http://www.nanoctr.cn/e_index.jsp
	Key Laboratory of Molecule Nanostructure and Nanotechnology, Chinese Academy of Sciences	http://www.icas.ac.cn/5_keyanxitong/stm/english/home.htm
Taiwan	Industrial Technology Research Institute (ITRI)	http://www.itri.org.tw/eng/research/nano/index.jsp
	NanoTechnology Research Center (NTRC)	http://www.ntrc.itri.org.tw/eng/index.jsp
	nanoMark	http://www.nanomark.itri.org.tw/Eng
	National Nano Device Laboratories (NDL)	http://www.ndl.org.tw
	National Science and Technology Program for Nanoscience and Nanotechnology	http://nano-taiwan.sinica.edu.tw/newsen.asp

TABLE 3 *(Continued)*

Korea	Korean Food & Drug Administration (KFDA)	http://www.kfda.go.kr
	National NanoFab Center (NNFC)	http://www.nnfc.com/index.html
	Korea Institute of Science and Technology Information (KISTI)	http://www.kisti.re.kr
	Center for Nanostructured Materials and Technology (CNMT)	http://cnmt.kist.re.kr
	National Center for Nanomaterials Technology (NCNT)	http://www.nano.or.kr/english/index.asp
Australia	CSIRO Manufacturing and Materials Technology (CMMT)	http://www.cmit.csiro.au/brochures/tech/nanotech
	Department of Industry, Tourism and Resources (ITR)	http://www.industry.gov.au/nano
	Nanostructural Analysis Network Organisation, Major National Research Facility (NANO-MNRF)	http://www.nano.org.au

10.6 CONCLUSION

The future development of novel bio-nanocomposites with improved properties and functionality can be foreseen as an emerging, open field of research, with plenty of possibilities. Bio-nanocomposites may replace the existing petroleum based plastic polymers in near future as they succeeded in protecting environmental interests as well as food. Ideally any food packaging material should have good barrier properties against moisture, gas, flavor, and microbial attack. By impregnating nano sized reinforcement materials into the biopolymers, bio-nanocomposites obtain the benefit of variety of property profiles, which are displayed only at nano range. In addition, barrier properties of the biopolymer film will increase significantly and makes it suitable to pack food material. However, the future prospects of bio-nanocomposites are unpredictable because of its high cost, poor moisture resistance and most importantly, and other barrier properties are not sufficient enough to compete with plastic polymers. More research work needs to be performed in this area to come out with a solution to the problems involved in use of bio-nanocomposites as food packaging material. Nevertheless, the concept of safety in using bio-nanocomposites should not be neglected in the interest of getting new systems.

ACKNOWLEDGMENT

Authors wish to thank Prof. Ram Rajasekharan, Director, CSIR-CFTRI for his encouragement and support and also Ms. Ezhilarasi, Research Fellow, Department of Food Engineering, CFTRI, Mysore for her help.

KEYWORDS

- **Bio-nanocomposite**
- **Biopolymer**
- **Nanotechnology**
- **Polymeric material**
- **Pullulan.**

REFERENCES

1. Abdollahi, M., Rezaei, M., and Farzi, G. A novel active bio-nanocomposite film incorporating rosemary essential oil and nanoclay into chitosan. *J. Food Eng.*, **111**, 343–350 (2012).
2. Adame, D. and Beall, G. W. Direct measurement of the constrained polymer region in polyamide/clay nanocomposites and the implications for gas diffusion. *Appl. Clay Sci.*, **42**, 545–552 (2009).
3. Ahearn, D. G., May, L. L., and Gabriel, M. M. Adherence of organisms to silver-coated surfaces. *Ind. Microbiol.*, **15**, 372–376 (1995).
4. Alexandre, M. and Dubois, P. Polymer-layered silicate nanocomposites: Preparation, properties, and uses of a new class of materials. *Mater. Sci. Eng.*, **28**, 1–63 (2000).
5. Ali, S. S., Tang, X., Alavi, S., and Faubion, J. Structure and physical properties of starch/polyvinyl alcohol/sodium montmorillonitenanocomposite films. *J. Agric. Food. Chem.*, **59**, 12384–12395 (2011).
6. Angellier-Coussy, H., Torres-Giner, S., Morel, M., Nathalie Gontard, N., and Gastaldi, E. Functional properties of thermoformed wheat gluten/montmorillonite materials with respect to formulation and processing conditions. *J. Appl. Polym. Sci.*, **107**, 487–496 (2008).
7. Appendini, P. and Hotchkiss, J. H. Review of antimicrobial food packaging. *Innovat. Food Sci. Emerg. Tech.*, **3**, 113–126 (2002).
8. Arora, A. and Padua, G. W. Nanocomposites in Food Packaging. *J. Food Sci.*, **75**, R43–R49 (2010).
9. Avella, M., Vlieger, J. J. D., Errico, M. E., Fischer, S., Vacca, P., and Grazia Volpe, M. G. Biodegradable starch/clay nanocomposite films for food packaging applications.*Food Chem*., **93**, 467–474 (2005).
10. Azeredo, H. M. C. Nanocomposites for food packaging applications. *Food Res. Int.*, **42**, 1240–1253 (2009).
11. Bae, H. J., Park, H. J., Hong, S. I., Byun, Y. J., Darby, D. O., Robert, M., Kimmel, R. M., and Whiteside, W. S.Effect of clay content, homogenization RPM, pH, and ultrasonication on mechanical and barrier properties of fish gelatin/montmorillonitenanocomposite films. *LWT-Food Sci. Technol.*, **42**, 1179–1186 (2009).
12. Balakrishnan, H., Hassan, A., Imran, M., and Wahit, M. U. Toughening of polylactic acid nanocomposites: A short review. *Polym. Plast. Technol. Eng.*, **51**, 175–192 (2012).
13. Banker, G. S. Film coating theory and practice. *J. Pharm. Sci.*, **55**, 81–89 (1966).
14. Barron, C., Varoquaux, P., Guilbert, S., Gontard, N., and Gouble, B. Modified atmosphere packaging of cultivated Mushroom (AgaricusBisporus L) with hydrophobic films. *J. Food Sci.*, **67**, 251–255 (2002).
15. Bharadwaj, R. K. Modeling the barrier properties of polymer-layered silicate nanocomposites. *Macromolecules*, **34**, 9189–9192 (2001).
16. Bhattacharya, S. Biomems and nanotechnology based approaches for rapid detection of biological entities. *J. Rapid Methods Auto. Microb.*, **15**, 1–32 (2007).

17. Bondeson, D. and Oksman, K. Polylactic acid/cellulose whisker nanocomposites modified by polyvinyl alcohol. *Composites: Part A: Applied Science and Manufacturing*, **38**, 2486–2492 (2007).
18. Bouwmeester, H., Dekkers, S., Noordam, M. Y., Hagens, W. I., Bulder, A. S., and de Heer, C. Review of health safety aspects of nanotechnologies in food production. *Regul. Toxicol. Pharmacol.*, **53**, 52–62 (2009).
19. Bower, C. K., Avena-Bustillos, R. J., Olsen, C. W., Mchugh, T. H., and Bechtel, P. J. Characterization of fish-skin gelatin gels and films containing the antimicrobial enzyme lysozyme. *J. Food Sci.*, **71**, 141–145 (2006).
20. Bradley, E. L, Castle, L., and Chaudhry, Q. Applications of nanomaterials in food packaging with a consideration of opportunities for developing countries. *Trends Food Sci. Technol.*, **22**, 604–610 (2011).
21. Brandenburg, A. H., Weller, C. L., and Testin, R. F. Edible films and coatings from soy protein. *J. Food Sci.*, **58**, 1086–1089 (1993).
22. Brody, A. L. Nanocomposite technology in food packaging. *Food Tech.*, **61**, 80–83 (2007).
23. Cabedo, L., Feijoo, J. L., Villanueva, M., Lagaron, J. M., and Gimenez, E. Optimization of Biodegradable nanocomposites based on a PLA/PCL blends for food packaging applications. *Macromol. Symp.*, **233**, 191–197 (2006).
24. Cagri, A., Ustunol, Z., and Ryser, E. T. Antimicrobial edible films and coatings. *J. Food Protect.*, **67**, 833–848 (2004).
25. Carrado, K. A. Synthetic organo- and polymer-clays: Preparation, characterization, and materials applications. *Appl. Clay. Sci.*, **17**, 1–23 (2003).
26. Castello, R., Ferreira, A. R., Costa, N., Fonseca, I. M., Alves, V. D., and Coelhoso, I. M. Nanocomposite films obtained from carrageenan/pectin biodegradable Polymers. Presented at International conference on food innovation. *Food Innova.*, pp. 25–29 (2010).
27. Cerrada, M. L., Serrano, C., Sanchez-Chaves, M., Fernandez-Garcia, M., Fernandez-Martin, F., Andres, A. D., Rioboo, R. J. J., Kubacka, A., Ferrer, M., and Fernandez-Garcia, M. *Adv. Funct. Mater.*, **18**, 1949–1960 (2008).
28. Cha, D.S. and Chinnan, M. S. Biopolymer-based antimicrobial packaging: Review. *Crit. Rev. Food Sci. Nutr.*, **44**, 223–237 (2004).
29. Chaleshtori, M. Z., Masud, S. M. S., and Saupe, G. B. Using new porous nanocomposites for photocatalytic water decontamination. *Mater. Res. Soc. Symp. Proc.*, **1145**, 75–80 (2008).
30. Chau, C., Wu, S., and Yen, G. The development of regulations for food nanotechnology. *Trends. Food Sci. Technol.*, **18**, 269–280 (2007).
31. Chawengkijwanich, C. and Hayata, Y. Development of TiO_2 powder-coated food packaging film and its ability to inactivate Escherichia coli in vitro and in actual tests. Int. *J. Food Microbiol.*, **123**, 288–292 (2008).
32. Chen, G. X., Hao, G. J., Guo, T. Y., Song, M. D., and Zhang, B. H.structure and mechanical properties of poly(3-hydroxybutyrateco-3-hydroxyvalerate) (PHBV)/clay nanocomposites. *Journal of materials science letters*, **21**, 1587–1589 (2002).
33. Cheng, Q., Li, C., Pavlinek, V., Saha, P., and Wang, H. Surface-modified antibacterial TiO_2/Ag^+ nanoparticles: Preparation and properties.*Appl. Surf. Sci.*, **252**, 4154–4160 (2006).
34. Chung, Y., Ansari, S., Estevez, L., Hayrapetyan, S., Giannelis, E. P., and Lai, Hsi-Mei. Preparation and properties of biodegradable starch-clay nanocomposites. *Carbohydr. Polym.*, **79**, 391–396 (2010).
35. Coffin, D. R. and Fishman, M. L. Physical and mechanical properties of highly plasticized pectin/starch films. *J. Appl. Polym. Sci.*, **54**, 1311–1320 (1994).
36. Coles, R., McDowell, D., and Kirwan, M. J. *Food Packaging Technology*. Blackwell Publishing, Oxford, UK, pp. 346 (2003).
37. Conn, R. E., Kolstad, J. J., Borzelleca, J. F., Dixler, D. S., Filer, L. J., LaDu, B. N., and Pariza, M. W. Safety assessment of polylactide (PLA) for use as a food-contact polymer. *Food Chem. Toxicol.*, **33**, 273–283 (1995).

38. Cox, D. N., Koster, A., and Russell, C. G. Predicting intentions to consume functional foods and supplements to offset memory loss using an adaptation of protection motivation theory. *Appetite*, **43**, 55–64 (2004).
39. Cuq, B., Gontard, N., and Guilbert, S. Proteins as Agricultural Polymers for Packaging Production. *Cereal Chem*., **75**, 1–9 (1998).
40. Cushen, M., Kerry, J., Morris, M., Cruz-Romero, M., and Cummins, E. Nanotechnologies in the food industry-Recent developments, risks, and regulation. *Trends Food Sci. Technol*., **24**, 30–46 (2012).
41. Cussler, E. L., Hughes, S. E., Ward III, W. J., and Aris, R. Barrier membranes. *J. Membr. Sci*., **38**, 161–174 (1988).
42. Danilczuk, M., Lund, A., Saldo, J., Yamada, H., and Michalik, J. Conduction electron spin resonance of small silver particles. Spec. *Acta A-Mol. Biomol. Spec*., **63**, 189–191 (2006).
43. David, L. S. and Gupta, R. K. *A finite element analysis of the influence of morphology on barrier properties of polymer-clay nanocomposites*. Excerpt from the Proceedings of the COMSOL Conference, Boston (2007).
44. Dean, K. M., Do, M. D., Petinakis, E.,and Yu, L. Key interactions in biodegradable thermoplastic starch/poly(vinyl alcohol)/montmorillonite micro-and nanocomposites. *Compos. Sci. Technol*., **68**, 1453–1462 (2008).
45. Dean, K. and Yu, L. Biodegradable protein-nanocomposites. R. Smith (Ed.), *Biodegradable polymers for industrial application*,CRC,Boca Raton, Florida,pp. 289–309 (2005).
46. Dibrov, P., Dzioba, J., Gosink, K. K.,and Häse, C.C. Chemiosmotic mechanism of antimicrobial activity of Ag^+ in *Vibrio cholerae*. *Antimicrob. Agents Chemother*, **46**, 2668–2670 (2002).
47. Directive 94/36EC(1994).On colors for use in foodstuffs. *Official Journal of European Community*, **L237**, 13–29(October9, 1994).
48. Du Plooy, W., Regnier, T.,and Combrinck, S. Essential oil amended coatings as alternatives to synthetic fungicides in citrus postharvest management. *Postharvest Biol. Tec*., **53**, 117–122 (2009).
49. Duncan, V. T.Applications of nanotechnology in food packaging and food safety: Barrier materials, antimicrobials, and sensors. *Colloid Interface Sci*. doi:10.1016/j.jcis.2011.07.017.
50. Edwards, H.W.and Petersen, R.P. Reflectivity of evaporated silver films. *Phys. Rev*., **9**, 871 (1936).
51. Espitia, P. J. P., Soares, N. F. F., Coimbra, J. S. R., de Andrade, N. J., Cruz, R.S.,and Medeiros, E. A. A. Zinc oxide nanoparticles: Synthesis, antimicrobial activity, and food packaging applications. *Food Bioprocess Tech*., 5, 1447–1464 (2012).
52. Faheemuddin. Clays, Nanoclays, and Montmorillonite Minerals. Metall. Mater. *Trans. A*, **39**, 2804–2814 (2008).
53. Fama, L., Rojoe, P. G., Bernal, C.,and Goyanes, S. Biodegradable starch based nanocomposites with low water vapor permeability and high storage modulus. *Carbohydr.Polym*., **87**, 1989–1993 (2012).
54. FDA. *Inventory of Effective Food Contact Substance (FCS) Notifications No.178*.(2002). http://www.accessdata.fda.gov/scripts/fcn/fcnDetailNavigation.cfm?rpt=fcsListing&id=178.
55. Foegeding, E. A., Davis, J. P., Doucet, D.,and McGuffey, M. K. Advances in modifying and understanding whey protein functionality. *Trends Food Sci. Technol*., **13**, 151–159 (2002).
56. Garcia, M. Electronic nose for wine discrimination. *SensorsActuat. B*, **113**, 911–916 (2006).
57. Ghandour, W., Hubbard, J.A., Deistung, J., Hughes, M.N.,and Poole, R.K.The uptake of silver ions by Escherichia coli K12: Toxic effects and interaction with copper ion. *Appl. Microbiol. Biotechnol*., **28**, 559–565 (1988).
58. Goettler, L. A., Lee, K. Y.,and Thakkar, H. Layered silicate reinforced polymer nanocomposites: Development and applications. *Polym. Rev*., **47**, 291–317 (2007).
59. Gomez-Guillen, M. C., Perez-Mateos, M., Gomez-Estaca, J., Lopez-Caballero, E., Gimenez, B.,and Montero, P. Fish gelatin: A renewable material for developing active biodegradable films. *Trends Food Sci. Technol*., **20**, 3–16 (2009).

60. Gontard, N., Guilbert, S.,and Cuq, J. L. Water and glycerol as plasticizers affect mechanical and water vapor barrier properties of an edible wheat gluten film. *J. Food Sci.*, **58**, 206–211 (1993).
61. Han, J.H. New technologies in food packaging: Overview. J.H.Han(Ed.), *Innovations in Food Packaging*, Elsevier Academic Press, London, UK, pp. 3–11 (2005).
62. IFST. Institute of Food Science and Technology. Nanotechnology (2006). Available from http://www.ifst.org/uploadedfiles/cms/store/ATTACHMENTS/Nanotechnology.pdf.
63. Jamshidian, M., Tehrany, E. A., Imran, M., Jacquot, M.,and Desobry, S. Poly-lactic acid: Production, applications, nanocomposites, and release studies. *Compr. Rev. Food Sci. Food Saf.*, **9**, 552–571 (2010).
64. Jongjareonrak, A., Benjakul, S., Visessanguan, W., Prodpran, T.,and Tanaka, M. Characterization of edible films from skin gelatin of brownstripe red snapper and big eye snapper. *Food Hydrocoll.*, **20**, 492–501 (2006).
65. Kim, B., Kim, D., Cho, D.,and Cho, S. Bactericidal effect of TiO_2photocatalyst on selected food-borne pathogenic bacteria.*Chemosphere.*, **52**, 277–281 (2003).
66. Kim, J. S., Kuk, E., Yu, K. N., Kim, J. H., Park, S. J., Lee, H. J., Kim, S. H., Park, Y. K., Park, Y. H., Hwang, C. Y., Kim, Y. K., Lee, Y. S., Jeong, D. H.,and Cho, M. H. Antimicrobial effects of silver nanoparticles. *Nanomedicine.*, **3**, 95–101 (2007).
67. Kim, Y., Jung, R., Kim, H. S.,and Jin, H. J. Transparent nanocomposites prepared by incorporating microbial nanofibrils into poly(l-lactic acid). *Curr. Appl. Phys.*, pp. S69–71 (2009).
68. Klemm, D., Heublein, B., Fink, H.P.,and Bohn, A. Cellulose:Fascinating biopolymer and sustainable raw material.*Acc. Chem. Res.*, **44**, 3358–3393 (2005).
69. Kristo, E.and Biliaderis, C. G.Physical properites of starch nanocrystalreinforced pullulan films. *Carbohydr. Polym.*, **68**, 146–158 (2007).
70. Krochta, J. M.and DeMulder-Johnston, C.Edible and biodegradable polymer films: Challenges and opportunities. *Food Technol.*, **51**, 61–74 (1997).
71. Kumar, P., Sandeep, K. P., Alavi, S.,and Truong, V. D. A review of experimental and modeling techniques to determine properties of biopolymer-based nanocomposites. *J. Food Sci.*, **76**, E2–E14 (2011).
72. Kumar, P.,Sandeep, K. P., Alavi, S., Truong, V. D.,and Gorga, R. E. Preparation and characterization of bio-nanocomposite films based on soy protein isolate and montmorillonite using melt extrusion. *J. Food Eng.*, **100**, 480–489 (2010).
73. Kurose, T., Urman, K., Otaigbe, J.U., Lochhead, R.Y.,and Thames, S.F. Effect of uniaxial drawing of soy protein isolate biopolymer films on structure and mechanical properties. *Polym. Eng. Sci.*, **47**, 374–380 (2007).
74. Kvitek, L., Panacek, A., Soukupova, J., Kolarj, M., Vecerjova, R., Prucek, R., Holecova, M.,and Zbooil, R. Effect of surfactants and polymers on stability and antibacterial activity of silver nanoparticles (NPs). *J. Phys. Chem. C.*, **112**, 5825–5834 (2008).
75. Labuza, T.P.and Breene, W.M. Applications of active packaging for improvement of shelf-life and nutritional quality of fresh and extended shelf-life foods. *J. Food Process. Preserv.*, **13**, 1–69 (1988).
76. Lagaron, J. M., Catala, R.,and Gavara, R. Structural characteristics defining high barrier polymeric materials. *Mater. Sci. Technol.*, **20**, 1–7 (2004).
77. Lagaron, J.M., Cabedo, L., Cava, D., Feijoo, J. L., Gavara, R.,and Gimenez, E. Improving packaged food quality and safety. Part 2: Nanocomposites. *Food Addit. Contam.*, **22**, 994–998 (2005).
78. Lange, D. Complementary metal oxide semiconductor cantilever arrays on a single chip: Mass-sensitive detection of volatile organic compounds. *Anal. Chem.*, **74**, 3084–3095 (2002).
79. Li, H., Li, F., Wang, L., Sheng, J., Xin, Z., Zhao, L., Xiao, H., Zheng, Y.,and Hu, Q. Effect of nanopacking on preservation quality of Chinese jujube (Ziziphusjujuba Mill. var. inermis (Bunge) Rehd). *Food Chem.*, **114**, 547–552 (2009).

80. Lok, C. N., Ho, C. M., Chen, R., He, Q. Y., Yu, W. Y., Sun, H., Tam, P. K. H., Chiu, J. F.,and Che, C. M. Proteomic analysis of the mode of antibacterial action of silvernanoparticles.*J. Proteome Res.*, **5**, 916–924 (2006).
81. Luo, J. J.and Daniel, I. M. Characterization and modeling of mechanical behavior of polymer/clay nanocomposites. *Compos. Sci. Technol.*, **63**,1607–1616 (2003).
82. Luo, P. G.and Stutzenberger, F. J. Nanotechnology in the detection and control of microorganisms. *Adv. Appl. Microbiol.*, **63**, 145–181 (2008).
83. Lu, Y., Weng, L.,and Zhang, L. Morphology and properties of soy protein isolate thermoplastics reinforced with chitin whiskers. *Bio-macromolecules*, **5**,1046–1051 (2004).
84. Mahmoudian, S., Wahita, M.U.,Ismailc, A.F.,and Yussuf, A.A. Preparation of regenerated cellulose/montmorillonitenanocomposite films via ionic liquids.*Carbohydr. Polym.*, **88**, 1251–1257 (2012).
85. Maksimov, R. D., Lagzdins,A., Lilichenko, N.,and Plume, E. Mechanical Properties and Water Vapor Permeability of Starch/Montmorillonite Nanocomposites. *Polymer engineering and science.*, pp. 2421–2429 (2009).
86. Marambio-Jones, C.and Hoek, E. M. V. A review of the antibacterial effects of silver nanomaterials and potential implications for human health and the environment.*J. Nanopart. Res.*,**12**, 1531–1551 (2010).
87. Marsh, K.and Bugusu, B. Food packaging: Roles, materials, and environmental issues. *J. Food. Sci.*, **72**, R39–R55 (2007).
88. Martucci, J. F.and Ruseckaite, R. A. Biodegradable bovine gelatin/Na^+-montmorillonite nanocomposite films. Structure, barrier and dynamic mechanical properties. *Polymer Plast. Tech. Eng.*, **49**, 581–588 (2010).
89. Mauricio-Iglesias, M., Peyron, S., Guillard, V.,and Gontard, N. Wheat Gluten nanocomposite films as food-contact materials: Migration tests and impact of a novel food stabilization technology (high Pressure). *J. Appl. Polym. Sci.*, **116**, 2526–2535 (2010).
90. Minelli, M., Baschetti, M. G.,and Doghieri, F. Analysis of modeling results for barrier properties in ordered nanocomposite systems. *J. Membr. Sci.*, **327**, 208–215 (2009).
91. Mitchell, J. R., Hill, S. E.,and Ledward, D. A. Functional properties of food macromolecules:2nd Edition. Aspen,Gaithersburg, Maryland(1998).
92. Morillon, V., Debeaufort, F., Blond, G., Capelle, M.,and Voilley, A. Factors affecting the moisture permeability of lipid-based edible films: A review. *Crit. Rev. Food Sci. Nutr.*, **42**, 67–89 (2002).
93. Morones, J.R., Elechiguerra, J.L., Camacho, A., Holt, K., Kouri, J.B., Ramírez, J.T.,and Yacaman, M.J. The bactericidal effect of silver nanoparticles. *Nanotechnology*, **16**, 2346–2353 (2005).
94. Mujica-Paz, H.and Gontard, N. Oxygen and Carbon dioxide permeability of WG film: Efffect of relative humidity and temperature. *J. Agric. Food Chem.*, **45**, 4101–4105 (1997).
95. Nielsen, L.E. Model for the permeability of filled polymer system. *J. Macromol. Sci. Part A: Chem.*, **1**, 929–942 (1967).
96. Okada, A. and Usuki, A. Twenty Years of Polymer-Clay Nanocomposites. *Macromol.Mater. Eng.*, **291**, 1449–1476 (2006).
97. Paiva, L. B., Morales, A. R.,and Diaz, F. R. V.Organoclays: Properties, preparation, and applications. *Appl. Clay Sci.*, **42**, 8–24 (2008).
98. Park, H.J. Development of advanced edible coatings for fruits. *Trends Food Sci. Technol.*, **10**, 254–260 (1999).
99. Paul, J. A. B., David, R.,and Stephan, H.The potential risks of nanomaterials: A review carried out for ECETOC. *Part.FibreToxicol.*, **3**, 1–35 (2006).
100. Percival, S.L., Bowler, P.G.,and Dolman, J. Antimicrobial activity of silver containing dressings on wound microorganisms using an in vitro biofilm model. *Int. Wound. J.*, **4**, 186–191 (2007).

101. Phan, T. D., Debeaufort, F., Luu, D., and Voilley, A. Moisture barrier, wetting, and mechanical properties of shellac/agar or shellac/cassava starch bilayer bio-membrane for food applications. *J. Membrane Sci.*, **325**, 277–283 (2008).
102. Ponce, A.G., Roura, S.I., DelValle, C.E., and Moreira, M.R. Antimicrobial and antioxidant activities of edible coatings enriched with natural plant extracts: In vitro and in vivo studies. *Postharvest Biol. Tec.*, **49**, 294–300 (2008).
103. Quarmley, J.and Rossi, A. Nanoclays: Opportunities in polymer compounds.*Ind. Minerals.*, **400**, 47–49; 52–53 (2001).
104. Rai, M., Yadav, A.,and Gade, A. Silver nanoparticles as a new generation of microbials. *Biotechnol. Adv.*, **27**, 76–83 (2009).
105. Ray, S., Easteal, A., Quek, S. Y.,and Chen, X. D.The potential use of polymer-clay nanocomposites in food packaging (Article 5). *Int. J. Food Eng.*, **2**(4)(2006).
106. Rhim, J. W. Potential use of biopolymer-based nanocomposite films in food packaging applications. *Food Sci. Biotechnol.*, **16**, 691–709 (2007).
107. Rhim, J.W. and Ng, P. K. W. Natural biopolymer-based nanocomposite films for packaging applications.*Crit. Rev. Food Sci. Nutr.*, **47**, 411–433 (2007).
108. Rhim, J.W. Increase in water vapor barrier property of biopolymer-based edible films and coatings by compositing with lipid materials. *Food Sci. Biotechnol.*, **13**, 528–535 (2004).
109. Rhim, J.W., Hong, S.I., Park, H.M.,and Ng, P.K.W. Preparation and characterization of chitosan-based nanocomposite films with antimicrobial activity.*J. Agric. Food Chem.*, **54**, 5814–5822 (2006).
110. Rinaudo, M. Chitin and chitosan: Properties and applications. *Prog.Polym. Sci.*, **31**, 603–632 (2006).
111. Rhim, J.W. Effect of clay contents on mechanical and water vapor barrier properties of agar-based nanocomposite films.*Carbohydr.Polym.*, **86**, 691–699 (2011).
112. Ruan, D., Zhang, L., Zhang, Z.,and Xia, X. Structure and properties of regenerated cellulose/tourmaline nanocrystal composite films. *J. Polym. Sci. Part B: Polym. Phys.*, **42**, 367–373 (2003).
113. Sanchez-Gonzalez, L., Chafer, M., Chiralt, A.,and Gonzalez-Martinez, C. Physical properties of edible chitosan films containing bergamot essential oil and their inhibitory action on Penicilliumitalicum. *Carbohydr. Polym.*, **82**, 277–283 (2010).
114. Schofield, J. D., Bottomley, R. C., Timms, M. F.,and Booth, M. R. The effect of heat on wheat gluten and the involvement of sulphydryl-disulphide inter change reactions. *J. Cereal. Sci.*, **1**, 241–253 (1983).
115. Schreurs, W. J. A.and Rosenberg, H. Effect of silver ions on transport and retention of phosphate by Escherichia coli. *J. Bacteriol.*, **152**, 7–13 (1982).
116. Shahidi, F., Arachchi, J.,and Jeon, Y. Food applications of chitin and chitosans. *Trends Food Sci. Technol.*, **10**, 37–51 (1999).
117. Shewry, P. R. and Tatham, A. S. Disulphide bonds in wheat gluten proteins. *J. Cereal. Sci.*, **25**, 207–227 (1997).
118. Shrivastava, S., Bera, T., Roy, A., Singh, G., Ramachandrarao, P.,and Dash, D. Characterization of enhanced antibacterial effects of novel silver nanoparticles. *Nanotechnology.*, **18**, 225103–225112 (2007).
119. Siegrist, M., Cousinb, M., Kastenholzc, H.,and Wiek, A. Public acceptance of nanotechnology foods and food packaging: The influence of affect and trust. *Appetite.*, **49**, 459–466 (2007).
120. Simkovic,I. What could be greener than composites made from polysaccharides?*Carbohydr. Polym.*, **74**, 759–762 (2008).
121. Simoncic, B.and Tomsic, B. Structures of novel antimicrobial agents for textiles-a review. *Text. Res. J.*, **80**, 1721–1737 (2010).
122. Sinha Ray, S.and Okamoto, M. Polymer/layered silicate nanocomposites: A review from preparation to processing. *Prog. Polym. Sci.*, **28**, 1539–1641 (2003).

123. Sondi, I.and Salopek-Sondi, B. Silver nanoparticles as antimicrobial agent: A case study on Escherichia coli as a model for Gram-negative bacteria. *J. Colloid Interface Sci.*, **275**, 177–182 (2004).
124. Sorrentino, A., Gorrasi, G.,and Vittoria, V. Potential perspectives of bio-nanocomposites for food packaging applications.*Trends Food Sci. Technol.*, **18**, 84–95 (2007).
125. Sothornvit, R., Hong, S., An, D. J.,and Rhim, J. Effect of clay content on the physical and antimicrobial properties of whey protein isolate/organo-clay composite films. *LWT-Food Sci. Technol.*, **43**, 279–284 (2010).
126. Sothornvit, R.and Krochta, J. M. Oxygen permeability and mechanical properties of films from hydrolyzed whey protein. *J. Agric. Food Chem.*, **48**, 3913–3916 (2000).
127. Sothornvit, R. and Krochta, J. M. Plasticizers in edible films and coatings.J. H. Han (Ed.), *Innovations in food packaging*,Elsevier Academic Press, London, pp. 403–433 (2005).
128. Sothornvit, R., Rhim, J.,and Hong, S. Effect of nanoclay type on the physical and antimicrobial properties of whey protein isolate/clay composite films.*J. Food Eng.*, **91**, 468–473 (2009).
129. Srinivasa, P.and Tharanathan, R. Chitin/chitosan: Safe, ecofriendly packagingmaterials with multiple potential uses. *Food rev. int.*, **23**, 53–72 (2007).
130. Suvorova, A. I., Tyukova, I. S., Smirnova, E. A.,and Peshekhonova, A. L. Viscosity of blends of pectins of various origins with ethylene-vinyl acetate copolymers. *Russ. J. Appl. Chem.*, **76**, 1988–1992 (2003).
131. Tankhiwale, R.and Bajpai, S. K. Silver-Nanoparticle-Loaded Chitosan Lactate Films with Fair Antibacterial Properties. *J. Appl. Polym. Sci.*,**115**, 1894–1900 (2010).
132. Thostenson,E. T., Li, C. Y.,and Chou, T. W. Nanocomposites in context. *Compos. Sci. Technol.*, **65**, 491–516 (2005).
133. Trovatti, E., Fernandes, S.C.M., Rubatat, L., Freire, C.S.R., Silvestre, A.J.D.,and Neto, C.P.Sustainable nanocomposite films based on bacterial cellulose and pullulan. *Cellulose.*, **19**, 729–737 (2012).
134. Uyama, H., Kuwabara, M., Tsujimoto, T., Nakano, M., Usuki, A.,and Kobayashi, S. Green nanocomposite from renewable resources: Plant oil-clay hybrid materials. *Chem. Mater.*, **15**, 2492–2494 (2003).
135. Vanin, F. M., Sobral, P. J. A., Menegalli, F. C., Carvalho, R. A.,and Habitante, A. M. Q. B. Effects of plasticizers and their concentrations on thermal and functional properties of gelatin-based films. *Food Hydrocoll.*, **19**, 899–907 (2005).
136. Vicentini, D. S., SmaniaJr., A., and Laranjeira, M. C. M. Chitosan/poly(vinyl alcohol) films containing ZnO nanoparticles and plasticizers.*Mater. Sci. Eng.*, **30**, 503–508 (2010).
137. Wang, S., Shen, L., Tong, Y., Chen, L., Phang, I., Lim, P.,and Liu, T. Biopolymer chitosan/ montmorillonitenanocomposites: Preparation and characterization. *Polym. Degrad. Stabil.*, **90**, 123–131 (2005).
138. Wang, S., Sue, H. J.,and Jane, J. Effects of polyhydric alcohols on the mechanical properties of soy protein plastics. *J. Macromol. Sci. Part A: Pure Appl. Chem.*, **33**, 557–569 (1996).
139. Waszkowiak, K. Antioxidative activity of rosemary extract using connective tissue proteins as carriers. *Int. J. Food Sci. Technol.*, **43**, 1437–1442 (2008).
140. Wei, H., Jun, Y. Y., Tao, L. N.,and Bing, W. L. Application and safety assessment for nanocomposite materials in food packaging. *Chinese Sci. Bull.*, **56**, 1216–1225 (2011).
141. Weiss, J., Takhistov, P.,and McClements, D. J. Functional materials in food nanotechnology. *J. Food Sci.*, **71**, R107–R116 (2006).
142. Wijnhoven, S. W. P., Peijnenburg, W. J. G. M., Herberts, C. A., Hagens, W. I., Oomen, A. G., Heugens, E. H. W., Roszek, B., Bisschops, J., Gosens, I.,Van de Meent, D., Dekkers, S.,De Jong, W. H.,Van Zijverden, M., Sips, A. J. A. M.,and Geertsma, R. E.Nanosilver: A review of available data and knowledge gaps in human and environmental risk assessment.*Nanotoxicology*, **3**, 109–138 (2009).
143. Wright, J. B., Lam, K.,and Burrell, R. E. Wound management in an era of increasing bacterial antibiotic resistance: A role for topical silver treatment. *Am. J. Infect. Control*, **6**, 572–577(1998).

144. Wu, T, S., Wang, K. X., Li, G.D., Sun, S.Y., Sun, J.,and Chen, J. S. Montmorillonite-supported Ag/TiO_2 nanoparticles: An efficient visible-light bacteria photodegradation material. *ACS Appl. Mater. Int.*, **2**, 544–550 (2010).
145. Xiong, H. G., Tang, S. W., Tang, H. L., and Zou, P. The structure and properties of a starch-based biodegradable film. *Carbohydrate Polymers*,**71**, 263–268 (2008).
146. Xu, Y., Ren, Y., and Hanna, M. A. Chitosan/clay nanocomposite film preparation and characterization. *J. Appl. Polym. Sci.*, **99**, 1684–1691 (2006).
147. Yi, J. B., Kim, Y. T., Bae, H. J., Whiteside, W. S.,a nd Park, H. J. Influence of transglutaminase-induced cross-linking on properties of fish gelatin films. *J. Food Sci.*, **71**, 376–383 (2006).
148. Zhao, G. J.and Stevens, S. E. Multiple parameters for the comprehensive evaluation of the susceptibility of *Escherichia coli* to the silver ion. *Biometals*, **11**, 27–32 (1998).

CHAPTER 11

BIO-BASED NANOCOMPOSITES: PROSPECTS IN GREEN PACKAGING APPLICATIONS

PRODYUT DHAR, UMESH BHARDWAJ, and VIMAL KATIYAR

ABSTRACT

This chapter puts emphasis on numerous bio-based nanocomposites used for green packaging, a step towards replacing non-biodegradable packaging. First section deals with the importance and present scenario of nanotechnology in the present Century. Thereafter, the need of nanotechnology for the packaging applications is discussed, wherein the role of fillers including nanofillers, has been included. Subsequent section includes classification, properties, and applications of nanocomposites followed by the role of bio-nanocomposites. Fabrication technologies of bio-nanocomposites are the key highlights for the chapter.

11.1 INTRODUCTION

In the twentieth Century, there have been significant advancements in the packaging where rapid increase of plastic uses has been noticed, especially for food packaging applications. However, during the storage of raw or minimal processed food for a longer period of time, it has a risk of biofilm formation due to microbial contamination, oxidation, surface dehydration, and so on. Nevertheless, plastics have enormous advantageous properties such as lightness, seal ability, flexibility, and stable mechanical properties over wide range of environmental conditions (for example, temperature, humidity, pH, and so on). Therefore, plastic can be exploited for its application in high value added components. Further, other than conventional plastics, metals, ceramics, aluminum, tinplated steels, and so on are also used for packaging, which cause severe pollution to the environment. In addition, there are many agro-based feedstocks other than conventional fossil feedstock which shows great potential value for their use in production of bio-plastics (Beccari et al., 2009). Such environment friendly packaging are generally synthesized by using feedstock from renewable resources which include aliphatic thermoplastic polyesters such as poly (lactic acid) (PLA), poly (butylene succinate) (PBS), poly (hydroxyl butyrate) (PHB), and so on (Amass et al., 1998). Synthesis of PLA from lactic acid precursor, produced by fermentation from waste products, attracted much attention for research having excellent mechanical properties, biodegradability, and biocompatibility at a competitive cost (Singh and Ray, 2007; Averous and Martin, 2001; Lin et al. 2007).

The potent threat towards safety of polymer-based packaged food may be contributed by significant permeation of the oxygen, water vapor, and other gases through it, migration of toxic materials which adds subsequent loss in food quality and safety standards. Barrier limitations generally occur in mono-layered packaging, however, it may overcome with the development of multi-component multilayer–packaging-based newer technologies. Complex multilayer packaging has significantly higher resistance towards oxygen and water vapor permeation to prevent deterioration during storage and handling of food. The most commonly used multilayer films include non-degradable polymer resins such as poly (ethylene terephthalate), polyamide, and ethylene-vinyl alcohol copolymers. It is noteworthy to report that with the concept of degradability and environmental protection, research towards environment friendly polymers is growing. Such polymers can be classified in four categories based on the resources which include bio-polymers directly from natural resources, synthesized from microorganisms, synthesized using biotechnology, and petroleum products (Figure 1). Biodegradable polymers are defined as those that undergo mineralization by microbial induced chain scission under specific conditions in term of pH, humidity, oxygenation, and so on. Such environment friendly polymers can be made from bio sources like corn, wood cellulose, and so on or synthesized by bacteria from small molecules like butyric acid or valeric acid or from different carbon sources that give polyhydroxybutyrate (PHB) and polyhydroxyvalerate (PHV). Other biodegradable polymers can be derived from petroleum sources or obtained from mixed sources of biomass and petroleum. The best known petroleum source-derived biodegradable polymers are aliphatic polyester or aliphatic-aromatic copolyesters. Biodegradable polymers derived from bio-based precursor such as lactic acid yields PLA with comparable thermophysical and barrier properties to some of petroleum-based conventional plastics such as polypropylene, polyethylene or polystyrene, and so on. Hence, PLA have potential to compete with commodity plastics but still some of the properties such as brittleness, low thermal stability, low heat distortion temperature, high gas permeability, and low melt viscosity for its processing restrict its uses for wide-range of applications. Therefore, modification of the biodegradable polymers through innovative technologies will come up as the research topic in the near future.

Recently, nanotechnology-based modern approaches have been introduced as a tool to overcome the material properties limitations such as gas barrier properties and improved mechanical properties of packaging plastics. It is well understood that the polymer with nano-scale dispersion of particles possess significant physical and chemical properties than their micro scale dispersion (Cho and Kim, 2011; Zhang et al., 2001). For example, organically modified, layered silicates when disperse in nanoscale into polymer matrices yields biodegradable nanocomposites with improved mechanical and barrier properties (Ray et al., 2003; Chang et al., 2003; Yang et al., 2007). Fabrication and processing of biodegradable polymer-based nanocomposites may be considered as green nanocomposites, the wave of the future, considered as the next generation materials.

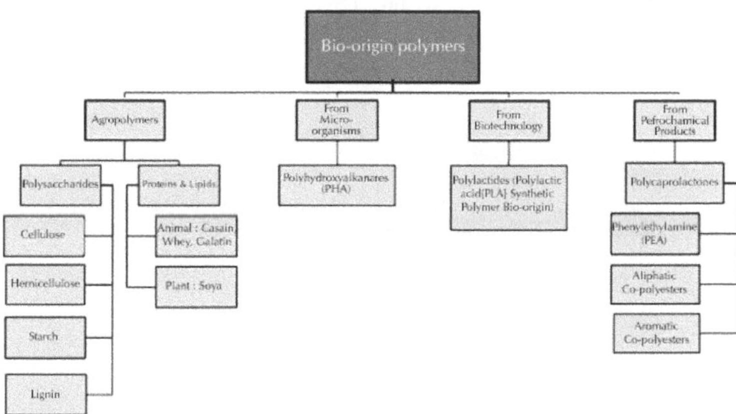

FIGURE 1 Schematic of bio-origin polymers.

Polymer composites are recently developed with different filler materials from bio-originated biopolymer to clays and metallic micro and nanoparticles. Recent decade has seen an enormous research with the use of cellulose whiskers, where cellulose binding domains serve as significant role in the improvement of the physical properties of polymers. Cellulose fibers derived from agricultural products (from cotton, ramie, hemp, flax, sisal, wheat, straw, palm, bleached softwood, beet pulp, hardwood pulps, and waste husks of cereals) and bacterial origin cellulose have been used for preparation of cellulose nanocrystals as one of the potential biofiller for enhancing polymer mechanical and gas barrier properties. Also, the applications of clay for example, montmorillonite, hectrite, and saponite have made polymers extremely light weighted and flame resistant material. Metallica-based micro and nano-structured materials are also incorporated into food contact polymer, to prevent the photo degradation and enhancement of the mechanical and barrier properties. Recent advances in technology have incorporated silver and heavy metal-based nano-engineered materials as food packaging material because of their antimicrobial activity. Several metal ions such as copper, silver, zinc, palladium, or titanium occur naturally and under certain threshold concentration does not have adverse effects on eukaryotic cell (Llorensa et al., 2012). Migration of cations from polymer surface is of utmost important parameter of study, which determines the antimicrobial effectiveness and applicability of the polymeric material as a food contact material. The antimicrobial, photo catalytic, oxidizing, and UV protecting properties of the innovative organic nanostructured materials are also investigated (Tunç and Duman, 2011).

Nanotechnology has experienced enormous development in recent years with the high demand in worldwide research and it will make significant contribution in the global economy by 2020 (Roco et al. 2010). Nanotechnology has applications in all the areas of food science, from agriculture to food processing, security to packaging, nutrition, and nutraceuticals, as mentioned in Figure 2. Despite the growth in these areas surrounding the nanotechnology, there has been a little subdue attention towards application in food nanotechnology because many toxicological information about nanofillers are still not documented so far. This chapter focuses on enhancement of knowledge on the current developments in nanotechnology application to food packaging related systems.

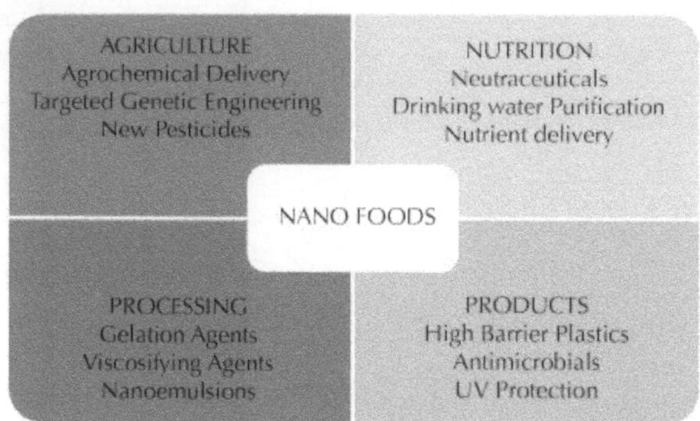

FIGURE 2 Nano food packaging and their applications.

The subsequent part of chapter highlights the use of nanocomposites for bio-based food packaging with the newly developed manufacturing technologies. Various techniques used to prepare biodegradable polymer nanocomposites, physicochemical characterization, mechanical and materials properties, biodegradability, and processing of biodegradable polymer-based nanocomposites are discussed further.

11.2 NEED OF NANOTECHNOLOGY FOR PACKAGING

In this modern world, priority to packaged food has increased enormously. Food packaging serves numerous functions such as from protecting the food from dirt or dust, oxygen, light pathogenic micro-organisms, moisture, and other varieties of destructive or harmful substances. The packaging materials must be able to withstand extreme conditions during processing, filling and transport conditions, and resistant to physical damage (Yang et al., 2007). Polymer processing has revolutionized from monomer to polycarbon chain processing but still a major drawback the inherent permeability to gas, water, and other small molecules exists. For example, PET provides a good barrier to oxygen (O_2 permeability = 6–8 nmolm^{-1}s^{-1}GPa), while high density polyethylene (HDPE) resists water vapor but permeates oxygen to large extent (O_2 permeability = 200–400 nmol m^{-1}s^{-1}GPa^{-1}) (Finnigan, 2009).

Permeability depends on several interrelated factors: polarity and structural features of polymeric side chains, hydrogen bonding characteristics, molecular weight and polydispersity, degree of branching or cross-linking, processing methodology, method of synthesis, and degree of crystalinity (Yang et al., 2007). Ethylene–vinyl alcohol (EVOH) under dry conditions exhibit excellent oxygen transmission rate (OTR) but at relative higher humid conditions (relative humidity >75%) EVOH swells and plasticizes the polymer (Zhang, 2001; Kollen and Gray, 1991). For this approach, EVOH is sandwiched in between polymer layers to form relatively hydrophobic polymer such as polyethylene (Zhang, 2001; Yam, 2009).

The PNCs preparation can help in solving the aforementioned problems for food packaging applications. The PNCs refer to the composite materials containing typically low additions of some nanoparticles blended with the polymers. These are generally used with a concentration limit of 1–7 wt% of modified nanoclays (Lagaron, 2005). The PNCs can also be fabricated by using clay and silicate nano particles, silica (SiO_2) nanoparticles (Wu et al., 2002; Vladimirov et al., 2006; Jia et al. 2007; Tang et al. 2008), carbon nanotubes (Zhou et al., 2004; Chen et al., 2005; Bin et al., 2006; Zeng et al., 2006; Morlat-Therias et al., 2007; Kim et al., 2007; Prashantha et al., 2009), graphene (Kim et al., 2009; Wakabayashi, 2008), starch nanocrystals (Chen, 2008; Kristo, 2007), cellulose-based nano fibers or nano whiskers (Azeredo et al., 2009; Azeredo et al., 2010; Bilbao-Sainz et al., 2010; Dufresne, 1996; Helbert, 1996; Podsiadlo, 2005; Samir, 2004; Oksmanet al., 2006 34-42; Cao et al., 2008), chitin or chitosan nanoparticles (Lu et al., 2004; Sriupayo et al., 2005; de Moura, 2009; de Moura, 2011), and other inorganics (Yang et al., 2008; Zhang et al., 2008; Ma et al. 2009; Yudin, 2007).

The major challenge in the development of nanocomposites is the interaction between matrix and filler which is highly desired. The matrix–filler interactions significantly improves with the reduction in size of the reinforcing agent having better compatibility to form homogeneous dispersions of fillers in the polymeric matrix (H. M. C. de Azeredo et al., 2009). Thus, shifting to the nano-sized particles from micro-sizes, leads to a better performance of composite materials and polymer matrices due to high surface to volume ratio. The PNCs not only improves the barrier properties but also enhance the mechanical strength (Chen, 2005; Li Schulz, 2009, Ray and Okamoto et al., 2003; Park, 2002; Cabedo et al., 2006; Gorrasi et al., 2004; Ray and Bousmina, 2005; Ogata et al., 1997; Osman et al., 2005; Kojima et al., 1993; Kojima et al., 1993; Powell and Beall, 2006; Zhang et al., 2007), resistant to flame (Ma et al., 2002; Ray and Okamoto, 2003; Zhang et al., 2007; Horrocks et al., 2007; Kandola et al., 2008; Smart et al., 2008; Porter et al., 2000; Gilman et al., 2000; Morgan et al., 2005), better thermal properties, (for example, melting points, degradation, and glass transition temperature) (Park et al., 2002; Kumar et al., 2003; Yoo et al., 2004; Bertini, 2006) surface wettability, and increased hydrophobicity (Zhou et al., 2011). For example, layer-by-layer assembly fabrication of PNC material composed of clay nanoparticles dispersed in cross-linked polyvinyl acetate (PVA), possessed a modulus stiffness of 106 ± 11 GPa, which is twice the order of magnitude than neat PVA (Podsiadlo et al., 2007).

The application of the nanotechnology in food packaging can be done through several ways in forms of nanocomposites, nano-coatings, surface biocides, active packaging, and intelligent packaging. In nano-coatings, nonmaterial are embedded onto the packaging surface (either inside surface, outside surface, or sandwiched as a layer in a laminate) especially to improve the barrier properties. Nanomaterials with antimicrobial properties, for the surface protection, are coated to form surface biocides. Active packaging incorporates nanomaterials with antimicrobial or other properties (for example, antioxidant) with intentional release into and consequent effect on the packaged food. Intelligent packaging is applied for smart products like nanosensors to monitor and report on the condition of the food.

11.2.1 ROLE OF FILLERS IN GAS BARRIER ENHANCEMENT

The permeability of gases in the polymeric material depends upon the diffusion rate of the diffusant through polymer matrix (Cussler, 1988). The adsorption rate depends upon the rate of formation of free volume holes in the polymer created due to random Brownian motion or thermal motion of the polymeric chains and diffusion caused by jumps of molecular gas molecules to neighboring empty holes during polymer formation (Yang, 2007). Permeability of polymer films is also dependent on the intrinsic polymer chemistry, polymer-polymer, and polymer-gas interactions.

Since, clay has such large surface energies, its nanoplatelets tend to stick together, particularly when dispersed in nonpolar polymer environments.

On the basis of the dispersion behavior of clay platlets in polymer matrix, structurally three different types of nanocomposites are thermodynamically achievable (Figure 3):

1. *Aggregated Microcomposites*: Agglomeration of clay platelets leads to tactoid structures (micro composites) with reduced aspect ratios and, according to the Nielsen model, reduced barrier efficiencies.
2. *Intercalated Nanocomposites*: Where insertion of polymer chains into the silicate structure occurs in a crystallographically regular fashion, regardless of polymer to OMLS ratio, and a repeat distance of few nanometers. Intercalated morphologies are characterized by moderate intrusion of polymer strands into the gallery volume and the shape of the layered stack is preserved.
3. *Exfoliated Nanocomposites*: In this case, individual silicate layers are separated in polymer matrix by average distances, totally depend on the OMLS loading. In fully exfoliated structures, on the other hand, individual platelets are well separated and have extremely favorable interactions with the polymer matrix (Ludueña, 2007).

The dispersion of nano-sized fillers into polymer matrix creates a tortuous path for the diffusant in comparison to neat polymer having straight diffusive path perpendicular to the film (Figure 3). In case of intercalated and exfoliated nanocomposites, path travelled by diffusant molecule would be greater than that of aggregate micro composite.

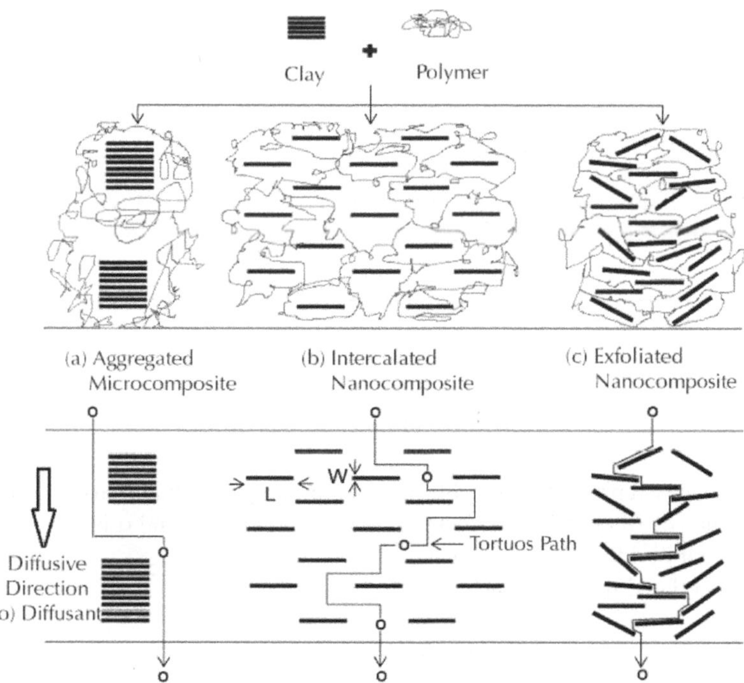

FIGURE 3 Tortuosity in the path travelled during permeability of water vapor and oxygen through the polymer micro composite and polymer nanocomposites.

Various models have been proposed to describe the behavior of the diffusive and mass transport through the nanocomposites as described in Table 1. Shape of the platelets is assumed to be either rectangular or circular pattern in most of the models to form a regular array in the space. The platelets act as a semi-permeable barrier to the diffusant, forcing it to follow a longer and more tortuous path (Figure 3). Nielsen proposed a simple theoretical model for gas diffusion, which assumes that the fillers are evenly distributed throughout the matrix and take the shape of rectangular platelets of finite width (L) and thickness (W) (Figure 3) of uniform size with tortuosity of path being the only factor influencing the gas diffusion rate (Nielsen, 1967; Choudalakis and Gotsis, 2009; Dunkerley, 2010).

TABLE 1 Models for permeability behavior of PNCs

Model	Filler Type	Orientation	2D/3D	Equation
Nielsen	Ribbon	Regular array, oriented	2D	$R_p = (1 - \varphi)/(1 + \varphi\ \alpha/2)$
Cussler – regular array	Ribbon	Regular array, oriented	2D	$R_p = (1 - \varphi)/(1 + (\varphi\alpha/2)^2)$

TABLE 1 *(Continued)*

Cussler – random array	Ribbon	Random array, oriented	2D	$R_p = (1 - \varphi)/(1 + (\varphi\ \alpha/3)^2)$
Gusev and Lusti (2001)	Disk	Random array,oriented	3D	$R_p = (1 - \varphi)/\exp[(\varphi\ \alpha/3.47)^{0.71}]$
Fred-rickson-Bicerano	Disk	Random array oriented	3D	$R_p = (1 - \varphi)/[4(1 + x + 0.1245x^2)/(2 + x)]^2$ $x = \frac{\pi \alpha \phi}{2\ ln\left(\frac{\alpha}{2}\right)}$ where, $x = \pi\varphi\alpha/(2\ \ln(\alpha/2))$

The gas permeability in Nielsen model is given by:

$$K_{composite}/K_{matrix} = (1 - \varphi)/(1 + \varphi\alpha/2)$$

where, K values represent permeabilities of the composite materials and of matrix in the absence of filler, φ is the volume fraction of the filter, and α is the aspect ratio (L/W) of the individual filler materials.

The φ is related with τ (tortuosity) as:

$$\tau = 1 + [L/(2W)]\ \varphi$$

Nielson model is valid for small loading percentage ($\varphi < 10\%$), while higher loading leads to agglomeration which affects the intercalation of polymer with nano-clay (Cussler et al., 1988). Nielsen model has been modified by taking several factors into considerations such as random positioning of the filler throughout the matrix (Brydges et al., 1975), shape of filler materials (Fredrickson et al., 1990; Gusev et al., 2001), degree of agglomeration or stacking (Nazarenko et al., 2007), temperature, and high nano-clay filler contents (Lape et al., 2004).

11.3 NANOCOMPOSITES: CLASSIFICATION, PROPERTIES, AND THEIR APPLICATIONS

Polymer composites with inorganic or organic additives can have different geometries like fibers, flakes, spheres, or particulates. A uniform dispersion of nanoparticles leads to ultra-large interfacial area between the constituents. It yields very large organic–organic or organic–inorganic interface, which alters the molecular mobility, polymer relaxation behavior, and the consequent thermal and mechanical properties of the resulting nano-composites material. This also offer extra benefits like low density, transparency, good flow, better surface properties, and recyclability in new generation polymer composites.

The PNCs are also classified on the basis of the filler materials used with the polymer, which leads to the enhancement of their physical properties.

There are several composites in nano and micro scale levels discussed as follows:

11.3.1 POLYMER CLAY NANOCOMPOSITE

Undoubtedly, the promising nanoscale fillers for PNCs are nanoplatelets composed of clays or other silicate materials. The popularity of nanoclays in food contact applications derives from their low cost, effectiveness, high stability, benignity, and anti-microbial activity. The prototypical clay utilized in PNC applications is montmorillonite (MMT) $[(Na,Ca)_{0.33}(Al,Mg)_2(Si_4O_{10})(OH)_{2n}H_2O)]$, a soft 2:1 layered phyllosilicate clay comprised of highly anisotropic platelets, separated by thin layers of water (Figure 4). The platelets have an average thickness of 1 nm and average lateral dimensions ranging between a few tens of nm to several μm. Each platelet contains a layer of aluminum or magnesium hydroxide octahedra sandwiched between two layers of silicon oxide tetrahedra. The faces of each platelet have a net negative surface charge due to isomorphous substitution of Si^{+4} for Al^{+3} or Al^{+3} for Mg^{+2} in the silicate layer, which counter by interstitial cations (Ca^{2+}, Mg^{2+}, Na^{+}, and so on) and allow the construction of multi-layer polymer assemblies under appropriate conditions, as shown in Figure 4. However, organophilic clay is obtained, if the sodium ions are replaced by long-chain organic cations, such as alkyl ammonium ions, alkylphosphonium ions, or protonated amino acids called as organically modified layered silicate (OMLS) nanocomposites. The OMLS demonstrates significant enhancement, relative to an unmodified polymer resin of a large number of physical properties including barrier, extremely light weight materials, flammability resistance, thermal and environmental stability, solvent uptake, and rate of biodegradability of biodegradable polymers attained at lower silicate concentration (< 5 weight%).

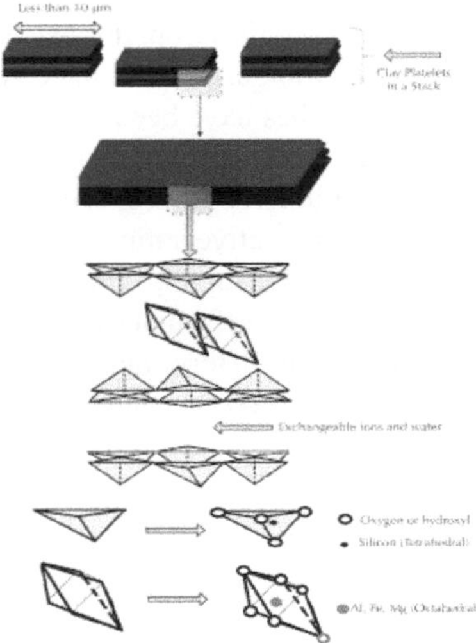

FIGURE 4 Structure of montmorillonite.

The structural characteristics of OMLS contributes excellent utility as a filler material for PNCs, typically giving rise to impressive increase in polymer strength and barrier properties with only a few weight percentage addition to the polymer matrix. For example, polyimide nanocomposites containing 2 wt% of hectrite, saponite, montmorillonite, and synthetic mica have water vapor permeability of 12.3, 10, 5.86, and 1.16 g mm /(m^{-2} day) compared to 12.9 g mm /(m^{-2} day) for the neat polymer (Yano et al., 1997). Theoretical modeling of this data shows that the variation in H_2O permeability and oxygen transmission rate (OTR) values, as a function of clay type, corresponds well to the natural platelet lateral dimensions for each clay. The first successful application of polymer–clay nanocomposites (PCNC) was a nylon-6 MMT hybrid material developed by the Toyota Corporation in 1986 (Dastjerdi and Montazer, 2010).

The efficient delamination of platelets to form fully exfoliated morphologies is hindered by the fact that clay particles are hydrophilic and many polymers of interest (PET, PE, PP, and so on) are hydrophobic. Good dispensability of nanoclays platelets in hydrophobic matrices is typically achieved by functionalizing the polar clay surface with organic ammonium ions bearing long aliphatic chains (Ray and Okamoto, 2003; Alexandre and Dubois, 2000).

11.3.2 METALLIC NANOPARTICLES BASED POLYMER COMPOSITES

Metallic-based micro and nano-structured materials in the form of salts, oxides, and colloids, are incorporated into food contact polymers to enhance mechanical properties, barrier properties, and to avoid the photo degradation of plastics. They not only serve as food preservatives but also have extended application to decontaminate surfaces in industrial environments. Other relevant properties in active food packaging are the capability for ethylene oxidation for oxygen scavenging which helps in extending food shelf life. Among the range of the metallic cations, the silver based nano-engineered materials having greatest antimicrobial capacity against broad range of gram positive and gram negative microorganisms. In the context of antimicrobial food packaging polymer, numerous researches have been carried out in recent past.

Copper, zinc, and titanium nanostructures are promising materials to be used in food safety and technology. The antimicrobial properties of above metals at the nanoscale provide affordable and safe innovative antimicrobial strategies (Table 2). Copper has been shown to be an efficient sensor for humidity, while titanium oxide has resistance to abrasion and UV-blocking performance. The migration of cations from the polymer matrices is the key point to determine their antimicrobial effectiveness, however, this migration of cations may affect the legal status of the polymer as a food contact material.

TABLE 2 Antimicrobial metallic based micro and nanocomposites in food packaging

Metal	Size	Carrier	Microorganism	Log reduction	References
Ag Metal					
Zeolites	Micro	SS	*Bacillus spp.*	$3log_{10}$CFU/mt	Cowan et al.2003
Cluster	90nm	PE	*A.acidoterrestris*	$2log_{10}$CFU/mt	Nobile et al.2004

TABLE 2 *(Continued)*

Clay	Nano	Chitosan	*E.coli*	Inhibition	Rhim et al.2006
Nanoparticles	5–35nm	Cellulose	*E.coli*	1-3$\log_{10}$CFU/mt	An et al.2008

Copper at low concentrations below 2 mg Cu^{+2}/kg serve as a co-factor for metalloproteinase and enzymes, hence, it has potent application of antimicrobial properties. Environmental Protection Agency (EPA) of the United States approved the registration of copper based alloys on the claim that they reduce microbial infections and confirmed the antimicrobial efficacy ((http:// www.epa.gov/pesticides/factsheets/copper-alloy-products.htm). Nano-sized ZnO particles have biocidal activity and several advantages over the silver nanoparticles. They have low cost, white appearance, high versatility, and doping ability with several inorganic carriers and UV blocking property.

Titanium oxide nano-sized particles have shown positive results for their application as food additives, having photo catalytic properties, self-cleaning, anti-bacterial property, and anti-UV light (Alexandre and Dubois, 2000). Silver and zinc ions are useful in zeolite type micro porous inorganic ceramics expanding the applications of silver in diverse fields. In Japan, diverse food contact polymers incorporate silver-substituted zeolites, which are the most common antimicrobials. Zeolite based technologies are listed under the FDA food contact substance notification in the USA for use in all types of food-contact polymers. A positive opinion concerning the use of two zeolites containing Ag^+ ions in food contact surfaces was released by the European Food Safety Agency in 2005. In addition, 50 µg Ag^+/kg of food is the authorized amount, therefore, silver migration into food matrices is highly restricted in Europe (Lorensa et al., 2012; FDA, 2007).

11.4 ROLE OF BIO-NANOCOMPOSITES: THEIR CLASSIFICATION AND IMPORTANCE

The extraordinary success of the synthetic PNCs concept has stimulated new research on the application of bio-based nanocomposites having biodegradable polymer as its matrix. The innovations in the development of economically and ecologically attractive green materials from biodegradable polymers led to the preservation of fossil-based raw materials, complete biological degradation of bioplastics in due period of time through natural cycle and protection of climate through reduction in carbon dioxide released. So far, the most studied bio-based polymers for the fabrication of nanocomposites are polylactides (PLA), starch, cellulose, polyhydroxy-butyrate (PHB), chitosan, and so on.

Perspective of these bio-based materials towards fabrication of bio-based nanocomposites for the food packaging applications will be discussed in the following section:

11.4.1 STARCH-BASED NANOCOMPOSITES

Starch is a promising raw material because of its cyclic availability from many plants, excessive production with regard to current needs, low cost, and complete biodegradability (Gonera and Cornillon, 2002; Smits et al. 1998). Moreover, it has an ability to promote degradability of the non-biodegradable polymers when blended at different composition in soil and water. Starch does not form films with suitable mechanical strength and properties unless it is plasticized or chemically modified. Hence, glycerol and other low molecular weight poly hydroxyl compounds, polyether, urea, and water are used as common plasticizer. When starch is treated in an extruder by application of both thermal and mechanical energy, it is converted to a thermoplastic material, in which plasticizers are expected to efficiently reduce intra-molecular hydrogen bonds and to provide stability to product properties. Thermoplastic starch (TPS) alone cannot meet all these requirements because of its hydrophobicity, performance changes during processing, and change in water content. To overcome this drawback, many different routes have been reported.

De Carvalho et al. provided a first insight in the preparation and characterization of thermo plasticized starch–kaolin composites by melt intercalation technique (2001). As films or bag, starch could be employed as packaging for fruits and vegetables, snacks, dry products, or as starch based adsorbent pads due to its hygroscopic nature for meat exudation.

Starch/PLLA nanobiocomposites is gaining interest due to the improvement in degradation properties. Thermal degradation of PLLA and PLLA/starch nanobiocomposites has been compared using thermo gravimetric studies, where the change in thermal degradation range has been observed. Onset temperature for PLLA degradation observed is 310°C and completed by 400°C, however, after addition of starch, the degradation temperature of Starch/ PLLA nanobiocomposites decreased to 220–230°C while most of the degradation completed between 280–340°C (Kim et al., 1998). Thermal degradation temperature could be lowered by increasing the corn starch content in the composite which also increases the moisture absorption capability (6–8% as compare to 1% in PLLA). Addition of starch to PLLA, had no effect on its glass transition temperature, however, increase in crystallization was observed indicating, starch as a nucleating agent (Park et al., 1999). Change in plasticizer used in PLLA/ Starch composite changes the mechanical properties like tensile strength, percentage elongation at break, and modulus of the nanobiocomposites (Figure 5).

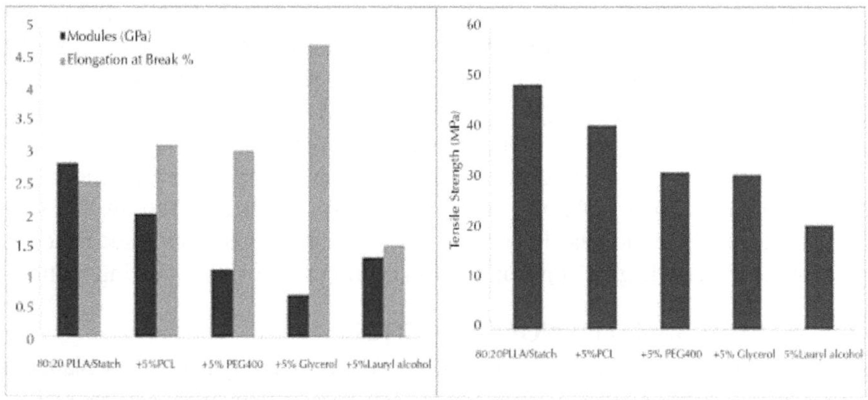

FIGURE 5 Comparison of the mechanical properties with the change in the plasticizers for the nanobiocomposite. (Kim et al., 1998).

Hydrophobicity of the starch-based films and its poor mechanical properties can also be improved by fabrication of nanobiocomposite of thermoplastic starch (TPS) and nanocellulose fibers (NCF). Tensile strength of the base polymer film increases by ~46% by addition of 0.4 wt% NCF and it starts deteriorating above 0.4 wt% (Savadekar and Mhaske, 2012). Tensile strength calculated for neat TPS films was 9.83 MPa, which increased to 10.35 MPa and 12.84 MPa with an addition of 0.1% and 0.5 % NCF. However, maximum tensile strength was found at 0.4% NCF/TPS film about 18 MPa totaling to 46.1% of increase. Significant decrease of WVTR is observed in 0.4% NCF/TPS films having value of 4.3×10^{-4} g/h/sq m while the highest WVCTR recorded is 7.8×10^{-3} g/h/sq m through the neat TPS films. However, The OTR reduced by 93% in 0.4% NCF/TPS films.

A great deal of research has been performed on the preparation of starch/clay nanocomposites by melt intercalation in details. There has been an increase in elongation at break and tensile strength by more than 20% and 25%, respectively, and a decrease in water vapor transmission rate by 35% for potato starch/MMT nanocomposites after addition of 5% clay (Park et al. 2002, 2003). Wilhelm et al. observed 70% increase in tensile strength of cara root starch/hectrite nanocomposites films at 30% clay level while the percentage of elongation decreased by 50% (2003). Avella et al. reported the preparation of starch/MMT nanocomposites films for food packaging applications, where the decomposition temperature is increased, indicating a better thermally stabile film along with increase of modulus and tensile strength (2005). Several investigations have shown that the order of adding different components affected the mechanical properties of the starch/MMT films.

11.4.2 CELLULOSE-BASED NANOCOMPOSITES

Cellulose is one of the most promising natural raw materials and constitutes the most abundant renewable polymer resource available today. With its inherent renewability and sustainability, it makes itself attractive towards the research area. Cellulosic materials, when subjected to acid hydrolysis, (Figure 6) yield defect-free, crystalline residues called as cellulose nanocrystals (CNCs) or whiskers (Kalia et al.). The CNCs possess many advantages, such as nanoscale dimension, high specific strength and modulus, high surface area, unique optical properties, and so on. These incredible physicochemical properties and wide application prospects have attracted significant interest from both research scientists and industrialists. These properties make them promising for different application areas like reinforcement of polymeric fillers in nanocomposite materials, chemical transformations, nanocomposite films, drug delivery, protein immobilization and metallic reaction template, and so on exploring their thermal and mechanical strengths. Polymer matrix gets transformed when they are pooled with cellulose nanocrystals and thereafter utilized to prepare nanobiocomposites having enhanced mechanical, thermal, barrier, and electrical properties along with degradability ease.

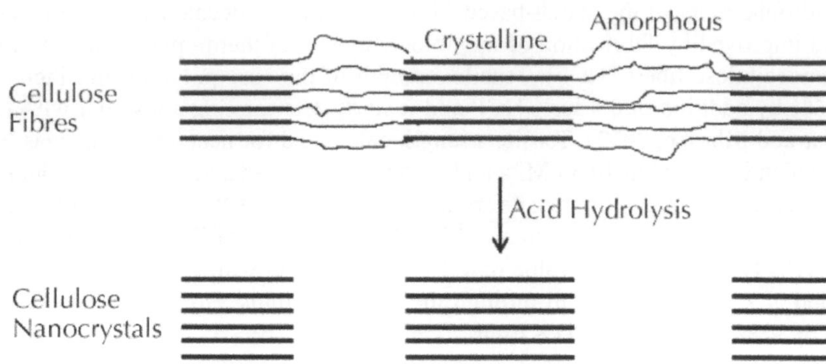

FIGURE 6 Acid hydrolysis to form cellulose nanocrystals.

The CNCs improve the barrier properties like oxygen transmission rate (OTR) and water vapor transmission rate (WVTR) as well as mechanical properties like young's modulus and strength of PLA matrix (Okasman et al., 2006).

Cellulose derived plastics such as cellulose acetate (CA), cellulose acetate propionate (CAP), and cellulose acetate butyrate (CAB) are thermoplastic materials produced through esterification of cellulose, which can be used as a substitute for petroleum feedstock. Among the different derivatives of cellulose, cellulose acetate (CA) is of particular interest because of its biodegradable nature, excellent optical clarity, and high toughness. Cellulose ester powders derived from different raw materials such as cotton, recycled paper, wood cellulose, and sugarcane in presence of different plasticizer and additives are melted processed *via* extrusion processes to produce commercial pelletized cellulose plastics.

Phthalate plasticizer can be replaced with ecofriendly triethyl citrate (TEC), organo clays, and citrates derivatives from vegetable oil. The fabricated nanocomposites from CA through extrusion–injection molding at 20 wt% TEC plasticizer showed significantly improved physical and mechanical properties with intercalation cum exfoliation mixed morphology. (Chen and Evans, 2005).

11.4.3 GELATIN-BASED NANOCOMPOSITES

Gelatin is prepared by the thermal denaturation and hydrolytic cleavage of collagen, isolated from the skin of animal and fish and bones using very dilute acid. Gelatin contains a large number of glycine (almost 1 in 3 residues, arranged every third residue), proline, and 4-hydroxyproline residues. Gelatin is a heterogeneous mixture of single or multi-strand polypeptides with extended left handed proline helix conformations of 300–400 amino acids with a typical structure–Ala–Gly–Pro–Arg–Gly–Glu–4 Hyp–Gly–Pro–, molecular structure of gelatine. The gelatin solution undergo coil-helix transition followed by aggregation of helixes to form triple helix(proline rich junctions) of type I collagen extracted from skin and bones, as a source for gelatin, is composed of two a1(I) and one a2(I) chains, each with molecular mass ~95 kD, width ~1.5 nm, and length ~0.3nm. Primarily Gelatin is a gelling agent, which dissolves at low temperature to give flavored products forming transparent elastic thermo reversible gels on cooling below 35°C. The polymer can be used as biopolymer in tissue engineering as well as edible coatings since it reduces oxygen, moisture, and oil migration and carry an antioxidant or antimicrobial but the major drawback is its poor mechanical properties and hence, a topic of challenge.

Grape fruit seed extracts (GFSE) derived from the seed and pulp of the grape fruit is a potential additive to gelatin-based polymer because it has been reported to be non–toxic and increases shelf life by inhibiting the growth of the food borne pathogens. Polymer films coated with GFSE layer with polyamide as binder showed antimicrobial activity against a variety of micro-organisms. These gelatin and polymeric-based GFSE composites have effective application for packaging in beef and fish products (Ha et al., 2005; Cho et al., 2004).

The barley bran (BB) is a byproduct of the barley powder manufacturing industries, a protein source (18.7 %) and can be used for protein film preparation because of its low cost. The barley bran protein and the gelatin-based composite films showed decrease in the tensile strength with the increase in the protein concentration but increased with the addition of the gelatin content. The BB films coated with GFSE and have potent application in the packing of the salmon, since GFSE decreased the peroxide value and thiobarbituric acid, respectively. In above cases, concentration of the GFSE added did not affect the quality of the beef product (Song et al., 2012). The property of thermo-reversible at melting point close to body temperature makes gelatin a base material for protein films but its large scale production possibilities are debatable due to its high cost.

11.4.4 PLA NANOCOMPOSITES

The PLA is a synthetic biopolymer made from agricultural raw material. The PLA has good mechanical (stiffness and strength) and thermal properties. It is more attractive towards packaging, medical, and automotive applications. Economical consideration of PLA nano-bio composites for food packaging at commercial scale is still a big challenge. Nano fillers are added to PLA to improve its properties like barrier properties, thermal stability, stiffness, strength, and so on. Best suitable example for the same is PLA/CNCs (cellulose nanocrystals) composite. Embedding cellulose nanocrystals to PLA matrix may increase many folds barrier properties and still the research work is in progress to improve it. Many other nanobiocomposites like PLA/Starch, PLA/PCL, PLA/layer silicate nanocomposites, and PLA/MMT are known with the improved mechanical properties. Biggest challenge with PLA nanocomposites is to improve the barrier properties mainly to oxygen, water vapor, and carbon dioxide, to use it for food packaging applications. Katiyar et al. developed an *in situ* polymerization method for simultaneous PLLA synthesis while dispersion of clay and characterized the morphology of poly(L-lactic acid) (PLLA) nanocomposites using TEM (2011). This study also demonstrated twinning of either intercalated or exfoliated morphology based on the functionality of clay modifier and found intercalated PLACN when used non-functionalized modifier based Cloisite® clay. Representative TEM micrograph is shown in Figure 7.

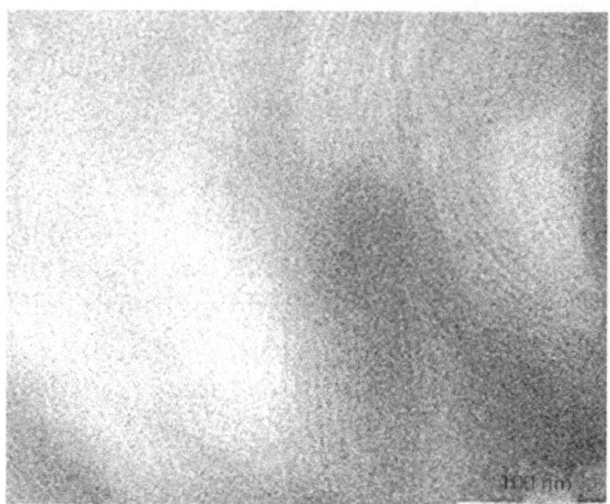

FIGURE 7 The TEM bright field representative image of intercalated PLACNs. The dark entities are the cross section of intercalated organoclay layers, and the bright areas are the matrix. (Katiyar et al., 2011); (Reprinted by the permission from authors).

Bright lines represent the PLLA while clay layers are identified by the dark lines of ~1 nm thickness, intercalated by the clay galleries. The TEM image shows the effect of 4 wt% clay loading depicting a well-ordered intercalated nanostructure of alternating layers of PLLA and clay with the polymer growth inside clay galleries.

Layered double hydroxides (LDHs) are used as clay nanofillers to improve the barrier properties of polylactides films. Ring opening polymerization has been used to ensure the good dispersion of LDH carbonate (LDH-CO_3) and laurate modified-LDH (LDH-C_{12}) in polylactides matrix. The TEM images clearly show the morphological difference between LDHs and PLA-LDH nanocomposites (Katiyar et al., 2011). Aggregates of LDH-CO_3 can be seen in Figure 8(a) giving the fibrous morphology. Irregular shaped sheet like LDH coexists of ~50–100 nm (Figure 8(a)). Hexagonal morphology of LDH-C_{12} is clearly visible in Figure 8(b) appearing as thin plates from ~ 50 nm to 200. Figure 9 shows the TEM images of PLA-LDH nanocomposites in comparison to Figure 8.

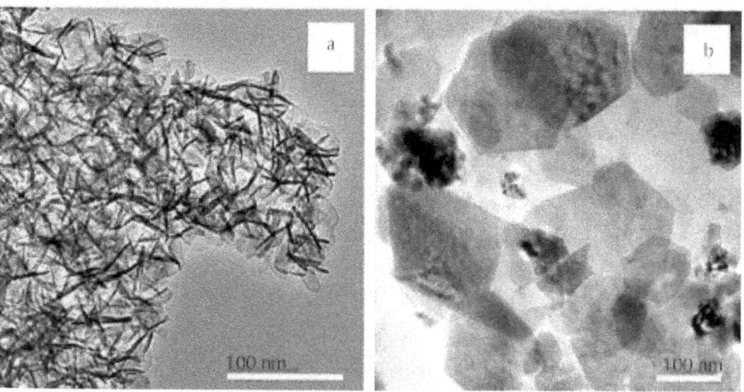

FIGURE 3.8 TEM images of LDH-CO_3 (a) and LDH-C_{12} (b). (Katiyar et al., 2011); Reprinted by the permission from authors.

Bright areas in the images show PLA matrix while dark areas indicate cross section of LDH platelets. Exfoliated LDH layers in PLA matrix can be seen in PLA/LDH-CO_3 5% (Figure 9(a)) while well distributed LDH platelets in PLA-LDH-C_{12} 5% (Figure 9(b)).

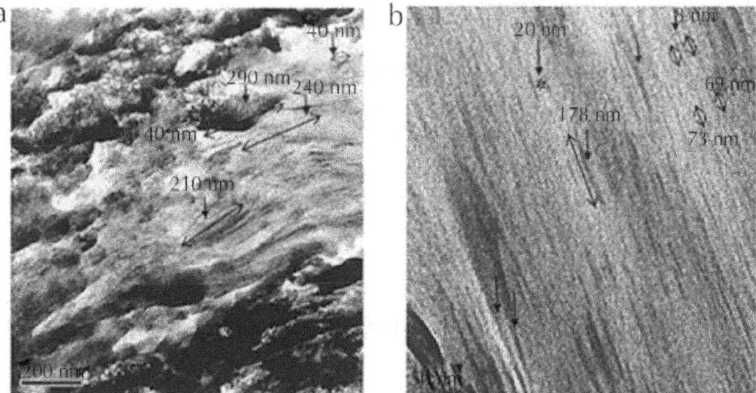

FIGURE 9 TEM brightfield images of PLA/LDH-CO_3 (5%) (a), PLA/LDH-C_{12} (5%) (b),(Katiyar et al., 2011), Reprinted by the permission from authors.

11.5 FABRICATION TECHNOLOGIES FOR NANOBIOCOMPOSITES

Embedding the nanoparticles in the polymer matrix is still a big challenge. Execution may be feasible at a bench scale but scaling up for commercial purpose faces many problems like homogeneously dispersing the nanoparticles in the polymer matrix, agglomeration of nanoparticles, and so on Moreover, interaction mechanisms between nanomaterial, polymer and, solvent play a crucial role.

We will be discussing few techniques for the fabrication of nanobiocomposites as following:

11.5.1 SOLVENT CASTING

This is a simple technique, which was developed hundred years ago for plastic film manufacturing. It attracts to be a fabrication technique of nanobiocomposites, since, it is capable of fabricating high quality films in terms of uniform thickness requirement, maximum optical purity, and extremely low haze. Basic requirement for this technology is the solubility of a polymer in volatile solvent or water, stability of the solution having less viscosity and removal of the film from the casting support. It includes mainly six steps that is, dope preparation, casting on the support, spreading homogeneously, drying, inline coating, and solvent recovery (Ulrich, 2005). López-Suevos et al. used solvent casting approach for the preparation of composites of refined beech pulp with poly (vinyl acetate) latex (2010). Nano-fibrillated cellulose in water was homogenized with the polymer solution using high shear blender. Increasing the amount of cellulose nano-fibrils, increased the reinforcing effects in the glassy state, and had superior heat resistance properties after board fabrication.

The NCF based thermoplastic starch films have been synthesized using solvent casting method followed by evaporation process. Dispersion of NCF in varied amounts was prepared in water in which TPS is added in respective weight percentage till complete dissolution. Glycerol and acetic acid are added for gelatinization. Average thickness of the films formed is 0.6 mm (Savadekar and Mhaske, 2012). Properties improved as discussed in section 11.1.

Ciobanu et al. investigated the formation of nanocrystalline hydroxyapatite layer on the surface of porous polyurethane for the promising scaffolding material in tissue engineering (2009). A solution of polyurethane and cellulose acetate is formed in N, N-dimethylformamide to form polyurethane membranes. Glass plate is used as a support for the solvent evaporation under caloric radiation to form thin membranes. Theses membranes are treated with supersaturated calcification solution to form three dimensional scaffolds of hydroxyapatite layer on the polyurethane membrane.

11.5.2 MELT INTERCALATION

In melt extrusion, polymer is forced to flow under shear along with helical screw direction, which divided into number of zones based on their functions and the requirement for processing specific combination such as feed section, mixing and melting section, and compression section and takes the shape of the die at the other end. In melt intercalation, biofiller can be directly mixed mechanically with the polymer melt to form a homogeneous mixture. Method is very widely used for thermoplastic nanocomposites and can be applied to nanobiocomposite. Melt intercalation is highly specific for a polymer, which may lead to new hybrids. Moreover, absence of solvent in the technique leads to industrially economical as well as environment friendly by waste point of view (Ray et al., 2005).

Krishnamachari et al. prepared nanobiocomposites of poly lactic acid and several nanoclays (organically modified montmorillonite) using twin screw extruder by melt compounding. They observed an increase in the thermal stability with a high loading level of 3% by weight (2009).

Oksman et al. applied an innovative approach of separating cellulose whiskers through swelling of microcrystalline cellulose (2006). A mixture of separated whiskers from the solution of *N,N*-dimethylacetamide (DMAc) containing lithium chloride (LiCl) was pumped through extrusion process to form a poly lactic acid/cellulose whisker nanocomposite melt. Subsequiently, the mechanical strength of the nanocomposites improved by 800% but TGA studies showed that polymer composite degraded at relatively lower temperature than neat PLA temperature.

In their further investigations, Bondeson and Oksman used anionic surfactant to uniformly disperse the cellulose whiskers (CNW) in the PLA matrix (2006). The tensile strength and elongation at break improved with the increase in concentration of surfactant but maximum modulus is achieved at 5 wt% surfactant concentration.

Further, transparent polymer composites have been made using CNWs when suspended in cellulose acetate butyrate (CAB) and plasticized with triethyl citrate (TEC). The mechanical characterization results indicate an increase in tensile modulus and strength by factor of 300% and 100%, respectively.

Iwatake et al. developed a sustainable green composite by reinforcing PLA with micro fibrillated cellulose (MFC) and needle-leaf bleached kraft pulp (NBKP), which improved the nano fiber network to large extent (2008). At 10 wt % nano fiber dispersion of MFCs in PLA, Young modulus and tensile strength improved by 40% and 25% without any reduction in yield strain, whereas NBKP reported reduction in yield strain and strength by 30% and 50% respectively.

Wang et al. extracted cellulose nanofibres from soybean stock by chemo-mechanical treatments with diameter ranging from 50–100 nm and size of ~1000 nm (2007). The cellulose nano fibers were dispersed in polymeric matrix of poly (vinyl alcohol) (PVA) and polyethylene (PE) by addition of dispersant ethylene-acrylic oligomer. The soya bean stock based PVA films showed an increase of 4–5 fold in tensile strength.

The melt intercalation method for preparation of bio-polymers can leads to better dispersion of fillers at large extent, which finds their space in manufacture of nanobiocomposite based packaging, however, there processing activity is limited at elevated temperature.

The PLA layered silicate nanocomposites were prepared by adding small amounts of the compatibilizer to form the randomly distributed intercalated silicate layers. Simple melt extrusion of PLA and organically modified montmorillonite lead to better parallel stacking of silicate layers and much stronger flocculation due to hydroxylated edge–edge interactions of silicate layers and consequently improved mechanical and barrier properties, which makes it suitable for food packaging application. Further, Bondeson et al. used melt extrusion to fabricate a transparent bio-based nanocomposite of 5 wt% cellulose nanowhiskers (CNW) and cellulose acetate butyrate (CAB), plasticized by triethyl citrate (TEC) (2007).

11.5.3 *IN-SITU* POLYMERIZATION

In-situ polymerization technique may be an effective approach for preparation of well-disperse nano or nanobiocomposites. Dispersion of such fillers at monomer level is quite easy due to significantly lower monomer viscosity than respective polymer. Hence, filler can be dispersed first into monomer followed by polymerization, which yields high molecular weight polymer with nano-scale filler dispersion. In case of layered clays, in which monomer can intercalate into clay galleries and subsequently upon polymerization, may yield either exfoliated or intercalated morphology depending on the selected organomodified clays. It can be clearly seen from Figure 10, where L-lactide monomer units are inserted into clay galleries and therefore the gallery gets expanded up to 20Å. Subsequently, on polymerization, it yields superior nanocomposite (Katiyar and Nanavati, 2011). Molecular level reinforcement is the main advantage of this technique as compared to others (Ray et al.). Kim et al. synthesized multi-walled carbon nanotube (MWCNT) and fibrin nanobiocomposites (hybrid structures) using transglutaminase catalyzed *in situ* polymerization (2009).

Y. Li et al. prepared nanobiocomposite of PLA and titanium dioxide by using *in situ* polymerization (2011). Covalent bonding between surface of titanium dioxide nanowire and PLA chains yields PLA nanocomposites through *in situ* melt condensation. Increases in the glass transition temperature and high thermal stability are found in the grafted PLA as compared to pure PLA. Grafting of PLA chains onto nanowire surfaces are confirmed by FTIR and TGA.

Urbanczyk et al. prepared polylactide/clay nanocomposites in supercritical carbon dioxide using *in situ* ring opening polymerization (2009). Polylactide is used for diluting nanocomposites with high clay contents (30–40%) by melt blending, so as result in exfoliated nanocomposites containing 3 wt% of clay. Mechanical properties like stiffness, toughness, and impact resistance of nanobiocomposites are improved significantly as compared to the pure matrix.

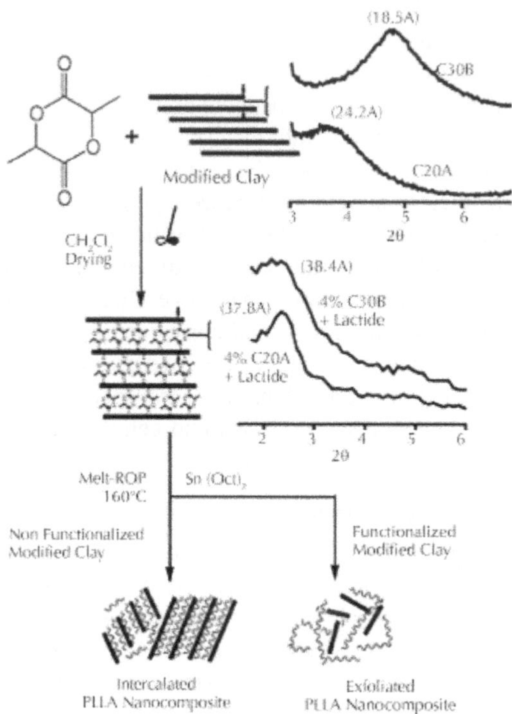

FIGURE 10 Schematic of *in situ* ROP process of L-lactide in the presence of 4 wt% organically modified clay.

11.5.4 ELECTROSPINNING

Electrospinning is a type of solution processing, based on the principle of inducing the electrostatic forces to form a jet of polymer solution. Biofillers or bionanoparticles are dispersed in the bio-based polymer matrix to form a polymer solution. A polymer solution of nanobiocomposite is thereafter added to a syringe attached to high voltage DC power supply. A fine spinneret is attached in front of the syringe maintained at positive or negative charge. A rotating or revolving collector is installed at the bottom of the spinneret. When electrostatic repelling force overcomes the surface tension force of the polymer solution then solution is forced to move through the spinneret forming fine filament formed by solvent evaporation and thus continuously leading to nanobiocomposite nanofibers. Above technique is widely used for embedding the substrate in the polymer matrix and forming the nanofibers. Electrospinning finds major applications in the tissue engineering, however, fabrication of the bio-nanocomposites with the fibrous network for food packaging can be done.

Vassalli prepared a nanobiocomposite material, which is helpful in fabricating biocompatible and durable hernia repair mesh. It is based on the crosslinking of bioabsorbable polymer fibers. A solution of poly-caprolactone (biodegradable) in ethylene diamine (aminolyzing agent) has been prepared for adding amine groups and are fabricated electro-spun fibers using electrospinning technique.

Pilon et al. used PLA and monetite cement infused with functional particulates of calcium phosphate cement to develop a nanofibrous scaffold for tissue engineering applications using electrospinning technique (2010). Similar work has been carried in tissue engineering such as fabrication of novel PLA/CDHA (carbonated calcium deficient hydroxyapatite) nanobiocomposite fibers (Zhou, 2011), preparation of PVP/PLLA ultrafine blend fibers (Xu, 2011), fabrication of biodegradable polyester nanocomposites (Zhou, 2011), nanobiocomposites from electrospun PVA/pineapple nanofibers/*Stryphnodendro nastringents* bark extract for medical applications (Costa et al., 2013], and so on.

11.6 CONCLUSON

Although we have non-biodegradable materials with good mechanical and thermal properties having slightly poor barrier properties for food packaging applications, nanobiocomposites may explore a new horizon to use them for the food packaging. Biodegradability of nanobiocomposites adds an advantage to save our environment from being polluted. Embedding the nanobio fillers to the polymer matrix improves its various properties mainly barrier towards oxygen and water vapor and mechanical properties. Thermal stability may also be improved at times. Though, fabrication techniques are not very well defined for nanobiocomposites but we can imply the techniques used for fabricating nanocomposites. Solution processing, melt intercalation, *in situ* polymerization are the major one used frequently. Characterizing the fabricated composite is much important to know about the improved properties. Bio-based nanocomposite research is developing rapidly. However, there is need to address issues such as toxicological, gas barrier, and processing limitation especially when these materials are focused to use in food contact applications such as food packaging.

KEYWORDS

- **Biodegradable polymers**
- **Brownian motion**
- **Cellulose fibers**
- **Nanostructured materials**
- **Polymer nanocomposites**

REFERENCES

1. Alexandre, M. and Dubois, P. *Mater. Sci. Eng.*, **28**, 1–63 (2000).
2. Amass, W., Amass, A., and Tighe, B. *Polym. Int.*, **47**, 89–144 (1998).
3. Avella, M., De Vlieger, J. J., Errico, M. E., Fischer, S., Vacca, P., and Volpe, M. G. *Food Chemistry*, **93**, 467–474 (2005)
4. Averous, L. and Martin, O., *Polymer*, **42**, 6209–6219 (2001).
5. Awarenet. *Handbook for the prevention and minimization of waste and valorization of byproducts in European agro-food industries*. Deposito legal: BI-223-04 (2004).
6. Azeredo, H. M. C., Mattoso, L. H. C., Avena-Bustillos, R. J., Filho, G. C., Munford, M. L., Wood, D., and McHugh, T. H. *J. Food Sci.*, **75**, N1–7 (2010)
7. Azeredo, H. M. C., Mattoso, L. H. C., Wood, D., Williams, T. G., Avena-Bustillos, R. J., and McHugh, T. H. *J. Food Sci.*, **74**, N31–35 (2009).
8. Beccari, M., Bertin, L., Dionisi, D., Fava, F., Lampis, S., and Majone, M. *Journal of Chemical Technology & Biotechnology*, **84**, 901–908 (2009).
9. Bertini, F., Canetti, M., Audisio, G., Costa, G., and Falqui, L. *Polym. Degrad. Stab.*, **91**, 600–605 (2006).
10. Bharadwaj, R. K. *Macromolecules*, **34**, 9189–9192 (2001).
11. Bin, Y., Mine, M., Koganemaru, A., Jiang, X., and Matsuo, M., *Polymer*, **47**, 1308–1317 (2006).
12. Bilbao-Sainz, C., Avena-Bustillos, R. J., Wood, D. F., Williams, T. G., and McHugh, T. H. *J. Agric. Food Chem.*, **58**, 3753–3760 (2010).
13. Bondeson, D. and Oksman, K. *Compos. Interface*, **14**, 617–630 (2007).
14. Bondeson, D., Syre, P., and Oksman, K. *J. Biobased Mat. Bioenergy*, **1**, 367–371 (2007).
15. Brydges, W. T., Gulati, S. T., and Baum, G. *J. Mater. Sci.*, **10**, 2044–2049 (1975).
16. Cabedo, L., Feijoo, J. L., Villanueva, M. P., Lagaron, J. M., and Gimenez, E. *Macromol. Symp.*, **233**, 191–197 (2006).
17. Cao, X., Chen, Y., Chang, P. R., Stumborg, M., Huneault, M. A. *J. Appl. Polym. Sci.*, **109**, 3804–3810 (2008).
18. Chang, J., An, Y. U, and Sur, G. S. *J. Polym. Sci. B: Polym. Phys.*, **41**, 94–103 (2003).
19. Chen, B. and Evans, J. R. G. *Carbohydrate Polymers*, **61**, 455–463 (2005).
20. Chen, Y., Cao, X., Chang, P. R., and Huneault, M. A. *Carbohydr. Polym.*, **73**, 8–17 (2008).
21. Chen, W., Tao, X, Xue, P., and Cheng, X. *Appl. Surf. Sci.*, **252**, 1404–1409 (2005).
22. Cho, S. M., Kwak, K. S., Park, D. C., Gu, Y. S., Ji, C. I., Jang, D. H., and Kim, S. B. *Food Hydrocolloids*, **18**, 573–579 (2004).
23. Cho, T. W. and Kim, S.W. *J. Appl. Polym. Sci.*, **121**, 1622–1630 (2011).
24. Choudalakis, G., Gotsis, A. D. *Eur. Polym. J.*, **45**, 967–984 (2009).
25. Ciobanu, G., Ignat, D., and Luca, C. Chemical Bulletin of "Politehnica" University of Timisoara, **54**, 1 (2009).
26. K. L. Yam (Ed.). Coextrusion for Semirigid Packaging. *The Wiley Encyclopedia of Packaging Technology*, John Wiley and Sons, Inc., New York, 297–299 (2009).
27. *Compos. Sci. Technol.*, **69**, 1756–1763 (2009).
28. Costa, L. M. M., Olyveira, G. M., Cherian, B. M., Leão, A. L., Souza, S. F., and Ferreira, M. *Industrial Crops and Products*, **41**, 198–202 (2013).
29. Cussler, E .L., Hughes, S. E., Ward, W. J., and Aris, R. *J. Membr. Sci.*, **38**, 161–174 (1988).
30. Dastjerdi, R. and Montazer, M. *Colloids and Surfaces B: Biointerfaces*, **79**, 5–18 (2010).
31. De Azeredo, H. M. C. *Food Research International.*, **42**, 1240–1253 (2009).
32. De Carvalho, A. J. F., Curvelo, A. A. S., and Agnelli, J. A. M. *Carbohydrate Polymers*, **45**, 189–194 (2001).
33. De Moura, M. R., Aouada, F. A., Avena-Bustillos, R. J., McHugh, T. H., Krochta, J. M., and Mattoso, L .H. C. *J. Food Eng.*, **92**, 448–453 (2009).
34. De Moura, M. R., Lorevice, M. V., Mattoso, L. H. C., and Zucolotto, V. *J. Food Sci.*, **76**, N25–29 (2011).
35. Dubois, P. *Polym. Degrad. Stab.*, **92**, 1873–1882 (2007).

36. Dufresne, A., Cavaille, J. Y., and Helbert, W. *Macromolecules*, **29**, 7624–7626 (1996).
37. Dunkerley, E. and Schmidt, D. *Macromolecules*, **43**, 10536–10544 (2010).
38. FDA. Inventory of Effective Food Contact Substance (FCS) notifications. http://www.cfsan.fda.gov/wdms/opa-fcn.html. FCN (2007).
39. Finnigan, B. Barrier polymers. K.L. Yam (Ed.), The Wiley Encyclopedia of
40. Fredrickson, G. H. and Bicerano, J. *J. Chem. Phys*., **110**, 2181–2188 (1999).
41. Lo´pez-Suevos, F., Eyholzer, C., Bordeanu, N., and Richter, K. *Cellulose.*, **17**, 387–398 (2010).
42. Gilman, J. W., Jackson, C. L., Morgan, A. B., Harris, R., Manias, E., Giannelis, E. P., Wuthenow, M., Hilton, D., and Phillips, S. H. *Chem. Mater*., **12**, 1866–1873 (2000).
43. Gonera, A. and Cornillon, P. *Starch/Stärke*, **54**, 508–516 (2002).
44. Gorrasi, G., Tortora, M., Vittoria, V., Pollet, E., Alexandre, M., and Dubois, P. *J. Polym.Sci., Part B: Polym. Phys*., **42**, 1466–1475 (2004).
45. Gu, S. Y., Ren, J., and Dong, B. *J. Polym. Sci., Part B: Polym. Phys*., **45**, 3189–3196 (2007).
46. Gusev, A. A. and Lusti, H. R. *Adv. Mater*., **13**, 1641–1643 (2001).
47. Ha, J. U., Kim, Y. M., and Lee, D. S. *Packaging Technology and Science*, **15**, 55–62 (2001).
48. Helbert, W., Cavaille, J. Y., and Dufresne, A. *Polym. Compos*., **17**, 604–611 (1996).
49. Horrocks, A. R., Kandola, B. K., Smart, G., Zhang, S., and Hull, T. R. *J. Appl. Polym. Sci*., **106**, 1707–1717 (2007).
50. Iwatake, A., Nogi, M., and Yano, H. *Compos. Sci .Technol*., **68**, 2103–2116 (2008).
51. Jia, X., Li, Y., Cheng, Q., Zhang, S., and Zhang, B. *Eur. Polym. J*., **43**, 1123–1131 (2007).
52. Kalia, S., Kaith, B. S., and Kaur, I. *Cellulose Fibers: Bio- and Nano-Polymer Composites*, Springer Heidelberg, Dordrecht, London, and New York.
53. Kandola, B. K., Smart, G., Horrocks, A. R., Joseph, P., Zhang, S., Hull, T. R., Ebdon, J., Hunt, B., and Cook, A. *J. Appl. Polym. Sci*., **108**, 816–824 (2008).
54. Katiyar, V., Nanavati, H. Society of Plastic Engineers, Plastics research Online, 10.1002/spepro.003699.
55. Katiyar, V. and Nanavati, H. Polymer Engineering and Science, 2066–2077 (2011).
56. Kim, J. Y., Han, S. I., Kim, D. K., and Kim, S. H. *Compos. Part A: Appl. Sci. Manuf*., **40** (2009).
57. Kim, J. Y., Han, S. I., and Kim, S. H. *Polym. Eng. Sci*., **47**, 1715–1723 (2007).
58. Kim, M., Jang, J. H., Han, J. H., Lee, Y. W., Cho, S. M., Son, S.Y., Hulme, J. Choi, I. S., Paik, H.J., and Soo, A. S. *Bulletin of the Korean Chemical Society*, **30,** 405–408 (2009).
59. Kim, S. H., Chin, I. J., Yoon, J. S., Kim, S. H. and Jung, J. S. *Korea Polymer Journal*, **6**, 422–427 (1998).
60. Kojima, Y., Usuki, A., Kawasumi, M., Okada, A., Fukushima,Y., Kurauchi, T., and Kamigaito, O.*J. Mater. Res*., **8**, 1185–1189 (1993).
61. Kojima, Y., Usuki, A., Kawasumi, M., Okada, A., Kurauchi, T., and Kamigaito, O. *J. Polym. Sci., Part A: Polym. Chem*., **31**, 1755–1758 (1993).
62. Kollen, W. and Gray, D. *J. Plast. Film. Sheet*., **7**, 103–117 (1991).
63. Krishnamachari, P., Zhang, J., Lou, J., and Yan, J., Uitenham, L. *International Journal of Polymer Analysis and Characterization*, **14**, 336–350 (2009).
64. Kristo, E. and Biliaderis, C. G. *Carbohydr. Polym*., **68**, 146–158 (2007).
65. Kumar, S., Jog, J. P., and Natarajan, U. *J. Appl. Polym. Sci*., **89**, 1186–1194 (2003).
66. Lagaron, J. M., Cabedo, L., Cava, D., Feijoo, J. L., Gavara, R., and Gimenez, E. *Improving packaged food quality and safety Food Addit Contam*, **22**, 994–998 (2005).
67. Lape, N. K., Nuxoll, E. E., and Cussler, E. L. *J. Membr. Sci*., **236**, 29–37 (2004).
68. Li, Y., Chenb, C., Lib, J., and Sun, X. S. *Polymer*., **52**, 2367–2375 (2011).
69. Li, Y. C., Schulz, J., and Grunlan, J. C. *ACS Appl. Mater. Int*., **1**, 2338–2347 (2009).
70. Lin, L., Liu, H., and Yu, N. J. *Appl. Polym. Sci*., **106**, 260–266 (2007).
71. López-Suevos, F., Eyholzer, C., Bordeanu, N., and Richter, K. *Cellulose*, **17**, 387–398 (2010).
72. Llorens, A., Lloret, E., Picouet, P. A., Trbojevich R., and Fernandez, A. *Trends in Food Science & Technology*., **24**, 19–29 (2012).
73. Lu, Y., Weng, L., and Zhang, L. *Biomacromolecules,* **5**, 1046–1051 (2004).
74. Ludueña, L. N., Alvarez, V. A., and Vazquez, A. *Mater. Sci. Eng., A*., **460**, 121–129 (2007)

75. Maksimov, R. D., Gaidukov, S., Zicans, J., and Jansons, J. *Mech. Compos. Mater*., **44**, 505–514 (2008).
76. *Mater. Manuf. Process*, **24**, 1053–1057 (2009).
77. Ma, X. D., Qian, X. F., Yin, J., and Zhu, Z. K. *J. Mater. Chem*., **12**, 663–666 (2002).
78. Morlat-Therias, S., Fanton, E., Gardette, J. L., Peeterbroeck, S., Alexandre, M.;
79. Moggridge, G. D., Lape, N. K., Yang, C., and Cussler, E. L. *Prog. Org. Coat*., **46**, 231–240 (2003).
80. Morgan, A. B., Chu, L. L., and Harris, J. D. *Fire Mater*., **29**, 213–229 (2005).
81. Nazarenko, S., Meneghetti, P., Julmon, P., Olson, B. G., and Qutubuddin, S. *J. Polym. Sci., Part B: Polym. Phys*., **45**, 1733–1753 (2007).
82. Neilsen, L. E. *J. Macromol. Sci. A*., **1**, 929–942 (1967).
83. Ogata, N., Jimenez, G., Kawai, H., and Ogihara, T. *J. Polym. Sci., Part B: Polym. Phys*., **35**, 389–396 (1997).
84. Oksman, K., Mathew, A. P., Bondeson, D., and Kvien, I. *Compos. Sci. Technol*., **66**, 2776–2784 (2006).
85. Osman, M. A., Rupp, J. E. P., and Suter, U. W. *Polymer*, **46**, 1653–1660 (2005).
86. Prashantha, K., Soulestin, J., Lacrampe, M. F., Krawczak, P., Dupin, G., and Claes, M. *Packaging Technology*, John Wiley and Sons, Inc., New York, 103–109 (2009).
87. Park, H. W., Lee, W. K., Park, C. Y., Cho, W. J., and Ha, C. S. *Journal of Materials Science*., **38**, 909–915 (2003).
88. Park, H. M., Li, X., Jin, C. Z., Park, C. Y., Cho, W. J., and Ha, C. S. *Macromolecular Materials and Engineering*., **287**, 553–558 (2002).
89. Park, H. M., Li, X., Jin, C. Z., Park, C. Y., Cho, W. J., and Ha, C. S. *Macromol. Mater. Eng*., **287**, 553–558 (2002).
90. Park, J. W., Lee, D. J., Yoo, E. S., Im, S. S., Kim, S. H., and Kim, Y. H. *Korea Polymer Journal*, **7**, 93–101 (1999).
91. Picard, E., Vermogen, A., Gerard, J. F., and Espuche, E. *J. Membr. Sci*., **292**, 133–144 (2007).
92. Pilon, A., Touny, A., Lawrence, J., and Bhaduri, S. SAE Technical Paper 2010-01-0423 (2010). doi:10.4271/2010-01-0423.
93. Podsiadlo, P., Choi, S. Y., Shim, B., Lee, J., Cuddihy, M., and Kotov, N. A. *Biomacromolecules*, **6**, 2914–2918 (2005).
94. Podsiadlo, P., Kaushik, A. K., Arruda, E. M., Waas, A. M., Shim, B. S., Xu, J. Nandivada, H., Pumplin, B. G., Lahann, J., Ramamoorthy, A., and Kotov, N. A. *Science*, **318**, 80–83 (2007).
95. Porter, D., Metcalfe, E., and Thomas, M. J. K. *Fire Mater*., **24**, 45–52 (2000).
96. Powell, C. E. and Beall, G. W. *Curr. Opin. Solid State Mater. Sci.,* **10**, 73–80 (2006).
97. Ray, S. S. and Bousmina, M. *Prog. Mater. Sci*., **50**, 962–1079 (2005).
98. Ray, S. S. and Okamoto, M. *Prog. Polym. Sci*., **28**, 1539–1641 (2003).
99. Ray, S. S. and Bousmina, M. *Progress in Materials Science*., **50**, 962–107 (2005).
100. Ray, S. S. and Okamoto, M. *Prog. Polym. Sci*., **28**, 1539–1641 (2003).
101. Ray, S. S., Yanada, K., Okamato, M., and Ueda, K. *J. Nanosci. Nanotechnol*., **3** 503–510 (2003).
102. M. C Roco, C. A. Mirkin, M. C Hersam (Eds.). World Technology Evaluation Center (WTEC) and the National Science Foundation (NSF), Springer (2010).
103. Samir, M. A. S. A., Alloin, F., Sanchez, J. Y., and Dufresne, A. *Polymer*, **45**, 4149–4157 (2004).
104. Savadekar, N. R. and Mhaske. S. T. *Carbohydrate Polymers*, **89**, 146–151 (2012).
105. Siemann, U. *Progress in Colloid and Polymer Science*., **130**, 1–14 (2005).
106. Singh, S. and Ray, S. S. *J. Nanosci. Nanotechnol*, **7**, 2596–2615 (2007).
107. Smart, G., Kandola, B. K., Horrocks, A. R., Nazare, S., and Marney, D. *Polym. Adv.Technol*., **19**, 658–670 (2008).
108. Smits, A. L. M., Ruhnau, F. C., and Vliegenthart, J. F. G. *Starch/Stärke*, **50**, 478–483 (1998).
109. Song, H. Y., Shin, Y. J, and Song, K. B. *Journal of Food Engineering*., **113**, 541–547 (2012).
110. Sriupayo, J., Supaphol, P., Blackwell, J., and Rujiravanit, R. *Polymer*, **46**, 5637–5642 (2005).
111. Tang, S., Zou, P., Xiong, H., and Tang, H. *Carbohydr. Polym*., **72**, 521–526 (2008).
112. Tunç, S. and Duman, O. *LWT: Food Science and Technology*., **44**, 465–472 (2011).

113. Siemann, U. *Progress in Colloid and Polymer Science*., **130**, 1–14 (2005).
114. Urbanczyk, L., Ngoundjo, F., Alexandre, M., Jérôme, C., Detrembleur, C., Calberg, C., *European Polymer Journal*, **45,** 643–648 (2009).
115. Vassalli, JT. https://mospace.umsystem.edu/xmlui/bitstream/handle/10355/5631/research.pdf?sequence=3.].
116. Vladimirov, V., Betchev, C., Vassiliou, A., Papageorgiou, G., and Bikiaris, D. *Compos. Sci. Technol*., **66**, 2935–2944 (2006).
117. Wakabayashi, K., Pierre, C., Dikin, D. A., Ruoff, R. S., Ramanathan, T., Brinson, L. C., and Torkelson, J. M. *Macromolecules*, **41**, 1905–1908 (2008).
118. Wang, B. and Sain, M. *Polym. Int*., **56**, 538–546 (2007).
119. Wilhelm, H. M., Sierakowski, M. R., Souza, G. P., and Wypych, F. *Carbohydrate Polymers*., **52**, 101–110 (2003).
120. Wu, C. L, Zhang, M. Q., Rong, M. Z., and Friedrich, K. *Compos. Sci. Technol*., **62**, 1327–1340 (2002).
121. Xing, Z. C., Han, S. J., Shin, Y. S., Kang, I. K. J. Nanom (2011). Article No. 929378, doi;10.1155/2011/929378.
122. Xu, J., Wang, J., Dong, X., Liu, G., and Yu, W. *Intern. J. of Chem*., **3**, 56–60 (2011).
123. Yang, D., Hu, Y., Song, L., Nie, S., He, S., and Cai, Y. *Polym. Degrad. Stab*, **93**, 2014–2018. (2008)
124. Yang, K., Wang, X., and Wang, Y. *J. Ind. Eng. Chem*., **13**, 485–500 (2007).
125. Yano, K., Usuki, A., and Okada, A. *J. Polym. Sci., Part A: Polym. Chem*., **35**, 2289–2294 (1997).
126. Yoo, Y., Kim, S. S., Won, J. C., Choi, K. Y., and Lee, J. H. *Polym. Bull*., **52**, 373–380 (2004).
127. Yudin, V. E., Otaigbe, J. U., Gladchenko, S., Olson, B.G., Nazarenko, S., Korytkova, E.N., and Gusarov, V. V. *Polymer*., **48**, 1306–1315 (2007).
128. Zeng, H., Gao, C., Wang, Y., Watts, P. C. P., Kong, H., Cui, X., and Yan, D. *Polymer*, **47**, 113–122 (2006).
129. Zhang, F., Zhang, H., and Su, Z. *Polym. Bull*., **60**, 247–251 (2008).
130. Zhang, S., Hull, T. R., Horrocks, A. R., Smart, G., Kandola, B. K., Ebdon, J., Joseph, P., and Hunt, B. *Polym. Degrad. Stab*., **92**, 727–732 (2007).
131. Zhang, Z., Britt, I. J., and Tung, M. A. *J. Appl. Polym. Sci*., **82**, 1866–1872 (2001).
132. Zhang, Z., Britt, I. J., and Tung, M. A. *J. Plast. Film Sheeting*, **14**, 287–298 (1998).
133. Zhou, H., Touny, A. H., and Bhaduri, S. B. *J. Mat. Sc.: Mat. Med*., **22**, 1183–1193 (2011).
134. Zhou, Q., Pramoda, K. P., Lee, J. M. Wang, K., and Loo, L.S. J. *Colloid Interface Sci*., **355**, 222–230 (2011).
135. Zhou, X., Shin, E., Wang, K. W., and Bakis, C. E. *Compos. Sci. Technol*., **64**, 2425–2437 (2004).

CHAPTER 12

UNDERSTANDING OF MECHANICAL AND BARRIER PROPERTIES OF STARCH, POLYVINYL ALCOHOL AND LAYERED SILICATE NANOCOMPOSITE FILMS UTILIZING MATHEMATICAL MODELS

X. Z. TANG and S. ALAVI

ABSTRACT

Nanocomposites consisting of starch and polyvinyl alcohol (PVOH) filled with layered silicate clays were investigated. Four types of clay (untreated sodium montmorillonite Na-MMT, Modified MMT Cloisite 10A and Cloisite 30B, and synthetic Laponite-RD LRD) were used. Films were prepared by solution casting method. The structure and morphology of the nanocomposites were studied using X-ray diffraction and transmission electron microscopy, which indicated that the compatibility of the multiphase system determined the formation of exfoliated or aggregated structures. The addition of Na-MMT and LRD led to significant enhancement in film mechanical and barrier properties. The content (0–20%) of both clays were correlated to property changes, which showed significant increase in tensile strength (up to 104% and 56% increase, respectively) and tensile modulus (up to 323% and 132% increase, respectively) and decrease in water vapor permeability (up to 115% and 40% decrease, respectively). Mathematical models for mechanical and barrier properties were utilized to understand the impact of the clay content on nanocomposite structure and properties. Comparison of experimental data with model predictions indicated that the models could serve as indirect tools for quantification of the degree of exfoliation/intercalation /aggregation in the production of nanocomposite products.

12.1 INTRODUCTION

The environmental deterioration by solid waste from petroleum based packaging materials (Jayasekara and others, 2004). Consumer awareness about environmental damage by non-degradable packaging sources is playing an importance role in the increased use of biodegradable plastics (Elizondo and others 2009). Starch has unique advantage of being cheap, abundant, renewable and biodegradable material for replacing petroleum based packaging materials (Averous 2004). Starch can be used to form edible or biodegradable films. However, starch films are very brittle in nature with poor water barrier properties (**Dean and others 2007; Tang and others 2008 a, b),** which limit their use as packaging films. Researchers are trying different methods to improve barrier and mechanical properties of starch based biodegradable films.

One successful option is to produce starch blends with other polymers such as polycaprolactones (Koenig and Huang 1995; Averous and others 2000), polylactic acid (Chang 2004; Jang and others 2007), and polyvinyl alcohol (PVOH) (Mao and others 2002; Jayasekara and others 2004; Yang and Huang 2008; **Zou and others 2008)**. Starch and PVOH blends are of particular interest because they are highly compatible and form hydrogen bonds easily (Zhou and others 2009; Sin and others 2010). Films from starch and PVOH blends show improved mechanical properties over starch alone and are biodegradable.

Nanocomposits of starches with layered silicate clays is another approach to improve the physical properties of starch based films (Tang and others 2008a, b). Smectite clays are the type mostly used in polymer clay nanocomposites due to their swelling properties and capacity to host water and organic molecules between their platelet layers. These clays have high cation exchange capacity, large surface area and high aspect ratio (**Chen and others 2008).**

Nanocomposites based on starch/PVOH/silicate clay are not widely studied (Dean and others 2008; Vasile and others 2008; Majdzadeh-Ardakani and Nazari 2010). Dean and others (2008) produced a series of thermoplastic starch/PVOH/montmorillonite (MMT) micro and nanocomposites which exhibit intercalated and exfoliated structures through extrusion processing. A small amount of PVOH (up to 7 wt%) and MMT (up to 5 wt%) were used and the improvement of tensile strength (up to 67% increase) and tensile modulus (up to 85% increase) were reported. Vasile and others (2008) prepared starch/PVOH/MMT composites by the melt mixing method. They revealed the morphology and thermal behavior changes for the nanocomposites and reported a slight increase in thermal stability for starch.

In this work, the starch/PVOH/silicate clay blend films were prepared by a solution casting method. Four different types of layered silicate clays were used to investigate the compatibility of multiphased system. Special attention is driven to study the effect of the clay content on the film properties based on barrier and mechanical modeling predictions of nanocomposite films.

12.2 MATERIALS AND METHODS

12.2.1 MATERIALS

Four types of silicate clays were obtained from Southern Clay Products (Austin, TX). Laponite-RD (LRD) is a synthetic hectorite-type clay with each platelet having a diameter of 25–30nm and a thickness of 1nm (aspect ratio 25–30). Cloisite Na^+ (Na-MMT) is naturally occurring sodium montmorillonite with aspect ratio of 75–100 (diameter of 75–100nm and thickness of 1nm). Cloisite 10A (C10A) and Cloisite 30B

(C30B) are MMTs modified by the addition of quaternary ammonium salts. Cloisite 10A contains dimethyl, benzyl, hydrogenated tallow, quaternary ammonium, whereas Cloisite 30B contains methyl, tallow, bis-2-hydroxyethyl, quaternary ammonium. The level of hydrophobicity is: C10A>C30B>Na-MMT. Regular corn starch was obtained from Cargill Inc. (Cedar Rapids, IA). Polyvinyl alcohol (Elvanol 71–30) was purchased form DuPont (Wilmington, DE). Elvanol 71–30 is a medium viscosity, fully hydrolyzed grade of polyvinyl alcohol. Glycerol (Sigma, St. Louis, MO) was used as a plasticizer for all studies.

12.2.2 PREPARATION OF NANOCOMPOSITES AND FILM CASTING

Aqueous starch/PVOH/clay/glycerol solutions were prepared by dissolving 6.7g starch, 3.3g PVOH, clay (0–20% based on base polymers starch and PVOH) and 3g glycerol in 300ml water at 95°C for 30 minutes. The heated solution was cooled to 55°C and equal amounts (35 g) were poured in Petri dishes. The water was allowed to evaporate while drying for 36–48 hours at room temperature and then the resulting films were peeled off and stored at room temperature in air tight bags for further tests.

X-RAY DIFFRACTION ANALYSIS

X-ray diffraction (XRD) studies of the films were carried out using a Bruker D8 Advance X-ray diffractometer (40kV, 40mA) (Karlsruhe, Germany). Samples were scanned in the range of diffraction angle $2\theta = 1–10°$ at a step of 0.01° and a scan speed of 4 sec/step. The x-ray radiation was generated from Cu-Kα source with a wavelength (λ) of 0.154 nm. D-spacing was estimated from the XRD scans by using Bragg's Law

$$D = \frac{\lambda}{2Sin\theta} \tag{1}$$

where λ = wavelength of X-ray beam, θ = the angle of incidence

TRANSMISSION ELECTRON MICROSCOPY

Transmission electron microscopy (TEM) was performed using a Philips CM100 Electron microscope (Mahwah, NJ), operating at 100kV. Solution prepared for film casting was put on a carbon-coated copper grid and dried for scanning.

12.2.3 WATER VAPOR PERMEABILITY

Water vapor permeability (WVP) was determined gravimetrically according to the standard method E96-00 (ASTM 2000). The films were fixed on top of test cells containing a desiccant (silica gel). Test cells then were placed in a relative humidity chamber with controlled temperature and relative humidity (25°C and 75% RH). After steady-state conditions were reached, the weight of test cells was measured every 12 hours over three days. The water vapor transmission rate (WVTR) was determined using equation (2).

$$WVTR = \frac{\left(\frac{G}{t}\right)}{A} \text{ g/h•m}^2 \quad (2)$$

where G = weight change (g), t = time (h) and A = test area (m^2)

WVP was then calculated using equation (3):

$$WVP = \frac{WVTR \times d}{\Delta p} \text{ g•mm/kPa•h•m}^2 \quad (3)$$

where d = film thickness (mm) and Δp = partial pressure difference across the films (kPa).

12.2.4 TENSILE PROPERTIES

Tensile properties of the films were measured using a texture analyzer (TA-XT2, Stable Micro Systems Ltd., UK), based on standard method ASTM D882-02 (ASTM 2002). Films were cut into 2 cm wide and 8 cm long strips and conditioned at 23°C and 50% RH for three days before testing. The crosshead speed was 1 mm/min.

BARRIER MODELS

Nielsen (1967) developed a simple model to describe the permeability of filled-polymers on the basis of tortusity argument. In an ideal case where clay particles are fully exfoliated and uniformly dispersed along a preferred orientation ($\theta = 0°$) in a polymer matrix, the tortuous factor τ (defined as the ratio of the detour distance d′ to the thickness d of the specimen) becomes:

$$\tau = 1 + \frac{l}{2w} V \quad (4)$$

where l is length of silicate platelets, w is the thickness of silicate platelets (aspect ratio a is l/w), V is volume fraction of the silicate platelets.

Thus, Relative permeability coefficient is given by equation (5):

$$\frac{P_c}{P_p} = \left(1 + \frac{l}{2w} V\right)^{-1} \quad (5)$$

where P_c and P_p are a permeability coefficient of composite and polymer matrix respectively.

However, mostly, the ideal case of $\theta = 0°$ is not achieved. To consider the influence of platelet orientations on permeability, Bharadwaj (2001) introduced an order parameter S' into the above equation. S' is defined as:

$$S' = (3\cos^2\theta - 1)/2, \tag{6}$$

where θ represents the angle between the direction of preferred orientation and the sheet normal unit vectors. This function can range from 1 (θ = 0°), indicating perfect orientation, and a value of 0 (θ = 90°), indicating random orientation. Hence, the relative permeability can be written as:

$$\frac{P_c}{P_p} = \left(1 + \frac{l}{2w}V\frac{2}{3}(S' + \frac{1}{2})\right)^{-1} \tag{7}$$

These simple models have been widely used to describe the barrier properties of polymer layered silicate nanocomposites (Yano and others 1997; Gusev and Lusti 2001; Lu and Mai 2007).

MICROMECHANICAL MODELS

The Halpin-Tsai equations (Halpin 1969; Halpin and Kardos 1976) have long been popular for predicting the properties of short fiber composites. Tucker and Liang (1999) reviewed the application of several composite models for fiber-reinforced composites. They reported that the Halpin-Tsai theory offered reasonable predictions for composite modulus.

The elastic modulus (E) of composite materials of Halpin-Tsai model is expressed in the following equation:

$$\frac{E}{E_m} = \frac{1 + \xi\eta V}{1 - \eta V} \tag{8}$$

where, E_m is matrix Young's modulus, ξ is a shape parameter dependent upon filler geometry, orientation and loading direction, and η is given by

$$\eta = \frac{E_f / E_m - 1}{E_f / E_m + 2\xi}$$

where E_f is filler's Young's modulus.

For perfectly oriented lamellar shape filler reinforcements, such as rectangular platelets, ξ is equal to 2a (a = *l*/*w*), where a, *l* and *w* are aspect ratio, length and thickness of the dispersed phase, respectively.

On the basis of equation (8), Wu and others (2004) further considered the morphology difference between the plate-like filler and the fiber-like filler phase and modified equation (8) into the following equation:

$$\frac{E}{E_m} = \frac{1 + 0.66\xi\eta V}{1 - \eta V} \tag{9}$$

12.2.5 EXPERIMENTAL DESIGN AND STATISTICAL ANALYSIS

Four different silicate clays with varied contents (0–20% based on base polymers) were evaluated. All experiments are replicated three times. All the data were analyzed using OriginLab (OriginLab Corporation, Northampton, MA) scientific graphing and statistical analysis software.

12.3 RESULTS AND DISCUSSION

12.3.1 DISPERSION OF CLAYS AND ORGANOCLAYS IN STARCH, PVOH MATRIX

Direct evidence of the intercalation or exfoliation of silicate plates is provided by the XRD patterns of the obtained hybrids (Tang and others 2008 a, b). Figure 1 through Figure 4 show XRD patterns of the hybrids with different clay and clay contents. It can be seen that the dispersion states of clays in the polymer matrix depends on the type of clay and clay content used.

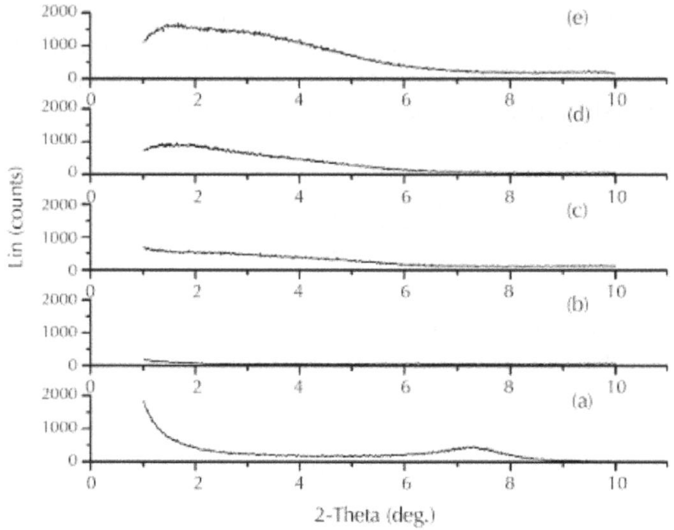

FIGURE 1 XRD patterns of (a) natural montmorillonite (Na-MMT), (b) starch/PVOH polymer matrix with no clay addition (0% Na-MMT), and (c) to (e) starch/PVOH/Na+-MMT hybrids with 5, 10, 20% Na-MMT, respectively.

Figure 1 shows the XRD patterns of Na-MMT (a), starch/PVOH polymer matrix with no clay addition (b) and starch/PVOH/Na-MMT hybrids (c) to (e). The Na-MMT exhibits a single peak at $2\theta = 7.2°$, which corresponds to d-spacing 1.23nm. The Na-MMT has a very ordered clay platelet structure. In the case of starch/PVOH/Na-MMT hybrids, this original peak of clay at $2\theta = 7.2°$ diminished or disappeared, indicating the morphology and d-spacing of silicate layers varied with the incorporation of Na-MMT. In the lowest clay content at 5 %, the XRD curve shows no peaks, which implies that the periodicity of silicate layer is not observed and the silicate plates have the disordered and exfoliated structure in polymer matrix. When clay content increased to 10%, the XRD curve shows a single plane peak at 2θ = 1.6°, corresponding to d-spacing 5.51nm. This disappearance of the original peak, appearance of new broad peak and increase of d-spacing suggests that the silicate plates have the intercalated and disordered structure in the polymer matrix. When the clay content is 20%, there are two plane peaks observed. The lower angle peak is at 2θ = 1.8°, which corresponds to d-spacing 4.90nm, while the higher angle peak is at 2θ =3.2°, which corresponds to d-spacing 2.76nm. Compared to 10% Na-MMT, High clay content leads to lower d-spacing, meaning the clay plates are less separated from each other. There are more ordered intercalated structures in the polymer matrix. It should also be noted from Figure 1 (b), that as a compare, the starch/ PVOH polymer matrix without clay addition shows a featureless flat curve.

In Figure 2, the starch/PVOH/C30B hybrids show the weak peaks at around 2θ = 4.9°, While C30B exhibits an intensive peak at 2θ = 4.8°. There are almost no changes of d-spacing, indicating hardly to say the formation of intercalated hybrids. Also in the case of starch/PVOH/C10A hybrids, the original peaks of the C10A (at 2θ = 4.9°) remain in the hybrids, indicating little or no intercalation has occurred (Figure 3).

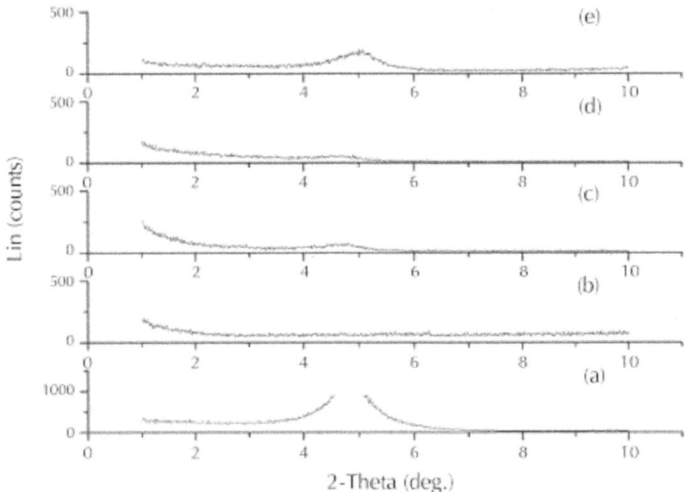

FIGURE 2 XRD patterns of (a) C30B, (b) starch/PVOH polymer matrix with no clay addition (0% C30B), and (c) to (e) starch/PVOH/Na+-MMT hybrids with 5, 10, 20% C10A, respectively.

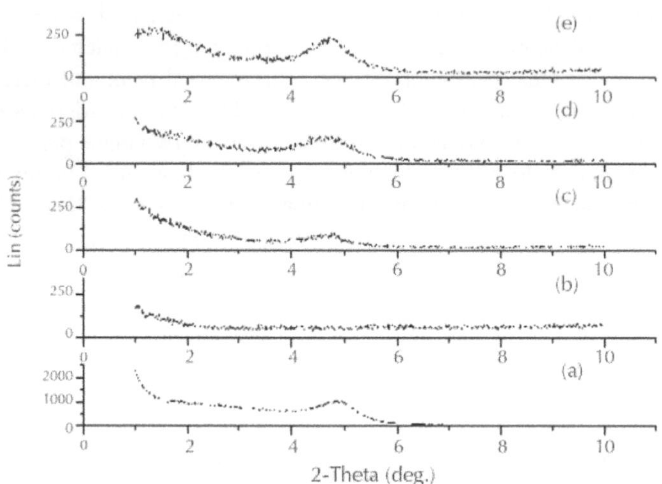

FIGURE 3 XRD patterns of (a) C10A, (b) starch/PVOH polymer matrix with no clay addition (0% C10A), and (c) to (e) starch/PVOH/Na+-MMT hybrids with 5, 10, 20% C10A, respectively.

Figure 4 shows the XRD patterns of LRD (a), starch/PVOH polymer matrix with no clay addition (b), and starch/PVOH/LRD hybrids (c) to (e). The LRD exhibits a very broad peak at around $2\theta = 6.4°$, meaning the arrangement of silicate plates in LRD itself is not in the ordered state. Similar to starch/PVOH/Na-MMT hybrids, starch/PVOH/LRD hybrids also show exfoliated and intercalated structure, and the degree of exfoliation and intercalation decreased as the LRD content increased. For clay content at 5 and 10%, the XRD curves show very weak peaks at 2θ around 1.2°, and 1.4°, respectively, indicating the silicate plates have higher degree of disordered and exfoliated structure in the polymer matrix. At 20% LRD, a very intensive peak at $2\theta = 2.3°$ was found, suggesting that highly ordered and aggregated layered structure can be induced at this content, rather than the random distribution of clay platelets in the original LRD.

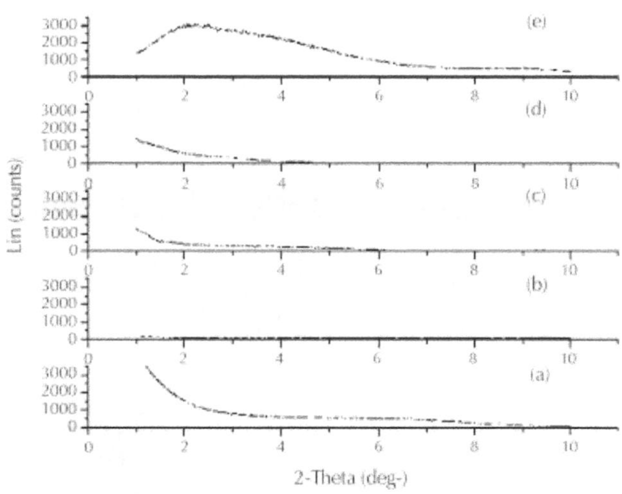

FIGURE 4 XRD patterns of (a) LRD; (b) starch/PVOH polymer matrix with no clay addition (0% LRD), and (c) to (e) starch/PVOH/LRD hybrids with 5, 10, 20% LRD, respectively.

A typical transmission electron micrograph of starch/PVOH/Na-MMT and starch/PVOH/LRD at 10 and 20% is shown in Figure 5. The TEM image of starch/PVOH/Na-MMT hybrid at 10% clearly exhibits multilayer nanostructure with alternating polymeric and inorganic silicate layers (Figure 5-a), whereas TEM image of starch/PVOH/Na-MMT hybrid at 20% exhibits more tightly stacked layer structure (Figure 5-b). Figure 5-c shows the TEM image of starch/PVOH/LRD hybrid at 10%. One can observe that individual silicate plates are randomly dispersed in the polymer matrix, supporting that 10% hybrids have higher degree of disordered and exfoliated structure in polymer matrix, although there are still slight amount of unexfoliated silicate plates remaining in the structure. Figure 5-d shows the TEM image of starch/PVOH/LRD hybrid at 20%. It is very clear that more aggregated layered structures are seen over the whole sample. All the TEM results correspond very well with the XRD patterns.

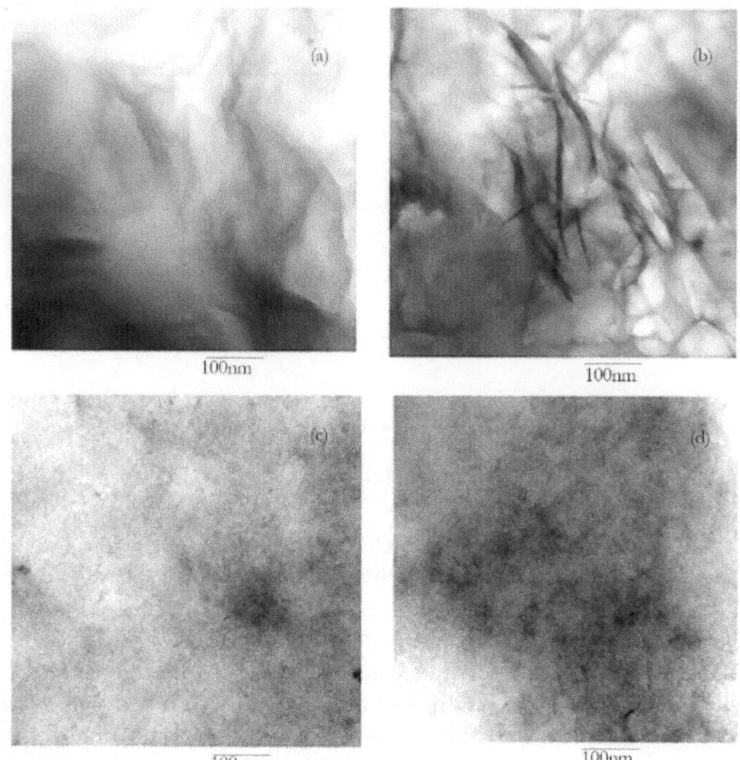

FIGURE 5 TEM images of (a) starch/PVOH/10% Na-MMT, (b) starch/PVOH/20% Na-MMT, (c) starch/PVOH/10% LRD, and (d) starch/PVOH/20% LRD nanocomposites.

The above results clearly show that compatibility and optimum interactions between polymer matrix, organic modifiers (if any) and the silicate layer surface are crucial to the formation of intercalated or exfoliated nanocomposite structure. Starch and PVOH are more compatible with Na-MMT and LRD than C30B and C10A because of their hydrophilicity. On the other hand, with increase of clay content, the clay layers are harder to be dispersed due to the weak intermolecular forces. In such case, more mechanical energy input instead of molecular interactions may be needed to peel off the silicate plates to produce exfoliated nanocomposite structure.

12.3.2 BARRIER PROPERTIES OF STARCH/PVOH/LAYERED SILICATES NANOCOMPOSITE FILMS

WVP results were presented in Figure 6. The overall trend of WVP for starch/PVOH/Na-MMT films is that the incorporation of Na-MMT decreased WVP of hybrid films from 1.706 g•mm/kPa•h•m^2 (0% Na-MMT) to 0.793 g•mm/kPa•h•m^2 (20% Na-MMT). When the intercalated/exfoliated structure is formed, the impermeable clay layers mandate a tortuous pathway for water molecules to traverse the film matrix, thereby increasing the effective path length for diffusion (Yano and others 1997; Tang and others 2008 a). The decreased diffusivity due to formation of nanocomposite nanostructure, in the case of starch/PVOH/Na-MMT hybrids, reduced the WVP continuously. For starch/PVOH/LRD hybrid films, the WVP decreased from 1.706 g•mm/kPa•h•m^2 (0% LRD) to 1.219 g•mm/kPa•h•m^2 (15% LRD), then increased to 1.376 g•mm/kPa•h•m^2 (20% LRD). Based on XRD and TEM analysis, hybrids with 20% LRD have an aggregated structure, which possibly provides more channels in the membrane that allow for more rapid permeation. As for starch/PVOH/C30B and starch/PVOH/C10A hybrid films, the WVP first decreased slightly with 5 % of clay (and also 10 % for C30B) due to the existence of some individual silicate plates in the polymer matrix. Then WVP increased with the increase of clay content and the final WVP with 20% C10A was even higher than that of starch/PVOH films without clay addition. This may be attributed to the fact that an increase in the spacing between the polymer chains (because of bad dispersion of C30B and C10A as already shown in the XRD analysis) due to some aggregation of clay particles, may promote water vapor diffusivity through the film and hence accelerate the WVP. It should be noted that starch/PVOH/Na-MMT hybrid films show the lowest WVP, meaning the best barrier property among all the hybrids. This enhanced barrier characteristics result from higher aspect ratio (compared to LRD) and good compatibility of Na-MMT with starch/PVOH polymer matrix (compared to C30B and C10A). It is also notable that starch/PVOH/C30B films have relatively better WVP than starch/PVOH/C10A films because C30B is less hydrophobic than C10A, which leads to relatively better dispersion of C30B in the polymer matrix. Figure 7 shows comparison of theoretical predictions of equation (7) with experimental values for WVP in starch/PVOH/Na-MMT and starch/PVOH/LRD nanocomposites. According to this model, the obvious expectations can be quantified: higher aspect ratio filler provide substantial lower permeabilites for given filler volume fraction (for example Na-MMT has higher aspect ratio than LRD), and aligned fillers are much more effective barriers for a given aspect ratio and filler loading. In Figure 7, it is observed that WVP of starch/PVOH/Na-MMT follows well with case for a=100 and random orientation. It should be noted that WVP of lower clay loadings was slightly lower than theoretical predictions, while WVP of higher clay loadings was slightly higher than theoretical predictions. This represents that the extent of exfoliation or intercalation decreased with the increase of clay content, which corresponds well with the XRD and TEM results. As for starch/PVOH/LRD, the experimental data didn't follow very well with case for a = 30 and random orientation, meaning LRD having a different dispersion capabilities from Na-MMT in the starch/PVOH polymer matrix. With lower loadings of LRD, the data point is close to perfect orientation and with higher loadings of LRD, clay layers are much harder to be dispersed and aligned in the polymer matrix. This is possibly because at lower loadings, LRD is easily dispersed due to its lower aspect ratio and random orientation, whereas at higher loadings, LRD can form the relatively higher strength gel during solution mixing and the clay platelets need more mechanical energy to be peeled off.

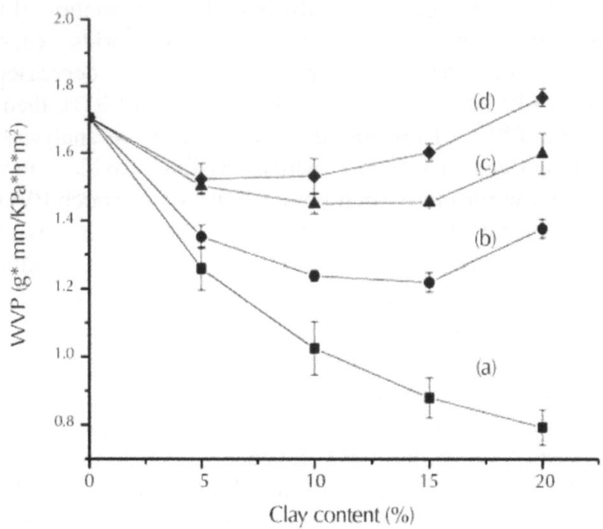

FIGURE 6 Effects of clay type and clay content on water vapor permeability (WVP) of composite film: (a) starch/PVOH/Na-MMT, (b) starch/PVOH/LRD, (c) starch/PVOH/C30B, and (d) starch/PVOH/C10A. Error bars indicate the standard deviation.

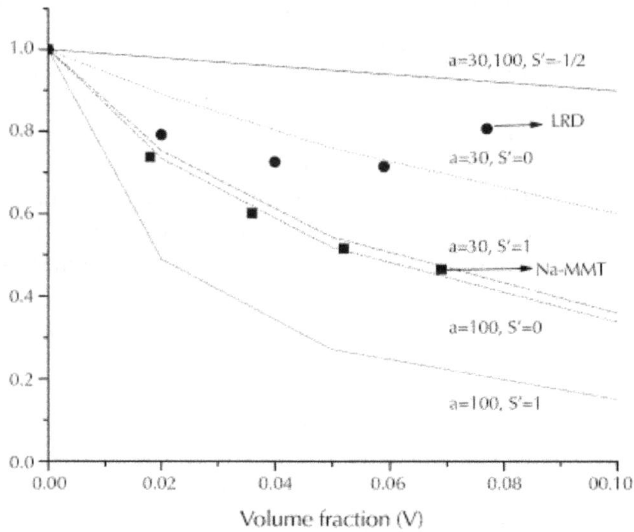

FIGURE 7 Comparison of theoretical predictions with experimental values for WVP in starch/PVOH/Na-MMT and starch/PVOH/LRD nanocomposite films.

Comparison of experimental observations with barrier model predictions clearly indicated that increase of clay content has an impact on effective clay aspect ratio and alignment, which in turn affect the nanocomposite barrier properties. At the same time, the barrier models could serve as indirect tools for quantification of the degree of exfoliation/intercalation/aggregation in the production nanocomposite products.

12.3.3 MECHANICAL PROPERTIES OF STARCH/PVOH/LAYERED SILICATES NANOCOMPOSITE FILMS

Table 1 displays the mechanical properties of starch/PVOH/layered silicate nanocomposite films. For starch/PVOH/Na-MMT, an increase in tensile strength and Young's modulus is observed and correlated to the clay content. With the increase of clay content from 0 to 20%, tensile strength increase from 8.53 to 17.42 MPa and modulus increased from 160.35 to 677.71 MPa. This increase is commonly observed in nanocomposite materials and is related to higher strength and modulus of inorganic clay layers (Luo and Daniel, 2003; Chivrac and others 2010). The coupling between the tremendous surface area of the clay and the polymer matrix facilitates stress transfer to the reinforcement phase, allowing for such tensile and toughening improvements. Conversely, Elongation at break did not exhibit much improvement. It decreased with the increasing of Na-MMT content. This was coincident with the report by Lee and others (2005), which suggested that good dispersion of clay platelets in the polymer reduced tensile ductility and impact strength compared to neat polymer. For starch/PVOH/LRD, similar trend is found for Young's modulus but different trends are found for tensile strength and Elongation at break. With the increase of clay content from 0 to 20%, modulus increased from 160.35 (0%) to 372.39 (20%) MPa, but tensile strength increase from 8.53 (0%) to 13.32 MPa (10%), then decreased to 11.1 MPa (20%). Elongation at break increased from 155.52% (0%) to 229.21% (10%), then decreased to 144.26% (20%). This is an interesting result in that the incorporation of LRD at 10% significantly improves the elongation of hybrid films without sacrifice of tensile strength. To the best of the authors' knowledge, there is no report on the effect. The reason for this is possibly because the interfacial interactions of filler and matrix play an important role. LRD has a smaller size and is easier to interact with hydroxyl groups of starch, PVOH and glycerol. The interactions will help the polymer matrix well connected and more flexible. In such case, LRD not only serves as reinforcement agent but also as compatibilizer. When LRD is increased to 20%, LRD is aggregated rather than dispersed. The aggregation discontinues the polymer matrix, leading to the decrease of tensile strength and elongation at break. As for starch/PVOH/C10A and starch/PVOH/C30B hybrid films, there is no significant improvement of tensile strength and modulus. There is even decrease of modulus with increase of clay content. Similar to WVP results, starch/PVOH/Na-MMT hybrid films show the highest tensile strength and Young's modulus because of higher aspect ratio and good compatibility of Na-MMT within starch/PVOH polymer matrix.

TABLE 1 Summary of tensile mechanical properties for starch/PVOH composites with different types of clay and clay content.

Sample	Tensile strength (MPa)	Modulus (MPa)	Elongation at break (%)
Na-MMT			
0% Na-MMT	8.53 (1.48)[a]	160.35 (17.14)	155.52 (17.89)
5% Na-MMT	10.25 (0.82)	325.37 (18.45)	106.54 (24.6)
10% Na-MMT	12.7 (1.2)	464.09 (26.17)	68.39 (10.04)
15% Na-MMT	14.09 (0.94)	557.47 (23.69)	60.46 (9.68)
20% Na-MMT	17.42 (1.12)	677.71 (25.46)	37.43 (15.43)
LRD			
0% LRD	8.53 (1.48)	160.35 (17.14)	155.52 (17.89)
5% LRD	10.65 (0.22)	222.76 (15.61)	172.5 (26.89)
10% LRD	13.32 (0.61)	282.51 (12.14)	229.21 (21.92)
15% LRD	13.2 (1.56)	335.99 (23.66)	183.98 (6.64)
20% LRD	11.1 (0.76)	372.39 (19.3)	144.26 (13.98)
C10A			
0% C10A	8.53 (1.48)	160.35 (17.14)	155.52 (17.89)
5% C10A	8.26 (0.14)	148.71 (14.61)	132.51 (2.12)
10% C10A	8.81 (0.29)	154.37 (26.70)	117.16 (5.95)
15% C10A	8.39 (0.39)	153.14 (4.72)	131.95)19.15)
20% C10A	7.74 (0.25)	83.96 (11.75)	148.22 (14.37)
C30B			
0% C30B	8.53 (1.48)	160.35 (17.14)	155.52 (17.89)
5% C30B	9.44 (0.41)	193.62 (16.35)	107.35 (10.04)
10% C30B	8.98 (0.45)	199.56 (43.24)	90.06 (8.88)
15% C30B	8.47 (0.46)	146.81 (8.18)	87.28 (6.79)
20% C30B	8.84 (0.22)	158.83 (22.54)	97.34 (15.64)

[a] Standard deviation in data shown in brackets.

Figure 8 shows the comparison of theoretical predictions of equation (9) with experimental values for Young's modulus in starch/PVOH/Na-MMT and starch/PVOH/LRD nanocomposites. Similar to WVP models, some obvious expectations can be quantified: higher aspect ratio filler provide substantial improvement of Young's modulus for given filler volume fraction. In Figure 8, it is observed that modulus of starch/PVOH/Na-MMT does not follow well with model predictions with aspect ratio a = 100. The reason for this is because most mechanical model predictions are based on complete layer exfoliation and perfect orientation. However, a state of delamination intermediate to intercalation and exfoliation is observed more often. For example, a situation wherein stacks of intercalated aggregates are dispersed throughout the matrix may be observed. In that case, aggregation through intercalation serves to increase the effective width of the nanocomposite. From Figure 8, experimental data with lower clay loadings is close to predictions with aspect ratio 50 and experimental data with higher clay loadings is close to predictions with aspect ratio 30, indicating the effective aspect ratio of Na-MMT in starch/PVOH polymer matrix is between 30 and 50. Higher clay loadings lead to lower effective aspect ratio, which represents that the extent of exfoliation or intercalation decreased with the increase of clay content. As for starch/PVOH/LRD, the effective aspect ratio of LRD in starch/PVOH matrix is between 10 and 15.

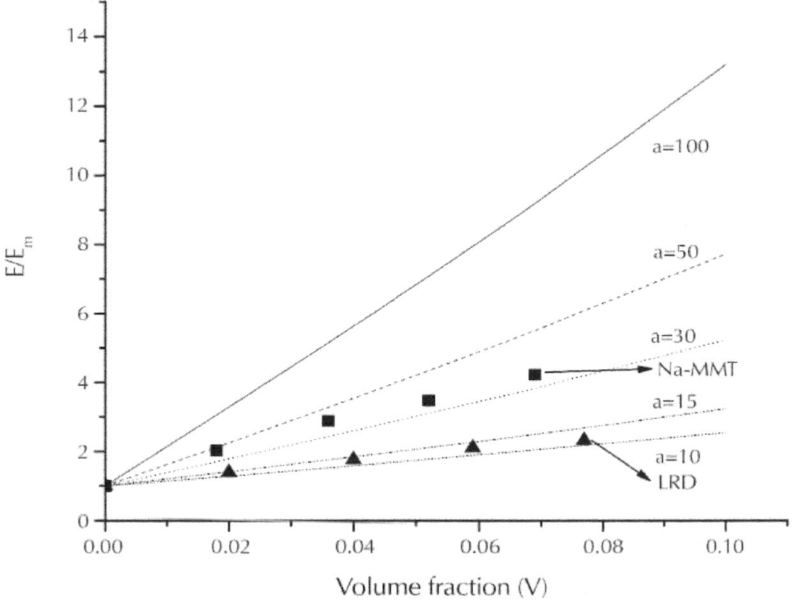

FIGURE 8 Comparison of theoretical predictions with experimental values for Young's modulus in starch/PVOH/Na-MMT and starch/PVOH/LRD nanocomposite films.

12.4 CONCLUSION

In this work, the effect of four types of clay (untreated montmorillonite Na-MMT, Modified MMT Cloisite 10A and Cloisite 30B, and synthetic Laponite-RD) on properties of starch/PVOH nanocomposite films was evaluated. The results show that compatibility and optimum interactions between polymer matrix, organic modifiers (if any) and the silicate layer surface are crucial to the formation of intercalated or exfoliated nanocomposite structure. Starch and PVOH are more compatible with Na-MMT and LRD than C30B and C10A. With the increase of clay content, the clay layers are harder to be dispersed using the solution method. In such case, more mechanical energy instead of intermolecular interactions may be needed to peel off the silicate plates to produce exfoliated nanocomposites. The addition of Na-MMT and LRD led to significant enhancement in film mechanical and barrier properties due to the formation of intercalated or exfoliated nanocomposite structure. Film properties are very closely related with the degree of exfoliation, which can also be observed from the comparison of experimental data with mathematic model predictions. Therefore, the models could serve as indirect tools for quantification of the degree of exfoliation/ intercalation /aggregation in the production of nanocomposite products.

ACKNOWLEDGEMENT

We would like to thank *United States Department of Agriculture* (USDA) for funding this project through USDA-NRI COMPETITIVE GRANT 2008 (Grant no. 20081503). This is Contribution Number 11-138-J from the Kansas Agricultural Experiment Station, Manhattan, Kansas 66506.

KEYWORDS

- **Bio-based nanocomposite films**
- **Starch**
- **Polyvinyl alcohol**
- **Layered silicates**
- **Mathematical modeling**
- **Barrier and mechanical properties**

REFERENCES

1. ASTM. Standard test methods for water vapor transmission of materials. E 96–100. ASTM, Philadelphia, p. **878** (2000).
2. ASTM. Standard test method for tensile properties of thin plastic. D 882–902, ASTM, Philadelphia, p. **161** (2002).
3. Averous, L, Moro, L, Dole, P, and Fringant C. Properties of thermoplastic blends: starch–polycaprolactone. *Polymer* **41** (11), 4157–4167 (2000).
4. Averous, L. 2004. Biodegradable multiphase systems based on plasticized starch: A review. *Journal of Macromolecular Science-Polymer Reviews* C**44** (3), 231–274 (2004).
5. Bharadwaj, R. K. Modeling the barrier properties of polymer-layered silicate nanocomposites. *Macromolecules*, **34**, 9189–3192 (2001).
6. Chang, J. L. Reactive blending of biodegradable polymers: PLA and starch. *Journal of polymers and the environment*, **8**(1), 33–37 (2004).
7. Chen, B., Evans, J. R. G., Greenwell, C, H., Boulet, P., Coveney, P. V., Bowden, A. A., and Whiting, A. A critical appraisal of polymer-clay nanocomposites. *Chem. Soc. Rev*, 37, 568–594 (2008).
8. Chivrac, F., Pollet, E., Dole, P., and Averous, L. Starch-based nano-biocomposites: plasticizer impact on the montmorillonite exfoliation process. *Carbohydrate Polymer*, **79**, 941–947 (2010).

9. Dean, K., Yu, L., and Wu, Y. D. Preparation and characterization of melt-extruded thermoplastic starch/clay nanocomposites. *Composite Science and Technology* 67, 413–421 (2007).
10. Dean, K. M., Do, M. D., Petinakis, E., and Yu, L. Key interaction in biodegradable thermoplastic starch/poly (vinyl alcohol)/montmorillonite micro and nanocomposites. *Composite Science and Technology*, **68**, 1453–1462 (2008).
11. Elizondo, N. J., Sobral, P. J. A., and Menegalli, F. C. Development of films based on blends of Amaranthus cruentus flour and poly (vinyl alcohol). *Carbohydrate Polymers*, **75**, 592–598 (2009).
12. Gusev, A. A. and Lusti, H. R. *Rational design of nanocomposites for barrier applications*. Adv Mater, **13**, 1641–1643 (2001).
13. Hapin, J. C. Stiffness and expansion estimates for oriented short fiber composites. *J compos Mater*, **3**, 732–734 (1969).
14. Hapin, J. C. and Kardos, J. L. The Halpin-Tsai equations: a review. *Polym Eng Sci*, **16**, 344–352 (1976).
15. Jang, W. Y., Shin, B. Y., Lee, T. J., and Narayan, R. Thermal properties and morphology of biodegradable PLA/starch compatibilized blends. *J. Ind. Eng. Chem*, **13**(3), 457–464 (2007).
16. Jayasekara, R., Harding, I., Bowater, I., Christie, G. B. Y., and Lonergan, G. T. Preparation, surface modification and characterization of solution cast starch PVA blended films. Polymer Testing, **23**, 17–27 (2004).
17. Koenig, M. F. and Huang, S. T. Biodegradable blends and composites of polycaprolactone and starch derivatives. *Polymer*, **36**, 1877–1882 (1995).
18. Lee, J. H., Jung, D., Hong, C. E., Rhee, K. Y., and Advani, S. G. Properties of polyethylene-layered silicate nanocomposites prepared by melt intercalation with a PP-g-MA compatibilizer. *Composites Science and Technology*, **65**, 1996–2002 (2005).
19. Lu, C. S. and Mai, L. W. Permeability modeling of polymer-layered silicate nanocomposites. *Composites Science and Technology*, **67**, 2895–2902 (2007)
20. Luo, J. J. and Daniel, I. M. Characterization and modeling of mechanical behavior of polymer/clay nanocomposites. *Composites Science and Technology*, **63**, 1607–1616 (2003).
21. Mao, L., Imam, S., Gordon, S., Cinelli, P., and Chiellini, E. Extruded cornstarch-glycerol-polyvinyl alcohol blends: Mechanical properties, morphology and biodegradability. *Journal of Polymers and the Environment*, **8**, 205–211 (2002).
22. Majdzadeh_Ardakani, K. and Nazari, B. Improving the mechanical properties of thermoplastic starch/poly (vinyl alcohol)/clay nanocomposites. *Composites Science and Technology*, **70**, 1557–1563 (2010).
23. Nielsen, L. E. Models for the permeability of filled polymer systems. *J Macromol Sci Chem*, A1(5), 929–942 (1967).
24. Sin, L. T., Rahman, W. A. W. A., Rahmat, A. R, and Khan, M. I. Detection of synergistic interactions of polyvinyl alcohol-cassava starch blends through DSC. *Carbohydrate Polymers* **79**, 224–226 (2010).
25. Tang, X. Z., Alavi, S., and Herald, T. J. *Barrier and Mechanical Properties of Starch-Clay Nanocomposite Films*. Cereal Chemistry, **85**, 433–439 (2008 a).
26. Tang, X. Z., Alavi, S., and Herald, T. J. Effect of plasticizers on the structure and properties of starch-clay nanocomposite films. Carbohydrate Polymers, **74**, 552–558 (2008 b).
27. Tucker, C. L. and Liang, E. Stiffness predictions for unidirectional short-fiber composites: Review and evaluation. *Composites Science and Technology*, **59**, 655–671 (1999).
28. Vasile, C., Stoleriu, A., Popescu, M., Duncianu, C., Kelnar, I., and Dimonie, D. Morphology and thermal properties of some green starch/poly (vinyl alcohol)/montmorillonite nanocomposites. *Cellulose Chemistry and Technology*, **42**, 549–568.
29. Wu, Y. P., Jia, Q. X., Yu, D. S., and Zhang, L. Q. Modeling Young's modulus of rubber-clay nanocomposites using composite theories. *Polymer Testing*, **23**, 903–909 (2004).

30. Yang, S. Y. and Huang C. Y. Plasma treatment for enhancing mechanical and thermal properties of biodegradable PVA/starch blends. *Journal of Applied Polymer Science*, **109**, 2452–2459 (2008).
31. Yano, K., Usuki, A., and Okada, A. Synthesis and properties of polyimide-clay hybrid films. *J. Polym Sci, PartA: Polym Chem*, **35**, 2289–2294 (1997).
32. Zhou, Y. X., Cui, F. Y., Jia, D. M., and Xie, D. Effect of complex plasticizer on the structure and properties of the thermoplastic PVA/Starch blends. *Polymer-Plastics Technology and Engineering*, **48**, 489–495 (2009).
33. Zou, G. X., Qu, J. P., and Zou, X. L. Extruded starch/PVA composites: water resistance, thermal properties and morphology. *Journal of Elastomers and Plastics*, **40, 303–316 (2008).**

CHAPTER 13

ANALYTICAL TECHNIQUES FOR STRUCTURAL CHARACTERIZATION OF BIOPOLYMER-BASED NANOCOMPOSITES

P. KUMAR, K. P. SANDEEP, S. ALAVI, and V. D. TRUONG

ABSTRACT

In recent years, biopolymer-based packaging materials have experienced a renewed industry and research focus. Such biopolymers include naturally occurring materials such as proteins, cellulose, and starches and synthetic materials such as polylactic acid that are manufactured from naturally derived monomers. Biopolymer-based packaging has the potential to replace plastics, as the latter are not only dependent on non-renewable resources such as petroleum for production but are also toxic to the environment. However, properties and performance of biopolymer-based packaging materials need much improvement. A new class of materials represented by biopolymer based nanocomposites or bio-nanocomposites have proven to be a promising option in this direction. Selection of proper techniques for characterization of these bio-nanocomposites is very critical in assessing their performance. This chapter reviews analytical techniques for the structural characterization of biopolymer-based nanocomposites, including X-ray diffraction, scanning electron micrsocopy, tunneling electron microscopy, atomic force microscopy, and spectroscopic techniques such as Fourier-transform infra-red and nuclear magnetic resonance. The relative strengths and drawback of these techniques are provided, and their application in understanding the interaction between nanoparticles and biopolymer matrix and synthesis of improved bio-nanocomposites is discussed.

13.1 INTRODUCTION

The non-biodegradable and non-renewable nature of plastic packaging has led to a renewed interest in packaging materials based on biopolymers derived from renewable sources. Such biopolymers include naturally occurring proteins, cellulose, starches, and other polysaccharides and those synthesized chemically from naturally derived monomers such as lactic acid. Commercialization of these bio-based polymers has already begun. Natureworks, LLC (Minnetonka, MN) manufactures polylactide from corn sugar. The polymer can be hydrolyzed back to lactic acid. Wal-Mart stores Inc. is already using polylactide to package fresh-cut produce (Marsh and Bugusu, 2007). However, biopolymers cannot meet the requirements of a cost-effective film with mechanical and barrier properties matching those of plastics. Recently, a new class of materials represented by bio-nanocomposites has proven to be a promising option in improving the properties of these biopolymer-based packaging materials. Bio-nanocomposites consist of a biopolymer matrix reinforced with particles (nanoparticles) having at least one dimension in the nanometer range (1–100 nm). Bio-nanocomposites exhibit much improved properties as compared to biopolymers due to the high aspect ratio and high surface area of nanoparticles. Therefore, efforts have been geared towards developing bio-nanocomposites for food packaging films with improved mechanical, barrier, rheological, and thermal properties.

The most common class of materials used as nanoparticles are layered silicates such as clay minerals, graphite, and metal phosphates. Clay minerals such as montmorillonite (MMT), hectorite, sapnotite, and laponite have proven to be very effective due to their unique structure and properties. These clay minerals belong to the general family of 2:1 layered silicates indicating that they have 2 tetrahedral sheets sandwiching a central octahedral sheet (Zeng et.al., 2005). Paiva and et al. (2008) describe the properties, preparation, and applications of layered clays in detail. Biopolymers can be reinforced with these layered clays in order to enhance their properties while maintaining their biodegradability.

For preparation of a bio-nanocomposite, polymer chains must diffuse into the galleries between silicate layers to produce structures ranging from intercalated to exfoliated (Figure 1). There are four possible arrangements of layered clays dispersed in a polymer matrix – phase separated or immiscible (microcomposite), intercalated, exfoliated, and disordered intercalated (partially exfoliated). In an immiscible arrangement, platelets of layered clays exist as tactoids (stack of platelets separated by about 1 nm) and the polymer encapsulates these tactoids. Intercalation occurs when a monolayer of extended polymer chain penetrates into the galleries of the layered silicates. Intercalation results in finite expansion (2–3 nm) of the silicate layers. However, these silicate layers remain parallel to each other. Extensive penetration of polymer chains into the galleries of layered silicate leads to exfoliation or delamination of silicate layers. Clay platelets are separated by 10 nm or more during exfoliation. An exfoliated nanocomposite consists of nanoparticles distributed homogeneously throughout the polymer matrix (Dennis et.al., 2001; Zeng et al., 2005).

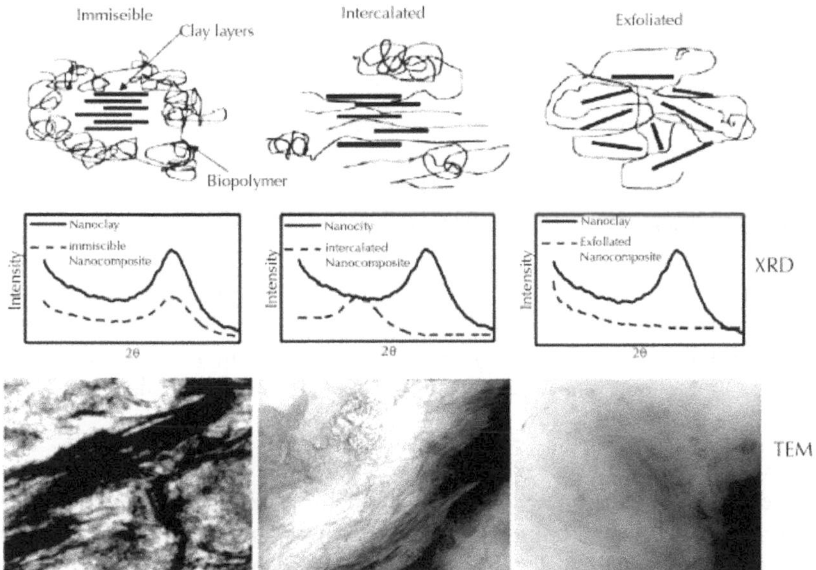

FIGURE 1 Possible arrangements of layered silicates in biopolymers with corresponding XRD and TEM results.

Bio-nanocomposites can be prepared by several methods which include *in situ* polymerization, solution exfoliation, and melt intercalation. In the *in situ* polymerization method, monomers are migrated into the galleries of layered silicates and subsequently polymerized via heat, radiation, or catalyst. In solution exfoliation, layered clays are exfoliated into single platelets. Exfoliation is achieved by dispersing the layered clays in a solvent. The polymer is adsorbed onto the platelets by mixing in the clay suspension. The solvent is removed either by evaporation or by precipitation. In melt intercalation, layered clays are mixed with the polymer matrix in molten state (Zeng et al., 2005).

Several review papers discuss the preparation, characterization, properties, and applications of bio-nanocomposites (Pandey et al., 2005; Ray and Bousmina, 2005; Yang et al, 2007; Rhim and Ng, 2007; Sorrentino et al., 2007; Zhao et al., 2008; Bordes et al., 2009). However, there is a lack of comprehensive review on various analytical techniques for the structural characterization of bio-nanocomposites. Selection of proper technique for characterization of these bio-nanocomposites is very critical in assessing their performance. A number of analytical techniques have been used to characterize the structure of bio-nanocomposites. These techniques include X-ray diffraction (XRD), microscopy transmission electron microscope (TEM), scanning electron microscope (SEM), scanning probe microscope (SPM), and confocal scanning laser microscope (CSLM), Fourier transform infra-red (FTIR) spectroscopy, and nuclear magnetic resonance (NMR). Each of the above mentioned techniques has its own benefits and limitations.

This chapter presents a review of analytical techniques for the structural characterization of bio-nanocomposites. A brief introduction of different techniques for the preparation of bio-nanocomposites is given. Analytical techniques for characterizing the structure of these bio-nanocomposites are discussed in terms of principle of operation, preparation of sample, interpretation of data, and limitations.

13.2 ANGLE X-RAY DIFFRACTION (WA-XRD)

X-rays were discovered in 1895 by W.C. Roentgen. When incident on a crystalline material, X-rays interfere with each other. This phenomenon is known as XRD. The WA-XRD is the most commonly used method to characterize the structure of bio-nanocomposites because of its ease of use and availability. The WA-XRD has been used to characterize dispersion of layered clays in nanocomposites based on protein (Chen and Zhang 2006; Shabeer et al., 2007; Yu et al., 2007) and starch (Dimonie et al., 2008; Tang et al., 2008).

PRINCIPLE OF OPERATION

X-rays are part of the electromagnetic spectrum with wavelength ranging from 0.01 to 10 nm. A schematic of an XRD instrument is shown in Figure 2. X-rays are generated by striking a pure anode of a metal (such as copper) with high-energy electrons in a sealed vacuum tube (X-ray tube). When a parallel and monochromatic X-ray beam with a wavelength of λ is incident on a sample (single crystal or powder ground to size less than 50 μm), the beam is transmitted, absorbed, refracted, scattered, and diffracted. A detector is used to measures the intensity of the diffracted X-rays in counts per second (cps). The sample may be set at any desired angle (θ) to the incident beam. For WA-XRD, the value of 2θ is greater than 1°. A diffraction plot detects the intensity of diffracted X-rays as a function of 2θ. Sharp peaks, corresponding to the interlayer spacing, are seen on the diffraction data. These peaks are characteristic to the material and the structure of the crystal. From the diffraction data, interlayer spacing (d-spacing or d) between clay layers is estimated from Bragg's equation (Kasai and Kakudo, 2005):

$$d = \frac{\lambda}{2\sin\theta} \quad \text{* MERGEFORMAT} \qquad (1)$$

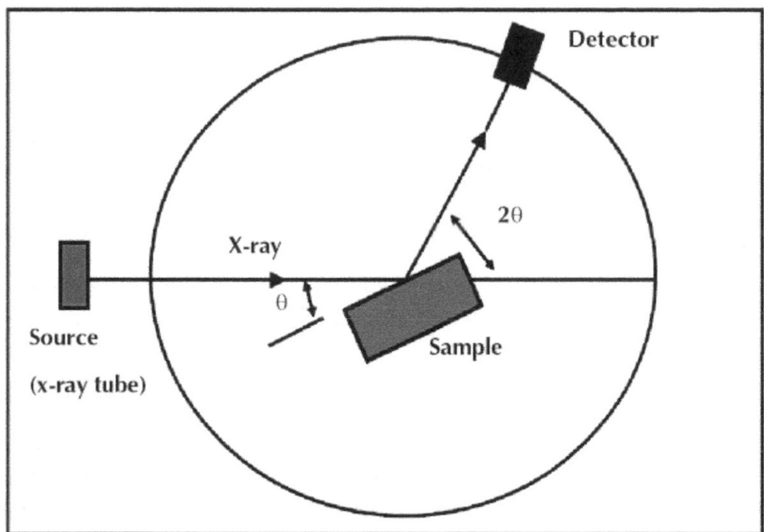

FIGURE 2 A schematic of an XRD instrument.

PREPARATION OF SAMPLES

Bio-nanocomposite samples for XRD analysis should be well ground to a particle size of less than 50 μm. Uniform particle size and random orientation of crystals are very critical for accurate XRD analysis. A sample with well-defined orientation of crystals (preferred orientation) can give a different diffraction pattern than that of a sample with randomly oriented crystals. Preferred orientation is particularly a problem with plate like crystals because they tend to lie horizontal, rather than perpendicular, on the sample holder. For a film sample, the thickness of the film should be at least 0.5 mm in order to detect the intensity of diffracted X-rays. The XRD analysis should be sensitive enough to detect the crystalline structure of layered silicates present in small amounts in bio-nanocomposites (Kasai and Kakudo, 2005).

COLLECTION AND INTERPRETATION OF DATA

Diffraction pattern for a particular sample is obtained by plotting the intensity of diffracted X-rays as a function of 2θ. Bio-nanocomposites are characterized by monitoring the intensity, width, and position of the peak in the diffraction pattern. Intensity of the peak provides information about the location of atoms in the unit cell. Higher intensity corresponds to higher electron density around the atom. Peak width provides information on the size of crystal and imperfections in the crystal. Peak width increases as the size of the crystal decreases. Position of the peak is used to estimate interlayer spacing by using the Bragg's equation as mentioned above.

The XRD patterns for various possible arrangements of layered silicates in biopolymers are shown in Figure 1. For an immiscible arrangement (microcomposite) of layered clays in a biopolymer matrix, the structure of the layered silicate in the composite is not affected. Thus, XRD pattern for the microcomposite should remain same as that obtained for the pure layered silicate. Intercalation of the polymer chains increases the interlayer spacing and according to Bragg's law, it causes a shift of the diffraction peak towards lower 2θ angle. As more polymers enter the interlayer spacing, the clay platelets become disordered, thus causing broader peaks and a wider distribution of such peaks. For exfoliated nanocomposites, there is no peak in the diffraction pattern because of the much larger interlayer spacing (>8–10 nm) between the silicate layers (McGlashan and Halley, 2003; Pavlidou and Papaspyrides, 2008; Paul and Robeson, 2008).

LIMITATIONS

The XRD is a simple and convenient method to determine d-spacing for immiscible or intercalated arrangements of layered silicates in bio-nanocomposites. However, it may be insufficient to characterize exfoliated nanostructures. Absence of peak in diffraction pattern is often misinterpreted as an indication of exfoliation. Other than exfoliation, dilution, and preferred orientation of clay in nanocomposites might result in a diffraction pattern with no peak. The XRD cannot be used to determine the spatial distribution and dispersion of layered silicates in bio-nanocomposites (Morgan and Gilman 2003). Therefore, XRD should always be used in conjunction with some other techniques such as TEM, SEM, or AFM.

Wide-angle XRD is not useful to study intercalation once the d-spacing exceeds 6–7 nm. However, small-angle ($2\theta < 1°$) X-ray diffraction (SA-XRD) analysis in combination with WA-XRD can be useful for characterization of such intercalated nanocomposites (Ray and Bousmina, 2005).

13.3 MICROSCOPY

Microscopy is the study of objects that are smaller than the spatial resolution (~75 μm) of human eye. Spatial resolution is the minimum possible length of an object which could be seen and separately identified from an adjacent and similar object. Light microscopes were developed in the early 17th Century. The spatial resolution (δ) of a light microscope can be approximated by the classical Rayleigh criterion as (Williams and Carter, 1996):

$$\delta = \frac{0.61\lambda}{\mu \sin\beta} \quad \text{* MERGEFORMAT (2)}$$

where λ is the wavelength of the radiation, μ is the refractive index of the viewing medium, and β is the semi-angle of collection of the magnifying lens. Wavelength for the light in the middle of the visible spectrum is ~ 550 nm. This gives light microscopes a resolution of approximately 300 nm. Early in the 20th Century, the wavelike nature of electrons was discovered. The electrons have a much lower wavelength as compared to that of light. Therefore, it is possible to achieve a resolution of as low as 0.2 nm by using electrons as a radiation source.

13.3.1 TRANSMISSION ELECTRON MICROSCOPE (TEM)

Transmission electron microscope was developed by Max Knoll and Ernst Ruska in 1931. In addition to XRD, TEM images provide further evidence for the occurrence of intercalation or exfoliation in nanocomposites. The TEM allows a qualitative understanding of the internal structure, spatial distribution, and dispersion of the nanoparticles within polymer matrices that are thin (< 100 nm) enough to transmit electrons.

PRINCIPLE OF OPERATION

Transmission electron microscope operates on the same basic principles as the light microscope but uses electrons instead of light. The TEM uses electrons as the radiation source and their much lower wavelength (in the order of one-tenths of a nanometer) makes it possible to achieve a resolution (0.2 nm) which is thousand times better than that of a light microscope. A schematic representation of a transmission electron microscope is shown in Figure 3. A TEM instrument can be divided into three sections: illumination system, specimen stage, and imaging system. The illumination system comprises of an electron gun, two or more condenser lens, and a condenser aperture. The specimen stage allows samples to be held for the analysis. The imaging system consists of an objective lens, an objective aperture, one or more intermediate lens, a projector lens, and a screen (Egerton, 2005).

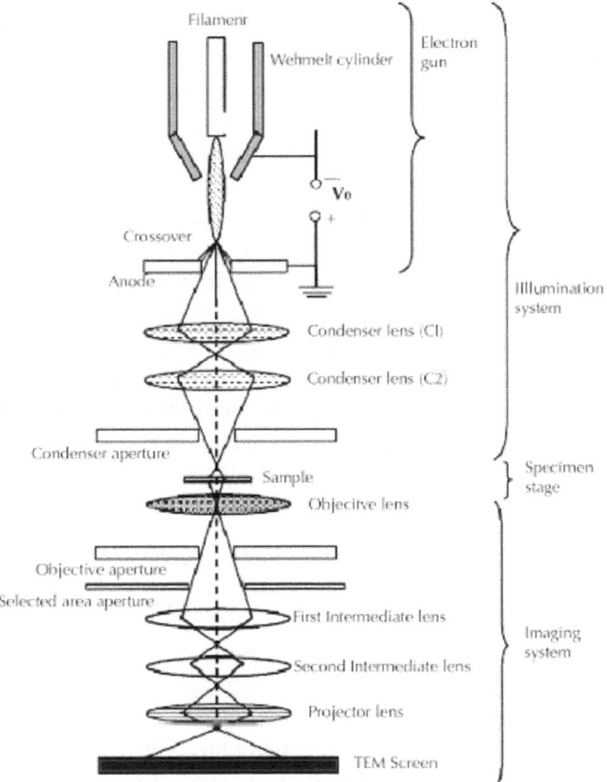

FIGURE 3 A schematic representation of a transmission electron microscope (adapted from Egerton 2005).

A fine beam of electrons with precisely controlled energy is produced in the electron gun. The gun consists of an electron source (filament for thermionic emission) and an accelerating chamber. The filament is made from a high melting point material such as tungsten. Once heated to a temperature above 2700K, the tungsten filament emits electrons into the surrounding vacuum. Wehnelt cylinder is a metal electrode which surrounds the filament except for a small opening through which the electron beam emerges. Wehnelt cylinder controls the emission current of the electron gun. After emission, electrons are accelerated by applying a potential difference (accelerating voltage), V_0 (100–500 kV) between the cathode and anode. Accelerating voltage determines the velocity and wavelength of the electron beam. Only 1% of the accelerated electrons are able to pass through the anode plate. Crossover is the effective source of illumination for the TEM. The first condenser lens (C1), which is a strong magnetic lens, creates a demagnified image of the electron beam. It also controls the minimum spot size available in the rest of the condenser system. The second condenser lens (C2), which is a weak magnetic lens, is used to vary the diameter of illuminated area of the sample. Condenser aperture controls the fraction of electron beam which illuminates the sample. Once the electron passes through the sample, the objective lens forms an inverted image. Objective aperture improves the contrast of the final image by passing only those electrons which will contribute to the final image. It also limits the blurring of image that arises from spherical and chromatic aberration. The purpose of the intermediate lens is to magnify the initial image formed by the objective lens. The intermediate lens can also be used to produce an electron diffraction pattern on the TEM screen. Projector lens produces an image or a diffraction pattern on the TEM screen. The TEM screen is used to convert the electron image to a visible form. It consists of a metal plate with a coating of $ZnSO_4$, which emits visible light under electron bombardment (Egerton, 2005).

The combination of TEM with other analytical techniques such as energy dispersive X-ray spectroscopy (EDS) and electron energy loss spectroscopy (EELS) can generate information for image analysis and chemical composition of the sample. The EELS can also provide information on the electronic structure of the sample (Koo, 2006). Some new developments in TEM for nanotechnology have been reviewed by Wang (2003). Transmission electron microtomography (TEMT) is an emerging three-dimensional imaging technique with a sub-nanometer resolution (~0.5 nm). Three-dimensional image obtained from TEMT can help in understanding the relationship between the structure and property of bio-nanocomposites (Jinnai and Spontak, 2009).

PREPARATION OF SAMPLES

Samples should be thin (< 100 nm) enough to transmit electrons and stable to electron bombardment under high vacuum. Typically, samples are made circular with a diameter of 3 mm. The sample must be inserted into the vacuum of the TEM column without introducing air. Samples are most commonly prepared by either ultramicrotomy or focused ion beam (FIB).

The TEM samples are prepared by cutting ultra thin (< 100 nm) sections from a small block of embedded material using ultramicrotomy. Ultramicrotomy is high precision cutting method using glass or diamond knife. The cut sections are floated onto a water surface. A fine-mesh copper grid (diameter of 3 mm) is then introduced below the surface of water. The grid is slowly raised and the cut section is supported by the grid. After drying in air, the ultra thin section remains attached to the grid. To produce scattering contrast, the cut section is stained by immersing in a solution of heavy metal (Egerton, 2005).

Focused ion beam (FIB) milling is a technique that uses a beam of accelerated ions to modify materials with nanometer precision. A schematic representation of a focused ion beam instrument is shown in Figure 4. A FIB instrument operates similar to an SEM. The FIB utilizes a liquid metal ion source (LMIS) at the top of its column to produce ions (usually Ga^{+}). The ions are focused into a beam by an electrostatic lens. After passing through the aperture and objective lens, the ion beam scans the surface of the sample. A beam diameter of 5–0.5 μm can be achieved in a FIB. The collision of ions with the atoms of the sample can be either elastic or inelastic in nature. An elastic collision results in removal of surface atoms. Removal of atoms, which is known as sputtering, modifies the surface of the sample. During inelastic collision, ions transfer a part of their energy to either surface atoms or electrons. Inelastic collision produces secondary electrons along with X-rays. The intensity of secondary electrons at each scan position can be used to create an image of the sample. Over an extended period of time, the sputtering process leads to noticeable removal of material. This is used for milling and probing applications. The small beam size in FIB makes it ideal for preparation of samples for TEM and SEM (Giannuzzi and Stevie, 1999; Yao, 2005).

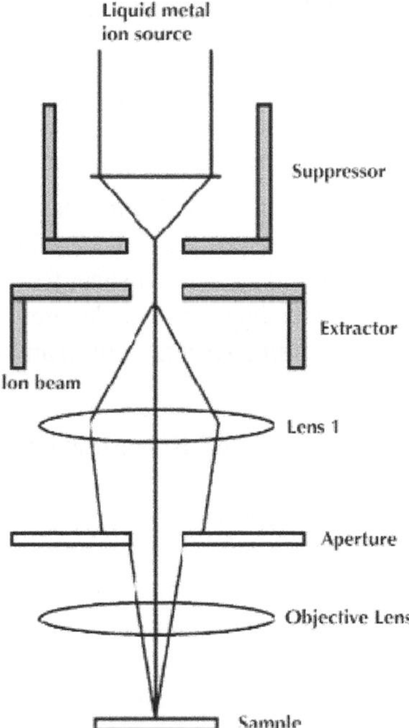

FIGURE 4 A schematic representation of a focused ion beam instrument (adapted from Guiannuzzi and Stevie 1999).

Samples of nanocomposite powders can be prepared by suspending the powder in a volatile solvent such as acetone. A drop of the suspension is then put on a fine-mesh copper grid. After drying in air, the nanocomposite powder remains attached to the grid and can be viewed under the TEM.

COLLECTION AND INTERPRETATION OF DATA

An ideal TEM image should have a good resolution and contrast. For bio-nanocomposites, contrast arises from differences in the atomic number of the layered silicates and the biopolymer matrix. Layered silicates with high atomic number appear dark in TEM image. The TEM results are quantified by the TEM particle density (number of platelets per unit area). The extent of exfoliation is better for samples with higher TEM particle density. Particle density per unit nanoparticles concentration is known as the specific particle density (Chavaria el al., 2004). However, it is difficult to calculate the TEM particle density for intercalated and partially exfoliated nanocomposites. Nam et al., (2001) calculated the length, thickness, and correlation length of the dispersed clay particles from TEM images. Correlation length was defined as the inter-particle distance in the direction perpendicular to that of the length of clay layers. Vermogen et al., (2005) proposed a method for quantifying TEM micrographs by measuring the length (L), thickness (t), and aspect ratio of clay layers. Inter-particle distance ($\varepsilon_{\parallel}$) in the direction parallel to that of the length of clay layers and inter-particle distance ($\varepsilon_{\perp}$) in the direction perpendicular to that of the length of clay layers were also measured.

The number (N) of platelets in an intercalated layered silicate was calculated as:

$$N = \frac{t_p + d - t_{platelet}}{d} \quad \text{* MERGEFORMAT} \tag{3}$$

where t_p is the thickness of the particle, d is the interlayer spacing, and $t_{platelet}$ is the thickness of a single platelet. Six classes of tactoids based on thickness were defined. These tactoids ranged from individual exfoliated sheet to micron size aggregate. The TEM has been used to characterize dispersion of layered clays in nanocomposites based on protein (Chen and Zhang 2006; Yu et al., 2007) and starch (Tang et al., 2008).

LIMITATIONS

A TEM instrument is more expensive than an XRD instrument. Methods for sample preparation are tedious. Only a small portion of the sample can be viewed. Samples containing water cannot be viewed under TEM because of the high vacuum in the sample chamber. The TEM results are difficult to quantify because TEM provides two-dimensional images of a three-dimensional sample. If the sample is not made very thin, electrons can be scattered or absorbed, rather than transmitted (Williams and Carter, 1996). Proper experimental conditions such as low exposure times and high accelerating voltage are necessary to prevent decomposition of the clay structure in the nanocomposites (Monticelli et al., 2007).

13.3.2 SCANNING ELECTRON MICROSCOPE (SEM)

The SEM is a type of microscope that uses high energy beam of electrons instead of light to scan surface of a relatively thick sample. SEM has a resolution of 1–10 nm.

PRINCIPLE OF OPERATION

A schematic representation of a scanning electron microscope is shown in Figure 5. Similar to TEM, an SEM incorporates an electron gun to generate a beam of electrons. The maximum accelerating voltage in SEM is typically 30 kV, which is lower than that in TEM (100–500 kV). The electron beam is demagnified by the first condenser lens (C1). The condenser aperture eliminates high-angle electrons by constricting the beam. The second condenser lens forms a thin and coherent beam. An objective aperture further eliminates high-angle electrons from the beam. The objective lens focuses the electron beam into a small-diameter (1–10 nm) electron probe. Electrons (primary) incident on a sample supply enough energy to the atomic electrons (secondary) present on the surface of the sample so that they can be released. A small fraction of primary electrons are elastically back-scattered (angle of deflection >90°) with only a small loss in energy. Due to their high energy, these back-scattered electrons (BSEs) can re-enter the surrounding vacuum. However, the secondary electrons can be distinguished from the backscattered electrons because of their much lower energy. A set of coils scans the electron beam over the sample in two perpendicular directions and covers a square or rectangular area of the sample. This procedure is known as raster scanning. An image of the raster is formed by sequential collection of secondary electrons from each scanned point. Thus, the image in SEM is generated sequentially rather than simultaneously as in TEM (Egerton, 2005).

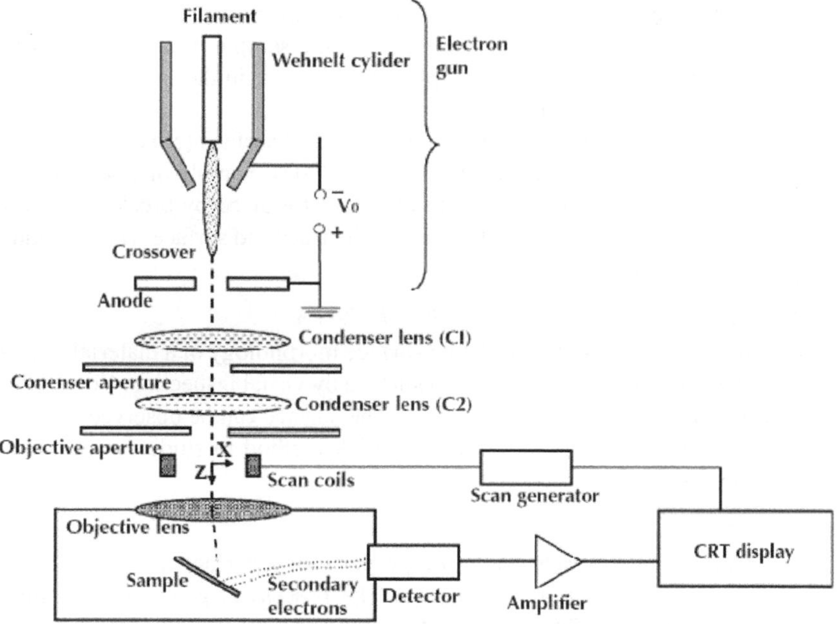

FIGURE 5 A schematic representation of a SEM (adapted from Egerton 2005).

PREPARATION OF SAMPLES

One major advantage of SEM over TEM is the ease of sample preparation. A sample does not need to be made thin. Conducting materials require no special sample preparation for SEM. Samples of insulating materials become charged when exposed to the electron probe. Negatively charged samples repel incident electrons, resulting in image distortion. Therefore, insulating materials are coated with a thin (15–40 nm) layer of metal (gold, chromium, and palladium) or conducting carbon. This process of coating is known as sputter coating. In sputter coating, the specimen chamber is exposed to very low vacuum (~0.1 Pa). An inert gas such as Argon (Ar) is introduced in the chamber. Gas molecules are ionized into ions and electrons because of the high voltage applied to the chamber. When struck with ions, the target metal ejects atoms. Atoms collide with residual gas molecules and deposit on the sample. Some of the limitations of the sputter coating process are thermal damage and surface contamination of the specimen. Sputter coating also reduces the penetration depth of SEM because of the thickness of the coating (Bozzola and Russell, 1999).

When coating is undesirable or difficult, low voltage should be used to avoid charging of the sample. An alternative to overcome charging of the samples is to surround the sample with a gaseous environment (low vacuum of up to 4000 Pa) rather than high vacuum ($\sim 10^{-4}$ Pa). This type of SEM is known as low-vacuum SEM or environmental SEM (ESEM). However, vacuum must be maintained in environmental SEM column for the operation of electron gun, acceleration of electrons, and focusing of electrons. After being focused by the objective lens, primary electrons ionize gas molecules (usually water vapor) before reaching the sample. Ionized gas molecules neutralize the surface charge of the sample. The pressure in the sample chamber can be as high as 5000 Pa (Egerton, 2005).

The SEM has also been used to study the fractured surface (cross-section) of bionanocomposite films (Vaz et al., 2002; Yu et al., 2007). Sample of fractured surface is prepared by freezing the film in liquid nitrogen, followed by breaking the film to expose the fracture surface, and sputter coating the fractured surface with a conducting coating (Yu et al., 2007).

COLLECTION AND INTERPRETATION OF DATA

The SEM images can be used to study the surface morphology of a material. A smooth surface can be distinguished from a rough surface by visual inspection. Comparatively larger (~1 μm) aggregates of nanoparticles at the surface can be observed in the SEM image. The SEM images can be used as a quick method to study the formation of nanocomposite structure.

LIMITATIONS

An image from SEM represents only the property of a surface and not the internal structure that is visible with a TEM image. The process of generating an image in SEM is slower than that in TEM. The SEM cannot be used to study detailed structure of nanocomposites because it has a poorer resolution as compared to TEM. Therefore, SEM analysis should always be used in combination with other microscopic techniques such as TEM or AFM.

SCANNING PROBE MICROSCOPE (SPM)

The use of a probe can resolve atoms as the resolution is no longer restrained by the wavelength of light or electrons. A probe can also be used to manipulate the structures along with scanning them. Some examples of SPM are scanning tunneling microscope (STM) and atomic force microscope (AFM).

13.3.3 SCANNING TUNNELING MICROSCOPE (STM)

Gerd Binnig and Heinrich Rohrer invented the scanning tunneling microscope in 1982, which gives three-dimensional images of conducting objects down to the atomic level. A spatial resolution of 0.01 nm can be achieved with a STM.

PRINCIPLE OF OPERATION

A schematic representation of a scanning tunneling microscope is shown in Figure 6. A very sharp metallic tip (M) is fixed on top of a piezodrive (P_z) to control the height (s) of the tip above the surface of a sample. The tip is brought within about 0.3–1 nm of the sample and a small potential difference (V_t ~ 0.01–1 V) is applied. Electrons move between the tip and the conducting sample by the process of quantum-mechanical tunneling. Quantum- mechanical tunneling is an effect of the wavelike nature of electrons which allows them to pass through small barriers such as the vacuum barrier between the tip and the sample. The amount of electrical current (tunneling current, J_t) flowing between the tip and the sample is measured. The tunneling current is very sensitive to the distance between the tip and the surface. A feedback control unit (CU) is used to adjust the height of the tip so that the tunneling current (J_t) remains constant. Two other piezodrives (P_x and P_y) are used to scan the tip in the lateral dimensions. During scanning of the surface, voltage (V_p) supplied to P_z is recorded as an image. This mode of operation is known as constant current mode and is mostly used to scan rough surfaces. In the constant height mode, the tip height (s) above the surface is kept constant and the variation of tunneling current reflects the corrugation of the surface (Bhushan 2004; Jia et al., 2005; Egerton 2005).

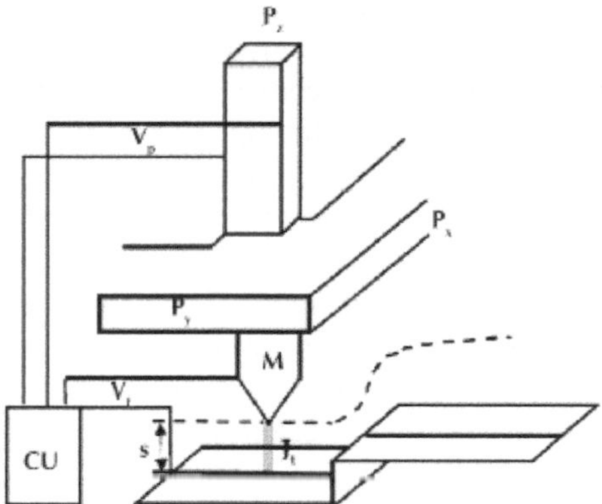

FIGURE 6 A schematic representation of a scanning tunneling microscope (adapted from Jia et al., 2005).

PREPARATION OF SAMPLES

Samples must be conductive enough to allow sufficient tunneling current (0.2–10 nA), which can be measured. Insulating samples can be coated with a thin layer of a conducting material and visualized with an STM (Bhushan, 2004).

COLLECTION AND INTERPRETATION OF DATA

The STM images can be interpreted in a manner similar to that for AFM, as discussed in the section under atomic force microscope.

LIMITATIONS

The main limitation of STM is that it can only take images of conducting or semiconducting surfaces under ultra high vacuum. It requires great mechanical precision to maintain a tip within 1 nm of a surface. An STM can provide structural information of only the surface of a sample. Therefore, it should be combined with some other technique such as TEM.

13.3.4 ATOMIC FORCE MICROSCOPE (AFM)

Gerd Binnig, Calvin Quate, and Christoph Gerber invented the atomic force microscope in 1985. Similar to an SEM, AFM can produce very high resolution three-dimensional images of the surface of a sample. The advantages of AFM include low cost, ease of use, and ease of sample preparation. The AFM does not require the sample to be conducting. Therefore, AFM can be used to take images of any type of surfaces including polymers, ceramics, composites, glass, and biological samples.

PRINCIPLE OF OPERATION

A schematic representative of an atomic force microscope is shown in Figure 7. A probe (cantilever with a sharp tip) is brought very close to the surface of a sample and the interatomic force between the probe tip and the sample is monitored. The interatomic force is repulsive if the tip is in direct contact with the sample. At a small distance above the sample, the tip senses an attractive van der Waals force. The interatomic force is determined by precisely measuring the displacement of the cantilever. A laser beam, which is deflected from the back of the cantilever, is directed to a photodetector. The photodetector measures the normal and lateral deflections of the cantilever. The deflections of the cantilever are converted into an image. A feedback control unit (CU) is used to adjust the height of the tip so that the interatomic force remains constant. A scanner is used for three-dimensional movement of the probe or sample in AFM. The scanners are made of piezoelectric materials to provide precise positioning in the nanometer range. There are two main modes of operation in AFM: contact mode and tapping or intermittent contact mode. The tip remains in permanent contact with the sample in contact mode. The force between the tip of the cantilever and the sample is measured at each point on the surface. A feedback loop keeps the force at a constant value by adjusting the height of the surface relative to the cantilever. The change in height is used to determine the topography of the surface. In tapping mode, the cantilever tip is oscillated above the surface of the sample at its resonant frequency. Oscillation of the cantilever is reduced because contact between tip and the surface results in loss of energy. The reduction in amplitude of oscillation is used to determine the topography of the surface. The tapping mode works well with soft materials that might be damaged in contact mode (Bhushan 2004; Magonov and Yerina, 2005).

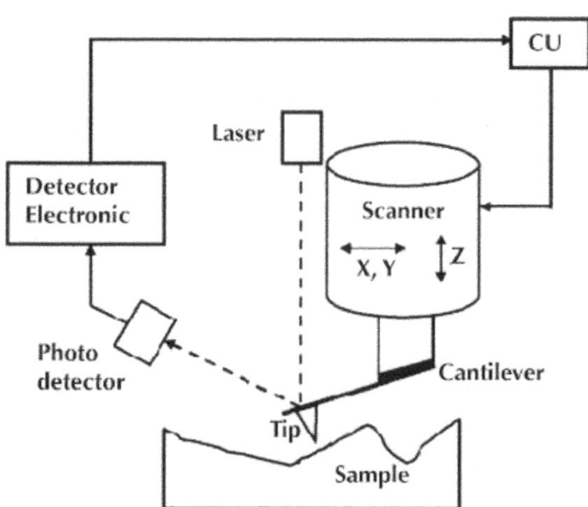

FIGURE 7 A schematic representation of an atomic force microscope (adapted from Magonov and Yerina 2005).

PREPARATION OF SAMPLES

Unlike other microscopy techniques, preparation of sample is quite simple in AFM. Samples do not require staining as in TEM or conductive coating as in SEM. Surface of the sample should be free of any scratches and foreign particles. Flat surfaces of samples can be prepared by ultramicrotoming (Yalcin and Cakman, 2004).

The AFM is typically conducted at ambient conditions. However, AFM can also operate with its tip and sample immersed in a liquid such as water. This makes AFM very useful in taking images of biological samples (Egerton, 2005). The AFM measurements at elevated temperatures of 150–300°C are now possible and AFM analysis can also be performed below 0 °C. The AFM imaging at elevated temperatures can be used to study structural changes in materials related to different thermal transitioning such as melting, crystallization, recrystallization, and glass transition (Ivanov et al., 2001; Magonov and Yerina, 2005).

COLLECTION AND INTERPRETATION OF DATA

The AFM images can be used to study detailed structure of the surface of a sample. The basic measurement in AFM is deflection of the cantilever over each x and y coordinates of the surface. This data can be used to define an average roughness (R_a) of the surface as (Ghanbarzadeha and Oromiehib 2008):

$$R_a = \frac{\sum_{i=1}^{N}\left(Z_i - \bar{Z}\right)}{N} \text{ * MERGEFORMAT} \quad (4)$$

where Z_i is the deflection value, $\bar{Z}\bar{Z}$ is the arithmetic mean of deflection values, and N is the number of x and y coordinates of the surface.

The AFM can also be used to determine adhesive and mechanical properties of a surface on nano-scale. The force required to pull the tip from the surface can be used as a quantitative measure of the adhesion between tip and surface. The mechanical response of the surface can be determined by using AFM in the contact mode at different force levels (Magonov and Renekar, 1997).

LIMITATIONS

The AFM can be used to study only the surface of a sample. Mechanical scanning of a large area of a sample in AFM is very time consuming. The probing of materials with AFM is based on mechanical interactions. This prevents AFM from being used for studying the chemical nature of materials. Another limitation of AFM is due to the finite size and shape of the probe. The AFM cannot be used to analyze a surface which has nanoparticles with dimensions smaller than the probe itself.

13.3.5 CONFOCAL SCANNING LASER MICROSCOPE (CSLM)

The CSLM is a non-destructive optical microscopic technique for the study of a material surface or internal structure of semi-transparent samples. The CSLM can be used to form a three-dimensional image of any sample.

PRINCIPLE OF OPERATION

A schematic representation of a confocal scanning laser microscope is shown in Figure 8. A beam of laser light is reflected by a dichroic mirror and focused onto a spot within the sample by the objective lens. Dichroic mirrors reflect light of one wavelength and transmit that of another wavelength. A laser beam illuminates the sample at the focus of the objective lens and excites fluorescence in the focused spot. The sample then emits light at a lower wavelength, which goes through the objective, passes through the dichroic mirror, and is focused down to another diffraction-limited spot, which is surrounded by a narrow pinhole. The pinhole is said to be confocal to the focal point of the objective lens on the sample because it is positioned at a point conjugate to the focal point of objective lens. The pinhole spatially filters out light originating from parts of the sample not focused by the laser beam. Light from the focused spot of the sample is then passed to the detector. The laser beam is focused at different portions of the sample to cover a range of depths. This allows point-by-point construction of the three-dimensional image of the sample (Lu, 2005).

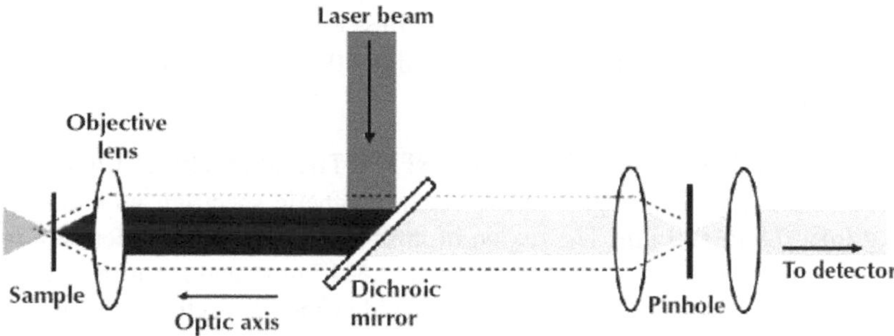

FIGURE 8 A schematic representation of a confocal scanning laser microscope (adapted from Lu, 2005).

PREPARATION OF SAMPLES

Preparation of samples for CSLM requires use of a staining agent for differentiating different phases and ingredients. Staining agent can be applied by either covalent or non-covalent labeling. Covalent labeling involves covalent linking of a fluorescent dye to the desired sample. Non-covalent labeling involves addition of fluorescent dye to the sample. For non-covalent labeling, fluorescent dye with affinity for either hydrophilic or hydrophobic regions can be used. Rhodamine B, a dye with affinity for hydrophilic region, is used to stain aqueous solution of proteins. Dyes with affinity for hydrophobic regions include Nile red and Nile blue A (Tromp et al., 2004).

Bio-nanocomposites based on layered clay can be stained with Safranine O by ion-exchange. Safranine O, a fluorescent dye, contains aluminum ion. Aluminum ion enables this dye to intercalate in clay galleries during ion-exchange. Confocal microscopy can then be used to determine the extent of dispersion of clay during mixing or sonication (Yoonessi et al., 2004).

COLLECTION AND INTERPRETATION OF DATA

The CSLM image can be used to evaluate the extent of mixing on micrometer scale. The CSLM image can be reconstructed to yield an image giving a three-dimensional impression of the sample. The three-dimensional image provides better insight into the extent of mixing on micrometer scale in the sample.

LIMITATIONS

One of the main limitations of CSLM is the requirement for a florescent dye which can bind to the region of interest. The resolution (~130 nm) of CSLM is poorer than that of TEM (Yoonessi et al., 2004). Therefore, CSLM analysis should always be used in combination with other microscopic techniques such as TEM or AFM.

13.4 SPECTROSCOPY

Spectroscopy is the study of the interaction between electromagnetic radiation and a material as a function of wavelength. Spectroscopy is used to determine functional groups, structural conformation, and concentration of different components in a material. The main spectroscopy techniques used for structural characterization of bionanocomposites are Fourier transform infra-red (FTIR), Raman, and nuclear magnetic resonance (NMR) spectroscopy.

13.4.1 FOURIER TRANSFORM INFRA-RED (FTIR) SPECTROSCOPY

Infra-red (IR) radiation is a part of the electromagnetic spectrum with wavelength ranging from 0.8 to 100 μm. The region of infra-red radiation is further sub-divided into near infra-red (0.8–2.5 μm), mid infra-red (2.5–15 μm), and far infra-red (15–100 μm). Typical wavenumber (1/wavelength) for mid infra-red spectroscopy ranges from 400–4000 cm^{-1}. The energy associated with absorption of IR radiation promotes rotation and vibration of chemical bonds. This vibrational energy is related to the strength of the bond and the molecular mass. The IR spectroscopy measures the absorption of specific frequencies of IR radiation by a sample. The measured absorption can be used to identify chemical bonds or functional groups in a particular sample. The IR spectroscopy uses high resolution diffraction monochromator for generating the absorption spectrum. This results in long measurement time because the absorption at each wavelength is measured sequentially. Another alternative form of IR analysis is Fourier transform infra-red spectroscopy. The FTIR spectroscopy uses an interferometer instead of a monochromator to measure absorption at all wavelengths simultaneously. The FTIR spectrometer is also fast and more sensitive as compared to a conventional IR spectrometer (Reh, 2001).

PRINCIPLE OF OPERATION

A schematic representation of an FTIR spectroscope is shown in Figure 9. The IR radiation is produced by heating an inert solid such as silicon carbide to 1,000–1,800°C. An interferometer consists of a beam splitter and two mirrors (fixed mirror and moving mirror). The beam splitter, which is a semi-reflecting device, is made by depositing a thin film of germanium onto a flat potassium bromide (KBr) substrate. The beam splitter divides the IR radiation in two beams: one beam is transmitted to the fixed mirror and the other beam is reflected to the moving mirror. When the two beams recombine after being reflected from the two mirrors, they undergo constructive and destructive interference due to the optical path difference between them. The optical path difference is created by varying the relative position of the moving mirror with respect to the fixed mirror. This recombined IR radiation is passed through the sample. Fluctuation in the intensity of energy at the detector is digitized into an interferogram. An interferogram contains time-domain spectrum (intensity *vs.* time) over the entire IR region. The interferogram is converted into a conventional frequency domain spectrum (intensity *vs.* frequency) using Fourier transform (Sherman-Hsu, 1997).

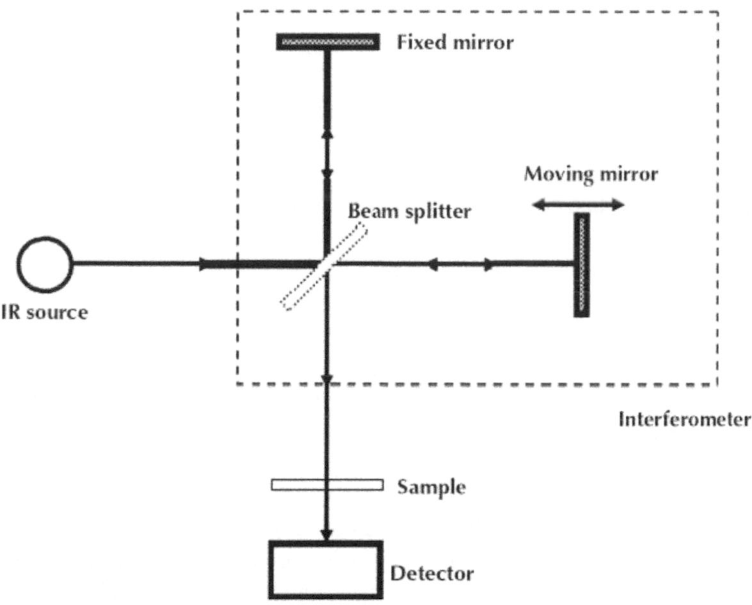

FIGURE 9 A schematic representation of an FTIR spectroscope (adapted from Sherman-Hsu 1997).

PREPARATION OF SAMPLES

Any solid, liquid, or gas sample can be analyzed using FTIR spectroscopy. The most common method to prepare solid samples involves use of a matrix to disperse the ground sample. The matrix can be a liquid (mineral oil) or solid (potassium bromide). For a liquid matrix, the paste of the ground sample and the liquid is spread between two IR transparent windows. For a solid matrix, a mixture of ground sample and potassium bromide (KBr) is pressed under high pressure (~80 MPa) for a few minutes. Recrystallization of KBr results in a clear disk, which can be inserted into the optical beam with a special sample holder. Liquids are analyzed as thin films in a cell with two IR transparent windows. A Teflon spacer is generally used to produce a film of the desired thickness (Reh, 2001).

Another method to prepare samples is based on attenuated total reflectance (ATR). The ATR device measures the total reflected energy from the surface of a sample in contact with an IR transmitting crystal. The refractive index of the crystal is significantly higher than that of the sample. This ensures total internal reflection of the radiation in the crystal. Infra-red radiation penetrates a small distance into the sample before reflecting back to the crystal. Therefore, there should be good optical contact between the sample and the crystal. Many solid samples give very weak spectra because the contact is confined to small areas (Reh, 2001).

COLLECTION AND INTERPRETATION OF DATA

The spectrum for a particular sample is obtained by plotting the intensity of % transmitted radiation as a function of wavenumber. Many functional groups absorb IR radiation in a very narrow range of wavenumber. The spectrum obtained from an FTIR analysis can be used to either determine the presence of a functional group in a sample or compare the spectrum of unknown material to that of known reference material (ASTM Standards, 1998).

Strong absorption bands appear in the region of 4000–2500 cm^{-1} because of the stretching vibrations between hydrogen and atoms with molecular weight (M) less than 19. Absorption bands in the region of 2700–1850 cm^{-1} usually appear because of triple bonds (C≡N) and other functional groups such as S-H, P-H, and Si-H. Many double bonded functional groups show absorption bands in the region of 1950–1450 cm^{-1}. Some of those double bonded functional groups include carbonyl groups, ketones, aldehydes, carboxylic acids, amides, and esters (Sherman-Hsu 1997). Absorption bands corresponding to C=O stretching (amide I), N-H bending (amide II), and C-N stretching (amide III) occur at 1630 cm^{-1}, 1530 cm^{-1}, and 1230 cm^{-1} respectively (Schmidt et al., 2005).

Absorption bands from stretching of silicon oxygen bonds (Si-O) occur in the region of 950–1100 cm^{-1}. This is particularly important for layered clay-based bio-nanocomposites because Si-O bonds are present in clay. The absorption band for agglomerated clay layers (tactoids) is broad. The absorption band becomes narrower as the degree of exfoliation increases. Thus, FTIR spectroscopy can be used to determine the extent of exfoliation in bio-nanocomposites (Klein et al., 2005). Absorption bands associated with different bonds of montmorillonite are shown in Table 1.

TABLE 1 Absorption spectra associated with montmorillonite (Chen et al., 2001)

Wavenumber (cm^{-1})	Functional groups
3627	Stretching vibration of-OH bond
3429	Stretching of interlayer H_2O
1635	Deformation of interlayer H_2O
1091 and 1039	Stretching of Si-O
519 and 466	Stretching of Al-O and bending of Si-O

LIMITATIONS

The FTIR spectroscopy can only be used to analyze samples, which are transparent to IR radiation. Multiple functional groups can absorb IR radiation in the same range of wavenumber. Therefore, this technique should be complemented with other analytical techniques such as Raman or nuclear magnetic resonance (NMR) spectroscopy (Sherman-Hsu, 1997).

Raman spectroscopy involves radiating a sample with monochromatic visible of near infra-red light from a laser. While IR spectroscopy detects the change in dipole moment, Raman spectroscopy detects change in polarizability of a sample. Some of the advantages of Raman spectroscopy over IR spectroscopy include ease of sample preparation, faster analysis, and better spatial resolution. However, heat generated by the laser in Raman spectroscopy may change the characteristics of the sample during measurement (Thygesen et al., 2003).

13.4.2 NUCLEAR MAGNETIC RESONANCE (NMR) SPECTROSCOPY

Nuclear magnetic resonance (NMR) is a phenomenon which occurs when the nuclei of certain atoms, under a static magnetic field (B_0), are exposed to a second oscillating magnetic field (B_1). The NMR spectroscopy can be performed on materials which have a net nuclear spin due to unpaired protons or neutrons. The unpaired proton or neutron with a spin generates a tiny magnetic field with a magnetic moment, μ. The alignment of the spin with the external magnetic field can result in either a low energy or a high energy configuration. There can be a transition between the two energy configurations by absorption (low to high energy configuration) or emission (high to low energy configuration) of a photon. The energy of the photon is equal to the difference between the two energy levels. The signal in NMR spectroscopy is obtained from the difference in energy absorbed and emitted. Thus, the NMR signal is proportional to the population difference between the two energy states. At room temperature, the number of spins in low energy level is slightly greater than those in high energy level. The NMR spectroscopy is sensitive enough to detect this small difference between the number of spins in low and high energy levels (Keeler, 2006).

PRINCIPLE OF OPERATION

A schematic representation of an NMR spectroscope is shown in Figure 10. A superconducting magnet (superconducting solenoid immersed in liquid helium) generates a static magnetic field (B_0) within the sample in the z-direction. Shim coils are used to maintain the homogeneity of the magnetic field. The sample tube is a cylindrical metal tube inserted into the bore of the magnet. The magnetic moments (μ) of all the spins in a sample can be represented by a net magnetization vector ($M = \sum\mu = M_x i + M_y j + M_z k$). At equilibrium, the net magnetization vector ($M_z = M_{0,}$ $M_x = M_y = 0$) lies along the direction of B_0 (z-direction). The value of B_0 is much greater than M_0. Therefore, it is not possible to detect the net magnetization vector of the sample. If the magnetization vector is tilted away from the z-axis by an angle β, it precesses around the z-axis and sweeps out a cone with an angle β.

The frequency (ν in rad s^{-1}) of precession is given by (Keeler 2006):

$$\nu = \gamma B_0 \text{ * MERGEFORMAT} \quad (5)$$

where γ, unique to each isotope, is the gyromagnetic ratio.

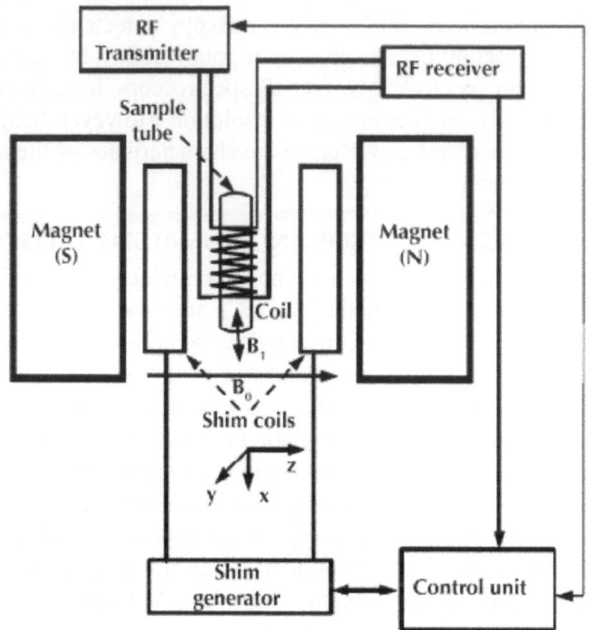

FIGURE 10 A schematic representation of an NMR spectroscope (adapted from Keeler 2006).

The values of γ for proton (^1H) and Carbon-13 (^{13}C) are 42.58 and 10.71 MHz/Tesla respectively. The frequency ν is also known as the resonance or Larmor frequency. Tilting of the magnetization vector is achieved by applying a very small oscillating magnetic field (B_1) along the x-axis. This oscillating secondary magnetic field is generated by transmitting radio frequency (RF) waves through a radio frequency (RF) coil. The response of the magnetization vector to this alternating field is used to receive signals in NMR spectroscopy. Resonance is achieved when the frequency of B_1 is equal to the Larmor frequency. Once resonance is achieved, the magnetization vector will precess about the x-axis with a frequency (ν_1) given by (Keeler, 2006):

$$\nu_1 = \gamma B_1$$

For a typical NMR experiment, B_1 is much smaller than B_0. Therefore, ν_1 is smaller than the Larmor frequency (ν). A pulsed magnetic field along the x-axis is produced by switching on and off the RF field in the coil. Pulse time (τ) is the duration of time the field is on. The pulse field rotates the magnetization vector about the x-axis by a certain angle, known as the flip angle (β). Flip angle is given by (Keeler 2006):

$$\beta = 2\pi\nu_1\tau \text{ \\* MERGEFORMAT} \quad (7)$$

A pulse field giving a flip angle of β is known as a β pulse. The most common flip angles are 90° and 180°. A 90° pulse rotates the magnetization vector from its equilibrium position to the negative y-axis. A 180° pulse, known as the inversion pulse, rotates net magnetization vector from positive z-axis to negative z-axis (Keeler, 2006). Rotation axis of the pulse can be changed by changing the phase offset (φ) of the RF field.

After the magnetization vector has been flipped by the RF pulse, transverse magnetization (M_{xy}) relaxes back to zero and longitudinal magnetization (M_z) relaxes back to its equilibrium value (M_0). Longitudinal or spin-lattice relaxation constant (T_1) is the time required for the longitudinal magnetization vector (M_z) to return to 63% of its equilibrium value (M_0). T_1 is also the time required by the spins to exchange energy with the surrounding lattice. The longitudinal relaxation time is most commonly measured by using an inversion recovery pulse sequence. In an inversion recovery pulse sequence, a 180° pulse is followed by a delay (τ) and a 90° pulse. Duration of the delay is longer than the duration of either the 180° pulse or 90° pulse. If the magnetization vector is placed in x-y plane, it will precess about the z-axis at the Larmor frequency. Along with precession, M_{xy} also starts to decay to its equilibrium value of zero because of the heterogeneity of magnetic field and molecular interactions. Transverse or spin-spin relaxation time (T_2) is the time required to reduce M_{xy} by a factor of e (~2.718). In the absence of a heterogeneous magnetic field, T_2 can be estimated by a spin echo pulse sequence. In a spin echo pulse sequence, a 90° pulse is followed by a delay (τ), a 180° pulse, and another delay (τ). The signal is acquired after the second delay. T_2 is always less than T_1 (Keeler, 2006).

The NMR signal is detected as an induced oscillating voltage (μV) in the RF coil by the RF receiver. The same RF coil is used for both exciting the spins and detecting the signal. The RF transmitter and receiver are separated by a device, known as diplexer. A diplexer is a fast acting switch which connects the receiver to the coil only when the transmitter is disconnected. The NMR signal is detectable only for a short duration because it decays due to the above mentioned relaxation processes. The decaying time-domain signal is known as the free induction decay (FID) signal. An analog to digital convertor (ADC) converts the signal from a voltage to a binary number. Fourier transform of the FID signal produces a spectrum with intensity as a function of frequency (Keeler, 2006).

PREPARATION OF SAMPLES

During sample preparation for NMR spectroscopy, the sample tubes should be clean and free of any contaminants. The sample should be dry and free of any solvent. The NMR samples are prepared by dissolving the sample in a solvent containing deuterium. The most common solvent is deuterated chloroform. The sample should be completely dissolved in the solvent because undissolved particles can distort the homogeneity of the magnetic field.

COLLECTION AND INTERPRETATION OF DATA

The NMR spectrum is a plot of intensity as function of frequency. A typical range for frequency is from 10 to 800 MHz. The NMR results are also expressed on a chemical shift scale. The chemical shift scale is set up by defining the peak frequency (ν_{ref}) from tetramethylsilane (TMS) as zero.

The chemical shift (δ) is defined as (Keeler 2006):

$$\delta = 10^6 \times \frac{(\nu_p - \nu_{ref})}{\nu_{ref}} \text{ \\* MERGEFORMAT} \quad (8)$$

where ν_p is the frequency of the peak intensity for a sample. The chemical shift is often expressed in parts per million (ppm).

The FID signal decays quicker in solids as compared to liquids. Thus, the shape of the FID signal can be used to distinguish between solid and liquid components of a sample. The relaxation times T_1 and T_2 provide information about the molecular structures of compounds (Choi et al., 2003). The values of T_1 and T_2 depend on the magnetic field, type of nucleus, temperature, and presence of other larger molecules. The value of T_1 for nuclei bound to larger molecules such as proteins is much lower than that for nuclei bound to pure water.

Some studies on the use of NMR for structural characterization of bio-nanocomposites have been reported (Avella et al., 2005; Bruno et al., 2008). Bruno et al., (2008) investigated the molecular structure and intermolecular interaction between the components of bio-nanocomposites based on poly(3-hydroxybutyrate) (PHB) and modified MMT using NMR. Values of T_1 were determined for the bio-nanocomposites at different MMT content. Results showed that NMR can be used as an efficient and rapid method to characterize the structures of bio-nanocomposites.

LIMITATIONS

The main limitation of NMR spectroscopy is its sensitivity. Atomic nuclei which have high sensitivity to NMR can only be analyzed by NMR spectroscopy. The NMR spectroscopy can only be used for a non-paramagnetic sample.

CONCLUDING REMARKS

This chapter reviews analytical techniques for the structural characterization of biopolymer-based nanocomposites. Selection of proper technique for characterization of these bio-nanocomposites is very critical in assessing their performance. The XRD is generally used as the preferred technique to determine immiscible or intercalated arrangements of nanoparticles in bio-nanocomposites. The SEM can be used as a quick method to study the dispersion of larger aggregates (~1 μm) of nanoparticles at the surface of a bio-nanocomposite. The TEM images can provide a qualitative understanding of the internal structure, spatial distribution, and dispersion of nanoparticles in bio-nanocomposites. The TEM in conjunction with XRD can provide information on the occurrence and degree of intercalation or exfoliation in bio-nanocomposites. The AFM can be used to study the structure as well as mechanical properties of a surface at the nano-scale. Spectroscopy techniques (FTIR and NMR) can be used in conjunction with microscopy techniques to obtain information on the interaction between nanoparticles and the biopolymer matrix.

Future studies on the characterization techniques need to focus on development of:

1. Microscopy techniques capable of obtaining three-dimensional images,
2. Better and faster methods of sample preparation,

3. Better image analysis techniques for better interpretation of results, and
4. Microscopy at elevated temperature.

These future developments will help in better understanding of the interaction between nanoparticles and biopolymer matrix. This will aid in the synthesis of bio-nanocomposites with improved properties.

KEYWORDS

- **Biopolymer-based nanocomposites**
- **X-ray diffraction**
- **Scanning electron micrsocopy**
- **Tunneling electron microscopy**
- **Atomic force microscopy**
- **Fourier-transform infra-red spectroscopy**
- **Nuclear magnetic resonance**

REFERENCES

1. ASTM Standards. E1252-98. General techniques for obtaining infrared spectra for qualitative analysis. Philadelphia, PA (1998).
2. Avella, M., Vlieger, J. J. D., Errico, M. E., Fischer, S., Vacca, P., and Volpe, M. G. Biodegradable starch/clay nanocomposite films for food packaging applications. *Food Chemistry*, **93**, 467–474 (2005).
3. Bhushan, B. Scanning probe microscopy: Principle of operation, instrumentation, and probes. B. Bhushan, (Ed.), *Springer Handbook of Nanotechnology* Springer, New York, pp. 325–369 (2004).
4. Bordes, P., Pollet, E., and Averous, L. Nano-biocomposites: biodegradable polyester/nanoclay systems. *Progress in Polymer Science*, **34**(2), 125–155 (2009).
5. Bozzola, J. J. and Russell, L. D. *Electron microscopy: principles and techniques for biologists*. Jones & Bartlett Publishers, Boston (1999).
6. Bruno, M., Tavares, M. I. B., Motta, L. M., Miguez, E., Preto, M., and Fernandez, A. O. R. Evaluation of PHB/clay nanocomposites by spin-lattice relaxation time. *Materials Research*, **11**(4), 483–485 (2008).
7. Chavaria, F. and Paul, D. R. Comparison of nanocomposites based on nylon 6 and nylon 66. *Polymer*, **45**(25), 8501–8515 (2004).
8. Chen, G., Liu, S., Chen, S., and Qi, Z. FTIR spectra, thermal properties, and dispersibility of a polystyrene/montmorillonite nanocomposite. *Macromolecular Chemistry and Physics*, **202**, 1189–1193 (2001).
9. Chen, P. and Zhang, L. Interaction and properties of highly exfoliated soy protein/montmorillonite nanocomposites. *Biomacromolecules*, **7**, 1700–1706 (2006).
10. Choi, S. G, Kim, K. M., Hanna, M. A., Weller, C. L., and Kerr, W. L. Molecular dynamics of soy-protein isolate films plasticized by water and glycerol. *Journal of Food Science*, **68**(8), 2516 2522 (2003).
11. Dennis, H. R., Hunter, D. L., Chang, D., Kim, S., White, J. L., Cho, J. W., and Paul, D. R. Effect of melt processing conditions on the extent of exfoliation in organoclay-based nanocomposites. *Polymer*, **42**, 9513–9522 (2001).

12. Dimonie, D., Constantin, R., Vasilievici, G., Popescu, M. C., and Garea, S. The dependence of the XRD morphology of some bio-nanocomposites on the silicate treatment. *Journal of Nanomaterials*, (2008). doi:10.1155/2008/538421.
13. Egerton, R. F. *Physical principles of electron microscopy: an introduction to TEM, SEM, and AEM*. Springer, New York (2005).
14. Guiannuzzi, L. A. and Stevie, F. A. A review of focused ion beam milling techniques for TEM specimen preparation. *Micron*, **30**(3), 197–204 (1999).
15. Jia, J. F, Yang, W. S, and Xue, Q. K.. Scanning tunneling microscopy. N. Yao and Z. L. Wang (Eds.). *Handbook of Microscopy for Nanotechnology*, Springer, New York, pp. 55–112 (2005).
16. Jinnai, H. and Spontak, R. J. Transmission electron microtomography in polymer research. *Polymer,* **50**, 1067–1087 (2009).
17. Ivanov, D. A., Amalou, Z, and Magonov, S. N. Real-Time evolution of the lamellar organization of poly(ethylene terephthalate) during crystallization from the melt: high-temperature atomic force microscopy study. *Macromolecules*, **34**, 8944-8955 (2001).
18. Kasai, K. Kakudo, M. *X-ray diffraction by macromolecules, Springer*, New York, p. 504 (2005).
19. Keeler, J. *Understanding NMR spectroscopy*, John Wiley & Sons Inc., New Jersey, p.459 (2006).
20. Klein, R. J., Benderly, D. Kemnetz, A., and Ijdo, W. L. A new methods to determining the quality of smectite clay/plastic nanocomposites: degree of nanodispersion and clay layer alignment. Nanocomposites 2005 fifth world congress, San Franciscso, California, USA (Aug 22–24, 2005)
21. Koo, J. H. Polymer nanocomposites: processing, characterization, and applications, McGraw-Hill, New York, p. 272 (2006).
22. Lu, P. J. Confocal scanning optical microscopy and nanotechnology. N, Yao and Z. L Wang (Eds.). *Handbook of Microscopy for Nanotechnology*. Springer, New York, pp. 3–24 (2005).
23. Magonov, S. N. and Yerina, N. A. Visualization of nanostructures with atomic force microscopy. N. Yao, Z. L. Wang, (Eds.). *Handbook of Microscopy for Nanotechnology*, Springer, New York, pp. 113–155 (2005).
24. Magonov, S. N. and Reneker, D. H. Characterization of polymer surfaces with atomic force microscopy. *Annual review of materials science*, **27**, 175–222 (1997).
25. Marsh, K. and Bugusu, B. Food packaging: roles, materials, and environmental issues. *Journal of food science*, **72**(3), R39–R55 (2007).
26. McGlashan, S. A. and Halley, P. J. Preparaion and characterization of biodegradable starch-based nanocomposite materials. *Polymer international*, **52**, 1767–1773 (2003).
27. Monticelli, O. Musina, Z. Russo, S., and Bals, S. On the use of TEM in the characterization of nanocomposites. *Materials letters*, **61**, 3446–3450 (2007).
28. Morgan, A. B. and Gilman, J. W. Characterization of polymer-layered silicate (clay) nanocomposites by transmission electron microscopy and X-ray diffraction: a comparative study. *Journal of applied polymer science*, **87**, 1329–1338 (2003).
29. **Nam, P. H., Maiti, P., Okamoto, M. Kotaka, T. Hasegawa, N., and Usuki, A.** A hierarchical structure and properties of intercalated polypropylene/clay nanocomposites. Polymer, 42(23), 9633–9640 (**2001).**
30. Pandey, J. K., Kumar, A. P, Misra, M., Mohanty, A. K, Drzal, L. T., and Singh, R. P. Recent advances in biodegradable nanocomposites. *Journal of nanoscience and nanotechnology*, **5**, 497–526 (2005).
31. Paul, D. R. and Robeson, L. M. Polymer nanotechnology: nanocomposites. *Polymer*, **49**, 3187–3204 (2008)
32. Pavlidou, S, Papaspyrides, C. D. A review on polymer-layered silicate nanocomposites. *Progress in polymer science*, **33**, 1119–1198 (2008).
33. Ray, S. S. and Bousmina, M. Biodegradable polymers and their layered silicate nanocomposites: In greening the 21st century materials world. *Progress in materials science*, **50**, 962–1079 (2005).
34. Reh, C. In-line and off-line FTIR measurements. E. Kress-Rogers and C. J. B Brimelow (Eds). *Instrumentation and Sensors for the Food Industry*, CRC Press, Boca Raton, pp. 213–232 (2001).

35. Rhim, J. W. and Ng, P. K. W. Natural biopolymer-based nanocomposite films for packaging applications. *Critical reviews in food science and nutrition*, **47**(4), 411–433 (2007).
36. Schmidt, V. Giacomelli, C. and Soldi, V. Thermal stability of films formed by soy protein isolate-sodium dodecyl sulfate. *Polymer degradation and stability*, **87**, 25–31 (2005).
37. Shabeer, A. Chandrashekhara, K. and Schuman, T. Synthesis and characterization of soy-based nanocomposites. *Journal of composite materials*, **41**(15), 1825–1849 (2007).
38. Sherman-Hsu, C. P. Infrared spectroscopy. F. Settle (Ed.). *Handbook of Instrumental Techniques for Analytical Chemistry*, Prentice Hall PTR, New Jersey, pp. 247–283 (1997).
39. Sorrentino, A. Gorrasi, G., and Vittoria, V. Potential perspectives of bio-nanocomposites for food packaging applications. *Trends in food science and technology*, **18**, 84–95 (2007).
40. Tang, X., Alavi, S., and Herald, T. J. Effects of plasticizers on the structure and properties of starch–clay nanocomposite films. *Carbohydrate polymers*, **74**, 552–558 (2008).
41. Thygesen, L. G, Lokke, M. M. Micklander, E. and Engelsen, S. B. Vibrational microspectroscopy of food. Raman vs. FTIR. *Trends in food science and technology*, **14**, 50–57 (2003).
42. Tromp, R. H., Nicolas, Y., Van de Velde, F., and Paques, M. Confocal scanning laser microscopy (CSLM) for monitoring food composition. I. E. Tothill (Ed.). *Rapid and On-line Instrumentation for Food Quality Assurance*, CRC Press, Boca Raton, pp. 306–323 (2003).
43. Vaz, C. M., Mano, J. F., Fossen, M., Van Tuil, R. F., De Graaf, L. A., Reis, R. L., and Cunh, A. M.. Mechanical, dynamic-mechanical, and thermal properties of soy protein-based thermoplastic with potential biomedical applications. *Journal of macromolecular science – physics*, **B41**(1), 33–46 (2002).
44. Vermogen A. Masenelli-Varlot, K. Sgula, R Duchet-Rumeau, J. Boucard, S., and Prele, P. Evaluation of the structure and dispersion in polymer-layered silicate nanocomposites. *Macromolecules*, **38**(23), 9661–9669 (2005).
45. Wang, Z. L. New developments in transmission electron microscopy for nanotechnology. *Advanced materials*, **15**(18), 1497–1514 (2003).
46. Williams, D. B. and Carter C. B. T*ransmission electron microscopy: a textbook for materials science*. Springer, New York. p. 721 (1996).
47. Yalcin B and Cakmak, M. The role of plasticizer on the exfoliation and dispersion and fracture behavior of clay particles in PVC matrix: a comprehensive morphological study. *Polymer*, **45**, 6623–6638 (2004).
48. Yang, K. K., Wang, X. L. and Wang, Y. Z. Progress in nanocomposite of biodegradable polymer. *Journal of industrial and engineering chemistry*, **13**(4), 485–500 (2007).
49. Yao, N. Focused ion beam systems – a multifunctional tool for nanotechnology. N. Yao and Z. L. Wang (Eds.), *Handbook of Microscopy for Nanotechnology*, Springer, New York, pp. 247–286 (2005).
50. Yoonessi, M. Toghiani, H. Kingery, W. L., and Pittman, C. U. Preparation, characterization, and properties of exfoliated/delaminated organically modified clay/dicyclopentadiene resin nanocomposites. *Macromolecules*, **37**(7), 2511–2518 (2004).
51. Yu, J, Cui, G, Wei, M, and Huang, J. Facile exfoliation of rectorite nanoplatelets in soy protein matrix and reinforced bio-nanocomposites thereof. *Journal of applied polymer science*, **104**, 3367–3377 (2007).
52. Zeng, Q. H., Yu, A. B., Lu, G. Q., and Paul, D. R. Clay-based polymer nanocomposites: research and commercial development. *Journal of nanoscience and nanotechnology*, **5**, 1574–1592 (2005).
53. Zhao, R, Torley, P., and Halley, P. J. Emerging biodegradable materials: starch- and protein-based bio-nanocomposites. *Journal of materials science*, **43**, 3058–3071 (2008).

Part IV: Modified Atmosphere Packaging for Foods and Other Applications

CHAPTER 14

MODIFIED ATMOSPHERE PACKAGING OF FOOD

SANJAYA K. DASH

ABSTRACT

Modified atmosphere packaging (MAP) is a well established technology, which in combination with proper temperature management helps in extending the shelf life and maintaining quality of perishable produce. As the technology relies on modification of package atmosphere by the respiration of the commodity and the permeability of the packaging material, the different types of polymers with a range of O_2, CO_2 and water vapor transmission rates have turned into the most preferred packaging material for MAP. The common plastic films used in MAP are LDPE, LLDPE, HDPE, PP, PVC, PET, PVDC, and PA (nylon). PVDC, PVC, PET and nylon have low gas permeabilities and can be used only for slow respiring commodities. However, suitable perforations in the films can expand their use to many commodities. The film properties can also be modified by combining the individual films with one another or with other materials such as paper or aluminum through processes such as coextrusion, lamination, coating and metallization. Additives as antioxidants, heat stabilizers, UV stabilizers, anti-slip agents, as well as color pigments also improve the polymer characteristics. Besides, antifogging agents, nucleating agents, antistatic agents, plasticizers, oxygen scavengers, antimicrobials are also used as additives, which have facilitated the development of active food packaging technologies. Though MAP has gained considerable acceptance, particularly for the fresh fruits and vegetables and minimally processed products, there is often a mismatch between the commodity requirements and the polymer characteristics, defeating the basic purpose of MAP, and hence, further investigation to develop polymer films of recommended gas and moisture transmission characteristics are required. Other recent research interests in MAP have been in the areas of active packaging technologies, modification of film properties by use of nanocomposites and additives, and modeling to standardize the package materials and designs for specific commodities.

14.1 INTRODUCTION

Post-harvest quality loss of food, particularly the fresh fruits and vegetables, is primarily a function of respiration, ripening and senescence, water loss (transpiration), enzymatic discoloration of cut surfaces, decay (microbial), oxidation of fats causing rancidity and mechanical damage during preparation, shipping, handling and processing. Food preservation and post harvest management, thus, involve control of these factors with the objective of maintaining quality, delaying spoilage and reducing waste. Different types of preservation techniques such as canning, drying and dehydration, refrigeration and freezing are employed. However, storage in a modified gaseous atmosphere is one of the best methods for extending the shelf life of respiring foods and making these available to the consumer as fresh as possible.

The MAP technology basically relies on the modification of atmosphere inside the package, which is achieved by the natural interaction between the respiration of the commodity and the permeability of the packaging material (Mangaraj et al., 2009). It is generally used for respiring commodities and is especially important for fresh-cut produce because of their greater susceptibility to water loss, cut surface browning, higher respiration rates, enhanced ethylene biosynthesis and action, and microbial growth. The commodities are kept in a package, which develops an atmosphere of reduced oxygen (O_2) and increased carbon dioxide (CO_2) concentrations, and thus reduces the deteriorative activities. However, a completely airtight package may drop the O_2 level below a certain lower acceptable limit, at which anaerobic respiration starts and the produce spoils. Besides, there is also loss of moisture from the commodity by respiration and transpiration, and there is a necessity of moisture transfer through the packaging material. Thus the packaging material should allow exchange of gases and water vapor through it to maintain the desirable gas concentration and relative humidity (RH) within the package. Therefore different polymeric materials, due to their different gas and vapor exchange characteristics along with other factors related to cost, convenience and acceptability, have found wide use in MAP.

14.2 PRINCIPLE OF MODIFIED ATMOSPHERE PACKAGING

The main factors which influence the shelf life and quality of fresh perishable commodities are the respiration and metabolism, microbial spoilage and loss of water through transpiration. The fresh foods continue to respire even after harvest. The respiration rates vary, for example potato and onion are low respiring commodities, whereas sweet corn, mushroom, peas are high respiring commodities. If the changes are due to internal tissue constituents and enzymes only, it is known as autolysis and if both microorganisms and enzymes are active in food spoilage, the process is known as decay. Transpiration loss of water reduces the weight of commodity and softens the tissues. It also affects the appearance, texture and flavor.

The respiration and metabolism as well as the activity of microorganisms require oxygen, and these continue at optimal rates if the commodity is stored in normal atmosphere. The rate of respiration depends upon various intrinsic (crop) and extrinsic (crop environment) factors. It is extremely difficult to maintain/control the intrinsic factors but relatively easier to alter the extrinsic ones that is gaseous concentrations in the crop environment, surrounding temperature and RH (Singh et al., 2012). Thus, the basic concept of MAP is to control respiration and other biochemical activities within the food tissues by reducing the O_2 and increasing the CO_2 concentrations of the storage environment. Both O_2 and CO_2 levels exert independent depressant effects on respiratory reactions. The net effect may be additive or synergistic.

A significant reduction in respiration rate takes place as the oxygen level is reduced below 21%, especially below 10%. Besides, the production and action of ethylene (C_2H_4) that plays a central role in ripening of the fruits can be inhibited at low O_2 levels. MAP also tends to maintain high humidity around the commodity, thereby reducing the transpiration loss. Other systems like hypobaric storage, vacuum packaging, and so on use the similar principle of modifying the storage atmosphere for extending the shelf life of perishable commodities. The MAP has now been extended to durable commodities like rice also.

The MAP is based on enclosing a respiring produce in a package which is selectively permeable to gases and water vapor. The respiration of the produce changes the relative proportions of gases in the package. The microbial activity also affects the rate of change of atmosphere. The consequential reduced O_2 and increased CO_2 inside the package create gradients across the film, which act as the driving force for gas movement into and out of the package. The levels of O_2 and CO_2 within a package depend on the interaction between commodity respiration and the permeability properties of the packaging material.

Sometimes the normal shelf life of commodities is very less or the rate of respiration alone is not capable of developing the EMA quickly. In that situation the produce tends to spoil before attaining the EMA conditions. Thus the time required to reach the EMA condition is reduced by injecting a desired gas mixture into the package or by having an active agent (to absorb O_2, ethylene, and so on) in the package itself. If the development of the modified atmosphere is a function of product respiration alone, it is known as "passive modification" and if some specified gas mixture is injected into the chamber or an active agent is put inside, then the system is known as "active modification".

In MAP, once the package is sealed with commodity, no further control on gas composition is possible. However in controlled-atmosphere packaging (CAP) and controlled-atmosphere storage (CAS), the proportion of each gas is maintained (controlled) throughout the distribution cycle. The facilities are often equipped with nitrogen generators and oxygen and ethylene scavengers along with refrigeration units. Controlled atmosphere technology is usually applied in case of bulk storage and transportation. With advances in active packaging systems, the distinction between MAP and CAS is no longer clear (Fellows, 2000).

If the package is completely impermeable to gases and moisture, then the O_2 level may go to such a low level that there is onset of anaerobic respiration and spoilage. Further the moisture generated by respiration and transpiration also condenses within the package, which causes more damage than the advantage of reduced desiccation. Hence in MAP, the package material should allow permeation of gases and water vapor at a desirable rate.

14.3 GASES USED IN MODIFIED ATMOSPHERE PACKAGING

The three main gases which are modified in MAP are CO_2, O_2 and nitrogen (N_2). The CO_2 dissolves readily in water (1.57 g/kg at 20°C, 100 kPa) to produce carbonic acid (H_2CO_3) that increases the acidity of the solution and reduces the pH. It can affect the quality and shelf life of stored produce in MAP (Sandhya, 2010). The high solubility property of CO_2 can result in reduction of head space volume which may lead to pack collapse.

O_2 supports the microbial growth and many other types of deteriorative biochemical reactions. Its water solubility is low, at 20°C and 100 kPa, the value is about 0.040 g/kg of water.

The N_2 can inhibit the growth of aerobic bacteria, but not that of the anaerobic ones. Thus it is used to displace O_2 in packs to control the aerobic bacteria and to delay oxidative rancidity. The N_2 also has a low solubility in water (0.018 g/kg water at 20°C, 100 kPa) and other food constituents, hence inclusion of sufficient N_2 in the package atmosphere helps to prevent pack collapse (Sandhya, 2010).

For different types of food products, three types of gas mixtures, viz. inert blanketing (N_2), semi-reactive blanketing ($CO_2 + N_2$, $O_2 + CO_2 + N_2$) and fully reactive blanketing (CO_2 or $CO_2 + O_2$) are used (Farber, 1991; Moleyar and Narasimham, 1994). Selection of the gas mixture depends on the commodity, packaging material and storage temperature. For respiring commodities like fruits and vegetables, optimum levels of O_2 and CO_2 are only considered.

Noble or "inert" gases such as argon are in commercial use for commodities such as coffee and snack products. Experimental use of carbon monoxide (CO) and sulfur dioxide (SO_2) has also been reported (Sandhya, 2010). However, commercial application of CO is limited due to its toxicity and the tendency to form potentially explosive mixtures with air.

Both the permeability of the packaging material and the respiration of the commodity vary with the storage temperature at different rates. If the two do not match at the specific storage conditions, the desired levels of O_2 and CO_2 will not be attained and the deterioration may not be under control. A very low O_2 level induces fermentation and accumulation of ethanol and acetaldehyde. These result in development of off-flavors and/or tissue damage. The lower O_2 limits vary from 0.15% to 5% for different commodities and the values vary with temperature, commodity and cultivar. The CO_2 content may also change to a too low or too high level, which can injure the stored produce. The fruits and vegetables have also been classified as per their tolerances to low oxygen and elevated CO_2 concentrations (Beaudry, 1999; Beaudry, 2000; Fellows, 2000). In general, 2–5% O_2 and 3–8% CO_2 levels are suitable for most commodities. The optimum gas composition required for different types of commodities are given in Table 1.

TABLE 1 Recommended gas mixtures for MAP of selected commodities

Product	O_2 (%)	CO_2 (%)	N_2 (%)
Fruits			
Apple	1–2	1–3	95–98
Banana	2–5	2–5	90–96
Mango	3–7	5–8	85–92
Orange	5–10	0–5	85–95
Papaya	2–5	5–8	87–93
Pear	2–3	0–1	96–98

TABLE 1 *(Continued)*

Strawberry	5–10	15–20	70–80
Vegetables			
Beans, snap	2–3	5–10	87–93
Broccoli	1–2	5–10	88–94
Brussels sprouts	1–2	5–7	91–94
Carrot	5	3–4	91–95
Cauliflower	2–5	2–5	90–96
Chili peppers	3	5	92
Corn, sweet	2–4	10–20	76–88
Cucumber	3–5	0	95–97
Lettuce (leaf)	1–3	0	97–99
Mushrooms	3–21	5–15	65–92
Tomatoes	3–5	0	95–97
Onion	1–2	0	98–99
Other commodities			
bakery products	0	20–70	30–80
Cheese	0	0	100
Fish, white	30	40	30
Fish, oily	0	40	60
Fresh meat	60–85	15–40	0
Cooked meat	0	25–30	70–75
Dry snack foods	0	20–30	70–80
Pizza	0–10	40–60	40–60
Poultry	0	20–50	50–80
Sausage	40	60	0

Adapted from: Farber, 1991; Day, 1993; Exama et al., 1993; Moleyar and Narasimham, 1994; Smith and Ramaswamy, 1996; Sandhya, 2010; Fellows, 2000

As seen from the Table 1, proper levels of both O_2 and CO_2 are important for MAP. In fact, altering the O_2 level automatically alters CO_2 level, but it may not lead to the recommended CO_2 level, and hence, auxiliary controls are required.

The lower O_2 limit in MAP is usually the level of O_2 that induces fermentation, however, in commercial practice, sometimes still lower O_2 levels are maintained to obtain certain benefits that offset the loss in flavor or other quality parameters (Mattheis and Fellman, 2000).

Production of compounds which help in development of characteristic aromas of many fruits such as apple, banana and strawberries, and so on can be adversely affected by MAP. The high CO_2 and low O_2 may affect the oxidative processes and ethylene perception and suppress aroma production. However, most produce recover from moderate low O_2 suppression of volatile aroma production and eventually develop characteristic flavors. Still the aroma of the commodity, when it is just taken out of the low O_2 packages may be inferior to the fresh commodity, which can affect consumer appeal (Beaudry, 2000).

14.4 BENEFITS OF CONTROLLED/MODIFIED ATMOSPHERE PACKAGING

The CA/MA helps to maintain the quality of stored produce for a longer period by reducing respiration and ethylene production rates, retarding softening of fruits, slowing down changes related to ripening and senescence, reducing degradation of chlorophyll and other pigments and decreasing the rate of microbial growth (Ahvenainen, 1996; Saito and Rai, 2005; Lurie et al., 2006; Rodriguez-Aguilera and Oliveira, 2009; Sabir et al., 2011). All these effects together contribute to the improvement in the shelf life of food. Low O_2 and elevated CO_2 can significantly reduce the rates of ripening and senescence primarily by reducing the synthesis and perception of ethylene and the respiration. Modified atmospheres are generally effective for reducing ripening prior to the onset of ripening, rather than at a later stage, though it can also reduce the rate of ripening of some commodities such as tomato even during later stages (Yang and Chinnan, 1988).

There is better retention of cellular constituents like sugars, organic acids, cellular proteins and flavor compounds. The desirable effect of MA on plant tissues has also been attributed to lower pH, due to dissolution of CO_2 in tissues. MA can use comparatively higher temperatures than refrigerated storage, and hence reduces physiological disorders like chilling injury. MAP reduces moisture loss from the commodity and keeps it fresh like for longer period. Usually the need for adding chemical preservatives is also reduced.

The MAP also has the general functionalities of packaging as containment, protection, communication and marketing. The environmental impact is less, particularly if the packaging materials can be recycled. There are also associated benefits as better management of supply chain, facilitation of a semi-centralized manufacturing option, management of product quality, reduction of waste during distribution and on retail shelves, and so on. As the commodity has a longer shelf life, it helps the stores to avoid frequent product rotation, removal and restocking, which helps in reducing labor and waste disposal costs (Moleyar and Narasimham, 1994).

Fresh product quality can be better maintained by good temperature management than simply by modifying the atmosphere. As a thumb rule, if the shelf life of a commodity at 20–25°C is 1, then by MAP, it will be doubled, refrigeration can extend the shelf life to 3, and refrigeration combined with MAP can increase it to 4 (Kader et al., 1989). Controlled atmosphere (CA) and modified atmosphere (MA) have been found to extend the shelf life of different types of foods by 1.5–4.0 folds under refrigeration depending on the commodity. In fact the use of MAP as a supplement for proper temperature maintenance to delay ripening is generally beneficial for all fruits though the advantage is more for chilling sensitive fruits. Increase in CO_2 and decrease in O_2 levels have been shown to reduce chilling injury of avocados, grapefruit, peaches, nectarines, okra, pineapples, potatoes, and zucchini squash (Wang, 1993). There is a minimum lowest possible temperature for storage of individual commodities to prevent the chilling injury and in MAP, this lowest feasible temperature is maintained, since temperature has a much more significant influence on preserving quality than the use of low O_2 atmosphere alone.

Low O_2 has a very limited effect on the activity or survival of microorganisms causing decay at levels above the fermentation threshold of most commodities. A concentration of 10–15% CO_2 is required to control decay in fresh fruits and vegetables. Some crops can tolerate this level (strawberries and spinach), but most cannot.

As discussed later in details, the perforated and continuous films differ in their permeability characteristics for O_2 and CO_2. The perforated films will generate a higher partial pressure of CO_2 for a given concentration of O_2 in the package and can increase the CO_2 to levels within the fungistatic range. Chemical additives can also help to reduce decay.

One of the major effects of packaging is the maintenance of high humidity surrounding the products. However, the MAP of high moisture as well as highly respiring commodities usually cause high RH in the package which accelerates decay. The control of in-package RH by using desiccants can help in limiting decay.

14.5 OPTIMIZATION OF ATMOSPHERE IN MODIFIED ATMOSPHERE PACKAGING

The MAP is a dynamic system during which respiration and permeation occur simultaneously (Mangaraj et al., 2012) and it requires proper interaction of produce, packaging film and package parameters to arrive at optimum in-pack atmosphere. The design and generation of an optimum MA requires thorough understanding of the interaction between the various factors such as film characteristics (surface area, gas and water vapor permeability), temperature, free volume inside the package, weight and respiration rate of the produce and initial gaseous composition (Salvador et al., 2002).

The optimum atmosphere for storage of a specific commodity can be found out by monitoring its shelf life and quality in different controlled atmosphere chambers. Subsequently the rate of respiration under such atmosphere is studied. For the respiring commodities as fruits and vegetables, the permeability (for O_2 and CO_2) of the packaging material should match the commodity respiration to establish the desired EMA within the package, that is the packaging material should allow the exact amount of O_2 required for respiration into the package and the exact amount of CO_2 evolving during respiration out of it.

Alternatively, the samples are stored in polymeric film packages of known permeability characteristics. The ratio of film area to mass of product is varied to generate the desired O_2 and CO_2 concentrations. A CO_2 absorbent like hydrated lime is put in the package prior to sealing. The O_2 concentration in the package is monitored and when the atmosphere attains equilibrium, the rates of oxygen uptake are calculated.

However, this method does not give the information about the desirable levels of CO_2 (Al-Ati and Hotchkiss, 2003). Altering film permselectivity (that is, β, which is the ratio of CO_2 and O_2 permeation coefficients) can simultaneously optimize the levels of both CO_2 and O_2 in MAP systems. For both methods, it is desirable to determine the concentrations of oxygen at which the respiration of the product becomes anaerobic.

Permeation is the movement of gases, vapors, or liquids (called permeants) across a homogenous material. Molecules of gases, vapors, and other low molecular weight substances can dissolve in polymers, diffuse through the polymers and then travel to a contact surface. In MAP the permeation takes place due to the partial pressure gradient of the particular gas or water vapor across the film. The characteristics of both polymer material and the permeant molecules, as well as the environment parameters affect the permeability. Some of these are the physical and chemical structure of the polymer, permeant concentration, size of permeant molecules, temperature, humidity, and so on The larger permeant molecules generally have lower diffusivity than the smaller ones, but higher solubility. Of course the chemical similarity between the solvent and the permeant strongly influence the solubility. Van Krevelen suggests an empirical relationship for the relative values of permeability of a polymer to various gases as given in Table 2 (1997).

TABLE 2 Relative values of permeability coefficients

Gas	P
Nitrogen	1
CH_4	3.4
O_2	3.8
H_2	22.5
CO_2	24
H_2O	550

The effects of factors like package characteristics, gas concentrations, and so on on the MA vary with the type of commodity and have been extensively reported. However, the parameters as the plant species, variety, cultural practices, maturity level at harvest, harvesting mechanism, tissue type, and postharvest handling, and so on, also affect the response of the stored produce to the modified atmosphere, which are difficult to define.

Nowadays the design of MA packages is mostly done by simulation studies and many models have been developed. Package modeling can improve understanding of how package, plant and environmental factors interact and can be useful in package design (Fonseca et al., 2000; Fonseca et al., 2002b). These models can also help to choose packaging materials for specific food products.

The respiratory metabolism is governed by enzymatic reactions, and hence, many researchers have used the Michaelis-Menten type equation to describe the relation between respiration and gas concentration (Cameron et al., 1994; Lakakul et al., 1999; Mangaraj and Goswami, 2008). Respiratory models are then coupled with an equation describing the temperature sensitivity of film permeability to gases (known as the Arrhenius equation) to predict package O_2 partial pressure as a function of temperature, product mass, surface area and film thickness (Cameron et al., 1994; Lakakul et al., 1999; Del Nobile et al., 2007; Mahajan et al., 2007; Mangaraj and Goswami, 2008; Mangaraj et al., 2012).

Both diffusion (D) and solubility (S) are functions of temperature and follow an Arrhenius type of equation (Selke et al., 2004).

$$\tau = \tau o e^{Ea/RT} \tag{1}$$

where, τ represents either D or S, τ_0 represents a proportionality constant (known as pre-exponential term),T is the temperature in degree K. The equation also generally holds good for permeability (P), which is the product of D and S. The value of R is 8.31434 $Jmol^{-1}\ K^{-1}$ or 1.314 atm $ft^3mol^{-1}\ K^{-1}$.

The equation is valid over a small range of temperature and can predict permeability at any temperature from known values at specific temperatures. When the polymer passes through a transition such as the glass transition temperature, a new relationship is needed.

As a rule of thumb, a change of approximately 10°C will affect the permeability by a factor of 2. At around 0°C a small variation in temperature greatly affects the permeability and Arrhenius equation cannot be used in that situation.

The differential mass balance equations that describe the change in O_2 and CO_2 concentrations in a package having respiring products are given as

$$\frac{dY_{O_2}}{dt} = -\left(\frac{W_p}{V_{fp}}\right) R_{O_2} + \left(\frac{A_p P_{O_2}}{V_{fp}}\right)\left(Y_{O_2}^a - Y_{O_2}\right) \tag{2}$$

$$\frac{dZ_{CO_2}}{dt} = -\left(\frac{W_p}{V_{fp}}\right) R_{CO_2} + \left(\frac{A_p P_{CO_2}}{V_{fp}}\right)\left(Z_{CO_2} - Z_{CO_2}^a\right) \tag{3}$$

where A_p (m^2) is the area of the package through which the permeation takes place, $Y_{O_2}^a$ and $Z_{CO_2}^a$ are the O_2 and CO_2 concentrations (cm^3cm^{-3} of air) in the atmospheric air, Y_{O_2} and Z_{CO_2} are the O_2 and CO_2 concentrations (cm^3cm^{-3} of air) inside the package, P_{O_2} and P_{CO_2} are the permeabilities ($cm^3m^{-2}h^{-1}$[concentration difference of O_2 in volume fraction]$^{-1}$) of packaging material, W_p (kg) is the weight of commodity, R_{O_2} and R_{CO_2} ($cm^3kg^{-1}h^{-1}$) are the respiration rates for O_2 consumption and CO_2 evolution by the commodity, respectively, V_{fp} (cm^3) is the free volume in the package and t (h) is the storage time (Exama et al., 1993; Das, 2005; Mahajan et al., 2007; Mangaraj et al., 2009; Mangaraj et al., 2012).

Most of these models express the O_2 and CO_2 concentrations in MAP as a function of produce respiration rate, temperature, and the area, thickness, and O_2 and CO_2 permeability coefficients of the packaging material and thus help in design of MAP (Mahajan et al., 2007; Mangaraj et al., 2009). Commercial packaging films typically have CO_2/O_2 perm-selectivities of 4–8 (Al-Ati and Hotchkiss, 2003). This ratio may be larger than optimal, and perm-selectivity can be a limiting factor for certain MAP applications.

The Michaelis-Menten type equation assumes no inhibition of O_2 consumption by the evolved CO_2. But actually, the evolved CO_2 may or may not inhibit the respiration reaction depending upon the storage conditions and produce characteristics. So, Singh et al. used a model proposed by Peppelenbos and Van't Leven (1996), which is based on the combined type of enzyme kinetics relationship that incorporates both competitive and uncompetitive parts of inhibition by CO_2 and found it more appropriate for study of respiration of fresh baby corn (2012).

Complex dynamic models have also been developed to account for changes in package volume, resistance to heat flow, product respiration, and the temperature and RH of the environment and developmental changes in product physiology on head-space gases, though they are not as common as the steady state models. Web based software tools are available which help in designing MAP for fresh and fresh cut fruits and vegetables (Mahajan et al., 2007).

Proper temperature control is a major requirement for MAP system and supply chain. The produce and package characteristics respond differently with temperature, thus in MAP the rates of change of O_2 uptake by the commodity and the O_2 permeation through the film in response to any change in temperature will be different. In that situation, a suitable model can suggest a film which changes its permeability with change in temperature almost at the same rate as the change in respiration of the commodity. The package is normally designed for the maximum temperature that is to be encountered in the supply chain.

The packaging film and head space atmosphere also offer resistance to heat transfer from the commodity to the cold outside air. A concept of "safe radius" for the distance from the center of the package to the circulated air has been proposed based on the heat of respiration and the rate at which heat can be removed by the cooler air (Sharp et al., 1993).

14.6 POLYMERS USED IN FOOD PACKAGING

The MAP is generally used for respiring commodities and its basic requirement is a permeable packaging film that can develop the EMA in the shortest possible time and maintain that environment at the specific storage/transit temperature. The different types of polymers with a range of O_2, CO_2 and water vapor transmission rates are thus the most preferred packaging material for MAP. Mangaraj et al. state that the flexible plastic packaging materials comprise nearly 90% of the materials used in MAP with paper, paperboard, aluminum foil, metal and glass containers accounting for the remainder (2009).

The plastic packaging systems include individual film wrapping, horizontal and vertical form-fill-seal pillow packs, semi rigid preformed trays with lidding film systems, bag in box containers, and so on Flexible packaging is mostly preferred to rigid or semi-rigid packages for economy. It makes very efficient use of materials and space. Storage of unfilled package also requires less space. Forming packages is also rapid and simple. Main disadvantages of flexible packs are the lack of strength and the lack of convenience for the user.

14.6.1 DESIRABLE CHARACTERISTICS OF POLYMERS FOR MAP

The packaging film for MAP should allow the ingress of CO_2 and exit of O_2 through it at the rate matching the respiration of the commodity in steady state in the storage temperature. In addition, the knowledge of low O_2 and high CO_2 concentrations tolerated by the commodity is important. High barrier films can lead to anaerobic conditions and excessive CO_2 buildup, while low barrier films can result in less than optimal CO_2 concentrations.

The major factors to be taken into account while selecting the packaging materials are as follows.

- Type of package (that is flexible or rigid or semi-rigid, that is lidded tray),
- Barrier properties needed (that is permeability of individual gases and gas ratios when more than one gas is used, water vapor transmission rate, and so on),
- Fogging of the film as a result of product respiration,
- Resistance to chemical degradation, toxicity, chemical inertness,
- Physical properties such as machinability, tensile strength, puncture resistance, clarity and durability, they should have wet and dry impact strength, they should be suitable for high speed filling,
- Easy sealability and sealing reliability.

In addition to the above, the consumer appeal, transparency, printability, economic feasibility are also considered before selection of appropriate packaging material for MAP.

Packaging materials for MAP are classified according to their oxygen barrier characteristics as follows (Fellows, 2000)

1. Low barrier (more than 300 cc m^{-2})
2. Medium barrier (50–300 cc m^{-2})
3. High barrier (10–50 cc m^{-2})
4. Ultra high barrier (less than 10 cc m^{-2})

The low barrier films are used for over-wraps on fresh meat or other applications where oxygen transmission is desirable. Most polymer films are made by extrusion, though there are also other methods as calendaring and casting. In calendaring, the polymer is passed through heated rollers until the required thickness is achieved and in casting the extruded polymer is cooled by chilled rollers. The mechanical, optical, thermal and barrier properties of the plastic films can be varied by changing the film thickness, orientation of polymer molecules, amount and type of additives and by coating, lamination, and so on. Plasticizers can be added to soften the film to make it more flexible and suitable for cold climates or for frozen foods.

14.6.2 DIFFERENT TYPES OF FILMS AND THEIR CHARACTERISTICS FOR USE IN MAP

Thermo-plastic materials are those which can be repeatedly softened and melted when heated. Thermo-sets are another group of materials, which can be molded once by heat and pressure. They cannot be resoftened, as reheating causes the material to degrade. The latter is commonly not used for food packaging.

The important thermoplastic materials, which are used in food packaging are normally classified into different groups based on the polymerization process and molecular structure as follows:

- Polyolefins: polyethylene (PE) and polypropylene (PP),
- Polyvinyl group: Polyvinyl chloride (PVC),
- Condensation polymers: Polyester (for example polyethylene terephthalate or PET),
- Polyamide: (PA) (for example Nylon-6),
- Styrene polymers: Polystyrene (PS) and expanded polystyrene (EPS),
- Carbonate group : Polycarbonate (PC).

The polyethylenes can be further classified as branched polyethylenes and linear polyethylenes. The low density polyethylene (LDPE), ethylene vinyl acetate (EVA), ethylene acrylic acid (EAA) and ionomers come under the category of branched polyethylenes. The linear polyethylenes include high density polyethylene (HDPE), linear low density polyethylene (LLDPE), metallocene polymers, and so on. Many new PE types are also available, which are developed by combinations of copolymers or by controlling molecular geometry and molecular weight distribution.

The common plastic films used in MAP are low-density polyethylene (LDPE), linear low-density polyethylene (LLDPE), high-density polyethylene (HDPE), polypropylene (PP), polyvinyl chloride (PVC), poly-ethylene terephthalate (PET), polyvinylidene chloride (PVDC) and polyamide (Nylon). PVDC, PVC, PET and nylon have low gas permeability and can be used only for slow respiring commodities. However, perforating the films can extend their use to many commodities.

The three main characteristics of the common packaging films, viz. the water vapor transmission rate, oxygen permeability and CO_2 permeability are given in Table 3.

TABLE 3 Permeability property of different polymers

Plastic film	WVTR	oxygen permeability	CO_2 permeability	Water absorption
LDPE	375–500 g μm/m² d at 37.8°C, 90% RH	163,000-213,000 cm³ μm/m² d atm at 25°C	750,000–1,060,000 cm³ μm/m² d atm at 25°C	Less than 0.01%
HDPE	125	40,000-73,000 cm³ μm/m² d atm at 25°C	200,000–250,000 cm³ μm/m² d atm at 25°C	Less than 0.01%
Non-oriented PP	590 (100–300 at 25°C)	146,000		0.01–0.03%

TABLE 3 *(Continued)*

Oriented PP (OPP)	100–300	50,000–94,000	200,000–320,000	
Flexible PVC	6096	3342 cm³ μm/m² d kPa		
PVDC	7.9–240	7.9–2700 cm³ μm/m² d atm at 25°C	1250–17,300 cm³ μm/m² d atm at 25°C	0.1%
PS	1,750–3,900	98,000–150,000 cm³ μm/m² d atm at 25°C	350,000 cm³ μm/m² d atm at 25°C	0.01–0.03%
Ethylene vinyl alcohol (EVOH) co-polymers 32% ethylene	2500	4 (at 0% RH) 13 (65% RH)		
Ethylene vinyl alcohol (EVOH) co-polymers 44% ethylene	800	2.4 (at 0% RH) 45 (65% RH)		
Nylon–6	3900–4300	470–1020	3900–4700	1.3–1.9% in 24 h (for a 0.32 cm thick sample)
PET	390–510	1200–2400	5900–9800	0.1–0.2% in 24 h (for a 0.32 cm thick sample)
PC	1900–2300	110,000	675,000	0.15% in 24 h (for a 0.32 cm thick sample)

A general summary in respect of the properties of the packaging materials can be as given in Table 4. Out of these, the PVDC is the best universal barrier.

TABLE 4 General summary in respect of the properties of the packaging materials

Characteristics	Rank	Examples of plastics
Oxygen barrier	Best	EVAL, PVDC, acrylonitrile polymers
	Good	PVC, PET, PA, Polychlorotrifluoroethylene
	Poor	LDPE, HDPE, PP, PS, PC
Water vapor barrier	Excellent	HDPE, PP, PCTFE, PVDC
	Good	LDPE, PET

TABLE 4 *(Continued)*

	Fair/poor	PVC, PS, PC, PA, acrylonitrile polymers
Clarity	Clear	PVC,PET, PS,PC
	Good	PP
	Poor	HDPE
	Hazy	LDPE
Impact strength	Very good	PC, LDPE
	Good	PVC, HDPE
	Fair	PP
	Poor	PET, PS

The gas permeability coefficient is a measure of the volume of gas through a unit thickness of the film whereas the gas transmission rate refers to the volume of gas passed through the total thickness of film. For composite films, the gas transmission rate is a better parameter for comparison than permeability coefficient. Mangaraj et al. have reviewed the different methods for measurement of gas and water vapor permeabilities (2009). Nowadays, it is possible to specifically design films with required oxygen transmission rate (OTR). Normally a high surface area to volume ratio is maintained in MA packages of respiring commodities, that is the bags are oversized.

The following paragraphs discuss some other functional characteristics of the important plastic food packaging materials important for MAP considerations along with some common applications related to food packaging.

LOW-DENSITY POLYETHYLENE (LDPE)

The LDPE is the most commonly used packaging film in food industry mainly due to its chemical inertness, transparency, favorable barrier properties and cost. It is obtained by addition polymerization of ethylene. It can be sealed at lower temperatures and over a wider temperature range (121–170°C). It has good water vapor barrier properties, however, it is a poor barrier to O_2, CO_2 and many odor and flavor compounds. The gas and water vapor transmission rates in general decrease as the density of the PE increases. In LDPE low slip properties can be introduced for self stacking or conversely high slip properties for easy filling of packs in secondary packages. The cost is lower than most films and is widely used in MAP as well as in shrink and stretch wrapping. Stretch wrapping uses either thinner LDPE than shrink wrapping does (25–38 μm compared with 45–75 μm), or alternatively 17–24 μm thick LLDPE is used.

ETHYLENE VINYL ACETATE (EVA)

The EVA is prepared by controlled hydrolysis of ethylene vinyl acetate copolymer. For most food applications, VA ranging from 5 to 20% is recommended. The EVA has good flexibility (better than LDPE), excellent toughness at low temperatures and heat sealability. As the melting temperature is low, it has to be processed at relatively lower temperatures. The film has excellent transparency and in comparison to LDPE, it has higher permeability to water vapor and gases. Due to the very good heat sealing and adhesive properties, it is used in extrusion coating and as a coextruded heat seal layer. It is very widely used for MAP, and frequently used as a component of the sealant layer in both lidding and base films. EVA and LLDPE are the major stretch wrap materials. Other applications of EVA include aseptic packaging, packing of yoghurt, ketchup, and so on.

ETHYLENE ACRYLIC ACID (EAA)

FDA regulations permit use of up to 25% acrylic acid for copolymers of ethylene in direct food contact. The EAA flexible films have similar chemical resistance and barrier properties as LDPE. However, these have better strength, toughness, hot tack and adhesion than LDPE. As the content of acrylic acid increases, the crystallinity decreases and clarity increases. It is used in blister packaging and as an extruded tie layer between aluminum foil and other polymers. EAA films are also used in flexible packaging of meat, cheese, snack foods, skin packaging and in adhesive lamination.

IONOMERS

Ionomers are obtained by neutralization of EAA or a similar copolymer, for example EMAA (ethylene methacrylic acid) with cations such as Na^+, Zn^{++}, Li^+, and so on. The American company DuPont uses the trade name Surlyn for these materials. They have better transparency, toughness, flexibility and higher melt strength than the un-neutralized polymer. The barrier properties of ionomers are not satisfactory, and hence these are combined with PVDC, HDPE or aluminum foil to improve barrier properties. They stick very well to aluminum foil. They also tend to absorb water readily. The heat seal characteristics are excellent.

Ionomers have good formability, toughness and visual appearance. In addition to MAP they are also used for frozen foods, cheese, snack foods, wine, water, and so on. The cost is relatively higher compared to films such as EVA.

HIGH-DENSITY POLYETHYLENE (HDPE)

The HDPE is less transparent than LDPE and has a cloudy appearance. A small amount of white pigment is commonly added to HDPE to have an opaque white film. It is stronger, thicker, less flexible and more brittle than LDPE. It is a good barrier of water vapor and gas. It has a higher softening temperature (121°C) and hence it can be sterilized by heat.

Medium density polyethylene (MDPE), which lies between HDPE and LDPE, is mechanically stronger than LDPE. LDPE is coextruded with MDPE to have good sealability as well as strength.

The HDPE is used for packaging of milk, juice, snack foods and many other food applications including shrink wrapping. It can be used in MAP in spite of its low O_2 and CO_2 permeabilities. Foamed HDPE film is thicker and stiffer than conventional films and has dead folding properties. It can be perforated with up to 80 holes cm^{-1} for use with fresh foods and bakery products (Fellows, 2000)

LINEAR LOW-DENSITY POLYETHYLENE (LLDPE)

The LLDPE is a flexible and soft material with less clarity and gloss than LDPE. It has highly linear arrangement of molecules and the distribution of molecular weights is smaller than that for LDPE. So it has higher impact strength, tensile strength, puncture resistance and elongation than LDPE of same density and thickness. Like LDPE, it is a good water vapor barrier, but a poor barrier to O_2, CO_2 and many odor and flavor compounds. The melting point of LLDPE is 10–15°C higher than that of LDPE.

The main applications of LLDPE include stretch/cling film. The cling properties of both LDPE and LLDPE are set on one side to help adhesion between layers of the film and to reduce adhesion with adjacent packages. LDPE and LLDPE are often blended to get the advantages of both. Each of them can be used in blends with EVA to improve strength and heat sealing.

METALLOCENE POLYMERS

These polymers are based on metallocene catalysts, in which metallocene, or single site catalysts (replacing Ziegler-Natta catalysts) provide a way to produce PE with very narrow molecular weight spreads. Metallocene polymers are somewhat harder to process but are tougher and provide stronger heat seals than conventional PE. The films have in general very high OTR, low WVTR, and good clarity. Metallocene PE can have up to 40% greater tensile strength, up to 10 times higher impact strength and better hot tack, it gives better heat seals than LLDPE. Cyclic olefin copolymers (COC) are copolymers of ethylene and norbornene made by using metallocene catalysts. They have excellent clarity, strength and water vapor barrier properties, they can be used in food packaging, though the use is not very common (Selke et al., 2004).

POLYPROPYLENE (PP)

The PP is produced by addition polymerization of propylene. The oriented PP (OPP) is a clear glossy film and has good optical properties. It has high tensile strength and good puncture resistance. PP has the lowest density and highest melting point of all the high volume usage thermoplastics and has a relatively low cost. The density of PP is 0.89–0.91 g/cm^3. The capability of more stiffness and ease of orientation make PP homopolymer suitable for stretch applications. Its melting point is high and hence can be used in situations where there is a need for sterilisation of packed food. The sealing temperature of OPP is 145°C. PP is chemically inert and resistant to most commonly found chemicals, both organic and inorganic. It is a barrier to water vapor and has oil and fat resistance. PP is not subject to environmental stress cracking (Kirwan and Strawbridge, 2003).

PP and OPP are used for bottles, jars, crisp packets, biscuit wrappers, and boil-in-bag films and for many types of commodities like biscuits, crisps (chips), snack foods, chocolate, sugar confectionery, ice cream, frozen food, tea and coffee (Manikantan et al., 2012). OPP is used for packaging of salad cut vegetables. In many MAP applications, biaxially oriented film (BOPP) is preferred which is stronger than OPP.

POLYVINYL CHLORIDE (PVC)

The PVC films are formed by combining PVC resin, produced by addition polymerization of vinyl chloride, with plasticizers and other additives. The films are generally soft and flexible and have excellent self-cling property, excellent resistance to chemicals, resilience and have good clarity. They are tough and easy to seal by heat. The sealing temperature of flexible PVC is 100–160°C. The permeability is relatively high.

PVC films are available in both oriented and non-oriented form. Some of these (for example plasticized PVC) have high CO_2 to O_2 permeability, thereby making these suitable for MAP. The flexible PVC film is also used for wrapping fresh fruits and vegetables and for packaging fresh red meat.

POLYVINYLIDENE CHLORIDE (PVDC)

The PVDC is an addition polymer of vinylidene chloride. The PVDC copolymer is a good barrier of moisture, odor and gases. It is fat resistant and heat sealable. In food packaging, the PVDC resins are used as barriers against moisture, gases, flavor and odor, either in a single layer or multi layers. Monolayer films are used in wrapping household items. They are also used in semi rigid thermoformed containers. The structures usually containing 10–20% of PVDC copolymer are commonly used as shrinkable films. Generally the best barrier films do not have the best heat sealability and vice versa. So two differently formulated PVDC copolymer coatings are applied when both heat sealability and barrier properties are desired. PVDC is also used as a coating for films and bottles to improve the barrier properties.

Polyvinyl chloride-vinylidene chloride copolymer is very strong and is used as thin films. It has very low gas and water vapor permeabilities, is heat shrinkable and heat sealable. However, it has a brown tint for which it is sometimes not liked for food applications.

POLYSTYRENE (PS)

It is an addition polymer of styrene. It is clear and has a high tensile strength. PS alone is brittle, but can be blended or generally biaxially oriented to achieve required properties. It has high gas and moisture permeability and hence can be used for highly respiring commodities. It may be oriented to improve the barrier properties. In heavier gauges, PS is used for transparent thermoformed trays.

ETHYLENE-VINYL ALCOHOL (EVOH)

Ethylene-vinyl alcohol is a copolymer of ethylene and vinyl alcohol. The film is excellent barrier to gases (especially O_2), odor and fl, od, hence it is most often used as an oxygen barrier. However, the hydrogen bonds make it a moisture-sensitive material. Its barrier properties decrease at high humidity.

POLYAMIDE (NYLON)

Nylons are condensation polymers, linear thermoplastic polyamides that contain the amide group as a recurring part of the chain. They have good mechanical properties over a wide temperature range (-60°C–200°C). They are excellent odor and flavor barriers, reasonably good oxygen barriers and very poor water vapor barriers. Normally they tend to lose some barrier performance when exposed to high humidity. They require high temperatures to form a hot seal. As the cost is higher, nylon films are used for specialty applications. They are often coextruded with other plastics, to reduce cost and gain strength and toughness. Co extrusions with polyolefins provide heat sealability and moisture barrier. Ionomers and EVA are also used. PVDC copolymer coating on nylons improves barrier properties.

Nylon-6 tends to be the most used nylon packaging film. MXD6 is a special nylon that has better gas barrier and thermal properties than Nylon 6 and has better moisture resistance. Usually nylons are used in similar situations as PET. Multilayer films containing a nylon layer are used in vacuum packaging of meat and other products.

POLYESTERS

Polyethylene terephthalate (PET), polycarbonate and polyethylene naphthalate (PEN) are polyesters, which are polymers containing ester linkages. The most commonly used polyester in food packaging is PET. It is commonly used in biaxially oriented form. It is a transparent glossy film and has excellent strength and good moisture and gas barrier properties. The barrier properties of PET can be further improved by coating with PVDC. PET is flexible at temperatures from -70°C to 135°C and undergoes little shrinkage with variations in temperature and humidity. Amorphous PET is clear and used for bottles and films. Crystalline PET is opaque and used for microwavable trays and semi-rigid containers.

POLYCARBONATE (PC)

The PC is a mostly amorphous polymer that has very good clarity with a very slight yellowish tinge. It is very tough and rigid and has good impact strength. The heat resistance and low temperature performance are also good. It has good resistance to water, fruit juices and oil. PC films are used in skin packaging and for food packaging where there is requirement of treating at high temperatures.

FLUOROPOLYMERS

Fluoropolymers are a family of polymers containing C-F bonds. Polytetrafluoroethylene (PTFE) is commonly known as Teflon. Polychlorotrifluoroethylene (PCTFE) is an extremely good water vapor barrier, a good gas barrier and is highly inert. PCTFE is used as a component in a laminated structure especially for moisture sensitive drugs. Their application in food is limited due to the high cost.

14.6.3 BIODEGRADABLE POLYMERS

Biodegradable films are also gaining interest in the food packaging sector with the objective to reduce the adverse pressure on the environment. Biodegradable polymers are generally made from cellulose and starches. The raw materials can be agricultural feed stocks, animal sources, food processing wastes or microbial sources.

Cellulose-based plastics such as cellulose acetate, cellulose butyrate, cellulose propionate and copolymers are used in food packaging to a relatively smaller extent, most often as sheet rather than film. Their high price and water sensitivity limit their use.

Plain cellulose is a glossy transparent film which is odorless, tasteless and biodegradable in approximately 100 days. It is tough and puncture resistant though it tears easily. However, it is not heat sealable and the dimensions and permeability of the film vary with humidity. It is used for films that do not require complete gas or moisture barrier. Cellophane is the most common cellulose-based biopolymer.

The starch-based polymers include amylose, hydroxyl-propylated starch and dextrin, polylactides (PLA), polyhydroxyalkanoate (PHA), polyhydroxybuterate (PHB), and a copolymer PHB and valeric acids (PHB/V). In addition, biodegradable films can also be formed from chitosan, which is derived from the chitin of crustacean and insect exoskeletons. Chitin is a biopolymer with a chemical structure similar to cellulose. A biodegradable laminate of a chitosan-cellulose has proved to be a suitable packaging material for MAP and storage of broccoli (Yoshio and Takashi, 1997).

Polyhydroxybutyrate/valerate (PHBV), one of the truly biodegradable non-cellulose based plastics, is a bacterially grown polyester with properties similar to PP. It was sold under the name Biopol. PHPB is one member of family of polyhydroxyalkanoates (PHAs) produced by microbes from sugars or other biobased materials. Metabolix PHAs are reported to be better water vapor barriers than most biodegradable plastics.

Del Nobile et al. used four different films, viz. two polyester-based biodegradable films, a multi-layer film made by laminating an aluminum foil with a PE film and an OPP film to compare their performance on storage of minimally processed lettuce at 4°C. Results suggest that the biodegradable films offer a longer shelf life than the PP package and that the gas permeability of the films is important in determining the quality of the packed produce (2008a).

The use of polylactides is gaining interest in food packaging. The lactic acid is obtained from corn or other biobased materials by a fermentation process and then chemical synthesis is used to produce the polymer from lactic acid or lactide monomers. Many companies are now producing PLA. It does not degrade when exposed to moisture. After the original use, the polymer can be hydrolyzed to recover lactic acid. Shoji and Shi compared the storage quality of green peppers using PLA-based biodegradable film packaging with LDPE and perforated LDPE film packaging and found that the biodegradable film with higher water vapor permeability could maintain the quality and sanitary conditions of freshly harvested green peppers (2007). Highbush blueberries could be stored for 18 days at 10°C and for 9 days at 23°C in PLA containers. The studies on physicochemical and microbiological qualities of the product indicated that PLA containers could prolong shelf life of blueberry at different temperatures (Almenar et al., 2008).

Edible films are another form of biodegradable polymer. These films act as moisture and aroma barriers and protect the food's mechanical integrity or handling characteristics. Emphasis is also being given for development and use of protein based plastics, polysaccharides, and wood derived plastics. The higher cost of the biodegradable plastics than petroleum-based polymers has been the major factor for their limited use in food industry.

14.6.4 CONTINUOUS FILMS AND PERFORATED FILMS

In the MAP, there are two strategies for maintaining the desired rates of gas and water vapor permeation. Either, there may be a continuous film which controls movement of gases and water vapor into or out of the package through the micro perforations in the structure of the polymer, or the film may have small holes or macro perforations. The relative permeabilities to gases and water vapor differ substantially between continuous and perforated films. In both types, the rates of O_2 and CO_2 exchanges are usually directly proportional to the differences in gas concentrations across the film. The EMA conditions are attained when the O_2 uptake by the commodity equals the O_2 permeating into the package and the CO_2 produced by the commodity is taken out. Of course this situation will happen only when the rate of respiration of the commodity is constant.

It is earlier mentioned that in continuous films packages, the permeability of the package to CO_2 is usually 4–8 times that of O_2 permeability. A common example is LDPE, where CO_2 permeability is four times that of O_2. They can be used for less CO_2 tolerant commodities such as banana, mango, apples, and so on. In continuous films, the gas permeability is temperature dependent and is described by Arrhenius-type equations. In perforated films, the perforations serve as parallel paths for movement of gas and moisture. The rate of gas transfer is the sum of gas movement through the perforations and that through the film. For such films, an apparent permeability term is used, which depends on the number and size of perforations as well as the continuous film area and its permeability. The equation which gives the permeability of macroscopic films with equal size perforations is as follows (Mahajan et al., 2007)

$$P^a = \left[P + \frac{\pi R_h^2 \, 16.4 \times 10^{-6}}{x + R_h} N_h \right] \tag{4}$$

where P^a is the apparent permeability of the macroscopic perforated film, R_h is the radius of the perforations, N_h is the number of holes, x is the film thickness and P is the permeability of the non-perforated film.

The macroperforations provide a high rate of gas exchange, a 1-mm perforation in a 0.0025 mm (1 mil) thick LDPE film has nearly the same gas flux as ½ m^2 film area (Fonseca et al., 2002a). When a continuous film is perforated, the O_2 flow increases by about 40-fold more than it increases the flow of water vapor. Thus such films can reduce the moisture loss without the chance of anaerobiosis.

In perforated films the ratio of CO_2 to O_2 permeability is almost close to unity. Therefore they are used for MAP of highly respiring commodities and fresh cut produce. In normal situations the rate of O_2 consumption by a commodity is almost same as the production of CO_2, and hence, the CO_2 gradient will be much lower than the O_2 gradient. So a LDPE package having an EMA of 6% O_2 will roughly have (21-6)/4 or 3.75% CO_2 level. However, in a perforated film, the CO_2 level will increase by the same extent as the reduction of O_2 level (so that the sum of O_2 and CO_2 partial pressures is usually in the range of 18–20%).

Micro-porous films with very high permeabilities have been developed and are also available commercially. Films using micro perforations, where the diameter of micro perforations generally range from 40 to 200 μm, can offer very high gas transmission rates (Sandhya, 2010).

These films can be prepared either by perforating with very minute tools or by incorporating some additives during the film manufacture which make the film porous. Inert inorganic minerals such as crushed calcium carbonate ($CaCO_3$) or talc or SiO_2 are the common additives used for the purpose. The type of filler material, particle size and the degree of stretching affect the gas exchange characteristics through the film. Ceramic filled LDPE films have higher O_2, CO_2 and C_2H_4 permeabilities, higher CO_2 to O_2 permeability ratio, and higher C_2H_4 to O_2 permeability ratios as compared to the conventional LDPE films (Lee et al., 2006).

The effect of change in temperature on the diffusion of gases is more pronounced in the continuous films than that in the perforations. The O_2 permeation through continuous LDPE film and perforations increases by 200% and 11%, respectively, when the temperature increases from 0–15°C (Beaudry, 2000). A combination of perforation and continuous films can be used to obtain permeability values in between those of the continuous and perforated films.

Perforations may also enable MAP for produce that is sensitive to even small changes in concentrations of O_2, CO_2 and C_2H_4.

There is another category of modified packages known as perforation mediated packaging or diffusion channel system, which is basically a rigid impermeable container, into which tubes are inserted for gas exchange. Thus, only the tubes serve for the development of MAP within the container. They are used for bulk packaging and for commodities sensitive to mechanical damage (Montanez et al., 2005; Mahajan et al., 2007).

14.6.5 MODIFICATION OF FILM CHARACTERISTICS

The gas and water vapor permeabilities and other characteristics of individual films differ widely and different commodities also differ in their requirement, hence the individual films often do not meet the requirement of MAPs for individual commodities. Therefore, the film properties are modified by combining the individual films with one another or with materials such as paper or aluminum through processes such as coextrusion, lamination, coating and metallization.

When a modified film is prepared by blending two or three different polymers, it is designed in such a way that each polymer performs a specific function such as strength, transparency and improved gas transmission to meet the desired characteristics. For instance, LDPE and LLDPE may be blended to give a better stretch film. LDPE and LLDPE can be used in blends with EVA to improve strength and heat sealing. A considerable research has been carried out on the properties of blends. In some applications, non-plastic substances are incorporated to reduce the cost and/or improve the performance, though they are not very common in food packaging (Mangaraj et al., 2009).

COEXTRUSION

In coextrusion, two or more layers of molten plastics are combined during the film manufacture. The copolymers used in coextruded films should have similar chemical structures, flow characteristics and viscosities when melted. The method can incorporate very thin layers of a material, more than that can be achieved by lamination. Coextruded films offer good barrier properties, similar to multilayer laminates. They are produced at a lower cost than laminates, they are also thinner than laminates, which makes their use easy on forming and filling equipment. The layers also do not separate. Therefore, if buried printing is not the objective, coextrusion is generally preferred to lamination. Blown film coextrusion and flat sheet coextrusion are the two main methods for producing co extrusions. The advances in coextrusion technology, coupled with single site catalyst-based plastic resins, have led to development of films with better gas and water vapor exchange and sealing characteristics.

Recent developments also include polymer based nanocomposite films. Many nanocomposite food packages are either already in the marketplace or being developed. The majority of these are targeted for beverage packaging (Brody et al., 2008). Improvements in fundamental characteristics of food packaging materials such as strength, barrier properties, antimicrobial properties, and stability to heat and cold are being achieved using nanocomposite materials. Polymer nanocomposites consist of resins (either thermoset or thermoplastics) and nano fillers which enhance the strength, stiffness, dimensional stability, optical properties, heat resistance and barrier properties (Sinha and Okamoto, 2003). Manikantan and Varadharaju (2011) observed a reduction of OTR, WVTR and increase in the tensile strength and percent elongation with decrease in the amount of both nanoclay and compatibilizer and increase of film thickness of PP based nanocomposite films. In a recent study, PP resin packaging material of 100 μm thickness consisting of 5% compatibilizer and 2% nanoclay and of 120 μm thickness having 10% compatibilizer and 4% nanoclay were used effectively for storage of banana chips (Manikantan et al., 2012).

LAMINATION

Lamination is the process of combining two webs of films together. In flexible packaging applications, a plastic film is combined usually with another film, paper or foil with the help of water, solvent-, or solid-based adhesive. LDPE is a common laminating adhesive, which is applied by extrusion. This process is known as extrusion laminating.

Lamination improves the barrier properties of the plastic. Lamination with paper is done to have strength and printability. Laminations enable reverse printing, in which the printing is buried between layers and thus not subjected to abrasion and can add or enhance heat sealability. Lamination using laser rather than adhesives are also used for thermoplastics.

Many types of laminated films and coextruded films have been found out to be effective in maintaining the modified atmosphere and extending the shelf life of commodities (Jacomino et al., 2001; Sra et al., 2011; Mangaraj et al., 2012). Common laminates used in food packaging are nylon-LDPE, nylon-PVDC-LDPE and nylon-EVOH-LDPE for non-respiring products. The nylon gives strength to the pack, EVOH or PVDC provides the gas and moisture barrier properties and LDPE gives heat sealability. PVC and LDPE are also commonly used for respiring products. There are many other examples of laminated films, for successful lamination the two films should have similar characteristics and the film tension.

It is difficult to recycle the films obtained by coextrusion and lamination. However, the amount of packaging materials is considerably reduced in case of coextruded and laminated films, and thus the environmental impact is also reduced (Sandhya, 2010; Mangaraj et al., 2009).

COATING

Coating of a thin layer of plastic on the surface of another plastic film or on a non-plastic substrate such as paper, cellophane or foil can improve the barrier properties. Coating also helps to avoid direct contact of the base material with the product, helps to impart heat sealability for plastics that are not heat sealed easily and protects the base materials as paper or cellophane from moisture. The coating is usually in the form of a solution, a suspension or a melt. Normally PVDC copolymer coatings are used to improve the barrier property and heat sealability.

Nitrocellulose, when coated on one side of cellulose film, can act as a moisture barrier, but does not affect O_2 permeability. A nitrocellulose coating on both sides of the film improves the barrier to oxygen, moisture and odor and enables the film to be heat sealed when broad seals are used. A PVC coating is applied to cellulose using either an aqueous dispersion (MXXT/A cellulose) or an organic solvent (MXXT/S cellulose). In each case the film is made heat sealable and the barrier properties are improved (Fellows, 2000).

METALLIZATION

Application of a thin metal layer on a plastic film can improve its barrier properties. In food packaging applications, aluminum is the most commonly used metal and usually low O_2 barrier films are metalized. Metallization makes the film good barrier to gases, moisture, odor and light. Metalized film is less expensive and more flexible than foil laminates, having similar barrier properties. A coating of vinyl chloride and VA gives a stiffer film with intermediate permeability. Metalized polyester has higher barrier properties than metalized PP, but PP is less expensive.

ADDITIVES TO IMPROVE THE POLYMER CHARACTERISTICS

The polymer characteristics like resin's stability to oxidation, impact resistance, hardness, surface tension, flame resistance, and so on can be improved by incorporating some additives with the resin during film manufacture. Additives can also facilitate extrusion and molding, controls blocking and reduces cost.

Several antioxidants, heat stabilizers, UV stabilizers, anti-slip agents, as well as color pigments are commonly added to polymers. Besides, antifogging agents, nucleating agents, antistatic agents, plasticizers, oxygen scavengers, antimicrobials are also used as additives, which have facilitated the development of active food packaging technologies. Clearance from regulating authorities dealing with food safety is necessary for using any such additive in packages for food products.

ANTIOXIDANTS

In food packaging, three resins, namely, PP, PE and high impact polystyrene (HIPS) have the maximum use of antioxidants. A combination of hindered phenol and phosphite antioxidants is used for PP. BHT, a phenolic antioxidant, is commonly used for LDPE. For HDPE and LLDPE, antioxidants less volatile than BHT, such as polyphenols, at higher concentrations, are used in combination with phosphites (Selke et al., 2004). α-tocopherol (vitamin E) is sometimes used as an antioxidant for polyolefins.

ANTIFOGGING AGENTS

Condensation of water molecules on the inner surface is often a problem for packages containing high moisture commodities. In those situations, antifogging compounds can be coated on the surface of the material or incorporated in the material during manufacture. Glycerol, sorbitol stearate, fatty alcohols, and ethyloxylates of nonylphenols are used as antifogging additives.

DESICCANT

Absorption of moisture is often required in MAP. It helps to prevent undue rise in humidity around the commodity and condensation. The water sensitive polymers can be protected by layers of desiccants, for example EVOH may be protected by PP. Incorporation of a desiccant in the tie layer between the EVOH and the main structural polymer in the package is also done (Selke et al., 2004).

Additives are also used in films for scavenging of ethylene and other compounds developed within the package space.

OXYGEN SCAVENGERS

Oxygen scavengers can be either in the form of coatings on the inner surface of the packages or may be added to the plastics during manufacture. Iron oxide is a commonly used oxygen scavenger. Metal free absorbents include mixtures of organic compounds such as phenolic, glycols, and quinones. Most of these are activated by moisture.

Oxygen scavenger films are also available in which scavenging activity is triggered by a UV light source, so that the initiation of scavenging activity is also under control. Another example of an oxygen scavenger consists of an oxidizable layer of MXD6 nylon, in amounts of 1–5% along with 50–200 ppm cobalt salt as a catalyst to scavenge oxygen permeating through the PET (Selke et al., 2004).

ANTIMICROBIALS

Most synthetic polymers are non-biodegradable, but there is chance of microbial attack if some low molecular weight additives are incorporated with them. In such situation antimicrobials may be added. However, in any case in food packaging, it is more important to protect the food than the polymer from microorganisms.

14.6.6 PERMEABLE LABELS FOR MAP SYSTEMS

Labels of different permeabilities can be fixed on the packages made of low permeable materials like rigid PVC to develop the desirable MA. New technologies allow development of films with very high OTR (greater than 15000 cc.m^{-2} day^{-1}). They can also be customized to provide almost any desired OTR, the OTR can be changed simply by varying the thickness of films, though the use of a very thin film creates problems in automated packaging machinery. The area of the label is another factor which controls the EMA within the chamber. Such films/ labels have found extensive applications for fresh cut fruits and vegetables.

Silicone membrane technology is another innovative technology in which a controlled ventilation system regulates gas levels of storage environment by relying on selective gas permeation (Raghavan et al., 1982; Gariepy et al., 1984; Naik, et al., 2007). The membrane material has the ability to permit selective passage of gases at different rates according to their physical and chemical properties, which is fixed on a rigid package for creating the modified atmosphere. The main advantages of silicone membrane system includes lower operating cost due to fewer controls and less maintenance during operation, lower refrigeration cost attributable to lower respiratory activity, membrane's high permeability to ethylene and low permeability to water vapor (Naik et al., 2012). The membrane area required is based on the level of CO_2 desired in the storage chamber. Raghavan et al. (1982) has proposed the following formula for calculation of the membrane area.

$$Area = \frac{RRxM}{P_{CO_2}\, xCO_2} \tag{5}$$

In the above equation, Area is the area of the silicone membrane (m^2), RR is respiration rate of the product stored under MA condition (liter of CO_2 $kg^{-1}day^{-1}$), M is mass of stored produce (kg), P_{CO2} is permeability of the silicone membrane to CO_2 (liter $day^{-1}m^{-2}$ atm^{-1}), and CO_2 is desired CO_2 partial pressure difference across the membrane (atm).

14.7 ENVIRONMENTAL PARAMETERS AFFECTING MODIFIED ATMOSPHERE PACKAGING

The external parameters like the temperature, RH, light, and so on affect the quality of commodity in storage as the package does not completely isolate the commodity from external influences. The characteristics of polymer films also change with time, temperature, humidity, and other environmental conditions.

14.7.1 TEMPERATURE

The modification of atmosphere in MAP depends upon several parameters such as film permeability to O_2 and CO_2, product respiration and the influence of temperature on both of these processes. The metabolic processes such as the rates of respiration and ripening are influenced by temperature. It is generally agreed that biological reactions increase 2–3 folds with every 10°C rise in temperature. The temperature also significantly affects film permeability and thereby the O_2 and CO_2 concentration in the package. The change in CO_2 permeability responds more than that of O_2 permeability. All these responses directly affect the storability of the commodity.

The rate of respiration increases at higher temperatures and thus the desired package conditions is attained in a shorter time (Singh et al., 2012), however, it also increases the physiological and biochemical processes within the food. Moreover, with increase in temperature, the produce demand for O_2 is increased, but the resultant increase in film permeability cannot keep pace with the increased O_2 demand, which cause a declined oxygen level. In perforated packages, this effect is more obvious because there is only a small increase in O_2 movement through perforations with increase in temperature. The effect of surrounding temperature on the respiratory parameters is evaluated mostly by using the Arrhenius relationship (Fonseca et al., 2002a; Benkeblia 2004; Charles et al., 2005; Kaur et al., 2013; Singh et al., 2012).

Singh et al. also indicate that all enzyme kinetics parameters viz. maximum oxygen consumption rate, and Michaelis-Menten constants for oxygen consumption, competitive inhibition of O_2 consumption by CO_2 and for uncompetitive inhibition of O_2 consumption by CO_2, and so on depend on temperature and minimal processing (2012). The rate of respiration, wound induced ethylene production, water loss, microbial growth, and so on are at a high level for fresh cut commodities and proper temperature management is very essential for such produce. A side-chain polymer technology that allows the film OTR to increase rapidly as temperature increases has also been developed. These polymers also have an adjustable CO_2/O_2 permeability ratio, and a range of vapor transmission rates.

At low temperatures, it takes longer time to attain EMA in the package and sometimes the produce may spoil before the steady-state conditions are reached. Initial purging of desired gas mixture into the package is recommended in such situations.

Fluctuations in the storage and transit temperatures are also quite normal in the fruits and vegetables supply chain, which do not allow the development of EMA in recommended time. Jacxsens et al. designed equilibrium MA packages for fresh-cut vegetables such as bell pepper, broccoli, carrots, chicory, cucumber, French beans, iceberg lettuce, mixed lettuce, mung bean sprouts, which would develop desired behavior in response to change in storage temperature, thus can be adopted in a distribution chain where temperature changes are expected (2000).

14.7.2 RELATIVE HUMIDITY

For fresh high moisture commodities, low RH in the surroundings increases the rate of transpiration and causes desiccation. High humidity in the package space causes condensation on the film that is driven by temperature fluctuations.

Condensation on the inner surface of the film is a common problem with MAP and it severely affects the quality of highly respiring commodities. Condensation takes place due to temperature fluctuation in the storage space, a temperature drop of only 0.2°C can cause condensation and temperature fluctuations in cold stores in this range is quite normal. Tano et al. observed that the quality of mushroom, broccoli, or mature green tomatoes in MAP was severely affected by storage temperature fluctuation (2007).

There are film surface treatments which result in droplet dispersion, so the condensing water forms a thin, uniform layer which is virtually invisible. Condensation can also be reduced in MAPs by adding some additives as salt in the package.

14.7.3 LIGHT

Some packages require display of the contents and entry of light into them is allowed, but many commodities including the fresh cut vegetables are susceptible to deterioration by light. There may be oxidation of lipids and degradation of pigments, for green vegetables light may enhance the consumption of CO_2 and produce O_2 through photosynthesis. Light should be prevented from entering in to these packages.

Some materials (as LDPE) transmit both visible and UV light to similar extent, whereas some others (for example PVDC) transmit visible light, but absorb UV light. Modification of the polymer films by incorporating pigments, overwrapping with polymer labels and printing the packages help to reduce the light transmission into packages. The other common practice is to keep the clear packages in fiber board boxes so that the primary packages are protected from light.

14.7.4 SHOCK AND VIBRATION

Shock and vibration may cause damage to cells, causing an increase in respiration. It may also lead to release of enzymes, which cause browning.

14.8 APPLICATIONS OF MODIFIED ATMOSPHERE PACKAGING

Examples of MAP products include raw and cooked meats, poultry, sea food, vegetables, fresh pasta, cheese, bakery products, sandwiches, sous vide foods, potato crisps, coffee, and tea, part baked bread, and so on. However, horticultural products are a main application for MAP, for which the MA reduces respiration rate, ethylene production, sensitivity, and texture losses, improves chlorophyll and other pigment retention, delays ripening and senescence and reduces the rate of microbial growth and spoilage (Rodriguez-Aguilera and Oliveira, 2009).

14.8.1 MAP OF FRESH FRUITS AND VEGETABLES

In recent years, several studies have confirmed the advantages gained by MAP on a variety of fruits as peach, melon, pumpkin, papaya, grapes, litchi, guava, strawberry, and many others (Baskaran et al., 2001; Aguayo et al., 2003; Pereira et al., 2004; Zhang et al., 2005; Deng et al., 2005; Azene and Nanos, 2005; Steiner et al., 2006; Berna et al., 2007; Sivakumar and Korsten, 2006; Sandhya, 2010; Azene et al., 2011; Sabir et al., 2011).

MAP of minimally processed fruits and vegetables has been an important area of development with the increase in consumer demand for such produce. Other terms that refer to minimally processed products are "lightly processed," "partially processed," "fresh processed," and "preprepared". The fresh cut processing increases respiration rates and causes major tissue disruption as enzymes and substrates get mixed. It also increases wound induced ethylene production and water activity, and exposes more surface area for the reactions, all these lead to quick spoilage under normal atmosphere. Discoloration of cut surfaces due to enzymatic browning, yellowing of green vegetables, and pale color of bright vegetables are the main defects of stored fresh-cut produce (Del Nobile et al., 2006). A low temperature regime, although insufficient alone, has been the principal way of overcoming this challenge so far. However, the use of low O_2 (1–5%) and high CO_2 (5–10%) concentrations, in combination with storage at low temperatures can maintain sensory as well as microbial quality of fresh-cut vegetables in a more effective way (Fonseca et al., 2002a). Sandhya has made a comprehensive review of recent developments in the application of MAP for different fruits and vegetables including the minimally processed products (2010).

PE packaging provides favorable atmospheres for improving the shelf life of apples even in ambient conditions (Geeson et al., 1994), it inhibits growth of P. expansum, thereby allowing patulin to be produced, regardless of gaseous environment (Moodley et al., 2002). Rocha et al. stored apples in PP (100 μm) for 6.5 months at 4°C and 85% RH and found that the MA stored apples had better color, firmness, and reduced weight loss than those stored in air (2004).

The use of plastic bags with O_2 permeability of 15 $cc.m^{-2}bar^{-1}.24\ h^{-1}$ and with initial atmosphere of 0 kPa O_2 extended the microbiological shelf life of Conference pear cubes for at least 3 weeks of storage (Soliva-Fortuny et al., 2005). The pear fruits could be stored under modified atmosphere in CFB boxes lined with HDPE liners for 60 days at 0–1°C and 90–95% RH with minimum loss in firmness (Kaur et al., 2012; Kaur et al., 2013).

The Tian et al. found that MAP was beneficial for storage of litchi as the high humidity prevents water loss and browning of litchi pericarp (2005). Sivakumar and Korsten observed that MAP of litchi fruits using BOPP film after suitable post harvest treatment minimized the rate of transpiration, and reduced weight loss and deterioration of food quality (2006). Artes-Hernandez et al. studied the quality of superior seedless table grapes under MAP using micro perforated and OPP films and reported that SO_2-free MAP maintained the quality of grapes close to the harvest conditions (2006).

Hailu et al. observed that modified atmosphere created by the polyethylene bags reduced weight loss of banana. The type of packaging significantly ($P \leq 0.05$) affected the physiological loss in weight, peel color, peel and pulp thickness, pulp to peel ratio, pulp firmness, dry matter of the pulp, decay loss and marketability of banana and recommended the use of HDPE and LDPE bags for extending the shelf life and maintaining the quality attributes of banana fruits (2012).

The polyethylene bag packaging combined with evaporatively cooled storage maintained the superior quality of papaya fruit for a period of 21 days (Azene at al. 2011). Guava packed with PVC, LDPE or PET and stored at 5 and 8°C hindered the change in peel color and the loss of firmness (Jacomino et al., 2001; Pereira et al., 2004). MA packaging of fresh guava in PET film had a strong influence on color preservation and weight loss of the guavas (Pereira et al., 2004). Jacomino et al. observed that multilayer coextruded polyolefin film with selective permeability (PSP) could prolong storage of guava up to 3 weeks, while LDPE film with incorporated minerals (LDPEm) was suitable for guava storage at 10°C with 85–90% RH. The PSP film and LDPE film with mineral incorporation could keep the fruit with good organoleptic characteristics for 28 days and 14 days, respectively (2001). Mangaraj et al. also observed that the MA packaging system increased the shelf life of guava by 128–200% compared to the unpacked fruits at various storage temperatures with a quality comparable with the freshly harvested commodity (2012).

The MAP has also proved very effective for a variety of vegetables, in whole as well as cut and shredded form in different types of package materials. Cucumber packed in perforated 31.75 μm LDPE bags had less chilling injury than unwrapped fruit in storage at 5°C and 90 to 95% RH (Wang and Qi, 1997). Karakas and Yildiz found that MAP increased the tissue hardness of cucumber for the first 3 days of storage, however the chilled tissues started to soften on the 6th day (2007). Cucumber stored in MAP having 2 perforations (0.33 mm dia each) combined with cold room condition (4 ± 1°C and 90 ± 2% RH) maintained good in-pack modified atmosphere and firmness, fruit weight, color, and sensory characteristics in acceptable limit (Manjunatha and Anurag, 2012).

Liu and Li studied the microbial proliferation and sensory quality aspects of sliced onions at different temperatures (2, 4 and 10°C) and atmospheric conditions (with or without 40% CO_2 + 59% N_2 + 1% O_2) and observed extended microbial shelf life in the MA (2006). The shelf life of the multiplier onion in the peel form could be increased from 4–5 days to 14 days in MA developed by the combined effect of silicone membrane (6 $cm^2 kg^{-1}$) and low temperature (5 ± 1°C) (Naik et al., 2012).

Del Nobile et al. observed enhanced storage of minimally processed lettuce in MAP at 4°C and found that polyester-based biodegradable films offer a longer shelf life than the PP package and that the gas permeability of the films was important in determining the quality of the packed produce (2008a). The storage life of fresh ginseng (*Panax Ginseng*) increased significantly by using film packages with low gas permeability coefficient, a shelf life of 5 months could be obtained at an O_2 concentration of 12.2–15.0 kPa and CO_2 concentration of 3.4–6.5 kPa with good quality and lower decay rate (Hu et al., 2013).

Packaging using PVC wrap and polyolefin films increased shelf life of mushroom by retarding cap opening and by reducing weight loss, respiration and internal browning, consequently resulting in superior quality (Kim, 2006). Many other work also reported beneficial effects of MAP on the postharvest storage potential of *Agaricus sp.* and *Pleurotus sp.* mushrooms (Ares et al., 2007; Villaescusa and Gil 2003; Ban et al., 2013). However, Parentelli et al. observed that shiitake mushrooms had a higher rate of deterioration under MA (active and passive) than those stored in PP macro perforated films, which was attributed to the mushroom's sensitivity to high CO_2 concentrations (2007).

14.8.2 MAP IN PERFORATED FILM PACKAGES

The use of perforations in otherwise less permeable packages are increasing in the recent years to store highly respiring commodities. Plestenjak et al. observed that a variation in CO_2 and O_2 concentrations and consequent accumulation of anaerobic metabolites adversely affected the sensory properties of MAP stored shredded cabbage (2008). Thus, some of the recent interests have been to identify suitable films and type of perforations to allow the desirable gas exchange and to standardize such package materials for specific commodities. The risk of developing injurious gas concentrations can be decreased by placing pin holes or micro perforations in the plastic film (Brar et al., 2011; Rai et al., 2011).

Wang and Qi also reported less severe chilling injury in cucumbers packaged in perforated or sealed 31.75 μm LDPE bags than un-wrapped fruit in storage at 5°C and 90–95% relative humidity (RH) (1997). Serrano et al. stored broccoli using macro perforated, micro perforated and non-perforated PP films and observed that the quality degradation was delayed in perforated films (2006). Rai et al. also observed that perforated PP packages (film area 0.1 m^2, 2 holes each of 0.3 mm diameter) could store broccoli (*Brassica oleracea italic*) for 4 days without degradation of chlorophyll and ascorbic acid (2008). Cauliflower curds packed in individual HDPE bags with 3% ventilation holes could retain white color and firmness with least loss in weight up to 21 days storage at 0 ± 1°C, 90–95% RH condition (Dhall et al., 2010).

However, Ramayya et al. observed that Alphonso mangoes stored in unperforated bags appeared to be better than those stored in perforated film bags under otherwise identical conditions. They suggest that an initial headspace composition of 50% v/v CO_2 is most effective in maintaining good keeping quality of mango over 21 days storage (2012). Combrink et al. also reported that non-perforated polyethylene bags maintained guava fruit quality better than that in perforated bags (2004). Nath et al. also recommend non-perforated PP (0.025 mm) as a better packaging material than LDPE (0.025 mm), LLDPE (0.0125 mm), and HDPE (0.025 mm) for extending the shelf life of pear fruits up to 15 days at ambient condition (2012).

14.8.3 SHRINK WRAP MAP

Individual shrink wrapping (ISW), a form of MAP, is another important technique which substantially increases the shelf life of the commodity. The principal advantage of shrink wrapping are: reduced weight loss, minimized fruit deformation, reduced chilling injury, and reduced decay by preventing secondary infection. LDPE and PVC have been the common traditional films used for shrink wrapping, though laminated films and blends are also used. Many studies have proved the beneficial effect of shrink wrapping on storage life and quality of different types of fruits and vegetables (Ben-Yehoshua 1985; Heaton et al. 1990; Sonkar and Ladaniya 1999; Rao and Rao; 2002; Singh and Rao, 2005). Wijewardane and Guleria observed that shrink wrapping extended the shelf life of Royal delicious apples and the apples precooled at 10°C and stored at 18–25°C (65–75% RH) with 2% neem oil coating or marigold extract application had a storage life of up to 45 days (2013). Rao et al. studied the effect of MAP and shrink wrapping on shelf life of cucumber and reported that shrink wrapping with PE film could extend the shelf life of cucumber up to 24 days at 10 °C (2000). Dhall et al. also observed that shrink wrapping of cucumber in Cryovac D955 (60 guage) film followed by storage at 12 ± 1°C, 90–95% RH extended the shelf life of commodity. Shrink wrap packaging reduced the weight loss, retained the freshness, color and firmness of cucumber without any decay. However the authors observed that the use of cling wrap films should be avoided as this led to accumulation of excessive moisture resulting in huge spoilage (2010, 2012).

14.8.4 MAP OF PROCESSED FOODS

The processed foods are non-respiring and the atmosphere can have as low O_2 and as high CO_2 as possible without causing the pack to collapse or changes to the flavor and appearance of the product. Ground coffee is protected against oxidation by using a CO_2/N_2 mixture or vacuum packing (Fellows, 2000). Studies on microbiology of processed foods as meat, poultry, fish, baked goods, and so on establish that the growth of pathogens in MAP products is no greater and frequently lower than the aerobic stored foods (Church, 1994; Davies 1995).

Sra et al. reported that the dried carrot slices pretreated with 6% KMS and packed in AFL pouches (polyethylene, aluminum foil and polyester) retained the physicochemical quality at an acceptable level up to 6 months under ambient conditions, the maintenance of quality was better in AFL pouches (32.5 μm) than that in HDPE pouches (56.0 μm) (2011). A study on storage of aloevera gel powder in three different packaging materials viz., laminated aluminum foil (AF), biaxially oriented polypropylene (BOPP) and PP suggest that packaging in AF is better than that in BOPP and PP in terms of color change (Ramachandra and Rao, 2011).

The deep fat fried banana chips could store better in laminated aluminum foil than the PP, OPP and LDPE packages under ambient conditions (27°C) in terms of crispness and odor (Ammawath et al., 2002). Bal et al. also reported that aluminum foil exhibited better moisture barrier than polyethylene and laminated paper. Aluminum foil gave a shelf-life of 90 days to the potato chips fried in sunflower oil under ambient conditions (2002). Silva et al. evaluated the lipid oxidation in potato chips and reported that an oxygen scavenger helped to avoid/delay lipid oxidation at room temperature storage (2004). Manikantan et al. used a PP based nanocomposite film for banana chips and observed that the increase in free fatty acid, breaking force and total color difference of stored banana chips was higher with increased proportion of both nanoclay and compatibilizer, but decreased by reducing the thickness of film (2012).

PET films maintained the quality and safety of pistachio nuts in a better way than the nylon, PA/PP or PVC during storage at ambient temperature (22–28°C) and 85–100% RH. All the packaging materials except LDPE delayed the moisture absorption and aflatoxin formation of the product. The shelf-life of pistachio could be extended from 2 months (control) to 5 months when PET was used as the packaging material (Shakerardekani and Karim, 2013).

14.9 ACTIVE AND SMART PACKAGING

Active packaging is the technology of improving the shelf life and maintaining quality of commodities for longer duration by incorporating some specific auxiliary substances into the packaging systems. These substances normally act by absorbing oxygen, C_2H_4, CO_2, flavors/odors, moisture and/or by releasing CO_2, antimicrobial agents, antioxidants and flavors.

Two approaches are normally used, either some specific substances are introduced into the package along with the commodity (in the form of label, sachets) to perform the desired activity or the property of the packaging material or a part of the package (for example cap, lining material) is modified to offer the desired function.

Another term, smart (or intelligent) packaging is also used interchangeably with active packaging. However, the terms are strictly not the same. Active packaging is that which changes the condition of the packed food to extend shelf-life or to improve safety or sensory properties and maintain the quality of the food, whereas intelligent or smart packaging monitors the condition within the package and gives information about the quality of the packed food (Dainelli et al., 2008; Dawange et al., 2010). Table 5 gives the different applications of active and intelligent packaging systems.

TABLE 5 Classification of active and intelligent packaging systems

Active packaging	Intelligent packaging
O_2 scavenging	Time-temperature history
Anti-microbial	Microbial growth indicators
CO_2 scavenging	Light protection (photochromic)
CO_2 emitting	Physical shock indicators
C_2H_4 scavenging	Leakage, microbial spoilage indicating
Odor and flavor absorbing/ releasing	
Moisture absorbing	
heating and cooling	

The active packaging systems may be further classified as active scavenging or active releasing types. The active scavengers normally absorb the O_2, CO_2, ethylene or moisture. The active emitters are mostly used to emit CO_2 or antimicrobial compounds in the package. Another broad classification, namely, non-migratory active packaging and active releasing packaging is based on whether or not there is migration of compounds from the active agent to the packaged produce. Non-migratory active packaging acts without the active component migrating from the packaging into the food, a common application is for the absorption of moisture. In active releasing system, there is controlled migration of non-volatile agents or an emission of volatile compounds in the atmosphere surrounding the food (Dainelli et al., 2008). The development of active packaging systems and their application for varieties of food have received considerable attention in the recent years. Many materials have been incorporated in active packaging applications as ethanol (Karabulut et al., 2004; Lichter et al., 2002; Sabir et al., 2006), mineral oils (Valero et al., 2006), chlorinated or hot water (Del Nobile et al., 2008b), ethephon (Jayasena and Cameron, 2009), chitosan (a natural polysaccharide) (Romanazzi et al., 2009; Zhan et al., 2011; Duan et al., 2011), and ozone (Sharpe et al. 2009). However, undesirable odors or bitterness developed in some cases have been the limitations in practical applications of such materials (Soliva-Fortuny and Martín-Belloso, 2003; Mishra et al., 2010). Table 6 list the different types of materials used/tested for different active packaging functions.

TABLE 6 Types of active packaging systems with mode of action

Active packaging system	Type of material used for the purpose
Ethylene absorbing	Activated carbon/potassium permanganate
Ethanol emitting	Micro-encapsulated ethanol
Moisture absorbing	Polyvinyl alcohol encapsulation, silica gel, clay-based materials, zeolite, cellulose and their derivatives, sorbitol, xylitol, sodium chloride, potassium chloride
Antimicrobial releasing	Organic acids (benzoic acid, sorbic acid, propionic acid, lactic acid), silver salts, sulfur and its compounds, bacteriocins (nisin and lacticin), zeolites, chlorine dioxide, grape seed extracts, lemon seed extracts, spice extracts (thymol, p-cymene, and cinnamaldehyde), enzymes (peroxidase and lysozyme), chitosan, chelating agents (EDTA), Plant essential oils
Antioxidant releasing	BHA/BHT, TBHQ, vitamin C or E. Cellulose acetate films
Flavor/odor absorbing	Activated carbon, sodium bicarbonate

The initial oxygen scavengers were in the form of self adhesive labels or other adhesives or in form of loose sachets placed inside the packages. Subsequently O_2-absorbing polymer was developed which involve the cobalt-catalyzed oxidation of MXD6 polyamide as a blend with PET (Robertson, 2006). Besides, oxygen scavenging films containing iron or organic compounds in their structure have been developed. The organic compounds absorb oxygen after being activated by UV light (Tewari et al., 2002). However, the potential toxicity of the added scavenger and chance of accidental ingestion have been the main hurdles in their wide spread adaptability (Damaj et al., 2009).

Moisture absorbers are mostly based on adsorption of water by zeolite, cellulose and their derivatives, sorbitol, xylitol, sodium chloride and potassium chloride (Dainelli et al., 2008, Azevedo et al., 2011). Silica gel is used as moisture absorbent for dry foods, which is incorporated in sachet and placed inside package. Moisture absorbents are more effective when incorporated into packaging material itself. Drip absorbing pads consisting of granules of a superabsorbent polymer sandwiched between two layers of a microporous or nonwoven polymer, which is sealed at the edges, have been used to absorb liquid water. Drip absorbent sheets are commercially available.

Ethylene absorbers can reduce the rate of senescence and keep the commodity fresh for a long period. Jacobson et al. observed that LDPE package that contained an ethylene absorber retained the overall quality of broccoli and extended the acceptable shelf life (2004). Kudachikar et al. stored Banana (Musa sp var. "Robusta") under active and passive MAP at 12 ± 1°C and 85–90% RH. The active packaging used green keeper (GK) ethylene absorbent. Results indicate that the shelf life of fruits packed under MAP and MAP + GK can be extended up to 5 and 7 weeks, respectively as compared to 3 weeks for openly kept control fruits (2011).

The antimicrobial food packaging is one most potential application of active packaging technology. The antimicrobial packaging can be categorized into five types, viz. volatile antimicrobial substances that are incorporated into a sachet/pad and kept in the package, antimicrobials that are coated or adsorbed onto polymer surfaces, volatile and non-volatile antimicrobial substances or agents that are directly incorporated in to polymers, antimicrobials immobilized into polymers by ion or covalent linkages, and the polymers that are inherently antimicrobial (Appendini and Hotchkiss, 2002). It is more effective when the antimicrobials are incorporated in the packaging film itself, it helps in giving a slow and continuous release of active materials from packaging film to package environment. Further it is beneficial in that the antimicrobial compounds are not directly added to the food product. Controlled release of the active ingredient is very important in these films. Antimicrobial polymers can also eliminate the need of peroxide treatment in aseptic packaging and reduce the chances of recontamination of processed products.

Many substances as organic acids (benzoic acid, sorbic acid, propionic acid, lactic acid), bacteriocins (nisin and lacticin), grape seed extracts, lemon seed extracts, spice extracts (thymol, p-cymene and cinnamaldehyde), enzymes (peroxidase and lysozyme), chelating agents (EDTA), metals (silver), essential oils, and so on can be used as antimicrobial agents in the packaging materials (Falcone et al., 2005; Conte et al., 2007; Mastromatteo et al., 2010). Burt et al. have reviewed the antimicrobial properties of essential oils and their applications in food (2004). An antimicrobial LDPE film has been developed by incorporating grapefruit seed extract (GFSE), which has a wide antimicrobial spectrum and high heat stability (Lee at al. 1998). Thymol and carvacrol, when incorporated in the PP film as antimicrobial additives, also modify the oxygen barrier and mechanical properties of the film (Ramos et al., 2012). Valero et al. developed active packaging by adding eugenol or thymol to table grapes and could store the commodity for 56 days under MA condition. The sensory, nutritional and functional property losses were significantly reduced in packages with added eugenol or thymol. In addition, there were lower microbial counts (2006).

Incorporation of sodium propionate into cellulose-based films also reduces mould growth in the food (Soarses et al., 2002). Treatment with emulsions of cinnamon oil combined with MAP could extend the shelf life of *Embul* bananas up to 21 days in a cold room and 14 days at 28 ± 2°C (Ranasinghe et al., 2005).

Silver ions entrapped in silicate network have been developed as antimicrobial agents in food packaging. A plastic film containing allyl isothiocyanate (AITC) extracted from mustard (*Brassica nigra*), brown mustard (*Brassica juncea*) and wasabi (*Eutrema wasabi* Maxim.) is commercially available in Japan. The AITC is entrapped in cyclodextrins during extrusion process of film preparation to protect it from thermal damage (Suppakul et al., 2003). When exposed to high moisture conditions, the cyclodextrins have the ability to change its structure and to release the antimicrobial agent into the atmosphere surrounding the food.

Cellulose acetate (CA) films with different morphological features are also used to control the release of low molecular weight natural antioxidants L-ascorbic acid and L-tyrosine (Gemili et al., 2010). Use of packaging films that release organic acids offer potential for reducing the effect of the growth of slime-forming bacteria on foods. Chitosan, derived from chitin is well known for its antimicrobial capability and excellent film forming properties (Appendini and Hotchkiss, 2002; Fernandez-Siaz et al., 2010). Water barrier novel blends of chitosan with EVOH copolymers have been developed by solution casting from water/ isopropanol solutions of acetic acid, which exhibit antimicrobial activity by the release of protonated glucosamine fractions. EVOH/ Chitosan [poly-β-(1,4) N-acetyl-D-glucosamine], due to its non-toxic nature, antibacterial activity and biocompatibility, has attracted much attention as a natural food additive (Dutta et al., 2009; Ravi Kumar, 2000). Chitosan (80/20 wt%) film can be used as antimicrobial packaging film in dry environments at very low RH and mild temperatures (4–23°C) so as to maintain antimicrobial activity of film itself (Fernandez-Siaz et al., 2010). Chitosan and its derivatives have been successfully used to prolong the shelf-life and maintain the quality attributes of many fresh-cut produce, such as litchi (Dong et al., 2004), mango (Chien et al., 2007) and mushroom (Eissa, 2007). Suppakul et al. has made a comprehensive review of the active packaging technologies with special emphasis on the antimicrobial packaging and its applications (2003).

14.10 GENERAL CONSIDERATIONS FOR MODIFIED ATMOSPHERE PACKAGING

For the proper outcome of MAP, it is very important to maintain appropriate headspace to delay senescence and maintain physico-chemical constituents. The response of plant materials to MAP differs with species, variety, growing conditions, conditions of harvest and cultural practices, and so on. Thus the improvements in shelf life for all commodities are not same in MAP. Similarly the technology can not benefit all types of produce.

The head space EMA in addition to other factors also depends on the temperature of storage and transit, and hence, special care is required to maintain proper temperature in the storage space, proper temperature is actually more important than the modification of atmosphere surrounding the commodity. Particularly for fruits, both ripening and C_2H_4 production rates increase with an increase in storage temperature. Often specific commodities require low O_2 or high CO_2 levels coupled with low temperature. In such situation, if the temperature increases even by a few degrees, the respiration of the commodity may further reduce the O_2 level and induce low O_2 injury in the commodity.

The polymer characteristics also greatly depend on temperature. The package designed for a particular storage temperature cannot work effectively when exposed to another temperature. Pre-cooling and proper temperature maintenance during preparation and transportation of the material also affect the MAP. Temperature should be brought close to the storage/shipping temperature as quickly as possible after packaging except in cases where a slightly elevated temperature is needed to assist in rapid atmosphere generation. The MAP recommendations for a commodity should also take into consideration the temperature profile likely to occur in the food supply chain and the corrective measures recommended under adverse situations.

The storage humidity is also important. Higher humidity though prevents desiccation, can increase decay. The gas and moisture vapor permeability characteristics depend on the type of polymer and the manufacturing parameters. Commercially available polymer based food packaging materials may not often match the commodity requirements, and hence, careful selection and development of suitable packaging material and/or designing perforations/window areas are very important for the success of MAP.

As is the case of almost all aspects of food production, the hazard analysis and critical control points (HACCP) system plays a major role in ensuring the safety of MAP foods (Fellows, 2000).

14.11 FUTURE POTENTIAL FOR MAP APPLICATIONS

MAP in general tend to maintain low O_2 (1–5%) and high CO_2 (5–10%) concentrations, and in combination with low temperature storage, maintains sensory as well as microbial quality of perishable commodities, it maintains the tissue structure and nutritional constituents and inhibits enzymatic browning reaction on the cut surfaces. The research, developments and application of MAP have increased in the last decade to meet the requirement of extending the shelf life of commodities, minimizing losses and providing wholesome and nutritious food to the people.

The use MAP for fruits and vegetables in retail stores and with local vendors has also been increasing in recent years and with increasing popularity of minimally processed fruits and vegetables, its application is surely to enhance further. As different types of polymers suit best as the packaging materials for MAP, the use of polymers will also rise analogous to the development of this sector.

The technology has been well accepted in the developed countries and a variety of modified atmosphere packed products are available in the retail stores. Nevertheless MAP has a great potential for developing countries also mainly due to the following reasons:

1. There is a huge quantity of product loss, particularly for the fruits and vegetables in the developing and underdeveloped countries and MAP, with its

capability to extend the shelf life of perishable commodities, has a good potential for reducing the losses,

2. The consumers are gradually shifting for minimally processed and convenience foods and the MAP has a great potential for maintaining quality and extending the shelf life of minimally processed or fresh cut produce,
3. The MAP technology is simple and does not require sophisticated machines,
4. The initial investment for the MAP is less and it can be easily integrated with a farm level pack house and precooling center. It will be particularly advantageous in situations where organized collection of raw materials is yet to be available,
5. MAP has further relevance in places where the cold chain facilities are not available. Even in the absence of cold chain, MAP is advantageous than storage in natural atmosphere, and permits the producer an additional time limit to sell and consume the produce,
6. Standard MAP practices do not use synthetic chemicals and there is nil residual toxicity,
7. MAP mostly uses polymer films of different types and the environmental impact will be less if the films are recycled (Mangaraj and Goswami, 2009).

14.12 CONCLUSION AND SUGGESTIONS FOR FUTURE RESEARCH WORK

The MAP, with its capability to keep the commodities in a fresh like form for an extended period, has been gaining importance with the processors and consumers. Studies have established the usefulness of MAP for a majority of produce, particularly the fresh fruits and vegetables, in whole or minimally processed form. As the different polymers offer a range of gas and water vapor permeability along with low cost, they are naturally the common packaging material for MAP.

The behaviors of the commodities in the MAP differ, the equilibrium gas concentrations within the package also depend both on the type of commodity and packaging material along with environmental parameters. Most of the recommendations on the use of MAP are based on studies with specific commodities undertaken on trial and error basis at specified low temperatures. Such recommendations for storage of a particular commodity cannot be considered universally acceptable, and thus the commercial storage does not behave in a similar way as the research findings. Thus a systematic design of MAP is required which can assist in developing the appropriate equilibrium O_2 and CO_2 conditions within the package depending on the package and commodity characteristics and the storage temperature.

It is very important to use the polymer films of recommended gas and vapor transmission characteristics, and this requirement is often not met, which defeat the basic purpose of MAP and the potential benefit is not realized. Therefore, more varieties of polymer blends, laminations, co-extrusions, and so on require to be investigated. The properties of the film within the same type also differ due to different uncontrollable factors during the manufacturing process. There is a need to develop and standardize a range of food grade microporous films for by incorporating suitable materials in the resins. The gas and vapor transmission values also depend upon the temperature fluctuations in storage, thus the development of packaging films to maintain the desired permeability and permeability ratios even if there is a change in storage temperature or an auxiliary device to take care of the changes in produce and polymer characteristics are very vital. Application of models in the packaging studies has proved useful in some cases, they can be further studied to cover wider range of package materials and conditions and commodities.

Active and smart packaging technologies can add to the reliability and acceptability of modified atmosphere packed produce, though they add to the cost. However, such packaging are sure to be popular in future, and hence, standardization of active packaging conditions for the varieties of commodities are essential with a clear identification of the type of packaging film as well as the additives/ active agents to be used for the commodity. The development of low cost sensors to indicate the safety of MAP product is also imperative. Nanotechnology has enough potential to influence the packaging sector and its application can improve the characteristics and active properties of the packages. With the growing population and increasing amount of food being handled, the use of polymers is continually on a rise and development of new biodegradable packages, which can be economically used for food, needs to be explored so as to reduce the pressure on the environment.

KEYWORDS

- **Metallocene polymers**
- **Modified atmosphere packaging**
- **Nanocomposite materials**
- **Plasticizers**
- **Polypropylene**

REFERENCES

1. Aguayo, E., Allende, A., and Artes, F. *European Food Res. Technol.*, **216**, 494–499 (2003).
2. Ahvenainen, R. *Trends in Food Sci. Tech.*, **7**, 179–186 (1996).
3. Al-Ati, T. and Hotchkiss, J. H. *J. Food Agric. Chem.*, **51**, 4133–4138 (2003).
4. Almenar, E., Samsudin, H., Auras, R., Harte, B., and Rubino, M. *Food Chem.*, **110**, 120–127 (2008).
5. Ammawath, W., Che Man, Y. B., Yusof, S., and Rahman, R. A. *J. Sci. Food Agric.*, **82**, 1621–1627 (2002).
6. Appendini, P. and Hotchkiss, J. H. *Innovative Food Sci. Emerging Technol.*, **3**, 113–126 (2002).
7. Ares, G., Lareo, C., and Lema, P. *Fresh Produce*. **1**, 32–40 (2012).
8. Artes-Hernandez, F., Tomas-Barberan, F. A., and Artes, F. *Post harvest Biol. Technol.*, **39**, 146–154 (2006).
9. Azene, M. and Nanos, G. D. *Postharvest Biol. Technol.*, **38**, 106–114 (2005).
10. Azene, M., Workneh, T. S., and Woldetsadik, K. *J. Food Sci. Technol.*, (2011) DOI 10.1007/s13197-011-0607-6.
11. Azevedo, S., Cunhaa, L. M., Mahajan, P. V., and Fonseca, S. C. *Procedia Food Sci.*, **1**, 184–189 (2011).
12. Bal, A., Sandhu, K. S., and Ahluwalia, P. *J Food Sci. Technol.*, **39**, 394–402 (2002).

13. Ban, Z., Li, L., Guan, J., Feng, J., Wu, M., Xu, X., and Li, J. *J. Food Sci. Technol.* (2012) DOI 10.1007/s13197-013-0935-9.
14. Baskaran, R., Prasad, R., Shivaiah, K. M., and Habibunnisa. *European Food Res. Technol.*, **212**, 165–169 (2001).
15. Beaudry, R. M. *Postharvest Biol. Technol.*, **15**, 293–303 (1999).
16. Beaudry, R. M. *Hort. Technol.*, **10**, 491–500 (2000).
17. Benkeblia, N. Braz. *J. Plant Physiol.*, **16**, 47–52 (2004).
18. Ben-Yehoshua, S. *Hortic. Sci.*, **20**, 32–37 (1985).
19. Berna, A. Z., Geysen, S., Li, S., Verlinden, B. E., Larnmertyn, J., and Nicolai, B. A. *Postharvest Biol. Technol.*, **46**, 230–236 (2007).
20. Brar, J. K., Rai, D. R., Singh, A., and Kaur, N. *J. Food Sci. Technol.* (2011) doi: 10.1007/s13197-011-0390-4.
21. Brody, A. L., Bugusu, B., Han, J. H., Sand, C. K., and McHugh, T. H. *JFS: Concise reviews and Hypothesis in Food Sci.*, **73**(8), R107–R116 (2008).
22. Burt, S. Analytical and Bioanalytical Chem. *Microbiol.*, **94**, 223–253 (2004).
23. Cameron, A. C., Beaudry, R. M., Banks, N. H., and Yelanich, M. V. *J. Amer. Soc. Hort. Sci.*, **119**, 534–539 (1994).
24. Charles, F., Guillaume, C., and Gontard, N. *Postharvest Biol. Technol.*, **48**, 22–29 (2008).
25. Chien, P. J., Sheu, F., and Yang, F. H. *J. Food. Eng.*, **78**, 225–229 (2007).
26. Church, N. *Trends Food Sci. Technol.*, **5**, 345–352 (1994).
27. Combrink, J. C., De-Kock, S. L., and Van-Ecden, C. J. *Acta. Hort.*, **275**, 539–645 (2004).
28. Conte, A., Speranza, B., Sinigaglia, M., and Del Nabile, M. A. *J. Food Prot.*, **70**, 1896–1900 (2007).
29. Dainelli, D., Gontard, N., Spyropoulos, D., Esther, Z. B., and Tobback, P. *Trends in Food Sci. Technol.*, **19**, S103–S112 (2008).
30. Damaj, Z., Naveau, A., Dupont, L., Hénon, E., Rogez, G., and Guillon, E. *Inorg. Chem. Comm.*, **12**, 17–20 (2009).
31. Das, H. *Food processing operations analysis*. Asian Books Private Limited, New Delhi (2005).
32. Davies, A. R. Advances in modified atmosphere packaging. G. W. Gould (Ed.), *New Methods of Food Preservation*, Blackie Academic and Professional, Glasgow, pp. 304–320 (1995).
33. Dawange, S., Dash, S. K., and Patil, S. B. *Indian Food Industry.*, **29**, 31–36 (2010).
34. Day, B. P. F. Fruits and vegetables. R. T. Parry (Ed.), *Principles and applications of MAP of foods*, Blackie Academic and Professional, New York, pp. 304–320 (1993).
35. Del Nobile, M. A., Biaiano, A., Benedetto, A., and Massignan. L. *J. Food Eng.*, **74**, 60–69 (2006).
36. Del Nobile, M. A., Conte, A., Cannarsi, M., and Sinigaglia, M. *J. Food Eng.*, **85**, 317–325 (2008).
37. Del Nobile, M. A., Sinigaglia, M., Conte, A., Speranza, B., Scrocco, C., Brescia, I., Bevilacqua, A., Laverse, J., La Notte, E., and Antonacci, D. *Postharvest Biol. Tech.*, **47**, 389–396 (2008).
38. Del Nobile, M. E., Licciardello, F., Scrocco, C., Muratore, G., and Zappa, M. *J. Food Eng.*, **79**, 217–224 (2007).
39. Deng, Y., Wu, Y., and Li, Y. *European Food Res. Technol.*, **221**, 392–397 (2005).
40. Dhall, R. K., Sharma, S. R., and Mahajan, B. V. C. *J. Food Sci. Technol.*, **47**, 132–135 (2010).
41. Dhall, R. K., Sharma, S. R., and Mahajan, B. V. C. *J. Food Sci. Technol.*, **49**, 495–499 (2012).
42. Dong, H., Cheng, L., Tan, J., Zheng, K., and Jiang, Y. *J. Food. Eng.*, **64**, 355–358 (2004).
43. Duan, J., Wu, R., Strik, B. C., and Zhao, Y. *Postharvest Biol. Technol.*, **59**, 71–79 (2011).
44. Dutta, P. K., Tripathi, S., Mehrotra, G. K., and Dutta, J. *Food Chem.*, **114**, 1173–1182 (2009).
45. Eissa, H. A. A. *J. Food Qual.*, **30**, 623–645 (2007).
46. Exama, A., Arul, J., Lencki, R. W., Lee, L. Z., and Toupin. C. *J. Food Sci.*, **58**(6), 1365–1370 (1993).
47. Falcone, P., Speranza, B., Del Nobile, M. A., Corbo, M. R., and Sinigaglia, M. *J. Food Prot.*, **68**, 1664–1670 (2005).
48. Farber, J. M. *J. Food Protect.*, **54**, 58–70 (1991).

49. Fellows, P. J. *Food Processing Technology Principles and Practice: 2nd Edition*. Woodhead Publishing Ltd., Cambridge, England (2000).
50. Fernandez-Saiz, P., Ocio, M. J., and Lagaron, J. M. *Carbohyd. Polym*., **80**, 874–884 (2010).
51. Floros, J. D., Dock, L. L., and Han, J. H. *Food, Cosmetics, and Drug Packag*., **20**, 10–17 (1997).
52. Fonseca, S. C., Oliveira, F. A. R., and Brecht, J. K. *J. Food Eng*., **52**, 99–119 (2002).
53. Fonseca, S. C., Oliveira, F. A. R., Frias, J. M., Brecht, J. K., and Chau, K. V. *J. Food Eng*., **54**, 299–307 (2002).
54. Fonseca, S. C., Oliveira, F. A. R., Lino, I. B. M., Brecht, J. K., and Chau, K. V. *J. Food Eng*., **43**, 9–15 (2000).
55. Gariepy, Y., Raghavan, G. S. V., and Theriault, R. *Canadian J. Agric. Eng*., **26**, 105–109 (1984).
56. Geeson, J. D., Genge, P. M., and Sharpies, R. O. *Postharvest Biol. Technol*., **4**, 35–48 (1994).
57. Gemili, S., Ahmet, Y., and Altınkaya, S. A. *J. Food Eng*., **96**, 325–332 (2010).
58. Hailu, M., Workneh, T. S., and Belew, D. *J. Food Sci. Technol*., (2012) DOI 10.1007/s13197-012-0826-5.
59. Heaton, E. K., Dobson, J. W., Lane, R. P., and Beuchat, L. R. *J. Food Prot*., **53**(7), 598–599 (1990).
60. Hu, W. Z., Jiang, A. L., and Qi, H. P. *J. Food Sci. Technol*., (2012) DOI 10.1007/s13197-012-0922-6.
61. Jacobsson, A., Tim, N., Ingegerd, S., and Karin, W. *Food Qual. Prefer*., **15**, 301–310 (2004).
62. Jacomino, A. P., Kluge, R. A., Sarantopoulos, C. I. G. L., and Sigrist, J. M. M. *Packaging Technol. Sci*., **14**, 11–19 (2001).
63. Jacxsens, L., Devlieghhere, F., De Rudder, T., and Debevere, J. *Lebensm.-Wiss. u.-Technol*., **33**, 178–187 (2000).
64. Jayasena, V. and Cameron, I. *Int. J. Food Sci. Tech*., **44**, 409–41 (2009).
65. Kader, A. A., Zagory, D., and Kerbel, E. L. *CRC Crit. Rev. Food Sci. Nutr*., **28**, 1–30 (1989).
66. Karabulut, O. A., Gabler, F. M., Mansour, M., and Smilanick, J. L. *Postharvest Biol. Tech*., **34**, 169–177 (2004).
67. Karakas, B. and Yildiz, F. *Food Chem*., **100**, 1011–1018 (2007).
68. Kaur, K., Dhillon, W. S., and Mahajan, B. V. C. *J. Food Sci. Technol*. (2012) DOI 10.1007/s13197-012-0773-1.
69. Kaur, K., Dhillon, W. S., and Mahajan, B. V. C. *J. Food Sci. Technol*., **50**, 147–152 (2013).
70. Kaur, P., Rai, D. R., and Paul, S. *J. Food Process Eng*. (2011) DOI:10.1111/j.1745-4530.2009.00508.x.
71. Kim, K. M., Ko, J. A., Lee, J, S., Park, H. J., and Hanna, M. A. *LWT-Food Sci. Technol*., **39**, 364–371 (2006).
72. Kirwan, J. M. and Strawbridge. J. W. Plastics in food packaging. C. Richard, M. Derek, and J. M. Kirwan (Eds.), *Food Packaging Technology*, Blackwell, CRC press, USA, pp. 174–240 (2003).
73. Kudachikar, V. B., Kulkarni, S. G., and Keshava Prakash, M. N. *J. Food Sci. Technol*., **48**(3), 319–324 (2011).
74. Lakakul, R., Beaudry, R. M., and Hernandez, R. J. *J. Food Sci*., **64**, 105–110 (1999).
75. Lee, D. S., Hagger, P. E., and Yam, K. L. *Packaging Technol. Sci*., **5**, 27–30 (2006).
76. Lee, D. S., Hwang, Y. I., and Cho, S. H. *Food Sci. Biotechnol*., **7**, 117–121 (1998).
77. Lichter, A., Zutkhy, Y., Sonego, L., Dvir, O., Kaplunov, T., Sarig, P., and Ben-Arie, R. *Postharvest Biol. Technol*., **24**, 301–308 (2002).
78. Liu, F. and Li, Y. *Postharvest Biol. Technol*., **40**, 262–268 (2006).
79. Lurie, S., Pesis, E., Gadiyeva, O., and Feygenberg, O., Ben-Arie, R., Kaplunov, T., Zutahy, Y., and Lichter, A. *Postharvest Biol. Technol*., **42**, 222–227 (2006).
80. Mahajan, P. V., Oliveira, F. A. R., Montanez, J. C., and Frias, J. *Innov. Food Sci. Emerg. Technol*., **8**, 84–92 (2007).
81. Mangaraj, S. and Goswami, T. K. *Fresh Prod*., **2**, 72–80 (2008).
82. Mangaraj, S., Goswami, T. K., Giri, S. K., and Joshy, C. G. *J. Food Sci. Technol*. (2012) DOI 10.1007/s13197-012-0860-3.

83. Mangaraj, S., Goswami, T. K., and Mahajan, P. V. *Food Eng. Rev.*, **1**, 133–158 (2009).
84. Manikantan, M. R., Sharma, R., Kasturi, R., and Varadharaju, N. *J. Food Sci. Technol.*, (2012) DOI 10.1007/s13197-012-0839-0.
85. Manikantan, M. R. and Varadharaju. N. *Packag Technol. Sci.*, **24**, 191–209 (2011).
86. Manjunatha, M. and Anurag, R. K. *J. Food Sci. Technol.*, (2012) DOI 10.1007/s13197-012-0840-7.
87. Mastromatteo, M., Conte, A., and Del Nobile, M. A. *Trends in Food Sci. Technol.*, **21**, 591–598 (2010).
88. Mattheis, J. P. and Fellman, J. P. *Hort. Technol.*, **10**, 507–510 (2000).
89. Mishra, B., Khatkar, B. S., Garg, M. K., and Wilson, L. A. *J. Food Sci. Technol.*, **47**(1), 109–113 (2010).
90. Moleyar, V. and Narasimham, P. *J. Food Sci. Technol.*, **31**, 267–278 (1994).
91. Montanez, J., Oliviera, F. A. R., Frias, J., Pinelo, M., Mahajan, P. V., Cunha, L. M., and Manso, M. C. *J. Proc. Fla. State Hort. Soc.*, **118**, 423–428 (2005).
92. Moodley, R. S., Govinden, R., and Odhav, B. *J. Food Prot.*, **65**, 867–871 (2002).
93. Naik, R., Ambrose, D. C. P., Raghavan, G. S. V., and Annamalai, S. J. K. *J. Food Sci. Technol.* (2012) DOI 10.1007/s13197-012-0898-2.
94. Naik, R. and Kailappan, R. *J. Food Sci. Technol.*, **44**(3), 301–306 (2007).
95. Nath, A., Bidyut, C. D., Singh, A., Patel, R. K., Paul, D., Misra, L. K., and Ojha, H. *J. Food Sci. Technol.*, **49**(5), 556–563 (2012).
96. Parentelli, C., Ares, G., Corona, M., Lareo, C., Gambaro, A., Soubes, M. et al. *J. Sci. Food Agric.*, **87**, 1645–1652 (2007).
97. Pereira, L. M., Rodrigues, A. C. C., Sarantopoulos, C. I. G. L., Junqueira, V. C. A., Cunha, R. L., and Hubinger, M. D. *J. Food Sci.*, **69**, 1107–1111 (2004).
98. Pereira, L. M., Rodrigues, A. C. C., Sarantopoulos, C. I. G. L., Junqueira, V. C. A., Cunha, R. L., and Hubinger, M. D. *J. Food Sci.*, **69**, 1107–1111 (2004).
99. Phillips, C. *Int. J. Food Sci. Tchnol.*, **34**, 463–479 (1996).
100. Plestenjak, A., Pozrl, T., Hribar, J., Unuk, T., and Vidrih, R. Food *Technol. and Biotechnol.*, **46**, 427–433 (2008).
101. Raghavan, G. S. V., Tessier, S., Chayet, M., Norris, E. R., and Phan, C. T. *Trans ASAE.*, **25**, 433–436 (1982).
102. Rai, D. R., Tyagi, S. K., Jha, S. N., and Mohan, S. *J. Food Sci. Technol.*, **45**, 247–250 (2008).
103. Rai, D. R., Chadha, S., Kaur, M. P., Jaiswal, P., and Patil, R. T. *J. Food Sci. Technol.*, **48**, 357–365 (2011).
104. Ramachandra, C. T. and Rao, P. S. *J. Food Sci. Technol.* (2011) DOI 10.1007/s13197-011-0398-9.
105. Ramayya, N., Niranjan, K., and Duncun, E. *J. Food Sci. Technol.*, **49**(6), 721–728 (2012).
106. Ramos, M., Jiménez, A., Peltzer, M., and Garrigós, M. C. *J. Food Eng.*, **109**, 513–519 (2012).
107. Ranasinghe, L., Jayawardena, B., and Abeywickrama, K. *Int. J. Food Sci. Technol.*, **40**, 97 (2005).
108. Rao, D. V. S. and Rao, K. P. G. *Packag. India.*, **34**, 27–32 (2002).
109. Rao, D. V. S., Rao, K. P. G., and Krishnamurthy, S. *Indian Food Packer.*, **54**, 65–71 (2000).
110. Ravi Kumar, M. N. V. *React. Funct. Polym.*, **46**, 1–27 (2000).
111. Robertson, G. L. Active and intelligent packaging. *Food packaging: Principles and Practices: 2nd Edition.* CRC Press, Boca Raton, Florida, USA, pp. 285–312 (2006).
112. Rocha, A. M. C. N., Barreiro, M. G., and Morais, A. M. M. B., *J. Food Control.* **15**, 61–64 (2004).
113. Rodriguez-Aguilera, R. and Oliveira, J. C. *Food Eng. Rev.*, **1**, 66–83 (2009).
114. Romanazzi, G., Gabler, F. M., Margosan, D., Mackey, B. E., and Smilanick, J. L. *Phytopat.*, **99**, 1028–1036 (2009).
115. Sabir, A., Sabir, F. K., and Kara, Z. *J. Food Sci. Technol.*, **48**, 312–318 (2011).
116. Sabir, A., Sabir, F. K., Tangolar, S., and Agar, I. T. *J. Cukurova. Uni. Agr. Fac.*, **21**, 45–50 (2006).
117. Saito, M. and Rai, D. R. *J. Food Sci. Technol.*, **42**, 70–72 (2005).

118. Salvador, M. L., Jaime, P., and Oria, R. *J. Food Sci.*, **67**, 231–235 (2002).
119. Sandhya, *LWT-Food Sci. Technol.*, **43**, 381–392 (2010).
120. Selke, S. E. M., Culter, J. D., and Hernandez, R. J. *Plastics Packaging, properties, processing, applications, and regulations: 2nd Edition*. Hanser Publications, Cincinati (2004).
121. Serrano, M., Martinez-Romero, D., Guillen, F., Castillo, S., and Valero, D. *Postharvest Biol. Technol.*, **39**, 61–68 (2006).
122. Shakerardekani, A. and Karim, R. *J. Food Sci. Technol.*, **50**, 409–411 (2013).
123. Sharp, A. K., Irving, A. R., and Morris, S. C. *Proc. 6th Intl Control Atmos. Res. Conf.* G. Blanpied, J. Bartsch, and J. Hicks (Eds.), Cornell Univ., Ithaca, New York, pp. 238–251 (1993).
124. Sharpe, D., Fan, L., Mc Rae, K., Walker, B., Mac Kay, R., and Doucette, C. *J. Food Sci.*, **74**, 250–257 (2009).
125. Shoji, K. and Shi, J. *Food Control.*, **18**, 1121–1125 (2007).
126. Silva, A. S., Hernandez, J. L., and Losada, P. P. *Anal. Chem. Acta.*, **524**, 185–189 (2004).
127. Singh, M., Kumar, A., and Kaur, P. *J. Food Sci. Technol.* (2012) DOI 10.1007/s13197-012-0735-7.
128. Singh, S. P. and Rao, D. V. S. *J. Food Sci. Technol.*, **42**(6), 523–525 (2005).
129. Sinha, S. R. and Okamoto, M. *Progress Polymer Sci.*, **28**, 1539–1641 (2003).
130. Sivakumar, D., Arrebola, E., and Korsten, L. *Crop Prot.*, **27**, 1208–1214 (2008).
131. Sivakumar, D. and Korsten, L. *Postharvest Biol. Technol.*, **41**, 135–142 (2006).
132. Smith, J. P. and Ramaswamy, H. S. Packaging of fruits and vegetables. L. Somogyi, H. S. Ramaswamy, and Y. H. Hui (Eds.), *Processing fruits: science and technology*, Technolomic Publ. Co., Lancaster, Pennsylvania (1996).
133. Soares, N. F. F., Rutishauser, D. M., Melo, N., Cruz, R. S., and Andrade, N. J. *Packaging Technol. Sci.*, **15**, 129–132 (2002).
134. Soliva-Fortuny, R. C. and Martin-Belloso, O. *Trends Food Sci. Tech.*, **14**, 341–353 (2003).
135. Soliva-Fortuny, R. C., Ricart-Coll, M., and Martin-Belloso, O. *Int. J. Food Sci. Technol.* **40**, 369 (2005).
136. Sra, S. K., Sandhu, K. S., and Ahluwalia, P. *J. Food Sci. Technol.* (2011) DOI 10.1007/s13197-011-0575-x.
137. Steiner, A., Abreu, M., Correia, L., Beirao-da-Costa, S., Leitao, E., Beirao-Costa, M. L. et al. *European Food Res. Technol.*, **222**, 217–222 (2006).
138. Suppakul, P., Miltz, J., Sonneveld, K., and Bigger, S. W. *J. Food Sci.*, **68**, 408–420 (2003).
139. Tano, K., Oule, M. K., Doyon, G., Lencki, R. W., and Arul, J. *Postharvest Biol. Technol.*, **46**, 212–221 (2007).
140. Tewari, G., Jayas, D. S., Jeremiah, L. E., and Holley, R. A. *Int. J. Food Sci. Technol.*, **37**, 209–217 (2002).
141. Tian, S. P., Li, B. Q., and Xu, Y. *Food Chem.*, **91**, 659–663 (2005).
142. Valero, D., Valverde, J. M., Martinez-Romero, D., Guillen, F., Castillo, S., and Serrano, M. *Postharvest Biol. Tech.*, **41**, 317–327 (2006).
143. Van Krevelen, D. W. *Properties of polymers: 3rd Edition*. Elsevier Science, Amsterdam (1997).
144. Villaescusa, R. and Gil, M. I. *Postharvest Biol. Technol.*, **28**, 169–179 (2003).
145. Wang, C. Y. *Hortic. Rev.*, **15**, 63–95 (1993).
146. Wang, C. Y. and Qi, L. *Postharvest Biol. Technol.*, **10**, 195–200 (1997).
147. Wijewardane, R. M. N. A. and Guleria, S. P. S. *J. Food Sci. Technol.*, **50**, 325–331 (2013).
148. Yang, C. C. and Chinnan, M. S. *Trans. ASAE.*, **31**, 920–925 (1988).
149. Yoshio, M. and Takashi, H. *Postharvest Biol. Technol.*, **10**, 247–254 (1997).
150. Zagory, D. and Kader, A. A. *Food Technol.*, **42**, 70–77 (1988).
151. Zhan, L., Hu, J., and Zhu, Z. *Int. J. Food Sci. Tech.*, **46**, 2634–2640 (2011).
152. Zhang, M., Xiaob, G., Peng, J., and Salokhe, V. M. *Int. J. Food Eng.*, **1**, 4 (2005)

CHAPTER 15

APPLICATION OF MODIFIED ATMOSPHERE PACKAGING FOR EXTENSION OF SHELF-LIFE OF FOOD COMMODITIES

S. D. DESHPANDE

ABSTRACT

Modified Atmosphere Packaging (MAP) helps in increasing the shelf life of fruits and vegetables and presents the consumer with distinct advantages of high quality product, clear view of the product, and nil or very little use of chemical preservative and the food producer with increased shelf life. Food distribution has undergone two major revolutions in the last Century, canning, and freezing. These gave consumers easy availability to most type of produce. However, energy crisis, ecological awareness, and demand for healthy and fresh food have created a need for a technology that allows distribution of fresh produce around the year. The MAP theoretically, offers a possibility of meeting these requirements. This new packaging concept is rapidly growing in the food packaging market. It improves the product quality, freshness, and increases the shelf life of the product as well as provides convenience to the consumer and adds value to the product. It is mainly used to extend the shelf life of fresh/perishable commodities maintaining freshness. The modified atmosphere concept for packaged produce consists of modifying the atmosphere surrounding a food product by vacuum, gas flushing, or controlled permeability of the pack thus, controlling the biochemical, enzymatic, and microbial actions so as to avoid or decrease the main degradations that might occur. This allows the preservation of the fresh state of the food product without the temperature or chemical treatments used by competitive preservation techniques, such as canning, freezing, dehydration, and other processes. It is one of the novel approaches to extend the self life of perishable commodity. Modified atmosphere packaging fits into the important area of preservation, where shelf life is extended without the loss of those important and elusive properties, which constitute fresheners in the consumer's mind, and therefore move the product into a premium bracket. At premium products mainly have short shelf lives process which extend this even by a few days are of great value to retailers by virtue of the reduced wastage, reduced costs for shelf filling, and in certain instances, the ability to offer products that might otherwise be obtainable only from a specialist outlet. One of the most important parameters to be considered for MAP studies is the need for proper packaging films.

15.1 INTRODUCTION

Packaging is a technique of using the most appropriate packaging media for the safe delivery of the contents from the centers of production to the site of consumption. Packaging serves as the vital link in the long line of production, storage, transportation, distribution, handling and marketing. The package must ensure the same high quality of the product to the consumer, as he is used to getting, in freshly manufactured products. It is not enough if more production is achieved through various means, what is important is, that all that is produced should reach the consumer in a usable condition. It is then that the efforts in any venture can become successful, meaningful, and purposeful.

Modern packaging systems for health foods are products from a synthesis of demands from producers, distributors, and consumers (Deshpande, 2012). The need for hygiene is the primary reason for retail packaging of perishable food products like milk. Although, this was realized more than a Century ago, packaging techniques for liquid milk were slow in developing (Deshpande, 2003). The advent of pasteurization in the 1920s made retail packaging of liquid essential, and the returnable glass bottle was soon to become universal.

The world production of fruits and vegetables has reached to the tune of 1.4 billion tons. Approximately 500 MT of fruits and 900 MT of vegetables were produced in the world annually during 2005–2006. India accounts for 10.1% of the total world production of fruits and ranks second with the production of 45.47 million tons in 2005 (Singhal, 2006). The estimated annual loss of fruits and vegetables due to inadequate facilities/infrastructure/marketing set up and improper technologies of handling, packaging, and storage is in the range of 25–30% of the total production (Mangaraj and Varshney, 2006).

Revolutions in food distribution in the last Century resulted in canning and freezing thus giving consumer's easy availability to most type of produce. However, energy crisis, ecological awareness, and demand for healthy and fresh food have created a need for a technology that allows distribution of fresh produce around the year. It uses minimum processing and little energy and maximizes nutrition and flavor. The MAP offers a possibility of meeting these requirements. This new packaging concept is rapidly growing in the food packaging market. It improves the product quality, freshness, and increases the shelf-life of the product as well as provides convenience to the consumer and adds value to the product. It is mainly used to extend the shelf life of fresh/perishable commodities maintaining freshness (Brody, 1989, Davies, 1995, Mangaraj and Goswami, 2008).

15.2 MODIFIED ATMOSPHERE PACKAGING (MAP)

The modified atmosphere concept for packaged produce consists of modifying the atmosphere surrounding a food product by vacuum, gas flushing, or controlled permeability of the pack thus, controlling the biochemical, enzymatic, and microbial actions so as to avoid degradations that might occur. This allows the preservation of the fresh state of the food product without the temperature or chemical treatments used by competitive preservation techniques, such as canning, freezing, dehydration, and other processes (Ooraikul and Stile, 1991, Saltveit, 1993, Chau and Talasila, 1994, Jayas and Jeyamkondan, 2002). It is one of the novel approaches to extend the self life of perishable commodity (Cameron 1989, Gorris and Tauscher, 1999, Ahvenainen, 2003).

Modified atmosphere packaging of fresh produce relies on modification of the atmosphere inside the package achieved by the natural interplay between two processes, the respiration of the product, and the transfer of gases through the packaging, that leads to an atmosphere richer in CO_2 and poorer in O_2, and it depends on the characteristics of the commodity and the packaging film (Jacxsens et al., 2000, Fonseca et al., 2002, Mahajan et al., 2007). Active MAP is the replacement of air in a pack with a single gas or mixture of gases, the proportion of each component is fixed when the mixture is introduced. No further control is exerted over the initial composition, and the gas composition is likely to change with time owing to the diffusion of gases into and out of the product, the permeation of gases into and out of the pack, and the effects of product and microbial metabolism (Church, 1994, Church and Parsons, 1995, Blakistone, 1997).

15.3 ADVANTAGES AND DISADVANTAGES OF MAP TECHNOLOGY

The benefits of MAP technology to the manufacturer, retailer as well as consumer far outweigh the drawbacks (Kader et al., 1989, Mahajan et al., 2007).

15.3.1 ADVANTAGES

- Extend product shelf life to meet the demands of long distance distribution.
- Maintenance of product quality through reduction of respiration rate, browning reactions, fruit softening, and decay,
- Increased shelf life allowing lesser frequency of loading of retail display shelves,
- Little or no need for chemical preservatives,
- Improved presentation – clear view of the product and all round visibility,
- Added product convenience through portion control and re-salable bags.
- Enhanced marketing through identification of brands and improved visibility of the product,
- Hygienic stackable pack sealed and free from product drip and odor,
- Potential shelf life increase,
- Reduction in production and storage costs due to better utilization of labor, space, and equipment,
- Reduction in retail waste, and
- Centralized packaging and portion control.

15.3.2 DISADVANTAGES

- Capital cost of gas packaging machinery,
- Cost of gases and packaging materials,
- Cost of analytical equipment to ensure that correct gas mixtures are being used,
- Cost of quality assurance systems to prevent the distribution of leakers, and so on,

- Increased pack volume which will adversely affect transport costs and retail display space,
- Potential growth of food borne pathogens due to non-maintenance of required storage temperature by retailers and consumers, and
- Benefits of MAP are lost once the pack is opened or leaks.

15.4 MODIFIED ATMOSPHERE PACKAGING OF FRUITS AND VEGETABLES

The MAP is a process by which the shelf life of a fresh product is increased significantly, by enclosing it in an atmosphere which slows down the degradative process such as the growth of microbial organisms, whilst enhancing some beneficial actions such as retaining of desirable color of fruits.

Flushing with nitrogen has long been employed in preserving fatty foods, vacuum packaging is a form of MAP, and the use of sterile air in asceptiic packaging machines might also be so described.

15.5 REASONS FOR USE OF MAP

It is the quality and the need to recognize that in preserving fresh fruits and vegetables we are dealing with products that have been parts of living organisms and retain at least some of the properties of living tissue for their entire existence in the "fresh" state. Equally, chemical changes in foods and spoilage by microbes are activities whose rates can be controlled or stopped entirely by preservation processes. Heat sterilization, as used in canning, is one example of a preservation technique which destroys all significant microbial action, whilst freezing slows down chemical change and microbial activity and is capable of stopping them virtually completely if the temperature is reduced to below 40°C.

From the consumer's point of view, however, these conventional preservation techniques have some disadvantages, chilling produces only very marginal effects on the shelf life. Freezing, even under the best conditions, will induce some changes in the product, particularly when thawing takes place.

Modified atmosphere packaging fits into the important area of preservation, where shelf life is extended without the loss of those important and elusive properties which constitute freshness in the consumer's mind, and therefore move the product into a premium bracket. At premium products mainly have short shelf lives process which extend this even by a few days are of great value to retailers by virtue of the reduced wastage, reduced costs for shelf filling, and in certain instances, the ability to offer products that might otherwise be obtainable only from a specialist outlet.

15.6 TECHNICAL REQUIREMENTS

The range of products is seemingly almost limitless, and extends well beyond the varieties of meat, fish, cheese, and pasta products that we have been customed to seeing.

Three principal gases are used in MAP- Carbon dioxide, Oxygen, and Nitrogen.

15.6.1 CARBON DIOXIDE

It has a powerful inhibitory effect on bacterial and mold growth when present in concentrations above about 20%, is the most important gas. Earlier studies showed that CO_2 inhibits growth of some anaerobic and facultative species, indicating that the effect is due to more than just the exclusion of oxygen. High concentrations of the gas may lead to discoloration and a sharp acid taste in some foods, because carbonic acid is produced by dissolution of CO_2 in the water contained in the food. A further problem consequent upon the solubility of the gas in both fatty and aqueous phases in the food is that of pack collapse, which is countered by adding a suitable gas.

15.6.2 OXYGEN

The presence of oxygen is generally to be avoided with many food products because its presence causes oxidation and enhances the growth of aerobic bacteria. However, it is a necessary means of sustaining the basic metabolism of serpining products.

15.6.3 NITROGEN

It is inert, tasteless, and is virtually insoluble in water. By excluding oxygen, it can inhibit the oxidation of fats and reduce the possibility of mold growth.

15.7 REQUIREMENT OF PACKAGING FILMS

One of the most important parameters to be considered for MAP studies is the need for proper packaging films. Commercially available packs are made from four basic polymers such as polyvinyl chloride (PVC), polyethylene terephthalate (PET), polypropylene, and polyethylene. An atmosphere of high CO_2 and low O_2 concentration in a package decreases product respiration and quality deterioration. Since, the maximum CO_2 and minimum O_2 concentration tolerated by different fruits and vegetables vary, even for different varieties of products, therefore, a film should be selected, which can maintain the desired gas concentration for a specific product for a desired length of storage. Kader et al. (1989) reported that for most available polymeric films, the ratios of CO_2 and O_2 permeabilities are 4–6. However, recently micro perforated films having similar or near similar gas permeability characteristics for O_2, CO_2, and N_2 have been developed and are being used in commercial packaging of highly perishable products like asparagus, mushroom, pea vines, strawberry, and so on.

15.8 INNOVATION IN MODIFIED ATMOSPHERE PACKAGING TECHNIQUES

Smart or intelligent or active packaging systems to improve food safety during packaging and storage have been classified depending upon the influence of food on packaging materials. In smart packaging, chemicals are incorporated into the packaging material. After the food is packaged, these chemicals interact with the atmosphere inside the film packages with the product characteristics or with both, ensuring that optimum conditions are maintained and shelf life of the products enhanced. As a result of the interaction, there is often a visible change in the characteristics of the chemical such as color, which allows the consumer to monitor the safety and shelf life of the product so that they are able to assess the freshness of the stored product (Deshpande, 2012).

TABLE 1 Factors to be considered in the selection of packaging films

Barrier properties	Permeability to various gases and water vapor
Machine capability	Capacity for trouble free operation
Sealing reliability	Ability to seal to itself and to the container
Anti-fog properties	Good product visibility
Special characteristics	Temperature sensitive, quality and freshness indicator

The active or intelligent packaging has various modes of action such as O_2 removal, O_2 barrier, water removal, gas indicator, ethylene removal, CO_2 release, antimicrobial properties, and so on. Recently, new films known as time temperature indicators have also been introduced for commercial application.

15.9 TECHNIQUES FOR CREATING MAP

The MAP can either be commodity generated/passive MA or active MA.

15.9.1 METHODS OF CREATING MODIFIED ATMOSPHERE CONDITIONS

Modified atmospheres can be created either passively by the commodity or intentionally by active packaging (Zagory and Kader, 1988, Kader et al., 1989, Mahajan et al., 2007).

PASSIVE MODIFIED ATMOSPHERE

In passive modification, the respiring product is placed in a polymeric package and sealed hermetically. Only the respiration of the product and the gas permeability of the film influence the change in gaseous composition of the environment surrounding the product. If the product's respiration characteristics are properly matched to the films permeability values, then a beneficial modified atmosphere can be passively created within a package. If a film of correct intermediary permeability is chosen, then a desirable equilibrium modified atmosphere is established when the rates of O_2 and CO_2 transmission through the package equal a product's respiration rate (Smith et al., 1987, Smith et al., 1988, Chau and Talasila, 1994, Jacxsens et al., 2000, Mahajan et al., 2007).

ACTIVE MAP

By pulling a slight vacuum and replacing the package atmosphere with a desired mixture of CO_2, O_2, and N_2, a beneficial equilibrium atmosphere may be established more quickly than a passively generated equilibrium atmosphere (Kader et al., 1989, Labuza, 1990, Zagory, 1998, Ahvenainen, 2003). In case of active modification, two basic techniques are employed to replace air in MAP that is Gas flushing and Compensated Vacuum.

GAS FLUSHING

In this, the air is replaced by passing a stream of gas. The gas flush technique is normally accomplished on a form fill-seal machine. The replacement of air inside a package is performed by a continuous gas stream. This gas stream dilutes the air in the atmosphere surrounding the food product. The package is then sealed. Since, the replacement of air inside the package is accomplished by dilution, there is a limit on the efficiency of this unit. Typical residual oxygen levels in gas flushed packs are 2–5% O_2. Therefore, if the food item to be packaged is very oxygen sensitive, the gas flush technique is normally not suitable. So, when considering a packaging system it is important to consider the oxygen sensitivity of the food product. The great advantage of the gas flush technique is the speed of the machine. Since, the action is continuous, the product rate can be very high (Labuza and Breene, 1989)

COMPENSATED VACUUM

Here, vacuum is first applied to remove the air and then the desired gas or gas mixture is incorporated. Since, the replacement of the air is accomplished in a two-step process, the speed of operation of the equipment is slower than the gas flush technique. However, since the air is removed by vacuum and not simply diluted, the efficiency of the unit with respect to residual air levels is better. Therefore, if the food product is extremely sensitive to oxygen, a compensated vacuum machine must be used (Labuza, 1990, Yam and Lee, 1995)

ACTIVE PACKAGING

Another active packaging technique is the use of O_2, CO_2, or ethylene scavengers/emitters. Such scavengers/emitters are capable of establishing a rapid equilibrium atmosphere within hermetically sealed produce packages (Labuza 1990).

Sometimes, certain additives are incorporated into the polymeric packaging film or within packaging containers to modify the headspace atmosphere and to extend shelf life. This is referred to as Active Packaging. The concept of active packaging has been developed to rectify the deficiencies in passive packaging. For example, when a film is a good barrier to moisture, but not to oxygen, the film can still be used along with an oxygen scavenger to exclude oxygen from the pack. Similarly, carbon dioxide absorbents/emitters, enthanol emitters, and ethylene absorbents can be used to control oxygen levels inside the MA pack. The appropriate absorbent materials are placed alongside with the food. By their activity, they modify the headspace of the package and thereby contribute to extend the shelf life of the contents (Ooraikul and Stiles, 1991, Blakistone, 1997, Ahvenainen, 2003).

15.10 REDUCING PACKAGING COSTS

To be able to identify areas where savings can be made in packaging costs, it is necessary to critically analyze current packaging practices and materials, look for alternatives, and have a meaningful dialogue with the supplier of packages and packaging materials to identify ways to economize without compromising the package quality and expected performance. It is also necessary to be familiar with the total packaging operation. The person responsible for purchasing packaging materials, packages, and accessories must have a thorough knowledge of all requirements for packaging design, equipment, and materials.

Some of the factors that can be analyzed in packaging operations for reducing expenses are given as:

15.10.1 CHOOSING A SUPPLIER

When choosing a supplier, check his facilities, track record, and competence. Do not shop around for marginal price benefits, which are often offset by irregular delivery, inferior or inconsistent quality, higher rejections, and lack of after-service after sales.

15.10.2 SPECIFICATION

Establish clear-cut technical specifications for the packages with respect to material, form, printing, and performance requirements such as compatibility, shelf-life, dispensing, drop test, printing, art work and color scheme, and so on. From an economical point of view, careful advance planning of packages specifications minimizes the risk of mistakes in purchase, reduces waste with effective quality control and better overall packaging management.

15.10.3 ASSESSING PRICE

When assessing price offers from different suppliers, keep in mind that the lowest unit price may not be the least expensive in the overall packaging operation. The price of packaging should always be looked at in terms of the overall performance of the package. Even large price differences might easily be offset by decreased shelf life of the product due to poor quality packaging, higher rejection or, in the case of packaging equipment of lower productivity.

A switch from glass to plastics bottles might have the same effect and in addition, one may be able to eliminate partitions used between the glass bottles and a secondary package, thereby reduce the size and cost of the transport package. In such a case, the total weight to volume ratio will also be reduced, which means lower freight costs. Another example would be a changeover from the blow molded container to a pouch of a multilayer film or laminate at about 20–30% weight and less price.

15.10.4 DETERMINING QUANTITY

In deciding what quantities of packages to order, check the relationship between the unit price of packaging materials and the various quantities that could be ordered. When asking for cost quotations, request prices for quantities corresponding to the expected consumption for 3, 6, and 12 months. A larger quantity is almost always quoted at a lower unit price. The cost savings from buying larger quantities should of course, be weighed against higher capital outlay, added storage costs for the larger volume, and similar other factors.

15.10.5 PACKAGE SIZE

Another way to save on packaging costs may be to increase package unit sizes. Intermediate bulk containers, such as “big bags” serve the purpose of reducing package costs per product unit. Another example is the use of large drums of say 200 1, in place of 5 and 10 I buckets.

15.10.6 PRINTING COSTS

To save on packaging production costs, one should also consider reducing the number of printed colors on the package. Multicolor printing usually means that the package or label has to be run on an expensive printing machine, which means extra production costs. It also means higher rejections in production.

For example, if a full-color reproduction of a product illustration requires four colors, the use of a fifth or even sixth color is only a waste of money. The basic question is, is it really necessary to use the extra color, and does the extra money spent have a decisive promotional effect? Many of the best quality products on the world market carry labels with very simple, one-color printed designs.

15.10.7 SCHEDULING SUPPLIES

Plan the procurement of packaging carefully and well in advance. The process of developing a new product often takes less time than developing an economical and effective packaging for that product. One should, therefore, start the search for, testing of, and packaging alternatives at an early stage. Delivery times may be long due to under-capacity, shortage of raw materials, and preoccupation. The first order almost always takes longer to prepare because extra time is required for making the cylinders, printing plates, and so on. Special attention should be paid to timely, coordinated deliveries of the necessary packaging accessories and auxiliary materials such as labels, closures, and adhesives. Plan any changes in packaging specifications well in advance to avoid expensive disposal of excess stock of obsolete packaging materials.

It is also possible to save on costs in the actual delivery process itself. For instance, one can send one's own truck to pick up the deliveries. Rigid packages, such as plastic bottles, can be delivered in corrugated boxes that can further be used for the transport of your own products after filling, thereby reducing the supplier's packing costs to your benefit, or alternately, boxes of the raw material supplies can be used for delivery of new production.

15.10.8 STORAGE AND HANDLING

Savings can be made by improving storage conditions, rotating stock, and improving internal handling of packaging materials. A lot of packaging materials are lost because of poor storage conditions and careless handling. Each type of packaging has its own storage requirements. For example, film rolls, laminate rolls, and so on can be very conveniently stacked on specially designed stands ensuring no pressure on the surface.

Stored materials can, at least, be protected against dust, mechanical damage, unnecessarily high heat or moisture, damage from water, and so on. Stocks should be sufficiently ventilated and placed on pallets and not directly on the floor, they should not be stacked too high. Personnel should not be allowed to walk on stacked material.

15.10.9 EQUIPMENT

An increasing degree of automation leads to decreasing labor costs but also to increased capital investment requirements. Therefore, it is important to find the most economical equilibrium between labor and automation costs. While computing the costs, in addition to direct labor costs, hidden costs of delays, lower outputs, higher wastage, damages and spilling over of expensive contents, and so on has to be precisely considered.

The packaging equipment must be well maintained for efficient performance. Clean and smoothly running equipment makes better use of packaging materials than machinery that has not been kept in good condition and also stops unnecessary wastage of the product itself.

15.10.10 MODIFYING THE PRODUCT

Packaging savings can also be achieved by making small modifications in the products being packed. Packaging economy actually starts with good product design. Space saving is an important consideration.

Package dimensions can be optimized for the most economical use of the materials. Constructions can be modified and unnecessary waste avoided. Most packaging materials are purchased by weight but the determining factor, from the economic point of view, is actually the yield, that is how many individual packages in numbers are available per kilogram of material.

15.10.11 COST

Sometimes, marketing success depends on how low the product is priced. The ultimate choice of the packaging material is often influenced by the cost at which the functions indicated above are achieved.

The criteria discussed above are some of the more important considerations. Depending upon the type of product and the market, many additional considerations may become important. However, by and large if above criteria are used while selecting the packaging material, the user is likely to arrive at an optimum choice of packaging material.

15.11 APPLICATIONS

Some applications of modified atmosphere packaging to illustrate possible uses are given below.

The use of seal wrapping for individual produce items is a good outlet. The technology of seal wrapping is one of forming a tight, uniform packaging skin over each produce item. Units can be passed through a hot air tunnel to shrink the wrapping film. In this form, produce can achieve twice the shelf life. The marginally, higher cost is offset by protecting the product loss by as much as 20% and the 100% extension of shelf life in case of most of the fruits. Wide application of seal wrapping could mean a large market for polyolefin films.

A second area of modified atmosphere packaging, which demands increased attention, is bakery products. Baked items are extremely reactive chemically at the end of the baking process. In addition, they are relatively free of microbial contamination. Extensive work has been carried out in Europe and in particular by David Seiler in England on the use of pure carbon dioxide atmospheres to extend the shelf life of bread. Findings to date have shown that carbon dioxide will inhabait the growth of the mold, thus allowing one-month storage at room temperature. The effect of carbon dioxide on the staling of bread is unclear and more work is required to determine whether a carbon dioxide atmosphere will reduce staling rates.

A third major development involving modified atmosphere applications is with red meat and poultry. Experiments and commercial tests using gas mixtures of 50% carbon dioxide and oxygen have proven this mixture useful in extending the refrigerated shelf life of prime cuts of beef. Color, which is dark on opening returns to normal red on exposure to air. Microbial growth is inhibited by the carbon dioxide, while the oxygen maintains the respiratory activity of the meat.

Poultry has been packed in multiple retail units with carbon dioxide flushed over the pack. The bulk package extends shelf life during distribution prior to retail display. Fish portions could be pre-packed and handled in a similar manner. In these cases, carbon dioxide inhabits microbial growth at refrigerated temperatures. It should be remembered that microbes are living plants cells. High concentrations of carbon dioxide inhabits the metabolic activity of many microbes.

In summary, modification of the gas atmosphere surrounding foods combined with selectively permeable films can result in significant shelf life extension, particularly at refrigerated temperatures. Applications include individual seal wrapping of produce, cuts of meat and bakery items. In the case of living foods, such as meat and produce, films or atmospheres must be selected to prevent the development of gas concentrations harmful to food quality. For products such as bread, pure carbon dioxide is useful in inhibiting mold growth. For products affected by oxygen, vacuum, or nitrogen atmospheres, perhaps with hydrogen, or oxygen scavenging systems, are desirable.

15.12 FUTURE CHALLENGES FOR MAP

The MAP undoubtedly helps in increasing the shelf life of fruits and vegetables and presents the consumer with distinct advantages of high quality product, clear view of the product and nil or very little use of chemical preservative and the food producer with increased shelf life. On the other hand, there is a viable added cost, which has to be borne by the consumer such as precise temperature control is required throughout the storage and all the benefits are lost once the pack is opened, the future thrust should be in the area of devising films, which are cost effective and able to maintain the quality of the packaged produce with less stringent temperature control or firms, that can change their gas permeability characteristics in response to temperature changes, thus compensating for frequent changes due to respiration of the packed product. The research in this area has already assumed some significance in terms of development of smart and intelligent films.

KEYWORDS

- **Active packaging**
- **Food safety**
- **Heat sterilization**
- **Pasteurization**
- **Raw material**

REFERENCES

1. Ahvenainen, R. *Novel food packaging technology*, Published in CRC Press, Boca Raton Boston New York Washinton, DC and Published by Woodhead Publishing Ltd., Cambridge, London (2003).
2. Blakistone, B. A. *Principles and Applications of Modified Atmosphere Packaging of Foods*. Laroisier Booksheller, Librairie, France (1997).
3. Brody, A. L. *Controlled/Modified Atmosphere/Vacuum Packaging of Meat, Controlled/Modified Atmosphere/Vacuum Packaging of Foods*. A. L. Brody (Ed.), Food and Nutrition Press, Trumbull, CT, USA, pp. 17–38 (1989).
4. Cameron, A. C. *Modified atmosphere packaging: a novel approach for optimizing package oxygen and carbon dioxide*. In: 5th international Controlled Atmosphere Conference, Washington, USA, **2**, 197–209 (1989).
5. Chau, K. V and Talasila, P. C. Design of modified atmosphere packages for fresh fruits and vegetables. R. P. Singh and F. A. R. Oliveira (Eds.), *Minimal processing of foods and process optimization*. CRC Press, Boca Raton, Florida, pp. 407−416 (1994).
6. Church, N. Development in modified-atmosphere packaging and related technologies. *Trends in Food Science and Tech*., **5**, 345–352 (1994).
7. Church, I. J. and Parsons, A. L. Modified atmosphere packaging technology: A review, *J. Sci. Food Agric*., **67**, 143–152 (1995).
8. Davies, A. R. *Advances in modified atmosphere packaging, new method of food preservation*, G. W. Gould (Ed.), Glasgow, UK, Blackie, pp. 304–320 (1995).
9. Deshpande, S. D. Packaging–Packaging of Liquids. (Ms: 0872). *Encyclopedia of Food Sciences and Nutrition*. B. Caballero, L. Trugo, and P. Finglas (Eds.) (Revised). Academic Press, pp. 4303–4309, April 2003 Second Edition **(UK)** (Book Chapter) (2003).
10. Deshpande, S. D. Recent Advances in Packaging of Health Foods. Book Chapter 32 in the Book published entitled '*Health Foods: Concept, Technology & Scope*. R. K. Gupta, Sangita Bansal, and Manisha Mangal (Eds.) CIPHET Ludhiana. Published by Biotech Books, New Delhi (2012).
11. Fonseca Susana, C., Oliveira Fernanda, A. R., and Brecht Jeffrey, K. Modeling of respiration rate of fresh fruits and vegetables for modified atmosphere packages: a review. *Journal of Food Engineering*, **52**, 99–119 (2002).
12. Gorris, L. and Tauscher, B. Quality and safety aspects of novel minimal processing technology. F. A. R. Oliveira and J. C. Oliveira (Eds.), *Processing of foods: Quality optimization and process assessment*, CRC Press, Boca Raton, USA, pp. 325–339 (1999).
13. Jacxsens, L., Devlieghhere, F., and Debevere, J. Validation of a systematic approach to design equilibrium modified atmosphere packages for fresh-cut produce. *Lebensmittel Wissenschaft und Technologie*, **32**, 425–432 (2000).
14. Jayas, D. S and Jeyamkondan, S. Modified atmosphere storage of Grain Meat Fruit and Vegetables. *Biosystem Engineering*, **82**(3), 235–251 (2002).
15. Kader, A. A. Regulation of fruits physiology by controlled and modified atmosphere. *Acta Horticulture*, **398**, 59–70 (1995).
16. Kader, A. A., Zagory, D., and Kerbel, E. L. Modified atmosphere packaging of fruits and vegetables. *CRC Critical Reviews in Food Science and Nutrition*, **28**, 1–30 (1989).
17. Labuza, T. P. Active food packaging technologies. *Food Science and Technology Today*, **4**(1), 53–54 (1990).
18. Labuza, T. P. and Breene, W. M. Applications of Active Packaging for improvement of shelf-life and nutritional quality of fresh and extended shelf-life foods. *Journal of Food Processing and Preservation*, **13**, 1–69 (1989).
19. Mahajan, P. V., Oliveira, F. A. R., Montanez, J. C., and Frias, J. Development of user-friendly software for design of modified atmosphere packaging for fresh and fresh-cut produce. *Innovative Food Science and Emerging Technology*, **8**, 84–92 (2007)
20. Mangaraj, S. and Goswami, T. K. Respiration rates modeling of Royal Delicious apple at different temperature. GSB publications, *Fresh Produce*, **2**(2), 72–80 (2008).

21. Mangaraj, S. and Varshney, A. C. Fruit grading in India: a review. *Indian Food Industry*, **25**(3), 46–52 (2006)
22. Ooraikul, B. and Stiles, M. E. *Modified Atmosphere Packaging of Food*. Ellis Horwood, UK (1991).
23. Saltveit, M. E. A summary of CA and MA requirements and recommendations for the storage of harvested vegetables. G. D. Blanpied, J. A. Barstch, and J. R. Hicks (Eds.), Proceedings of the Sixth International Controlled Atmosphere Research Conference, Ithaca, New York, USA, **2,** 800–81 (1993).
24. Singhal, V. *Indian Agriculture*. Indian Economic Data Research Center, New Delhi (2006).
25. Smith, S. M., Geeson, J. D., Browne, K. M., Genge, P. M., and Everson H. P. Modified atmosphere retail packaging of discovery apples. *J. Sci. Food Agric*., **40**, 161–167 (1987).
26. Smith, S. M, Geeson, J. D., and Genge, P. M. Effects of harvest date on the responses of discovery apples to modified atmosphere retail packaging, *Int. J. Food Sci. Technol*., **23**, 78–86 (1988).
27. Yam, K. L. and Lee, D. S. Design of modified atmosphere packaging for fresh produce. M. L. Rooney (Ed.), *Active food packaging*, Blackie Academic and Professional, New Zealand, p. 55 (1995).
28. Zagory, D. and Kader, A. A. Modified atmosphere packaging of fresh produce. *Food Technology*, **42**(9), 70–77 (1988).
29. Zagory, D. An update on Modified Atmosphere Packaging of Fresh Produce. *Packaging International*, http://www.davisfreshtech.com (1998).

[illegible]

CHAPTER 16

ACTIVE AND INTELLIGENT PACKAGING

PANUWAT SUPPAKUL

ABSTRACT

Nowadays, active and intelligent packaging is of increasing interest. It has been recognized and adopted to the agricultural, food and pharmaceutical industries. This chapter firstly defines the active and intelligent packaging concepts. Secondly, the present literature review outlines the active packaging including oxygen scavenging, moisture regulating, carbon dioxide generating and scavenging, ethylene scavenging and blocking, antimicrobial packaging, and antioxidant packaging, as well as thermally equilibrious modified atmosphere packaging. Thirdly, the chapter reviews the intelligent packaging including integrity indicator, freshness indicator, and time-temperature indicator, as well as radio frequency identification. The chapter then provides future trends to complete the chapter.

16.1 INTRODUCTION

Consumer demand for mildly preserved, minimally processed, easily prepared and ready-to-eat "fresher" foods – together with retail and distribution practices related to the globalization, new consumer product logistics, new distribution trends (for example internet shopping), automatic handling systems at distribution centers, and the stricter requirements regarding consumer health – pose major challenges for food quality and safety (Vermeiren et al., 1999, Sonneveld, 2000, Yam et al., 2005). A reduction in shelf life of foods as a result of microbial contamination leading to an increase in the risk of food-borne illness (Juneja and Sofos, 2010) and lipid oxidation resulting in a chronic toxicity to human cellular materials (Davies, 1995) are driving forces for innovation in food packaging. Active packaging (AP) technologies are being developed as a result of these driving forces (Suppakul et al., 2003). Moreover, food traceability is now a legal requirement, especially in the European Union. This establishes a chain of responsibility throughout the entire food supply chain. Consequently there is great interest among the food industry, retailers, consumers" rights watchdogs, and food safety controlling bodies in developing accurate, cost-effective, rapid, reliable, non-invasive and non-destructive methods or devices to evaluate real-time freshness of food products. An alternative concept to meet this requirement is the development of intelligent packaging (IP) (Nopwinyuwong et al., 2010). The Institute of Food Technologists (IFT) has issued the Scientific Status Summary to inform readers of innovations related to active and intelligent packaging materials (Brody et al., 2008). Recently, Pereira de Abreu et al. (2012) reviewed the literatures in accordance with the active and intelligent packaging for the food industry. Anyway, this chapter will define the terms of both "active packaging" and "intelligent packaging". Possible technologies to apply for foods, beverages and fresh produces in relation to polymers in packaging applications will be discussed.

16.2 DEFINITION

Generally, packaging is considered to provide the major functions: containment, protection, utility and convenience, communication and information, and machinability (Figure 1). Other minor functions often concerned are motivation, identification, resistance to processing, and reuse or recycle. Therefore, selection of appropriate packaging that is specified to foods, beverages or fresh produces should take into account all these functions. Certain trends in food packaging including AP and IP reflect increased understanding of factors that enhance the quality of foods (Robertson, 2012). These advances can enhance the protection and the communication and information function of packaging, respectively (Suppakul, 2012).

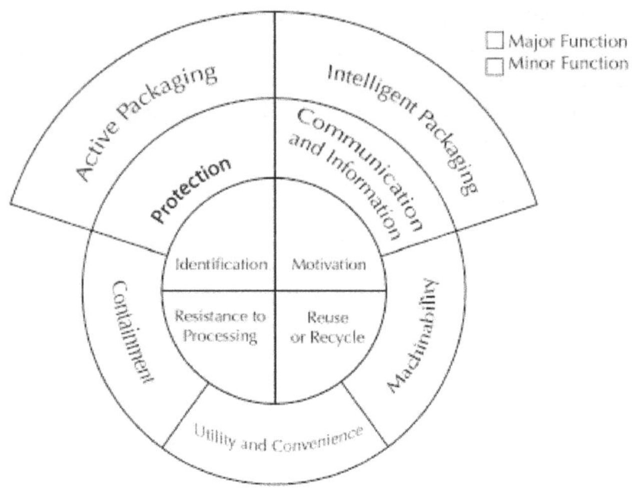

FIGURE 1 Packaging functions.

Active packaging is particularly important in the area of fresh and extended shelf life foods as originally described by Labuza and Breene (1989), and followed by Guilbert et al. (1996) and Rooney (1998). Floros et al. (1997) reviewed the products and patents in the area of AP. Labuza and Breene (1989) defined AP as a smart system that involves interactions between package components and food or internal gas atmosphere and complies with consumer demands for high quality, fresh-like, and safe products. Rooney (1998) defined AP as a packaging which does more than simply provide a barrier to outside influences. It can control, and even react to, events taking place inside the package in order to extend shelf-life and maintain quality and safety of food products. Active packaging is an innovative concept that can be defined as a mode of packaging in which the package, the product and the environment interact to prolong shelf life, and/or enhance safety and/or sensory properties, while maintaining the quality of the product (Suppakul et al., 2003). Definition stated in Regulation 1935/2004/EC and in Regulation 450/2009/EC considers *active materials and articles*: "materials and articles that are intended to extend the shelf-life or to maintain or improve the condition of packaged food". There are many different types of active materials and articles (Ozdemir and Floros, 2004, Rooney, 2005). The present literature review outlines the AP concepts including oxygen scavenging (OS), moisture regulating (MR) (that is liquid water absorption (LWA) and humidity buffering (HB)), carbon dioxide generating (CG), carbon dioxide scavenging (CS), ethylene scavenging (ES), ethylene blocking (EB), antimicrobial packaging (AMP), and antioxidant packaging (AOP), as well as thermally equilibrious modified atmosphere packaging (TEMAP).

Intelligent packaging refers to a package that can sense environmental changes, and in turn informs the changes to the users (Summers, 1992). Karel (2000) defined IP as a package containing sensor that notify consumers that the product is impaired, and it may begin to undo the harmful changes that have occurred in the products. IP is used to emphasize the role of the package as an intelligent messenger or information link in the Information Age (Yam, 2000). Clarke (2001) defined IP as a packaging system that possessed logic capability and smart packaging (SP) as one that communicated. Perhaps the most commonly encountered definition of IP is provided by the European study as "systems that monitor the condition of packaged foods to give information about the quality of the packaged food during transport and storage" (ACTIPAK, 2001). Rodrigues and Han (2003) defined IP as having two categories: simple IP (as defined by Summers (1992), and interactive or responsive IP (as defined by Karel (2000)). IP can also be defined as a mode of packaging that is capable of carrying out intelligent functions (such as detecting, sensing, recording, tracing, communicating, and applying scientific logic) to facilitate decision making to extend shelf life, enhance safety, improve quality, provide information, and warn about possible problems (Yam et al., 2005). Intelligent packaging is also an innovative concept that can be defined as a packaging system (or material) that use either internal (for example metabolites) or external (for example temperature, gas, humidity) package environment as "information" to monitor a status of product quality for enhancing product safety, biosecurity and/or convenience or capable of tracking a product for automatic product identification and traceability (Suppakul, 2012). Definition stated in Regulation 1935/2004/EC and in Regulation

450/2009/EC considers *intelligent materials and articles*: "materials and articles which monitor the condition of packaged food or the environment surrounding the food". The present literature review outlines the IP concepts including integrity indicator (II), freshness indicator (FI) (that is food spoilage indicator (FSI), ripeness indicator (RPI), and rancidity indicator (RCI)), time-temperature indicator (TTI) and radio frequency identification (RFID).

16.3 ACTIVE PACKAGING

16.3.1 OXYGEN SCAVENGING

The presence of O_2 in a packaged food is often a key factor in limiting the shelf life of food products. Oxidation can cause changes in flavor, color, and odor, as well as destroy nutrients and facilitate the growth of aerobic bacteria, moulds and insects. Therefore, the removal of O_2 from the package headspace and from the solution, in liquid foods and beverages, has long been a target of food-packaging scientists. The deterioration in quality of O_2-sensitive products can be minimized by recourse to O_2 scavengers that remove the residual O_2 after packing. Existing O_2 scavenging technologies are based on oxidation of one or more of the following substances: iron powder, ascorbic acid, photo-sensitive dyes, enzymes (for example glucose oxidase and ethanol oxidase), unsaturated fatty acids (for example oleic, linoleic and linolenic acids), rice extract, or immobilized yeast on a solid substrate (Floros et al., 1997). These materials are normally contained in a sachet. Details on O2 scavenging can be obtained from other reviews (Labuza and Breen 1989, Miltz et al., 1995, Miltz and Perry, 2000, Floros et al., 1997, Vermeiren et al., 1999).

Oxygen scavenging is an effective way to prevent growth of aerobic bacteria and moulds in dairy and bakery products and to retard loss of vitamins in beverages. Oxygen concentrations of 0.1% v/v or less in the headspace may be required for this purpose (Rooney 1995). Packaging of crusty rolls in a combination of CO_2 and N_2 (60% CO_2) has shown to be an effective measure against mould growth for 16-18 days at ambient temperature. However, the study also revealed that such an "anaerobic environment" is not totally effective without the incorporation of an oxygen scavenger into the package to ensure that the headspace O_2 concentration never exceeds 0.05%. Under such conditions the rolls remain mould-free even after 60 days (Smith et al., 1986).

Oxygen scavenging is advantageous for products that are sensitive to O_2 and light. One important advantage of AP over modified atmosphere packaging (MAP) is that the capital investment involved is substantially lower, in some instances, only sealing of the system that contains the oxygen absorbing sachet is required. This is of extreme importance to small and medium-sized food companies for which the packaging equipment is often the most expensive item (Ahvenainen and Hurme, 1997). An alternative to sachets involves the incorporation of the O_2 scavenger into the packaging structure itself. This minimizes negative consumer responses and offers a potential economic advantage through increased outputs. It also eliminates the risk of accidental rupture of the sachets and inadvertent consumption of their contents.

Since the share of polymers in primary packages for foods and beverages increases constantly, they have become the medium for incorporation of active substances such as antioxidants, O_2 scavengers, flavor compounds, pigments, enzymes, and AM agents (Hotchkiss, 1997). The BP Amoco Chemical (USA) is marketing Amosorb$^{\grave{O}}$ 2000 and 3000, which are polymer-concentrates containing iron-based O_2 scavengers. These can be used in polyolefins and in certain polyester packaging applications for wines, beers, sauces, juices and other beverages. During 2000, Ciba Specialty Chemical Corporation acquired the ferrous AmosorbÒscavengers. These can be used in polyolefins and in certain polyester packaging applications for wines, beers, sauces, juices and other beverages. During 2000, Ciba Specialty Chemical Corporation acquired the ferrous Amosorband inadvertent consumption of e OS2000Ôscavengers. These can be used in polyolefins and in certain polyester packaging app$_2$Ôscavengers. These can be used in polyolefins and in certain polyester packaging applications for wines, beers, sauces, juicesAustralia (Brody et al., 2001), as well as Oxy-GuardÔll as Oxy-Guardped by Sealed Air Corporation's Cryovac division and Süd-Chemie, the film features several layers to protect blistered tablets and capsules (Süd-Chemie, 2010). The outer two layers serve as barriers to moisture and oxygen ingress into the blister, whereas an inner layer actively scavenges oxygen in blister pack headspace as well as any fugitive oxygen molecules that enable to penetrate. The most inner layer serves as the contact layer. These are organic-based, UV light-activated O_2 scavengers that can be tailored to allow them to be bound into various layers of a wide range of packaging structures. OxbarÔscavengers that can be tailored to allow them to be bound into various layers of a wide range of packaging structures. Oxbarygen molecules that enable to penetrate. The most inner layer serves as the contact layer. These are organic-based, UV light-activaET) bottles for packaging of wine, beer, flavored alcoholic beverages and malt-based drinks (Brody et al., 2001). ActivSeal^{Ò} has been developed through collaboration with ColorMatrix and KTW/Husky. ActivSeal^{Ò} is based on the Hy-Guard technology developed by ColorMatrix and the closure has been designed with KTW/Husky. The ActivSeal® system consists of 2 elements, the bottle and the closure. Within the bottle wall, minute amounts of the palladium-based catalyst are present. A hydrogen activator is built in to the closure. Once the bottle is filled and the closure applied, hydrogen is released. As oxygen enters through the container wall, it binds with the hydrogen on the surface of the catalyst to form nano-particles of odorless and colorless water. Thus oxygen is prevented from entering the container and affecting the contents. The amount of water produced is at a part-per-million (ppm) level and does not affect the contents in any way (APPE, 2011).

Oxygen scavengers have opened new horizons and opportunities in preserving the quality and extending the shelf life of foodstuffs. However, much more information is needed on the action of O_2 scavengers in different environments before optimal, safe and cost-effective packages can be designed. The need for such information is especially acute on O_2 scavenging films, labels, sheets and trays that have begun to appear in the last two decades.

Aromatic pendant is synthesized by ethylene methylacrylate cyclohexeneylmethyl acrylate (EMCM) copolymerization, which is a resin enables to be oxidized by O_2. Because of 21% of O_2 in the atmospheric environment, 10% of oxygen scavenging polymer (OSP) is required to couple with a photoinitiator (for example 1,3,5-tris(4-benzoylphenyl) benzene) to prevent the reaction with O_2 prior to end use and catalyst (for example colbalt oleate) as a master batch. Then, this master batch will mix with 90% EMCM oxidizable resin in order to fabricate OS polymeric film. Before packing the food products, this OS film will be initiated with ultra-violet light in a narrow range of 250-320 nm (Rodgers and Compton, 2000) for oxygen scavenging from the food package headspace. These pendant cyclic olefins will scavenge oxygen molecules *via* both aromatization and ring opening but not dissociated from polymer backbone. This oxygen scavenging polymer is commercially available under the trademark of OS-PÔboth aromatization and ring opening but no

Damaj et al. (2008) revealed that copper (I) complexes with "glycoligands" containing a central saccharide scaffold can react with dioxygen, but not enough to be used as oxygen scavenger. Later on, Damaj et al. (2009) successfully developed Co(II) (L-proline)$_2$(H$_2$O)$_2$ solid complex as ligand which was incorporated into polymer matrix as oxygen scavenging films. Further studies onto the O_2 absorption kinetic, temperature and relative humidity influence, in order to predict and design active MAP are ongoing.

Supporting scavenging systems on nanoclays is a convenient strategy to develop novel engineered nanofiller with multiple functionalities (that is oxygen scavenger, antimicrobial, and antioxidant as well as drug control release, and so on). Iron in organo-modified momtmorillonite has been prepared and dispersed well into packaging plastics with efficient oxygen scavenging capacity and minimum impact on mechanical and optical properties. A feasible propriety technology is marketing under the trademark of O2Block$^{\grave{O}}$ (NanoBiomatters Ltd., Paterna, Spain) (Lagarón and Busolo, 2011).

Opposed to the currently available chemical oxygen scavengers, systems based upon natural and biological components have been developed. A model system for a new oxygen scavenging PET bottle is proposed using an endospore-forming bacteria genus *Bacillus amyloliquefaciens* as the active ingredient. A multilayer PET bottle consisting of a poly(ethylene terephthalate, 1,4-cyclohexane dimethanol) (PETG) middle layer containing bacterial spores surrounded by two outer PET layers. The inside of the bottle is in contact with the product, allowing moisture uptake of the bottle needed for spore germination. The system allows scavenging of residual oxygen from the in-bottle environment and scavenging from atmospheric oxygen permeating through the bottle wall (Anthierens et al., 2011).

16.3.2 MOISTURE REGULATING

In solid foods, a certain amount of moisture may be trapped during packaging or may develop inside the package due to generation or permeation. Unless it is eliminated, it may form a condensate with the attendant spoilage and/or low consumer appeal. Moisture problems may arise in a variety of circumstances, including respiration in horticultural produce, melting of ice, temperature fluctuations in food packs with a high equilibrium relative humidity (ERH), or drip of tissue fluid from cut meats and produce (Rooney, 1995). Their minimization via packaging can be achieved either by liquid water absorption or humidity buffering.

LIQUID WATER ABSORPTION

The major purpose of liquid water control is to lower the water activity, a_w, of the product, thereby suppressing the growth of microorganisms on the foodstuff (Vermeiren et al., 1999). Temperature cycling of high a_w foods has led to the use of plastics with an antifog additive that lowers the interfacial tension between the condensate and the film. This contributes to the transparency of the films and enables the customer to see clearly the packaged food (Rooney, 1995) although it does not affect the amount of liquid water present inside the package.

Several companies manufacture dripabsorbent sheets such as Thermarite® or Peaksorb® (Australia), or ToppanÔ(Australia), or Toppanacture dripabsorbent $_w$ foods such as meat, fish, poultry and fresh produce. Principally, these systems comprise a superabsorbent polymer incorporated between 2 layers of a micro-porous or nonwoven polymer. Such sheets are used as drip-absorbing pads placed under whole chickens or chicken cuts. Large sheets are also utilized for absorption of melted ice during air transportation of packaged seafood. The preferred polymers used for this purpose are polyacrylate salts and graft copolymers of starch (Rooney, 1995).

Desiccants have been successfully used for moisture control in a wide range of foods, such as cheeses, meats, chips, nuts, popcorn, candies, gums and spices. Silica gel, molecular sieves, calcium oxide (CaO) and natural clays (such as montmorillonite) are often provided in TyvekÔesiccants have been successfully used for moisture control in a wide range® and StripPax® packets, the DesiMax® (United Desiccants, U.S.A.) and the DesiPak®, Sorb-it®, Trisorb® and 2-in-1Ôand 2-in-1siccanorb Technologies Inc., U.S.A.).

HUMIDITY BUFFERING

This approach involves interception of moisture in the vapor phase by reducing the in-pack relative humidity and thereby the surface-water content of the food. It can be achieved by means of one or more humectants between two layers of a plastic film that is highly permeable to water vapor or by a moisture-absorbing sachet. An example of this approach is the PichitÔhis approach involves interception of moisture in the vapor phase by reducing theish and chicken and reduces the ERH in the vicinity of the product, but has not been evaluated experimentally. Pouches containing NaCl have also been used in the US tomato market (Rooney 1995).

In many cases, the commercialization of modified atmosphere packaging (MAP) is limited because of the condensed moisture accumulation in the package enhancing increased pathological and physiological disorders in fresh produces. Xtendò (XF), a hydrophilic plastic packaging film, was developed by StePac L.A. Ltd, Tefen, Israel, in cooperation with the Agricultural Research Organization-The Volcani Center, Israel. A series of plastic films has a higher water vapor permeability than most commercially available MAP products. A desired in-pack relative humidity was obtained using these films manufactured by co-extrusion of propriety blends consisting of different polyamides with other polymeric and non-polymeric compounds. The different blends permit these films with different water transmission rate (WVTR) in accordance with the requested relative humidity in a specific fresh produce packaging. The microperforated Xtendò packaging allowed the formation of a desirable modified atmosphere condition, suppressing ripening and senescence of the produce. Additional advantageous effects of Xtendò films included reduction in chilling injury, decay, leaf elongation, leaf sprouting, tissue discoloration, peel blemishes, and off-odor formation, and inhibition of microbial growth on the produce surface (Aharoni et al., 2007).

16.3.3 CARBON DIOXIDE GENERATING AND SCAVENGING

Carbon dioxide (CO_2) is known to exert a microbiological inhibitory effect in products such as meat, cheese, and baked goods. Relatively high CO_2 levels (60–80%) inhibit microbial growth on surfaces and, in turn, prolong shelf life. Pack collapse or the development of a partial vacuum can also be a problem for food packaged with an oxygen scavenger. To overcome this problem, generation of an equal volume of CO_2 is desired (Day, 2008). Although CO_2 often suppress microbial activity excess CO_2 may adversely affect the food products (that is anaerobic metabolism, pH reduction, color and flavor changes), or damage the package. Generally, the CO_2 permeability of plastic films is too low to evacuate the CO_2 excess produced by some products. So, a possible solution to that problem would be the design of packaging technologies capable of scavenging the excess of CO_2 that could affect the quality of certain foodstuffs (Brody et al., 2001).

CARBON DIOXIDE GENERATING

A complementary approach to O_2 scavenging is the impregnation of a packaging structure with a CO_2 generating system or the addition of the latter in the form of a sachet. Since the permeability of CO_2 is 3-5 times higher than that of O_2 in most plastic films, it must be continuously produced to maintain the desired concentration within the package (Suppakul et al., 2003). A CO_2 generator is only useful in certain applications such as fresh meat, poultry, fish and cheese packaging (Floros et al., 1997). In food products for which the volume of the package and its appearance are critical, an O_2 scavenger and CO_2 generator could be used together (Smith et al., 1995) in order to prevent package collapse as a result of O_2 absorption.

Nakamura and Hoshino (1983) reported that an oxygen-free environment alone is insufficient to retard the growth of *Staphylococcus aureus*, *Vibrio* species, *Escherichia coli*, *Bacillus cereus* and *Enterococcus faecalis* at ambient temperatures. For complete inhibition of these microorganisms in foods, the authors recommended a combined treatment involving O_2 scavenging with thermal processing, or storage under refrigeration, or using a CO_2 enriched atmosphere. They found that an O_2 and CO_2 absorber inhibited the growth of *Clostridium sporogenes* while an O_2 absorber and a CO_2 generator enhanced the growth of this microorganism, which is quite a surprising result. This result indicates the importance of selecting the correct scavenger to control the growth of *Clostridium* species in MAP foods.

The CO_2 generating sachet and label devices can either be used alone or combined with an oxygen scavenging system. An instance of the former is VerifraisÔenerating sachet and label devices can either be used alone or combined with an oxygen scavenging system. An i. This system comprises a standard MAP tray that has a perforated false bottom under which a porous sachet containing sodium bicarbonate/ascorbate is positioned. When juice exudates from MA packaged meat, poultry or fish drips onto the sachet, CO_2 is generated and this bacteriostatic gas can replace the CO_2 already absorbed by the fresh food, therefore avoiding pack collapse (Hogan and Kerry, 2008). In addition, CO_2 sachets and labels usually contain ferrous carbonate and a metal halide catalyst even though non-ferrous variants (for example ascorbate and sodium hydrogen carbonate) are available. Commercial manufacturers include AgelessÔascorbate and sodium hydrogen carbonate) are available. Commerc^{Ò} type C and type CW (Toppan Printing Co., Japan), Vitalon® type G (Toagosei Chemical Co., Japan), and FreshPax^{Ò} type M (Multisorb Technologies, Inc., USA).

CARBON DIOXIDE SCAVENGING

CO_2 concentration inside the package increases in some foods due to deterioration or respiration reactions. For example, packaged coffee beans may produce CO_2 during storage as a result of nonenzymatic browning reactions. High CO_2 levels may, however, cause changes in taste of products and the development of undesirable anaerobic glycosis in fruits (Suppakul et al., 2003). Fermented products such as pickles, sauces, kimchi (lactic acid fermented vegetables), and some dairy products can produce CO_2 after the packaging process. The CO_2 scavenging systems are quite useful for products that require fermentation and undergo aging processes after they have been packed.

The utilization of CO_2 scavengers is particularly applicable for fresh roasted or ground coffees, which create substantial volumes of CO_2. After hermetically sealing, the CO_2 released will build up within the packs and eventually cause them to burst. To overcome this problem, two solutions are used. The first is to use packaging with a patented one-way valves that will permit excess CO_2 built up in headspace to escape to outside. The other is to employ a CO_2 scavenging or a dual-action O_2 and CO_2 scavenging system. A mixture of calcium oxide and activated carbon has been used in polyethylene-lined coffee pouches to scavenge CO_2 but a dual-action sachet or lebel is more common and is commercially applied for canned and foil pouched coffees in Japan and the USA (Day, 2008). Commercially available dual-action scavengers are AgelessÔbut a dual-action sachet or lebel is more common and is commercially applied for c^{Ò} type CV (Toppan Printing Co., Japan).

An innovative dual-action CO_2 scavenger and O_2 emitter sachet has been developed by EMCO Packaging Systems Ltd. (Kent, UK) to counter respiration in high oxygen MAP of fresh-cut produce (Parker, 2002). Recently, Aday et al. (2011) investigated and documented the effectiveness of CO_2 and O_2 scavengers to maintain the quality attributes of fresh strawberries. The use of active packaging resulted in slowed accumulation of CO_2 and consumption of O_2. CO_2 scavengers were effective in preserving quality.

16.3.4 ETHYLENE SCAVENGING AND BLOCKING

Ethylene (C_2H_4) is a gaseous natural plant growth hormone. It acts at very low concentrations throughout the plant life cycle by stimulating or regulating seed germination, root initiation, flower development, fruit ripening, leaf abscission and senescence (Lin et al., 2009). Ethylene is biosynthesized from the amino acid methionine (MET) to S-adenosyl methionine (SAM) by methionine adenosyltransferase. SAM is then converted to 1-aminocyclopropane carboxylic acid (ACC) by ACC synthase. The final step requires oxygen and involves with ACC oxidase. In addition, ethylene biosynthesis can be induced by endogeneous or exogeneous ethylene. The principal sources of the low background levels of the latter are (i) climacteric fruits that ripen after harvesting and are characterized by a rise in respiration rate as well as a burst of ethylene production as they ripen, (ii) damaged or rotten produces, and (iii) exhaust gases from petrol combustion engines (Scully and Horsham, 2007). To prolong shelf-life and maintain acceptable appearance, textural property and organoleptic quality, accumulation of ethylene inside the package should be avoided (Vermeiren et al., 1999, López-Rubio et al., 2004).

ETHYLENE SCAVENGING

Ethylene scavenging can be defined as "When the packaging system acquires ethylene removal activity, the packaging system (or material) scavenges ethylene from the fresh fruit or vegetable environment then the ripening and deterioration processes of plant products are slowed, and so the storage life is prolonged." Packaging technologies designed to scavenge ethylene from the surrounding environment of packaged fresh produces have been developed. The most widely used ethylene scavenging technology is in the form of a sachet containing potassium permanganate ($KMnO_4$) immobilized on an inert porous support material with a large surface area such as perlite, alumina, silica, vermiculite, activated carbon or celite (Zagory, 1995), at a concentration of approximately 4–6 % w/w (Vermeiren et al., 1999, Vermeiren et al., 2003, Scully and Horsham, 2007). The ethylene is scavenged through an oxidation reaction with $KMnO_4$ to form manganese oxide, potassium hydroxide and carbon dioxide.

Although, these $KMnO_4$-based ethylene scavenging sachets, such as E.G.G. from Ethylene Control Inc., USA, are effective at removing ethylene, their use is sometime accompanied by undesirable effects including possible migration of $KMnO_4$ from the sachet onto the fresh produces, desirable aroma scalping and a general lack of user enthusiasm. Other types of sachet-based ethylene scavenging technologies employ activated carbon with a metal catalyst (for instance, palladium), such as SendoMate[Ô] from Mitsubishi Gas Chemicals Co. Ltd., Japan and NeupalonÔfrom Mitsubishi Gas Chemicals Co. Ltd., Japa

Whereas sachets have a limited consumer acceptance, ethylene scavenging packaging films are growing in popularity. In a decade, a number of ethylene scavenging polyethylene (PE)-based films incorporated with finely dispersed powdered material (mineral), such as zeolites, clays, and carbon, has become available. Instances of commercially available ethylene scavenging packaging films include Evert-Fresh from Evert-Fresh Co., USA, PeakfreshÔ(PE)-based films incorporated with finely dispersed powdered material (mineral), such as zeolites, clays, and carbon, has become available. Instances of commercially available ethylene scav, Day, 2008). The drawbacks of this technology are that most of the tested films are opaque and do not sufficiently absorb ethylene (Suslow, 1997). Moreover, the added minerals should be very finely divided to minimize alterations in the permeability and other physical properties of the package. The effectiveness of those films to increase the shelf-life of fresh produce is claimed to be caused by the ethylene-scavenging capacity of these minerals. However, an incorporation of mineral particles in polyolefinic films may greatly alter the permeability of such active films to oxygen, carbon dioxide, and ethylene, by discontinuities and pores (López-Rubio et al., 2004). Zagory (1995) speculated that the higher permeation to gases of these discontinuous materials might be the reason for the observed enhanced quality and longer shelf-life of highly perishable fresh produce, instead of the ethylene absorbance capacity of the zeolite. Biaxial oriented polypropylene (BOPP) sheets containing filler yield their microporous structure with the controllability of gas and aqueous vapor permeation (Mizutani et al., 1993), and therefore, they could also be used to extend shelf-life of fresh produces.

In Australia, researchers from Food Science Australia (FSA) developed ethylene scavenging technology, which is based on the irreversible and specific reaction between electron-deficient dienes and ethylene in what is known as an inverse electron-demand Diels-n Australia, researchers from Food Science Australia (FSA) developed ethylene scfor example benzenes, pyridines, diazines, triazines and particularly preferred tetrazines), having electron-withdrawing substituents, activated by groups such as fluorinated alkyl groups, sulphones and esters, and especially by the dicarboxyoctyl, dicarboxydecyl and dicarboxymethyl ester groups, reacted rapidly with ethylene. A promising development of electron-deficient nitrogen-containing trienes incorporated in ethylene-permeable films has been reported (Scully and Horsham, 2006, Scully and Horsham, 2007). However, since tetrazine is unstable in the presence of moisture, it must be incorporated in a hydrophobic polymeric film, like silicone-polycarbonate (PC), polystyrene (PS), PE and PP. Approximately 0.01-ure, it must be incorporated in a hydrophobic polymeric film, like silicone-polycarbonate (PC), polystyrene (PS), PE and PP. Approximately 0.01cully and Horsham, 2006 and 48 hours, respectively (Brody et al., 2001). As aforementioned tetrazine, the early limitation being sensitive to moisture under the high relative humidity (RH) conditions experienced in the storage and distribution of fresh produces, was overcome through the development of highly moisture stable tetrazine derivatives (Horsham et al., 2004). The scavenging capacity of films containing the earlier compound is decreased by 90% within 24 hours when kept at 93% RH, whereas films containing the latest compound are completely stable, within experimentally uncertainty, under these storage conditions (Scully and Horsham, 2007). Furthermore when the red tetrazine dicarboxyoctyl ester reacted with ethylene the color disappears which allowed the use of the film as both an ethylene indicator and an ethylene scavenger.

Apart from electron-deficient dienes or trienes, Terry et al. (2007) have reported on the discovery of a novel palladium (Pd)-based ethylene scavenging technology by the Johnson Matthey scientists. It shows a significant ethylene adsorption capacity at room temperature. A wide range of materials were synthesized and screened for activity. Pd gave by far the best performance of the promoter metals tested. The material is a Pd-impregnated zeolite giving finely dispersed palladium particles. The Pd-based material typically removed all measurable ethylene until breakthrough occurred. Under humid condition the Pd-promoted material was found to have an ethylene adsorption of 4162 $\mu L\ g^{-1}$ under *ca.* 100 % RH. This performance was increased to 45,600 $\mu L\ g^{-1}$ under dry conditions (Smith et al., 2009). Nowadays, the use of novel TiO_2 nanoparticles (TNPs) photocatalyst for ethylene degradation in fruits and vegetable cold storage has been developed with the assistance of the new fabricated photocatalytic reaction system. (Hussain et al., 2011). An efficient way of utilizing this photocatalyst, which has superior characteristics for the target application, has been investigated. The reaction mechanism of the TNP photocatalyst for ethylene photodegradation has been comprehensively described by Zorn et al. (2000). Both hydroxyl radical (•OH) and oxygen radical ($•O_2$) are photocatalytically produced, further react with ethylene to produce carbon dioxide and water. The proposed method with TNP for the ethylene photodegradation is simple and potentially economical to be applied commercially. Further research in the area of ethylene scavenging polyolefinic films would be useful. Sachets of $KMnO_4$ ethylene scavengers are in widespread commercial use in the United States even though their insufficient scavenging capacity is concerned. The development of the previously mentioned concepts could contribute, in the near future, to an increased shelf-life of the fresh produces, facilitating their commercialization in optimal conditions.

ETHYLENE BLOCKING

An alternative approach to minimize the impact of ethylene is the use of ethylene blockers such as 1-methylcyclopropene (1-MCP) due to its non-toxic mode of action and negligible residue. Therefore, ethylene blocking can be defined as "When the packaging system acquires ethylene blocking effect, the packaging system (or material) blocks ethylene via plant tissue receptor site then the ripening and deterioration processes of plant products are slowed, and so the storage life is prolonged." 1-MCP is a gaseous potent blocker of ethylene action and has been added to the options for extending the shelf life and maintaining the quality of fresh produce for which ethylene responses limit storability (Blankenship and Dole, 2003, Prange and De Long, 2003, Watkins, 2006). It is thought to interact with ethylene receptors and thereby prevent ethylene-dependent responses (Blankenship, 2001, Sisler and Serek, 2003). The use of cyclopropenes to inhibit ethylene action was patented by Sisler and Blankenship (1996). A commercial breakthrough in application of 1-MCP technology in which a 1-MCP is microencapsulated with gSisler and Serek, 2003). The use of cyclopropenes to inhibit ethylene action was patented by Sisler and Blen being dissolved by water vapor evaporating from the fresh produce. In 1999, the Environmental Protection Agency (EPA) approved 1-MCP for use on ornamentals. 1-MCP was marketed as EthylBloc® by Floralife, Inc. (Walterboro, SC). In addition, AgroFresh, Inc., a subsidiary of Rohm and Haas (Springhouse, PA), subsequently developed 1-MCP under the trade name SmartFreshÔby Floralife, Inc. (Walterboro, SC). In addition, AgroFresh, I

Commercial application of 1-MCP to horticultural products is based on release of gaseous 1-MCP from a aommercial application of 1-MCP to horticultural products is based on release of gaseous 1n. 1-MCP application is carried out in sealed rooms to ensure exposure of the crop to the chemical for several hours. This may be a limitation for some horticultural produces requiring for some capital investment to provide such rooms. Lee et al. (2006) reported that release of 1-MCP from a sachet can be influenced by a large number of factors such as, film or material characteristics, the adsorption abilities of the adsorbing agents, the concentration of acid salt, the size of the sachet and environmental factors. Moreover, the absorption capacity of 1-MCP by fresh produces (that is potato, parsnip, ginger, green bean, asparagus, tangerine, key lime, melon, apple, plantain, leaf lettuce, and mango) was investigated (Nanthachai et al., 2007). The use of 1-MCP in combination with controlled atmosphere storage (CAS) and/or MAP has been limited. Other strategies for 1-MCP application could include the combining of 1-MCP with an active packaging system. In an attempt to overcome aforementioned limitation, Hotchkiss et al. (2007) developed active packaging films (that is low density polyethylene (LDPE), polyvinylchloride (PVC), PP, PS and ethylene vinylalcohol (EVOH)) for the controlled release of an ethylene blocker during storage and transportation. It was found that higher holding temperatures result in a higher and faster release of the active compound. Additionally, the rate of release of 3-methylcyclopropene (3-MCP), an isomer of 1-MCP, increased when the packaging film is exposed to increasing RH (Hotchkiss, 2004). Nevertheless, 3-MCP is less effective than 1-MCP, at a certain extent (Sisler et al., 1999).

16.3.5 ANTIMICROBIAL PACKAGING

The quality of a food is one of the major determinants of its appeal to consumers and, consequently, sales of the product. Microbial contamination is one of the main factors that determine food quality loss and shelf-life reduction. Therefore, preventing microbial contamination is highly relevant to food processors. The growth of microorganisms in food products may cause spoilage or food-borne diseases which, in turn, contribute to the deterioration in safety, flavor, texture and color of the products (Man, 2002). Prevention of pathogenic and spoilage microorganisms in foods is usually achieved by using chemical preservatives. However, the increasing demand for minimally processed, extended-shelf-life foods, together with reports of chemical preservatives having potential toxicity, have required food manufacturers to find alternative means (Nychas, 1995). There is currently a strong debate about the safety aspects of chemical preservatives, since they are considered to have many carcinogenic and teratogenic attributes as well as residual toxicity. For these reasons, consumers tend to be suspicious of chemical additives. Thus the exploration of naturally occurring antimicrobials for food preservation has received increasing attention, due both to consumer awareness of natural food products and to a growing concern about microbial resistance toward conventional preservatives (Schuenzel and Harrison, 2002, Holley and Patel, 2005).

Quintavalla and Vicini (2002) have claimed that AMP is an extremely challenging technology and Appendini and Hotchkiss (2002) have reported that AMP is expected to grow in the next decade. Additionally, publications demonstrate AMP to be a rapidly emerging technology (Suppakul et al. 2003). Kerry et al. (2006) reported than AMP is gaining interest from researchers and industry due to its potential for providing quality and safety benefits. AMP technology, an innovative concept, can be defined as a version of AP in which the package, the product and the environment interact to extend the lag phase and/or reduce the growth rate of microorganisms. By this action, the shelf life of the product is prolonged and its quality and safety are better preserved (Suppakul et al., 2003). Alternatives to direct additives for minimizing the microbial load are canning, aseptic processing and MAP. However, canned foods cannot be marketed as "fresh". Aseptic processing may be expensive and hydrogen peroxide, which is restricted in level of use by regulatory authorities, is often used as a sterilizing agent. In certain cases, MAP can promote the growth of pathogenic anaerobes and the germination of spores, or prevent the growth of spoilage organisms which indicate the presence of pathogens (Farber, 1991).

Food packages can be made AM active by incorporation and immobilization of AM agents or by surface modification and surface coating. Present plans envisage the possible use of naturally derived AM agents in packaging systems for a variety of processed meats, cheeses, and other foods, especially those with relatively smooth product surfaces that come in contact with the inner surface of the package. This solution is becoming increasingly important, as it represents a perceived lower risk to the consumer (Nicholson, 1998). Nonetheless, the "natural" origins of the chemicals are likely to be a selling point, but this does not necessarily make them safer than artificial additives (Biever, 2003). Antimicrobial films can be classified in two types: (i) those that contain an AM agent that migrates to the surface of the food, and (ii) those that are effective against surface growth of microorganisms without migration.

MIGRATING SYSTEM

The direct incorporation of AM additives in packaging films is a convenient means by which AM activity can be achieved. Several compounds have been proposed and/or tested for AM packaging using this method. Han and Flores (1997) studied the incorporation of 1.0% w/w potassium sorbate in LDPE films. It was found that potassium sorbate lowered the growth rate and maximum growth of yeast, and lengthened the lag period before mould growth became apparent. The results of this study, however, contradict those obtained by Weng and Hotchkiss (1993) with LDPE films (0.05 mm thick) containing 1.0% w/w sorbic acid. In the latter case, the films failed to suppress mould growth when brought into contact with inoculated media. Devlieghere et al. (2000) investigated these contradictory results. Their results confirm that ethylene vinyl alcohol/linear low density polyethylene (EVOH/LLDPE) film (70 μm thick) impregnated with 5.0% w/w potassium sorbate is unable to inhibit the growth of microorganisms on cheese and to extend its shelf life. As suggested by Weng and Hotchkiss (1993), very limited migration of potassium sorbate into water as well as into cheese cubes occurs, probably because of the incompatibility of the polar salt with the non-polar LDPE. The choice of an AM agent is often restricted by the incompatibility of that agent with the packaging material or by its heat instability during extrusion (Weng and Hotchkiss, 1993, Han and Floros, 1997). While PE has been widely employed as the heat-sealing layer in packages, in some cases the copolymer polyethylene-co-methacrylic acid (PEMA) was found to be preferable for this purpose. Weng et al. (1999) reported a simple method for fabricating PEMA films (0.008-0.010 mm thick) with AM properties by the incorporation of benzoic or sorbic acids.

In Japan, the ions of silver and copper, quaternary ammonium salts, and natural compounds such as Hinokitiol are generally considered safe AM agents. Silver-substituted zeolite (Ag-zeolite) is the most common agent with which plastics are impregnated. It retards a range of metabolic enzymes and has a uniquely broad microbial spectrum. As an excessive amount of the agent may affect the heat-seal strength and other physical properties such as transparency, the normal incorporation level used is 1-3% w/w. Application to the film surface (that is increasing the surface area in contact with the food) is another approach that could be investigated in the future (Ishitani, 1995).

Another interesting commercial development is Triclosan-based antimicrobial agents such as Microban®, Sanitized® and Ultra-Fresh®. Vermeiren et al. (2002) reported that LDPE films containing 0.5 and 1.0% w/w triclosan exhibited AM activity against *S. aureus*, *Listeria monocytogenes*, *Escherichia coli* O157:H7, *Salmonella* Enteritidis and *Brocothrix thermosphacta* in agar diffusion assay. The 1.0% w/w Triclosan film had a strong AM effect in *in vitro* simulated vacuum-packaged conditions against the psychrotrophic food pathogen *L. monocytogenes*. However, it did not effectively reduce spoilage bacteria and growth of *L. monocytogenes* on refrigerated vacuum-packaged chicken breasts stored at 7 °on refrigerated vacuum-packaged chicken breasts stored at 7 d growt

Present plans envisage the possible use of naturally derived AM agents in packaging systems for a variety of processed meats, cheeses, and other foods, especially those with relatively smooth product surfaces that come in contact with the inner surface of the package. This solution is becoming increasingly important, as it represents a perceived lower risk to the consumer (Nicholson, 1998). Ha et al. (2001) studied GFSE incorporated (by co-extrusion or a solution-coating process) in multi-layered polyethylene films and assessed the feasibility of their use for ground beef. It was found that coating the PE film with grapefruit seed extract (GFSE) with the aid of a polyamide (PA) binder resulted in a more effective degree of AM activity on the agar plate medium that did its incorporation by a co-extrusion process. The film co-extruded with a 1.0% GFSE layer showed AM activity only against *Micrococcus flavus*, whereas the film coated with 1.0% GFSE showed activity against several microorganisms including *E. coli*, *S. aureus* and *Bacillus subtilis*. Both types reduced the growth rates of bacteria on ground beef stored at 3°C, as compared to plain PE film. The two investigated GFSE levels (0.5 and 1.0% w/w) did not differ significantly in the efficacy of the film in terms of its ability to preserve the quality of beef.

Suppakul et al. (2008) incorporated linalool or methylchavicol into LDPE-ethylene vinyl acetate (EVA) films of 45-50 μm thickness to minimize the loss of active agent. Cheddar cheese wrapped with AM incorporated LDPE-based films containing either 0.34% w/w linalool or methyl chavicol and stored at 4°C, revealed that an inhibitory effect of these AM films against mesophilic aerobic bacteria and coliform, as well as yeast and mould growths in naturally contaminated cheese. In addition, cheese samples inoculated with *E. coli* or *Listeria innocua*, wrapped with these AM films, and stored at refrigerated (4°C) or at abuse (12°C) temperatures, showed that the effect on suppression of *E. coli* and *L. innocua* growth was more pronounced at the abuse temperature. Methylchavicol-LDPE-based film exhibited a higher efficacy of inhibition than that of linalool-LDPE-based film. A sensory evaluation was performed with regards to possible taint in the flavor of the cheese. Taint in flavor as affected by linalool or methylchavicol was not significantly detectable by the panelists at the end of the storage period of 6 weeks. Later on, this work received a rare form of validation: a United States patent (Miltz et al., 2011). Recently, Sadaka et al. (2013) reviewed the literatures associated with antimicrobial packaging containing essential oils and their active biomolecules. They concluded that the use of bioactive packaging will probably rise in the world owing to the consumer's preference for naturally preserved food and also because of the industrial need in shelf life prolongation of packaged foods while preserving product quality and safety.

The combination of active technologies such as antimicrobials and nanotechnologies such as nanocomposites can synergistically lead to bioplastic formulations with balanced properties and functionalities for their implementation in packaging applications. The formulation of novel antimicrobial nanocomposites of polycaprolactone (PCL) is presented as a way to control solubility and diffusion of the natural biocide agent thymol. The enhancement in the antimicrobial solubility due to the presence of nanoplatelets was possibly due to retention of the apolar biocide agent over the engineered nanofiller surface. On the other hand, the thymol diffusion coefficient was seen to decrease (from *ca.* 2.8 ′ $2.^{-15}$ -11.1 ′ 10^{-15} m^2s^{-1}) with the addition of the nanoadditive in the biocomposite. This is likely the result of the larger tortuosity effect imposed to the diffusion of the biocide by the dispersed nanoclay. (Sánchez-García et al., 2008). This novel engineered nanofiller commercially releases as NanoBioTer® (NanoBiomatters Ltd., Paterna, Spain). However, a recent new patented technology that makes use of silver strongly stabilized on nanoclays either in the elemental nanoform or in ionic form and that is aimed at dispersing in food contact plastic materials has been developed, scaled up and commercially available under the trademark of Bactiblock® (NanoBiomatters Ltd., Paterna, Spain) (Lagarón and Busolo, 2011).

Persico et al. (2009) also investigated the AM activity of nanocomposite LDPE film containing 10% (w/w) carvacrol. It was found that this AM film showed a positive AM activity against *B. thermosphacta, L. innocua* and *Carnobacterium* spp., except *Pseudomonas fragi.* In addition, Tippayatum et al. (2009) studied the AM activity of EVA-coated LDPE film containing 2% or 4% (w/w) of thymol or eugenol or a combination of these AM compounds. The EVAflex150/LDPE incorporated with 4% thymol and eugenol was the most effective film against *L. monocytogenes*, *B. cereus*, *S. aureus* and *E. coli.* The EVAflex40w/LDPE films containing thymol or eugenol, or a combination of these, showed limited inhibitory effects against all bacteria tested. These results indicated that the effectiveness of the antimicrobial substances depends on their interactions with the packaging materials.

NON-MIGRATING SYSTEM

Besides diffusion and sorption, some AMP systems utilize covalently immobilized AM substances that suppress microbial growth. The technique used for the physical or chemical fixation of AM protein as enzymes (for example lysozyme, lactoperoxidase) or antimicrobial peptides (AMPs) (for example bacterial AMPs so called bacteriocins like nisin, brevicin, subtilin, pediocin, lactocin, lacticin and sakacin, amphibian AMPs like magainin, reptile AMPs like crocosin) or other proteins onto a solid support, into a solid matrix or retained by a membrane, in order to increase their stability and make possible their repeated or continued use, is termed immobilization. Goddard and Hotchkiss (2007) reviewed the advances in the covalent attachment of bioactive compounds to functionalized polymer surfaces including relevant techniques in polymer surface modification such as wet chemical, organosilanization, ionized gas treatments, and UV irradiation. These useful polymer modification techniques can be applied for polymeric film substrate preparation for AM immobilization for non-migrating AM system. The ionic or covalent immobilization of antimicrobials onto polymers requires the presence of functional groups on both the antimicrobial and the polymer (that is ethylene vinyl acetate (EVA), ethylene methyl acrylate (EMA), ethylene acrylic acid (EAA), ethylene metacrylic acid (EMAA), ionomer, polyamide (PA), polyvinylidene chloride (PVDC), polyvinyl chloride (PVC) copolymer, ethylene vinyl alcohol (EVOH), and PS. Additionally to functional antimicrobials and polymer supports, immobilization may require the use of "spacer" molecules that link the polymer surface to the bioactive agent. These spacers permit sufficient freedom of motion so the active portion of the agent can contact and inhibit microorganisms on the food surface. Spacers that could potentially be used for food antimicrobial packaging include dextrans, ethylenediamine, polyethyleneimine, and polyethylene glycol (PEG) due their low toxicity and common use in foods (Appendini and Hotchkiss, 2002). Appendini and Hotchkiss (1997) investigated the efficiency of lysozyme immobilized on different polymers. It is known that cellulose triacetate (CTA) containing lysozyme yields the highest AM activity. The viability of *Micrococcus lysodeikticus* was reduced in the presence of immobilized lysozyme on CTA film. Scannell et al. (2000) showed that PE/polyamide (70:30) film formed a stable bond with nisin in contrast to lacticin 3147. Nisin-adsorbed bioactive inserts reduced the level of *L. innocua* and *S. aureus* in sliced cheese and in ham.

16.3.6 ANTIOXIDANT PACKAGING

Lipid oxidation is one of the main factors that determine food quality loss and shelf-life reduction. Therefore, delaying lipid oxidation is highly relevant to food processors. Oxidative processes in food products lead to the degradation of lipids and proteins which, in turn, contribute to the deterioration in flavor, texture and color of the products (Decker et al., 1995). Oxidative deterioration of fat components in food products is responsible for off-flavors and rancidity which decrease nutritional and sensory qualities. The addition of antioxidants is required to preserve product quality. Synthetic antioxidants – for example butylated hydroxytoluene (BHT), butylated hydroxyanisole (BHA), tert-butylhydroquinone (TBHQ), and propyl gallate (PG) – are widely used as antioxidants in the food industry. Their safety, however, has been questioned. BHA has been revealed to be carcinogenic in animal experiments. At high doses, BHT may cause internal and external hemorrhaging, leading to death in some strains of mice and guinea pigs (Ito et al., 1986). There is much interest among food manufacturers in natural antioxidants to act as replacements for synthetic antioxidants currently used (Plumb et al., 1996). It is a paradox of aerobic life that aerobic organisms require oxygen for their normal metabolisms, yet oxygen can be toxic to cellular material (Davies, 1995). Detrimental events which may be caused by this lipid oxidation include membrane fragmentation, disruption of membrane-bound enzyme activity, swelling and disintegration of mitochondria, and lysosomal lysis.

Antioxidants can also be incorporated into plastic films for polymer stabilization in order to protect the films from degradation. It is well established that antioxidant concentrations in polymeric films decrease during storage due to oxidation but also because of diffusion through the bulk of the polymer towards its surface followed by evaporation (Miltz et al., 1989). For AOP, antioxidant is intentionally impregnated into polymeric packaging materials in order to perform antioxidant effectiveness against lipid oxidation. Hence, AOP is defined as the packaging system which acquires AO activity, the packaging system (or material) limits or prevents lipid oxidation by donating electron or hydrogen, quenching oxygen and/or scavenging free radical.

Oatmeal cereals, packaged in 0.32% BHT-incorporated high density polyethylene (HDPE), had a prolonged shelf-life in comparison with cereals packaged in 0.022% BHT-incorporated HDPE. After six weeks, the HDPE film was free of BHT and 19% of that originally present in the film remained in the cereal. The oat flakes packaged in the film with 0.32% BHT, showed less oxidation than the control (0.022% BHT). A considerable amount of the BHT was lost by outward migration. However, these outward losses can be controlled by the use of an extra layer with a low permeability for the antioxidant (Miltz et al., 1989). Vitamins E and C have been suggested for integration in polymer films to exert their antioxidative effects. Vitamin E has proved to be very stable under processing conditions and has an excellent solubility in polyolefins. It is confirmed, however, that vitamin E is a less mobile antioxidant in LDPE than

BHT, as vitamin E is a larger molecule. Since vitamin E migrated from the film to the fatty food simulant to a lesser extent than BHT, the success of the vitamin E releasing active packaging concept is difficult to assess (Wessling et al., 1998).

Nerín et al. (2008) developed and evaluated AO coated polypropylene (PP) film containing rosemary extract. It was found that 4.0% w/w rosemary extract coated PP film inhibited the oxidation extent up to 30.1% when compared to the control film at 15 days for iron (II) and 42.2% in comparison with additive-free PP film at 20 days for fatty acids from flax seed oil. Ascorbic acid has proven to be not adequate for antioxidant quantitative purposes owing to its prooxidant properties.

The major potential food applications of AM/AO packaging films generally include meats, seafoods, poultry, bakery goods, cheeses, fruits and vegetables, cereals, prepared meals or ready meals (Labuza and Breene, 1989, Suppakul et al., 2003). Very low doses of active agents can be possibly used to exhibit AM and/or AO activities. These powerful active natural plant extracts and their principal constituents should be considered and selected for food applications in order to avoid the pitfall of research in relation to detrimental effect on flavor. Alternatively, flavor matching between additive and food product should be employed. For instance, wasabi extract-incorporated sheet is commercially used for bento (Japanese lunch box).

16.3.7 THERMALLY EQUILIBRIOUS MODIFIED ATMOSPHERE PACKAGING

Fresh produces, when harvested, continue to consume O_2 and give out CO_2. In order to prolong the shelf-life of fresh produces *via* a reduction in respiration rate, cooling and lowering the O_2 should be acquired. The atmosphere in a package used to preserve fresh produce will reach equilibrium levels of O_2 and CO_2 depending on the respiration rate and weight of the produce, the available gas in the package, and the permeability of the package. There is a specific beneficial atmosphere for each fruit and vegetable, which together with good temperature control will help preserve the quality and freshness of produce.

The Intellipac^{ò} (Landec Corporation, CA, USA) membrane technology is a means of creating specific O_2 and CO_2 levels in a package (Clarke, 2001). This optimum atmosphere can be maintained, within limits, even as the temperature is changing. The highly permeable membrane covers a hole in the wall of the package. The membrane is made by coating a porous substrate with a proprietary side chain crystallizable (SCC) polymer. SCC polymers are polymers in which the side chain crystallizes independently of the main chain. Examples of such

polymers are siloxanes or acrylic polymers in which the side chain has eight or more carbon atoms. By varying the chain length of the side chain, the melting point of the polymer can be altered. An SCC polymer in a solid or crystalline state is relatively impermeable but when heated to the switch temperature becomes a molten fluid. This change in state has been used in fresh produce applications for providing a dramatic increase in permeability, when going from below to above the crystalline melting point. By altering the properties of the polymer, it is possible to achieve specific oxygen permeabilities, carbon dioxide to oxygen permeability ratio (bare siloxanes or acrylic polymers in which the side chain has eight or more carbon atoms. By varying the chain le have a fixed CO_2/O_2 ratio (Clarke, 2001). Thus, this independent control of permeability allows packaging of very large quantities of produce, while still providing the right gas atmosphere.

16.4 INTELLIGENT PACKAGING

16.4.1 INTEGRITY INDICATORS

Oxygen is an essential factor for either aerobic microorganisms or facultative anaerobic microorganisms and is involved in lipid oxidation. Protection from exposure to oxygen is one of major packaging functions. Consequently, packaging must provide a better oxygen barrier in order to prevent off-flavor development. However, the most common cause of integrity damage in flexible plastic packages is in association with leaking seals (Hurme, 2003). An alternative approach to establish destructive-package technique for determining oxygen permeated into package of food products or leaked via non-integrity seal of their packages, is the use of non-invasive indicator system in terms of integrity indicator.

A colorimetric oxygen indicator can be assembled using a semiconductor photosensitizer (SCP) coupled with a redox dye and a sacrificial electron donor (SED) in the form of an intelligent ink (Mills and Hazafy, 2008). In this system, the SCP (that is titanium dioxide, TiO_2) absorbs UV light in an activation stage, creating electron-hole pairs. The photogenerated holes react quickly with the SED (that is glycerol), leaving the photogenerated electrons to react with a redox dye in oxidized form (D_{ox}) (that is methylene blue (MB)), thereby generating a differently colored redox dye in reduced form (D_{red}), which is usually bleached and is oxygen-sensitive. The main chemicals including SCP powder particles, D_{ox} and SED are selected so that they are readily dispersed and dissolved in an aqueous solution containing a biopolymer (that is hydroxyethyl cellulose (HEC)), acting as an encapsulation medium when this intelligent ink is printed and dries to form a UV-activated oxygen-sensitive indicator film (Mills and Hazafy, 2008).

Activated by a pulse of UV light, the initially highly colored polymer/SED/D_{ox}/SCP film is rendered a different color, usually colorless, and oxygen-sensitive. In the absence of oxygen, this UV-activated film remains colorless indefinitely, but upon exposure to oxygen from atmospheric environment, it rapidly (usually within 10 min) regains its original color as D_{red} is reoxidized by oxygen to D_{ox} in a dark reaction stage. If the above SCP-sensitized-based indicator technology is to be used in frozen food products, the ink needs a more controllable route to activation, that is not by light commonly present in room or food cabinet lighting, but by UVB light (280-320 nm) which is largely absent from such sources. Mills and Hazafy (2009) developed a UVB-activated oxygen-sensitive indicator which uses nano-particulate tin (IV) oxide, *nc*SnO_2, as the SCP. It is clear that when *nc*SnO_2 is used as a photocatalyst in a HEC/glycerol/MB/SCP film, the resultant oxygen indicator is not UVA sensitive and, hence, not readily photobleached by the UVA component of white light fluorescent tubes (that is warm-white light (WWL)). This is not surprising given the bandgap of SnO_2 is 3.65 eV (°is 3.65 eV (ven the bandgap of SnOot surprising given the bandgap of SnOoxygen indicator is not UVA sensitive and, hence, not readily photobleached by t®.

16.4.2 FRESHNESS INDICATOR

According to Huis in't Veld (1996), the changes taking place in chilled food products can be categorized as (i) microbiological growth and metabolism leading to pH changes, formation of toxic substances, off-odors, off-flavors, gas and/or slime formation, (ii) oxidation of lipids and pigments resulting in undesirable flavors in terms of rancidity, formation of chemicals with adverse biological reactions or discoloration. The focus of these freshness indicators is on intelligent concepts indicating the changes associated with both categories. A freshness indicator determines directly the quality of food product using either metabolites from microbial growth or chemicals from lipid oxidation as "information" to monitor a status of the freshness of food product. Apart from microbial growth and lipid oxidation, freshness indicator can also detect the chemical ripeness index to monitor a status of fruit ripening.

FOOD SPOILAGE INDICATOR

An indicator that would demonstrate specifically the spoilage or the lack of freshness of the product, in addition to package leak or temperature abuse, would be ideal for the quality assurance of chilled food products. The number of publications on package indicators for spoilage or freshness of food product is still limited. The concepts for visible indicators sensitive to spoilage indicating metabolites, in turn, monitoring status of freshness, have been proposed by Mattila et al. (1990), Smolander et al. (2002), Pacquit et al. (2007), and Nopwinyuwong et al. (2010). Recently, Puligundla et al. (2012) comprehensively reviewed CO_2 sensors for intelligent food packaging applications. It can be concluded that the optical CO_2 gas sensors are suitable for food package indicator applications. Especially, dry optical sensors containing pH-sensitive dye indicators are better suited for CO_2 monitoring and/or as spoilage indicators in food packaging application.

Pacquit et al. (2007) developed a smart packaging for the monitoring of fish spoilage. Indicator spots are prepared by entrapping within a polymer matrix solution a selected pH sensitive dye (that is bromocresol green (BCG)) which is then spin-coated onto optically clear PET substrate discs as described previously (Pacquit et al., 2006). This colorimetric indicator offers the potential of developing dynamic "best-before" dates that may lead to important and exciting improvements in the quality assurance sector. In addition, Nopwinyuwong et al. (2010) developed a novel mixed pH dye indicator label for monitoring freshness of intermediate-moisture dessert spoilage. Indicator label was obtained by casting indicator coating onto nylon/LLDPE film. This on-package indicator contains mixed pH-sensitive dyes, bromothymol blue and methyl red, that respond through visible color change to carbon dioxide (CO_2) as a spoilage metabolite. A kinetic approach was used to correlate the response of the indicator label to the changes in intermediate-moisture dessert spoilage. Trials on golden drop have verified that the indicator response correlates with microbial growth patterns in dessert samples, thus enabling the real-time monitoring of spoilage either at various constant temperatures or with temperature fluctuation.

Apart from pH dye-based indicator, the Avery Dennison developed a food spoilage indicator on the basis of a platinum group metal (PGM) fluorophore (or optionally chromophore) complex which is delivered in a tag or label format. In the presence of gases containing sulphur or nitrogen (for example sulphides or amines) these complexes undergo a color change (in the case of the chromophore version) or a change in fluorescence, which can be seen under UV light (in the case of the fluorophore version). The color/fluorescent change is irreversible. Spoilage gases typically contain sulphur or nitrogen containing species and this technology has proven to be an excellent method for detecting many types of food spoilage (Hartman, 2003).

RIPENESS INDICATOR

Fruit ripening is a process in fruits that causes them to become more palatable. It results in loss of chlorophyll, production of carotenoids and anthocyanin, starches conversion into sugars, changes in organic acids, proteins and fats, aroma development, ethylene production, and reduction in tannins and fungistatic compounds. In some fruits, changes in color can be advantageous by consumers to decide when the fruit is ripe and ready to eat (for example strawberry, banana, mango, apricot). Nevertheless, some fruits do not exhibit obvious visual indication of ripening (for example apple, pear, kiwifruit, avocado, durian, cherimoya, sapodilla).

Based on hydrolysis of ester (McMurry, 2007), ripeness indicator targeted to aroma esters was firstly developed specifically for pears by HortResearch (New Zealand) and had market trials in 2003 (HortResearch, 2004). An indicator compound based on phenol red as a pH dye on alkalized solid phase indicator film can react with aroma esters to produce carboxylic acids, which results in a color change hence indicating ripening of fruits. The indicator changes color by detecting the aroma esters given off by the fruit as it ripens. Since phenol red shifts from basic form (red, pH 8.4) to acidic form (yellow, pH 6.8) (Srour and McDonald, 2008), then the indicator is initially red and gradually changes to orange-red orange and finally yellow. By matching the color of the indicator with their degrees of fruit ripening, customers can accurately recognize fruit as "crisp", "firm" and "juicy", respectively. Trade name of such device is namely ripeSense® as the world's first intelligent sensor label to indicate the ripeness of fruit.

On the basis of oxidation of ethylene, ripeness indicator targeted to ethylene was developed specifically for apples by the University of Arizona (USA) in 2005. (Riley, 2006). An indicator compound based on ammonium molybdates ($(NH_4)_2MoO_4$) as a visual dye (white) on palladium sulfate ($PdSO_4$)-catalyst indicator sticker can react with ethylene gas to produce molybdenum oxide (MoO_3) (dark blue), then the indicator is initially white and gradually changes to light blue and finally dark blue (Klein et al., 2006). Trade name of this sticker is namely RediRipe$^{\grave{O}}$.

RANCIDITY INDICATOR

Lipid oxidation is a complex process that proceeds upon a free radical process (Kochhar, 1996, Erickson, 1997). During the initiation stage, a hydrogen atom is removed from a fatty acid, leaving a fatty acid alkyl radical, being converted in the presence of oxygen to a fatty acid peroxyl radical. At propagation stage, the peroxyl radical abstracts a hydrogen from an adjacent fatty acid forming a hydroperoxide and a new fatty acid alkyl radical. Breakdown of the hydroperoxide is responsible for further free radical propagation. Decomposition of hydroperoxides of fatty acids to aldehydes and ketones is responsible for characteristic oxidative rancidity. Enzymatic pathway can also initiate lipid oxidation. Presented in many plants and animals, lipoxygenase is the major enzyme responsible for pigment bleaching and off-odors in frozen vegetables (Zaritzky, 2006). Moreover, in raw products, hydrolytic enzymes namely lipases and phopholipases, which catalyze the transfer of groups to water, may remain active during frozen storage. These enzymes hydrolyze ester linkages of triacylglycerols and phospholipids, respectively. If they are not in under control during storage, the hydrolysis of lipids can result in undesirable flavor and odor as hydrolytic rancidity (Zaritzky, 2006).

Thiobarbituric acid reactive substances (TBARS) are a measure of secondary oxidation products, mainly aldehydes, carbonyls or hydrocarbons, which contribute to off-odors and flavors in meat (Igene et al., 1985). TBARS values of 3 mg (kg meat^{-1}) are associated with oxidative rancidity of meat (Chouliara et al., 2008). Although TBARS is a common assay used to follow malondialdehyde (MDA) formation, other aldehyde products can also react with thiobarbituric acid to give the red chromogen. Hexanal is one of the most abundant volatiles formed and is produced from the oxidation of omega six-fatty acids. During refrigerated storage, hexanal content increased over time until day 10 where it peaked and later decreased. Its content ranged from 30 to 173 ppb. Hexanal content significantly increased from 3 to 60 ppb with time for frozen raw minced chicken breasts (MCB), similar to TBARS values (Chouliara et al., 2008).

Vo et al. (2007) developed pH sensitive indicator pad for detecting aldehydes. This indicator pad comprises a pH dye (that is methyl red) dissolved in an alkaline methanol solution, then applied onto cellulose-based pad. The indicator changes color by reacting with aldehydes. Since methyl red {2-[4-(dimethylamino) phenylazo]benzoic acid, sodium salt, $(CH_3)_2NC_6H_4N{=}Cc_6H_4CO_2Na$} shifts from basic form (yellow, pH 6.2) to acidic form (red, pH 4.5) (Sabnis, 2008). An alpha carbon atom is a carbon atom that is bonded directly to the carbon atom of a carbonyl group. A hydrogen atom bonded to an alpha carbon is called alpha hydrogen. The alpha hydrogens of a carbonyl compound are much more acidic (~103 times more acidic) than a typical C-H bond. Due to the existence of alpha hydrogens in glutaraldehyde, it can react with pH-sensitive dye. Similar to glutaraldehyde, hexanal containing alpha hydrogens can also react with methyl red.

16.4.3 TIME-TEMPERATURE INDICATOR

Temperature is a crucial factor affecting the quality and safety of food products during both distribution and storage. The difficulty in controlling and monitoring the temperature history of food products makes it difficult to precisely predict shelf life. Time-temperature indicators (TTIs) provide a visual summary of a product's accumulated chill-chain history, recording the effects of both time and temperature.

A modern quality and safety assurance system should prevent contamination through the monitoring, recording, and controlling of critical parameters such as temperature during a food product's entire life cycle. It includes the post-processing phase and extends to the time of use by the consumer (Koutsoumanis et al., 2005). Hence, monitoring and recording the temperature conditions during distribution and storage are of importance (Taoukis and Labuza, 2003). Time-temperature indicators or integrators (TTIs) are defined as simple, cost-effective and user-friendly devices to monitor, record, and cumulatively indicate the overall influence of temperature history on the food product quality from the point of manufacture up to the consumer (Taoukis and Labuza, 1989, Giannakourou et al., 2005).

Generally, the requirement that the TTI response matches the food quality losses must be fulfilled for the successful application of TTIs to food products. Consequently it is essential that the activation energy (E_a) (Table 1) (Vitsab International, 2007, Poças et al., 2008, Taoukis, 2008), indicating the temperature sensitivity of the TTI response, is similar to the Ea related to food quality losses (Table 2) (Labuza, 1982, Beaudry, 2007, Poças et al., 2008), and that the TTI endpoint should match with the end of a product's shelf life. Hence, the applicability of a particular TTI as a quality indicator requires a systematic kinetic study of both the food product's deterioration and the TTI response (Taoukis, 2001, Giannakourou et al., 2005).

TABLE 1 E_a Values of Commercial TTIs

TTIs	Activation Energy	
	kcal mol^{-1}	kJ mol^{-1}
Check Point® Type C	12.0	50.21
Check Point® Type M	26.8	112.13
Check Point® Type S	30.1	125.82
Check Point® Type L	46.5	194.56
3M MonitorMark® All Type	7.9-11.9	33.05-49.79
Fresh-Check®	19.1-21.5	79.91-89.96
OnVuÔ	21.5-35.8	89.96-149.79
TT SensorÔ	26.8-30.1	112.13-125.82
(eO)®	23.9-26.3	99.99-110.04

From: Vitsab International (2007); Poças et al. (2008); Taoukis (2008)

TABLE 2 Typical E_a Values for Food Quality Losses

Quality Loss	Activation Energy	
	kcal mol^{-1}	kJ mol^{-1}
Controlled diffusion	0-15	0-62.76

TABLE 2 *(Continued)*

Enzymatic reaction	10-15	41.84-62.76
Hydrolysis	15	62.76
Lipid oxidation	10-25	41.84-104.60
Nutrition loss	20-30	83.68-125.52
Non-enzymatic browning	25-50	104.60-209.20
Microbial growth	20-60	83.68-251.04
Spore destruction	60-80	251.04-334.72
Vegetative cell destruction	50-150	209.20-627.60
Plant respiration rate	29-79	121.34-330.54

From: Labuza (1982); Beaudry (2007); Poças et al. (2008)

The 3M Monitor Mark^{Ò} is based on diffusion of blue dyed ester. Fatty acid ester, having a melting temperature below a monitoring temperature, coupled with blue dye for vivid observation, is used as proprietary materials. The response of the indicator is the advance of a blue dyed ester diffusing along a wick (Manske, 1976). In addition, the Check Point® III is based on diffusion of pH-sensitive chemical. Diffusive reaction leads to pH change in printed dot on label face. For application and activation, activator labels are applied to the top of indicator label for merging into single active label at the activation step (Tiru and Tiru, 1979). For color change, the indicator is initially amber and graduates to orange, pink and finally pink-magenta. By matching the color of the indicator, customers can accurately recognize product quality in terms of expiration as "20% expired", "60% expired" and "100% expired", respectively. Other trade name of such device namely TT SensorÔ respectively. Other trade name of such device ® III.

The Check Point® I is based on enzymatic hydrolysis of a lipid substrate. Prior to activation the indicator comprises two separate compartments, in the form of plastic mini-pouches. One compartment is filled with an aqueous solution of a lipolytic enzyme. The other contains the lipid substrate incorporated in a pulverized polyvinylchloride (PVC) carrier, which is suspended in an aqueous phase and a pH indicator mix. Mixed esters of polyvalent alcohols and organic acids are included in substrates (Blixt et al., 1977, Blixt et al., 1980, Agerhem and Nilsson, 1081). Hydrolysis of the substrate causes acid release and the pH drop is translated into a color change, initially deep green to bright yellow, orange and finally red.

The TRACEO® is based on a particularly microbiological system. It is a transparent adhesive label in which selected strains of lactic acid bacteria (LAB) are trapped. Once put on the bar code of the package of the traced chilled food products and depending on the time temperature profile that the system goes through, TRACEO® delivers a clear twofold response: an irreversible change from colorless to pink and a simultaneous opacification reaction once the product has experienced critical temperature abuses or once it has reached its use by date. The opacification reaction prevents correct reading of the bar code. The product is therefore rejected by the scanner at check out, thus enabling automated and systematic detection of altered products in markets (Ellouze et al., 2008, Hogan and Kerry, 2008, Vaillant, 2010). Additionally, the (eO)® is also based on a microbiological system. It is an adhesive label in the form of a small gel pad shaped like the petals of a flower that change from green (good) to red (not good). The color change represents a pH change due to microbial growth of LAB within the gel itself (Hogan and Kerry, 2008).

The Fresh-Check® is based on a solid state polymerization. The TTI function is associated with the property of disubstituted diacetylene crystals (R-is based on a solid state polymerization. Tcontrolled solid-state reaction proceeding via 1,4-addition polymerization, which leads to a highly colored polymer (Patel et al., 1976, Patel and Yee, 1980). During polymerization, the crystal structure of the monomer is retained and the polymer crystals remain chain aligned and are effectively one-dimensional in their optical properties (Patel and Yang, 1983). The color response of the TTI can be observed in terms of a decrease in reflectance.

The OnVuÔhe OnVus on a photochromic solid state reaction. The pigmentary water ink changes color from colorless to blue upon UV light irradiation. As the organic photochromic pigment, dinitrobenzyl pyridine compound (DNBP) (Corval et al., 1996, Frank et al., 1996) exists in two states: state A is colorless and thermodynamically stable and state B is blue and metastable. In the dark, state B reverts to state A at a rate that is temperature-dependent. UV-light induces a coloration of the ink from colorless to blue (activation). After the label has been activated (charged) the label is covered with an optical filter in the form of a Thermal Transfer Ribbon (TTR), to protect it from re-charging by sunlight (Haarer and Eichen, 2006, Kreyenschmidt et al., 2010). The end of shelf life of the TTI is defined as the time that it takes the color of the blue photochromic spot of the label to reach a reference color, which is printed in the form of a ring surrounding the photochromic ink.

Recently, Nopwinyuwong et al. (2012a, 2013) presents a method to prepare a new class of polydiacetylene (PDA)/SiO_2 nanocomposites for colorimetric indicator applications. The silica concentration is 5 wt% of the diacetylene monomer, 10, 12-pentacosadiynoic acid (PCDA). Under ultraviolet light irradiation, PCDA vesicle/silica nanocomposites were formed and an intense blue solution was obtained. The silica nanoparticles can improve chromic properties of molecules. The color transition of PDA/silica nanocomposites can change from blue (λ_{max} ~ 640 nm) to red (λ_{max} ~ 540 nm), which is irreversible when stored at different temperatures and time. The PDA/5 wt% silica nanocomposites aqueous matrix would potentially be developed as a new time-temperature indicator. The E_a values of pure PDA and PDA/silica nanocomposites were 79.46 and 96.29 kJ mol^{-1}, respectively. According to the E_a values, the application of TTIs can monitor food quality losses in fruits and vegetables, and chilled foods, respectively (Nopwinyuwong et al., 2012b).

16.4.4 RADIO FREQUENCY IDENTIFICATION

Item identification through the use of numbering schemes has been established worldwide, mostly due to the introduction of barcode standards that serve as the foundation for optical scanning technology. Nowadays, nearly all products sold in supermarkets and department stores are equipped with a unique GS1 (formerly known as EAN/UCC) identification number containing information on its manufacturer and product type encoded in a one-dimensional barcode as a universal product code (UPC) (Thiesse and Michahelles, 2006). Later on, an electronic product code (EPC) has been conceived as a novel numbering scheme to identify all kinds of physical objects, not just traded goods. It must be sufficiently large to enumerate all objects, and to facilitate all current and future naming methods (Thiesse and Michahelles, 2006). EPCglobal Inc. is a joint venture of GS1 and GS1 US™, which administers the UPC bar code (Kleist et al., 2005). Like the UPC, the EPC is divided into numbers that identify the manufacturer, product serial number and version. The EPC ranges from 64 to 256 bits, with four distinct fields as depicted in Figure 2. What sets the EPC apart from UPC is its serial number, which permits individual item tracking (Kleist et al., 2005).

21	203D2A9	. 16E8B8	. 719BAE03C
Header (8 bits)	EPS Manager (28 bits)	Object Class (24 bits)	Serial Number (36 bits)

FIGURE 2 Electronic product code (EPC) format (96-bit version).

RFID is a data acquisition and storage method, providing accurate, real-time data without human intervention. With its advent, various business processes are poised for a new and rapid transformation (Li et al., 2006). RFID technology promises numerous advantages in supply chain management: decreased storage, handling and distribution expenses, increased sales through a reduction in stock outs, and improved cash flow through rise inventory turns (Käkkäinen, 2003), enhanced speed, accuracy, efficiency and security of information sharing across the supply chain (Jones et al., 2004). Prevention of product recalls is also considered a crucial role of RFID technology (Kumar and Budin, 2006). As one of major drivers behind RFID implementation in retailing, in June 2003 Wal-Mart announced a requirement to their top 100 suppliers to tag all pallets and cases they shipped to Wal-Mart distribution centers by January 2005. The next top 200 suppliers must implement by January 2006 and all suppliers by the end of 2006. Other early retail adopters of RFID Technology include Sainsbury's, 7-Eleven, Metro, Tesco, Carrefour, The Gap, Woolworth's, Prada, Benetton, and Marks & Spencer (Käkkäinen, 2003, Wilding and Delgado, 2004, Kerry et al., 2006, White et al., 2008).

Brewer and Sloan (1999) regarded RFID as an intelligent tracking technology in manufacturing to support logistics planning and execution. Käkkäinen and Holmström (2002) considered RFID as a wireless product identification technology to enable material handling efficiency, customization and information sharing in a supply chain. Wyld (2006) defined RFID as an automatic identification (auto-ID) technology, which identifies items and gathers data on items without human intervention or data entry.

A typical RFID system comprises tag and antenna, reader and communication infrastructure. At its basic level, an RFID tag consists of a tiny transponder, which is an integrated circuit with memory, being essentially a microprocessor chip, and antenna, which has a reading range both sideways and in front of itself. Antenna design and placement play a significant role in determining the coverage zone, range, and accuracy of communication of a tag, because the antenna both draws energy from the reader's signal to energize the tag and sends the data that are received from the reader. Having a unique number or alphanumerical sequence, the tag responds to signals received from a reader's antenna and transmits its number back to the reader (Kerry et al., 2006, Ngai et al., 2008).

The costs of RFID have reduced rapidly and significantly in recent years, and tag cost will probably be sufficiently competitive. The threshold value appeared to be 6-8 Euro cents for each tag (Regattieri et al., 2007). Importantly, the attractiveness of RFID tags derives from their low cost. In order to decrease further the cost of RFID, the use of organic semiconductor materials is apparently of interest for RFID antennas, being manufactured with printing technology. As conductive ink, metallo-organic compounds such as silver neodecanoate ($C_{10}H_{19}O_2Ag$) and (hexafluoroacetylacetonate)Cu(I)(vinyltrimethylsilane) dissolved in organic solvents do not have any solid particles (Vest, 1993, Rozenberg et al., 2002). Because this ink is free from particulate agglomerates and its viscosity tends to be low, it is suitable for inkjet printing. Shin et al. (2009) investigated the developed silver nanopaste in the range of 20 to 50 nm without the inclusion of microparticles and flakes which was sintered at 120 °Rozenberg et al., 2002). Because this ink is free from particulate agglomerates and its viscosity tends to be low, it is suitable for inkjet printing. Shin et al. (2009) investigated the developed silver nanopaste in the range of 20 to 50 nm without the iegrating it over time in order to determine the product's shelf life, and having a battery and an optional visual display that provides green (fresh), yellow (warning) and red (unsafe) indicators relying upon the status of its color, FreshtimeÔRozenberg et RFID tags enable to operate in the range of -Rozenberg et RFID tags enable to operate in the range of® RFID tag can function as both a temperature data logger and a supply-chain tracking tool. The range of operation for this data logger is from -RFID tag °RFID tag can function as bothet al. (2009) validated a RFID smart tag developed for real-time traceability and cold chain monitoring for an intercontinental cold logistic chain of fresh fish. The results proved that this system presents important advantages regarding conventional traceability tools and currently used temperature data loggers such as more memory, reusability, no human participation, no tag visibility needed for reading, possibility of reading many tags at the same time and more resistance to humidity and environmental conditions.

16.5 FUTURE TRENDS

Until 2004 in Europe there was a legislative lack for these kinds of packaging decreasing their penetration in the EU market. To face the problem Regulation 1935/2004/EC and more specifically Regulation 450/2009/EC set new legal basis for their correct use, safety and marketing. Nevertheless, due to its deliberate interaction with the food and/or its environment, the migration of substances could represent a food safety concern (Restuccia et al., 2010). A bright future may be anticipated for active and intelligent packaging. Regardingly, the Regulation 1935/2004/EC and new Regulation 450/2009/EC pose new basis for the general requirements and specific safety and marketing issues related to active and intelligent packaging. In fact, it should be concerned that complexity of systems create many variables into risk assessment. Pouches/sachets may introduce new migration substances and result in interactions between active agents and other packaging materials. A problem appears with the various qualifications of the specialists necessitated by the migration testing. There should be the specialists in polymers and polymer additives, the people responsible for the food, the experts in packaging, but also the bacteriologists as well as the analysts (Rosca and Vergnaud, 2007). The development and validation of migration testing system to reliably investigate new migration products could represent the prime challenge, as well as the risk assessment for nanomaterials (Imran et al., 2010, Restuccia et al., 2010). Since advances in nanotechnology and the improvement of nanomaterials will enable the development of better and new active and intelligent packaging (Pereira de Abreu et al., 2012). The use of the Threshold of Regulation and the Threshold of Toxicological Concern concepts may also play a critical role in the risk assessment of new food packaging technologies in the future. Its paradigm is sufficiently robust and flexible to be adapted to meet these technological challenges (Munro et al., 2009). Anyway, despite the hurdles that have to be overcome in the near future, there is a strong view that active and intelligent packaging will be a marketing tool with a high potential (Restuccia et al., 2010). In addition, the next technological revolution would be 3-BIOS, referring to Bioactive, Biodegradable, and Bionanocomposite. It is likely to be the smartest development yet to be been in modern packaging innovations (Imran et al., 2010).

Recognition of the advantages of active packaging technology by the brand owners, development of economically viable packaging systems and increased consumer acceptance opens new horizons. However, intelligent packaging technology will offer substantial potential as a marketing tool and the establishment of brand protection. It will become more commercially viable and common-place in the next decade.

In order to address the present imbalance between potential and realization of intelligent packaging concepts in comparison with active packaging, a number of research gaps need to be filled. A huge challenge in research area of the quality indicators is to find the indicators which are sensitive (ppm – ppb levels) and specific. Most of the current pH-based indicators lack these properties. To determine their accuracy and reliability, response function with time as rate constants is required to determine as exemplified in the study of Nopwinyuwong et al. (2010).

In summary, both the incorporation of active (oxygen scavenging, humidity buffering, antimicrobial, antioxidant, and so on) agents into food contact plastic materials and the diagnosis of intelligent (integrity and freshness, and so on) colorimetric materials has been reviewed. The next level of active and intelligent packaging, which will be the bioactive and biointelligent packaging, is thus a novel set of technologies designed to give response to a number of issues related to the feasibility, stability and bioactivity of functional ingredients as well as biosecurity for the food industry. This bioactive and biointelligent packaging will open new frontiers and opportunities in preserving the quality of food products and in monitoring a status of product quality, respectively.

KEYWORDS

- **Antimicrobial packaging**
- **Clostridium**
- **Liquid water absorption**
- **Microbial contamination**
- **Oxygen scavenging**

REFERENCES

1. ACTIPAK. Evaluating safety, effectiveness, economic-environmental impact, and consumer acceptance of active and intelligent packaging. FAIR Project CT-98-4170 (2001).
2. Aday, M. S., Caner, C., and Rahvalı, F. Effect of oxygen and carbon dioxide absorbers on strawberry quality. Postharvest Biol. Technol., 62, 179–187 (2011).
3. Agerhem, H. and Nilsson, H. J. Substrate composition and use thereof. United States Patent Publication US 4284719 (1981).
4. Aharoni, N., Rodoc, V., Fallik, E., Afek, U., Chalupowicz, D., Aharon, Z., Maurer, D., and Orenstein, J. Modified Atmosphere Packaging for Vegetable Crops using High-Water-Vapor-Permeable Films. C. L. Wilson (Ed.), Intelligent and Active Packaging for Fruits and Vegetables, CRC Press, Boca Raton, pp. 73–112 (2007).
5. Ahvenainen, R. and Hurme, E. Active and smart packaging for meeting consumer demands for quality and safety. Food Add. Contamin., 14, 753–763 (1997).
6. Anthierens, T., Ragaert, P., Verbrugghe, S., Ouchchen, A., De Geest, B. G., Noseda, B., Mertens, J., Beladjal, L., De Cuyper, D., Dierickx, W., Du Prez, F., and Devlieghere, F. Use of endospore-forming bacteria as an active oxygen scavenger in plastic packaging materials. Innov. Food Sci. Emerg. Technol., 12, 594–599.
7. APPE. ActivSeal®. [On-line]. Available: http://www.appepackaging.com/performance/activseal/. 2011 (Accessed on March, 2013).
8. Appendini, P. and Hotchkiss, J. H. Review of antimicrobial food packaging. Innov. Food Sci. Emerg. Technol., 3, 113–126 (2002).
9. Beaudry, R. MAP as a Basis for Active Packaging. C. L. Wilson (Ed.), Intelligent and Active Packaging for Fruits and Vegetables, CRC Press, Boca Raton, pp. 31–55 (2007).
10. Biever, C. Herb extracts wrap up lethal food bugs. New Scientist, 178, 26 (2003).
11. Blankenship, S. Ethylene effects and the benefits of 1-MCP. Perish. Hand. Quart., (November 2–4, 2001).
12. Blankenship, S. M. and Dole, J. M. 1-Methylcyclopropene: A Review. Postharvest Biol. Technol., 28, 1–25 (2003).
13. Blixt, K. G., Tornmarck, S. I. A., Juhlin, R., Salenstedt, K. R., and Tiru, M. Enzymatic substrate composition adsorbed on a carrier. United States Patent Publication US 4043871 (1977).
14. Blixt, K. G., Tornmarck, S. I. A., Juhlin, R., Salenstedt, K. R., and Tiru, M. Enzymatic substrate composition adsorbed on a carrier. United States Patent Publication US 4184920 (1980).
15. Brewer, A. and Sloan, N. Intelligent tracking in manufacturing. J. Intell. Manuf., 10, 245–250 (1999)
16. Brody, A. L., Bugusu, B., Han, J. H., Sand, C. K., and McHugh, T. Innovative food packaging solutions. J. Food Sci., 73, R107–R116 (2008).

17. Brody, A. L., Strupinsky, E. R., and Kline, L. R. Active Packaging for Food Applications. Technomic Publishing Company, Inc., Lancaster (2001).
18. Butler, B. L. Cryovac® OS2000™ Polymeric oxygen scavenging systems. World Conference on Packaging: Proceedings of the 13th Intl. Assoc. of Packaging Res. Insts., Michigan State Univ., CRC Press LLC, East Lansing, Michigan, Florida, pp. 157–162 (June 23–28, 2002).
19. Chouliara, E., Badeka, A., Savvaidis, I., and Kontominas, M. G. Combined effect of irradiation and modified atmosphere packaging on shelf-life extension of chicken breast meat: Microbiological, chemical, and sensory changes. Eur. Food Res. Technol., 226, 877–888 (2008).
20. Clarke, R. Radio frequency identification: smart or intelligent packaging? J. Packag. Sci. Technol., Japan, 10, 233–247 (2001).
21. Clarke, R. Temperature switchable membranes for creating and maintaining beneficial package atmospheres for fresh produce. J. Plast. Film Sheet., 17, 22–34 (2001).
22. Corval, A., Kuldová, K., Eichen, Y., Pikramenou, Z., Lehn, J. M., and Trommsdorff, H. P. Photochromism and thermochromism driven by intramolecular proton transfer in dinitrobenzylpyridine compounds. J. Physical Chem., 100, 19315–19320 (1996).
23. Damaj, Z., Cisnetti, F., Dupont, L., Hénon, E., Policar, C., and Guillon, E. Synthesis, characterization, and dioxygen reactivity of copper(I) complexes with glycoligands. Dalton Trans., 28, 3235–3245 (2008).
24. Damaj, Z., Naveau, A., Dupont, L., Hénon, E., Rogez, G., and Guillon, E. Co(II)(L-proline)2(H2O)2 solid complex: Characterization, magnetic properties, and DFT computations. Preliminary studies of its use as oxygen scavenger in packaging films. Inorg. Chem. Comm., 12, 17–20 (2009).
25. Davies, K. J. A. Oxidative Stress: the Paradox of Aerobic Life. C. Rice-Evans, B. Halliwell, and G. G. Lunt (Eds.), Free Radicals and Oxidative Stress: Environment, Drugs and Food Additives, Biochemical Society Symposium Series 61, Portland Press, London, pp. 1–31 (1995).
26. Day, B. P. F. Active Packaging of Food. J. Kerry and P. Butler (Eds.), Smart Packaging Technologies for Fast Moving Consumer Goods, John Wiley & Sons Ltd., West Sussex, pp. 1–18 (2008)
27. Decker, E. A., Chan, W. K. M., Livisay, S. A., Butterfield, D. A., and Faustman, C. Interactions between carnosine and the different redox states of myoglobin. J. Food Sci., 60, 1201–1204 (1995).
28. Devlieghere, F., Vermeiren, L., Bockstal, A., and Debevere, J. Study on antimicrobial activity of a food packaging material containing potassium sorbate. Acta Alimentaria, 29, 137–146 (2000).
29. Ellouze, M., Pichaud, M., Bonaiti, C., Coroller, L., Couvert, O., Thuault, D., and Vaillant, R. Modelling pH evolution and lactic acid production in the growth medium of a lactic acid bacterium: Application to set a biological TTI. Int. J. Food Microbiol., 128, 101–107 (2008).
30. Environmental Protection Agency (EPA). Federal Register 67. 48, 796–48 800 (2002).
31. Erickson, M. C. Lipid Oxidation: Flavor and Nutritional Quality Deterioration in Frozen Food. M. C. Erickson and Y. C. Hung (Eds.), Quality in Frozen Food, Chapman & Hall, New York, pp. 141–173 (1997).
32. Farber, J. M. Microbiological aspects of modified atmosphere packaging: A review. J. Food Prot., 54, 58–70 (1991).
33. Floros, J. D., Dock, L. L., and Han, J. H. Active packaging technologies and applications. Food Cosmet. Drug Packag., 20, 10–17 (1997).
34. Frank, I., Grimme, S., and Peyerimhoff, S. D. Quantum chemical investigations of the thermal and photoinduced proton-transfer reactions of 2-(2', 4'-dinitrobenzyl)pyridine. J. Physical Chem., 100, 16187–16194 (1996).
35. Giannakourou, M. C., Koutsoumanis, K., Nychas, G. J. E., and Taoukis, P. S. Field evaluation of the application of time temperature integrators for monitoring fish quality in the chill chain. Int. J. Food Microbiol., 102, 323–336 (2005).
36. Goddard, J. M. and Hotchkiss, J. H. Polymer surface modification for the attachment of bioactive compounds. Prog. Polym. Sci., 32, 698–725 (2007).

37. Guilbert, S., Gontard, N., and Gorris, L. G. M. Prolongation of the shelf-life of perishable food products using biodegradable films and coatings. Lebensm.-Wiss. u.-Technol., 29, 10–17 (1996).
38. Haarer, D. and Eichen, Y. Substrate for packaging perishable goods or for application onto same and method for determining the quality of said goods. United States Patent Publication US 7081364 (2006).
39. Ha, J. U., Kim, Y. M., and Lee, D. S. Multilayered antimicrobial polyethylene films applied to the packaging of ground beef. Packag. Technol. Sci., 14, 55–62 (2001).
40. Han, J. H. and Floros, J. D. Casting antimicrobial packaging films and measuring their physical properties and antimicrobial activity. J. Plastic Film Sheeting, 13, 287–298 (1997).
41. Hartman, B. A new technology for detecting food spoilage. Book of Abstracts of IFT Annual Meeting. July 12-16, 2003. Chicago, IL, 2003.
42. Hogan, S. A. and Kerry, J. P. Smart Packaging of Meat and Poultry Products. J. Kerry and P. Butler (Eds.), Smart Packaging Technologies, John Wiley & Sons, Ltd., West Sussex, pp. 33–59 (2008).
43. Holland, R. V. Absorbent Material and Uses Thereof. International Patent 1991, WO1991/04292 PCT/AU1990/00413.
44. Holley, R. A. and Patel, D. Improvement in shelf-life and safety of perishable foods by Plant essential oils and smoke antimicrobials. Food Microbiol., 22, 273–292 (2005).
45. Horsham, M. A., Murphy, J. K. G., and Santangelo, R. Absorbent Material for Use in Humid Conditions. International Patent 2004, WO2004/076545 PCT/AU2004/000251.
46. Hort Research. Ripesense ® [On-line]. Available: http://www.ripesense.com/pdf_store/Product Overview.pdf (Accessed on July, 2009) (2004).
47. Hotchkiss, J. H. Innovative approaches to active and intelligent packaging. The Proceedings of the BARD-Sponsored International Workshop on Active and Intelligent Packaging for Fruits and Vegetables. Shepherdstown, West Virginia (September, 2004).
48. Hotchkiss, J. H., Watkins, C. B., and Sanchez, D. G. Release of 1-methylcyclopropene from heat-pressed polymer films. J. Food Sci., 72, E330–E334 (2007).
49. Huis In't Veld, J. H. J. Microbial and biochemical spoilage of foods: An overview. International Journal of Food Microbiology, 33, 1–18 (1996).
50. Hurme, E. Detecting Leaks in Modified Atmosphere Packaging. R. Ahvenainen (Ed.), Novel Food Packaging Techniques, Woodhead Publishing Ltd., Cambridge, pp. 276–286 (2003).
51. Hussain, M., Bensaid, S., Geobaldo, F., Saracco, G., and Russo, N. Photocatalytic Degradation of Ethylene Emitted by Fruits with TiO2 Nanoparticles. Ind. Eng. Chem. Res., 50, 2536–2543 (2011).
52. Igene, J. O., Yamauchi, K., Pearson, A. M., Gray, J. I., and Aust, S. D. Evaluation of 2-thiobarbituric acid reactive substances (TBARS) in relation to warmed-over flavor (WOF) development in cooked chicken. J. Agric. Food Chem., 33, 364–367 (1985).
53. Imran, M., Revol-Junelles, A. M., Martyn, A., Tehrany, E. A., Jacquot, M., Linder, M., and Desobry, S. Active food packaging evolution: Transformation from micro- to nanotechnology. Crit. Rev. Food Sci. Nutr., 50, 799–821 (2010).
54. Ishitani, T. Active Packaging for Food Quality Preservation in Japan. P. Ackerman, M. Jagerstad, and T. Ohlsson (Eds.), Food and Packaging Materials-Chemical Interactions, Royal Society of Chemistry, Cambridge, pp. 177–188 (1995).
55. Ito, N., Hirose, M., Fukushima, H., Tsuda, T., Shirai, T., and Tatenatsu, M. Studies on antioxidants: Their carcinogenic and modifying effects on chemical carcinogens. Food Chem. Toxicol., 24, 1071–1092 (1986).
56. Jones, P., Clarke-Hill, C., Shears, P., Comfort, D., and Hillier, D. Radio frequency identification in the UK: opportunities and challenges. Int. J. Retail Distrib. Manag., 32, 164–171 (2004).
57. Juneja, V. K. and Sofos J. N. Pathogens and Toxins in Foods: Challenges and Interventions. ASM Press, Washington, DC (2009).
58. Käkkäinen, M. and Holmström, J. Wireless product identification: enabler for handling efficiency, customization, and information sharing. Supply Chain Manag., 7, 242–252 (2002).

59. Käkkäinen, M. Increasing efficiency in the supply chain for short shelf life goods using RFID tagging. Int. J. Retail Distrib. Manag., 31, 529–536 (2003).
60. Karel, M. Tasks of food technology in the 21st Century. Food Technol., 54, 5658; 60–62 (2000).
61. Kerry, J. P., O'Grady, M. N., and Hogan, S. A. Past, current, and potential utilisation of active and intelligent packaging systems for meat and muscle-based products: A review. Meat Sci., 74, 113–130 (2006).
62. Klein, R. A., Riley, M. R., DeCianne, D. M., and Srinavakul N. Non-invasive colorimetric ripeness indicator. United States Patent Application Publication US 2006/0127543 A1 (2006).
63. Kleist, R. A., Chapman, T. A., Sakai, D. A., and Jarvis, B. S. RFID Labeling: Smart Labeling Concepts and Applications for the Consumer Packaged Goods Supply Chain: 2nd Edition. Printronix, Inc., Irvene (2005).
64. Kochhar, S. P. Oxidative Pathways to the Formation of Off-flavors. M. J. Saxby (Ed.), Food Taints and Off-flavors: 2nd Edition, Blackle Academic and Professional, New York, pp. 168–225 (1996).
65. Koutsoumanis, K., Taoukis, P. S., and Nychas, G. J. E. Development of a safety monitoring and assurance system for chilled food products. Int. J. Food Microbiol., 100, 253–260 (2005).
66. Kreyenschmidt, J., Christiansen, H., Hübner, A., Raab, V., and Petersen, B. A novel photochromic time-temperature indicator to support cold chain management. Int. J. Food Sci. Technol., 45, 208–215 (2010).
67. Kumar, P., Reinitz, H. W., Simunovic, J., Sandeep, K. P., and Franzon, P. D. Overview of RFID technology and its applications in the food industry. J. Food Sci., 74, R101–106 (2009).
68. Kumar, S. and Budin, E. M. Prevention and management of product recalls in the processed food industry: A case study based on an exporter's perspective. Technovation, 26, 739–750 (2006).
69. Labuza, T. P. and Breene, W. M. Applications of active packaging for improvement of shelf-life and nutritional quality of fresh and extended shelf-life foods. J. Food Proc. Preserv., 13, 1–69 (1989).
70. Labuza, T. P. Shelf-life Dating of Foods. Food and Nutrition Press, Westport (1982).
71. Lagarón, J. M. and Busolo, M. A. Multifunctional Nanoclays for Food Contact Applications. J. M. Lagarón (Ed.), Multifunctional and Nanoreinforced Polymers for Food Packaging, Woodhead Publishing Ltd., Cambridge, pp. 31–42 (2011).
72. Lee, Y. S., Beaudry, R., Kim J. N., and Harte, B. Development of a 1-methylcyclopropene (1-MCP) sachet release system. J. Food Sci., 71, C1–C6 (2006).
73. Lin, Z., Zhong, S., and Grierson, D. Recent Advances in Ethylene Research. J. Exp. Bot., 60, 3311–3336 (2009).
74. Li, S., Visich, J. K., Khumawala, B. M., and Zhang, C. Radio frequency identification technology: applications, technical challenges and strategies. Sensor Rev., 26, 193–202 (2006).
75. López-Rubio, A., Almenar, E., Hernandez-Muñoz, P., Lagarón, J. M., Catalá, R., and Gavara, R. Overview of Active Polymer-Based Packaging Technologies for Food Applications. Food Rev. Int., 20, 357–387 (2004).
76. Man, D. Shelf Life. Blackwell Science, London (2002).
77. Manske, W. J. Selected time interval indicating device. United States Patent Publication US 3954011 (1976).
78. Mattila, T., Tawast, J., and Ahvenainen, R. New possibilities for quality control of aseptic packages: Microbiological spoilage and seal defect detection using head-space indicators. Lebens.-Wiss. und-Technol., 23, 246–251 (1990).
79. McMurry, J. E. Introduction to Organic Chemistry. Thomson Higher Education (2007).
80. Miltz, J., Bigger, S. W., Sonneveld, C., and Suppakul, P. Antimicrobial packaging material. United States Patent Publication US 8,017,667 B2 (2011).
81. Miltz, J., Hoojjat, P., Han, J., Giacin, J. R., Harte, B. R., and Gray, I. J. Loss of antioxidants from high density polyethylene-its effect on oatmeal cereal oxidation. Food and Packaging Interactions, ACS symposium Series 365, pp. 83–93 (1989).

82. Miltz, J., Passy, N., and Mannheim, C. H. Trends and Applications of Active Packaging Systems. P. Ackerman, M. Jagerstad, and T. Ohlsson (Eds.), Food and Food Packaging Materials-Chemical Interactions, Royal Society of Chemistry, Cambridge, pp. 201–210 (1995).
83. Miltz, J., and Perry, M. R. Active Packaging Technologies: Oxygen Scavenging. Proceedings of the 20th Intl. Assoc. of Packaging Res. Inst. Symposium, San Jose State Univ., San Jose, California, pp. 312–330 (June 14–18, 2000).
84. Mizutani, Y., Nakamura, S., Kaneko, S., and Okamara, K. Microporous polypropylene sheets. Industrial Eng. Chem. Res., 32, 221–227 (1993).
85. Munro, I. C., Haighton, L. A., Lynch, B. S., and Tafazoli, S. Technological challenges of addressing new and more complex migrating products from novel food packaging materials. Food Addit. Contam., 26, 1534–1546 (2009).
86. Nakamura, H. and Hoshino, J. Techniques for the preservation of food by employment of an oxygen Absorber. Tech Information, Ageless Division, Mitsubishi Gas Chemical Co., Tokyo, pp. 1–45 (1983).
87. Nanthachai, N., Ratanachinakorn, B., Kosittrakun, M., and Beaudry, R. M. Absorption of 1-MCP by fresh produce. Postharvest Biol. Technol., 43, 291–297 (2007).
88. Nerín, C., Tovar, L., and Salafranca, J. Behavior of a new antioxidant active film versus oxidizable model compounds. J. Food Eng., 84, 313–320 (2008).
89. Ngai, E. W. T., Moon, K. K. L., Riggins, F. J., and Yi, C. Y. RFID research: An academic literature review (1995-2005) and future research directions. Int. J. Prod. Econ., 112, 510–520 (2008).
90. Nicholson, M. D. The role of natural antimicrobials in food/packaging biopreservation. J. Plastic Film Sheeting, 14, 234–241 (1998).
91. Nopwinyuwong, A., Boonsupthip, W., Pechyen, C., and Suppakul, P. Formation Of polydiacetylene/silica nanocomposite as a colorimetric indicator: Effect of time and temperature. Adv. Polym. Technol. (2013) (In Press).
92. Nopwinyuwong, A., Boonsupthip, W., Pechyen, C., and Suppakul, P. Preparation of polydiacetylene vesicle and amphiphilic polymer as time-temperature indicator. Adv. Mat. Res. 506, 552–555 (2012(a)).
93. Nopwinyuwong, A., Boonsupthip, W., Pechyen, C., and Suppakul, P. Response modeling for polydiacetylene/silica nanocomposite as time-temperature indicator. The Book of Abstracts of the 5th Shelf Life International Meeting 2012. May 30–June 1, 2012. Changwon, South Korea, (2012(b)).
94. Nopwinyuwong, A, Trevanich, S., and Suppakul, P. Development of a novel colorimetric indicator label for monitoring freshness of intermediate-moisture dessert spoilage. Talanta, 81, 1126–1132 (2010).
95. Ozdemir, M. and Floros, J. D. Active food packaging technologies. Crit. Rev. Food Sci. Nutr., 44, 185–193 (2004).
96. Pacquit, A., Lau, K. T., McLaughlin, H., Frisby, J., Quilty, B., and Diamond, D. Development of a volatile amine sensor for the monitoring of fish spoilage. Talanta, 69, 515–520 (2006).
97. Pacquit, A., Frisby, J., Diamond, D., Lau, K. T., Farrell, A., Quilty, B., and Diamond, D. Development of a smart packaging for the monitoring of fish spoilage. Food Chem., 102, 466–470 (2007).
98. Parker, N. Innovative oxygen emitting/carbon dioxide scavenging technology for Fresh produce applications. Second International Conference on Active and Intelligent Packaging. Campden & Chorleywood Food Research Association, Chipping Campden, Gloucestershire, UK (September 12–13, 2007).
99. Patel, G. N., Preziosi, A. F., and Baughman, R. H. Time-temperature history indicators. United States Patent Publication US 3999946 (1976).
100. Patel, G. N. and Yee, K. C. Diacetylene time-temperature indicators. United States Patent Publication US 4228126 (1980).
101. Pereira de Abreu, D. A., Cruz, J. M., and Paseiro Losada, P. Active and intelligent packaging for the food industry. Food Rev. Int., 28, 146–187 (2012).

102. Persico, P., Ambrogi, V., Carfagna, C., Cerruti, P., Ferrocino, I., and Mauriello, G. Nanocomposite polymer films containing carvacrol for antimicrobial active packaging. Polym. Eng. Sci., 49, 1447–1455 (2009).
103. Plumb, G. W., Chambers, S. J., Lambert, N., Bartolomé, B., Heaney, R. K., Wanigatunga, S., Aruoma, O. I., Halliwell, B., and Williamson, G. Antioxidant actions of fruit, herb, and spice extracts. J. Food Lipid, 3, 171–188 (1996).
104. Poças, M. F. F., Delgado, T. F., and Oliveira, F. A. R. Smart Packaging Technologies for Fruits and Vegetables. J. Kerry and P. Butler (Eds.), Smart Packaging Technologies, John Wiley & Sons, Ltd., West Sussex, pp. 151–166 (2008).
105. Prange, R. K. and De Long, M. 1-Methylcyclopropene: The magic bullet for horticultural products. Chronica Hortic., 43, 11–14 (2003).
106. Puligundla, P., Jung, J., and Ko, S. Carbon dioxide sensors for intelligent food packaging applications. Food Control, 25, 328–333 (2012).
107. Quintavalla, S. and Vicini, L. Antimicrobial food packaging in meat industry. Meat Sci., 62, 373–380 (2002).
108. Regattieri, A., Gamberi, M., and Manzini, R. Traceability of food products: General framework and experimental evidence. J. Food Eng., 81, 347–356 (2007).
109. Restuccia, D., Spizzirri, U. G., Parisi, O. I., Cirillo, G., Curcio, M., Iemma, F., Puoci, F., Vinci, G., and Picci, N. New EU regulation aspects and global market of active and intelligent packaging for food industry applications. Food Control, 21, 1425–1435 (2010).
110. Riley, M. R. RediRipe® Sticker. [On-line]. Available: http://ag.arizona.edu/~riley/, 2006 (Accessed on July, 2009).
111. Robertson, G. L. Food Packaging: Principles and Practice: 3rd Edition. CRC Press, Boca Raton (2012).
112. Rodgers, B. D. and Compton, L. New polymeric oxygen scavenging system for coextruded packaging structures. The Proceedings of Oxygen Absorbers 2001 and Beyond Conference. June, 2000.
113. Rodrigues, E. T. and Han, J. H. Intelligent Packaging. D. R. Heldman (Ed.), Encyclopedia of Agricultural, Food and Biological Engineering, Marcel Dekker, New York, pp. 528–535 (2003).
114. Rooney, M. L. Active Packaging in Polymer Films. M. L. Rooney (Ed.), Active Food Packaging, Blackie Academic and Professional, Glasgow, pp. 74–110 (1995).
115. Rooney, M. L. Introduction to Active Food Packaging Technologies. J. H. Han (Ed.), Innovations in Food Packaging, Elsevier Academic Press, Oxford, pp. 63–79 (2005).
116. Rooney, M. L. New materials and methods in perishable transport. [On-line]. Available: http://www.atse.or…cations/symposia/proc-1998p15.html (Accessed on August, 2000) (1998).
117. Rosca, I. D. and Vergnaud, J. M. Problems of food protection by polymer packages. J. Chem. Health Safety, 14, 14–20 (2007).
118. Rozenberg, G. G., Bresler, E., Speakman, S. P., Jeynes, C., and Steinke, J. H. G. Patterned low temperature copper-rich deposits using inkjet printing. Appl. Phys. Lett., 81, 5249–5251 (2002).
119. Sabnis, R. W. Handbook of Acid-Base Indicators. CRC Press, Boca Raton (2008).
120. Sadaka, F., Nguimjeu, C., Brachais, C. H., Vroman, I., Tighzert, L., and Couvercelle, J. P. Review on antimicrobial packaging containing essential oils and their active biomolecules. Innov. Food Sci. Emerg. Technol. (2013) (In Press).
121. Sánchez-García, M. D., Gimenez, E., Ocio, M. J., and Lagaron, J. M. Novel polycaprolactone nanocomposites containing thymol of interest in antimicrobial film and coating applications. J. Plastic Film Sheeting, 24, 239–241 (2008).
122. Scannell, A. G. M., Hill, C., Ross, R. P., Marx, S., Hartmeier, W., and Arendt, E. K. Development of bioactive food packaging materials using immobilised bacteriocins Lacticin 3147 and Nisaplin. Int. J. Food Microbiol., 60, 241–249 (2000).
123. Schuenzel, K. M. and Harrison, M. A. Microbial antagonists of food borne pathogens on fresh, minimally processed vegetables. J. Food Prot., 65, 1909–1915 2002).

124. Scully, A. D. and Horsham, M. A. Active Packaging for Fruits and Vegetables. C. L. Wilson (Ed.), Intelligent and Active Packaging for Fruits and Vegetables, CRC Press, Boca Raton, pp. 57–71 (2007).
125. Scully, A. D. and Horsham, M. A. Emerging packaging technologies for enhanced food preservation. Food Sci. Technol., 20, 16–19 (2006).
126. Sisler, E. C. and Blankenship, S. M. Methods of counteracting an ethylene response in plants. US Patent 1996, US5,518,988.
127. Sisler, E. C., Serek, M., Dupille, E., and Goren, R. Inhibition of ethylene responses by 1-methylcyclopropene and 3-methylcyclopropene. Plant Growth Regulat., 27, 105–111 (1999).
128. Smith, A. W. J., Poulston, S., Rowsell, L., Terry, L. A., and Anderson, J. A. A new palladium-based ethylene scavenger to control ethylene-induced ripening of climacteric Fruit. Platinum Metals Rev., 53, 112–122 (2009).
129. Smith, J. P., Hoshino, J., and Abe, Y. Interactive Packaging Involving Sachet Technology. M. L. Rooney (Ed.), Active Food Packaging, Blackie Academic and Professional, Glasgow, pp. 143–173 (1995).
130. Smith, J. P., Ooraikul, B., Koersen, W. J., and Jackson, E. D. Novel approach to oxygen control in modified atmosphere packaging of bakery products. Food Microbiol., 3, 315–320 (1986).
131. Smolander, M., Hurme, E., Latva-Kala, K., Luoma, T., Alakomi, H. L., and Ahvenainen, R. Myoglobin-based indicator for the evaluation of freshness of unmarinated broiler cuts. Innov. Food Sci. Emerg. Technol., 3, 279–288 (2002).
132. Sonneveld, K. What Drives (Food) Packaging Innovation? Packag. Technol. Sci., 13, 29–35 (2000).
133. Srour, R. K. and McDonald, L. M. Determination of the acidity constants of methyl red and phenol red indicators in binary methanol- and ethanol-water mixtures. J. Chem. Eng. Data, 53, 116–127 (2008).
134. Süd-Chemie. Oxy-Guard™ Film Oxygen Scavenging Bilster Film. Süd-Chemie Inc. (2010).
135. Summers, L. Intelligent Packaging. Centre for Exploitation of Science and Technology, London (1992).
136. Suppakul, P. Intelligent Packaging. Part VII: Trends in Frozen Food Packing. D. W. Sun (Ed.), Handbook of Frozen Food Processing and Packaging: Second Edition, CRC Press, Boca Raton, pp. 837–860 (2012).
137. Suppakul, P., Miltz, J., Sonneveld, K., and Bigger, S. W. Active packaging technologies with an emphasis on antimicrobial packaging and its applications. J. Food Sci., 68, 408–420 (2003).
138. Suppakul, P., Miltz, J., Sonneveld, K., and Bigger, S. W. Efficacy of polyethylene-based antimicrobial films containing principal constituents of basil. LWT - Food Sci. Technol., 41, 779–788 (2008).
139. Suslow, T. Performance of zeolite based products in ethylene removal. Perishable Handling Q., 92, 32–33 (1997).
140. Taoukis, P. S. Application of time-temperature integrators for monitoring and management of perishable product quality in the cold chain. J. Kerry and P. Butler (Eds.), Smart Packaging Technologies, John Wiley & Sons, Ltd., West Sussex, pp. 61–74 (2008).
141. Taoukis, P. S. and Labuza, T. P. Applicability of time-temperature indicators as shelf life monitors of food products. J. Food Sci., 54, 783–788 (1989).
142. Taoukis, P. S. and Labuza, T. P. Time-temperature indicators (TTIs). R. Ahvenainen (Ed.), Novel Food Packaging Techniques, Woodhead Publishing Ltd., Cambridge, pp. 103–126 (2003).
143. Taoukis, P. S. Modeling the use of time-temperature indicators in distribution and stock rotation. L. M. M. Tijkskens, M. L. A. T. M. Hertog, and B. M. Nicolaï (Eds.), Food Process Modeling, CRC Press, Washington, DC, pp. 402–432 (2001).
144. Thiesse, F., Michahelles, F. An overview of EPC technology. Sensor Rev., 26, 101–105 (2006)
145. Tippayatum, P., Fuongfuchat, A., Jangchud, K., Jangchud, A., and Chonhenchob, V. Development of antimicrobial EVA/LDPE films incorporated with thymol and eugenol. Food Manufac. Effic., 2, 1–7 (2009).

146. Tiru, M. O. and Tiru, M. B. I. Thermochromic composition, method of making, and use. United States Patent Publication US 4149852 (1979).
147. Vaillant, R. Method, system, and component for controlling the preservation of a product. United States Patent Publication US 7691634 (2010).
148. Vermeiren, L., Devlieghere, F., and Debevere, J. Effectiveness of some recent antimicrobial packaging concepts. Food Add. Contamin., 19, 163–171 (2002).
149. Vermeiren, L., Devlieghere, F., van Beest, M., de Kruijf, N., and Debevere, J. Development in the Active Packaging of Foods. Trends Food Sci. Technol., 10, 77–86 (1999).
150. Vermeiren, L., Heirlings, L., Devlieghere, F., and Debevere, J. Oxygen, Ethylene, and Other Scavengers. R. Ahvenainen (Ed.), Novel Food Packaging Techniques, Woodhead Publishing Limited, Cambridge, pp. 22–49 (2003).
151. Vest, R. W. Electronics Films from Metallo-organic Precursors. J. B. Wachtman and R. A. Haber (Eds.), Ceramic Films and Coatings, Noyes Publications, Park Ridge, pp. 303–341 (1993).
152. Vitsab International. CheckPoint® III Product Details. Available Source: http://www.vitsab.com/Products.html (Accessed on May 31, 2010) (2007).
153. Vo, E., Murray, D. K., Scott, T. L., and Attar, A. J. Development of a novel colorimetric indicator pad for detecting aldehydes. Talanta, 73, 87–94 (2007).
154. Weng, Y. M., Chen, M. J., and Chen, W. Antimicrobial food packaging materials from polyethylene-co-methacrylic acid). Lebensm. Wiss. U. Technol., 32, 191–195 (1999).
155. Weng, Y. M. and Hotchkiss, J. H. Anhydrides as antimycotic agents added to polyethylene films for food packaging. Packag. Technol. Sci., 6, 123–128 (1993).
156. Wessling, C., Nielsen, T., Leufven, A., and Jägerstad, M. Mobility of tocopherol and BHT in LDPE in contact with fatty food stimulants. Food Add. Contamin., 15, 709–715 (1998).
157. White, A., Johnson, M., and Wilson, H. RFID in the supply chain: Lessons from European early adopters. Int. J. Phys. Distrib. Logist. Manag., 38, 88–107 (2008).
158. Wilding, R. and Delgado, T. RFID demystified company case studies. Logist. Transport Focus, pp. 32–44 (June, 2004).
159. Wyld, D. C. RFID 101: The next big thing for management. Manag. Res. News, 29, 154–173 (2006).
160. Yam, K. L. Intelligent packaging for the future smart kitchen. Packag. Technol. Sci., 13, 83–85 (2000).
161. Yam, K. L., Takhistov, P. T., and Miltz, J. Intelligent Packaging: Concepts and Applications. J. Food Sci., 70, R1–R10 (2005).
162. Zagory, D. Ethylene-Removing Packaging. M. L. Rooney (Ed.), Active Food Packaging, Blackie Academic and Professional, London, pp. 38–54 (1995).
163. Zaritzky, N. Physical-Chemical Principles in Freezing. D. W. Sun (Ed.), Handbook of Frozen Food Processing and Packaging, CRC Press, Boca Raton, pp. 3–31 (2006).
164. Zorn, M., Tompkins, D. T., Zeltner, W. A., and Anderson, M. A. Catalytic and Photocatalytic Oxidation of Ethylene on Titania-Based Thin Films. Environ. Sci. Technol., 34, 5206–5210 (2000).

CHAPTER 17

CONDUCTIVE POLYMERS FOR PACKAGING APPLICATIONS

S. B. KONDAWAR

ABSTRACT

Electromagnetic interference (EMI) is one of the most undesirable by-products of rapid proliferation of electronic products and telecommunications. Electromagnetic radiation also adversely affects human health. Hence, efforts are made to reduce its effect using EMI shielding materials as an intelligent packaging. The conducting paths of polymer due to nanoparticles embedded into polymer matrix increase the absorption of the electromagnetic wave to a large extent. The incorporation of magnetic and dielectric nanofillers in the polymer matrix leads to better absorbing materials which make them futuristic radar absorbing material. This chapter mainly focuses on nanostructure conductive polymers and conductive polymer nanocomposites as EMI shielding materials for packaging electronic gadgets. This chapter is divided into two sections. The first section describes the fundamentals of conductive polymers including doping, conduction mechanism, charge transport phenomenon, preparative strategy of nanostructure conductive polymers and nanocomposites. The second section of the chapter is on packaging applications which includes the introduction of electromagnetic shielding and microwave absorption, theory of electromagnetic shielding and the use of conductive polymers and nanocomposites for EMI shielding. Although only a few reports on the application of conductive polymer nanocomposites for microwave absorption and EMI shielding are available, it is believed that conductive polymer nanocomposites could be important as EMI shielding materials for packaging.

17.1 INTRODUCTION

Packaging is expected to not only to increase the economic value of products but also to preserve their qualities. Food packaging is traditionally used for protection, preservation, and storage of sensitive food items. Intelligent packaging and active packaging are two modern concepts that are making exciting advances compared to conventional packaging. The first technology (intelligent packaging) applies to materials able to respond to changing environmental condition and to signal changes, such as temperature indicators and leakage indicators. The second (active packaging) relates to changes that can be brought about in the condition of packaged food by altering the package properties, such as antimicrobial, odor absorbing and oxygen scavenging. Active packaging is of great interest because of the increasing demand for better quality and longer shelf life with food products. The most widely used active packaging technologies for foods are in the removal of undesirable substances through absorption or/and scavenging, as with the widely used oxygen-scavengers. Small molecule antioxidants have been commonly used as food additives to help protect food against oxidation because of their ability to scavenge free radicals. A few reports indicate that addition of an antioxidant, delays the oxidation of oil products. However, the health risk of synthetic antioxidants and a low thermal stability of natural antioxidants, have led to a consideration of alternative antioxidants. Moreover, addition of low molecular weight antioxidants may affect the taste and smell of a food. Antioxidant packaging materials can be considered for use as active packaging to prevent the oxidation of foodstuffs. The oxidation process normally occurs at the surface between air and the foodstuff. If the packaging surface can include an antioxidant material, oxidation might be retarded. If a packaging material (of a high molecular weight) can itself exhibit antioxidant activity, it may be excellent for food packaging applications. Conducting polymers with high molecular weight (and thus not liable to leach into the food product) and tailored antioxidant properties, can be considered as antioxidant packaging materials.

Electromagnetic interference (EMI) shielding belongs to the class of an intelligent packaging. EMI is one of the most undesirable by-products of rapid proliferation of electronic products and telecommunications. Mutual interference among devices such as TVs, computers, mobile phones, and radios degrade device performance. Electromagnetic radiation also adversely affects human health. Hence, efforts were made to reduce its effect using EMI shielding materials. Thus, attenuation of EMI has been investigated by many researchers (Wang, Tay, See, Sun, Tan, and Lua, 2009; Klemperer and Maharaj, 2009; Al-Saleh and Sundararaj, 2009; Saini, Choudhary, Singh, Mathur, and Dhawan, 2009). Traditionally, metals, in the form of thin sheets or sheathing, were used as EMI shielding materials. However, metals were expensive, heavy, prone to corrosion and difficult to process. Hence, conducting polymers and composites containing conductive fillers were developed as an alternative EMI shielding material. These materials are light weight, cheap, resistant to corrosion and easily processable. In the last four decades, conducting polymers (CPs) have gained a special status owing to wealth of applications. The EMI shielding and microwave absorption properties of these polymers can be explained in terms of electrical conductivity and presence of bound/localized charges (polarons/bipolarons) leading to strong polarization and relaxation effects. Polyaniline (PANI) has special status among other conducting polymers due to its non-redox doping, good environmental stability and economic feasibility. The properties can be further tuned by controlled polymerization conditions and using substituted anilines, specific co-monomers, dopants and fillers. PANI has low inherent specific strength and requires dispersion in some binding matrix to form composites for any commercially useful product. However, percolation threshold tends to be high due to low compatibilities, phase segregated morphology and low aspect ratio of the conducting polymer particles. Therefore, high concentration of conducting polymers is required in matrix for acceptable electrical properties which often affect the mechanical properties of resultant composites. This chapter is mainly focused on the materials used for EMI shielding as an intelligent packaging application.

17.2 CONDUCTIVE POLYMERS

The organic polymers which show semiconducting nature with electrical conductivity as those of conductors are called as conductive or more commonly conducting polymers. The biggest advantage of conductive polymers is their processability, mainly by dispersion. They can offer high electrical conductivity but do not show similar mechanical properties to other commercially available polymers. In most of the cases, polymers are insulators in their neutral state and they become conducting only after introduction of electron acceptors/donors by a process known as "doping". The conductivity of a polymer can be tuned by chemical manipulation such as the nature of the dopant, the degree of doping and blending with other polymers. Conductive polymers are rapidly gaining attraction in new applications with increasingly processable materials with better electrical and physical properties and lower costs. The emergence of conductive polymers has constituted a significant milestone in modern analytical science (Trojanowicz, 2003). These materials, for being conjugated systems possess considerable electron availability which provides them with a rigid structure and a better capacity to be adsorbed on metallic surfaces (Jadhav, Hundiwale, and Mahulikar, 2010). The most common forms of conductive polymers in their neutral states are insulators called conjugated polymers. However, these neutral conjugated polymers can be converted into semi-conductive or conductive states through chemical or electrochemical redox reactions. To extend the functions or to improve the performances of these polymers, conductive polymers are frequently doped with other functional materials to form composites. The general characteristics of these forms were reviewed recently (Holze, 2009). Due to their unique conductive properties, they are usually employed in significant applications in chemistry, physics, electronics, optics, materials, and biomedical science, which have prompted the need for analytical methodologies to characterize and to control the quality of these materials (Jang, 2006). Conductive polymers are highly delocalized π-electron system with alternate single and double bonds in the polymer backbone. The π-conjugation of the polymer chain generates high energy occupied molecular orbitals and low energy unoccupied molecular orbitals leading to a system that can be readily oxidized or reduced. Unlike traditional polymers, which are electrical insulators, conducting polymers are semiconducting and can be doped into regions of metallic conductivity (Stejskal, 2009; Laslaua, Zujovica, and Sejdica, 2010). This novel finding, at odds with what had been previously expected of polymers, yielded the 2000 Nobel Prize in Chemistry for Alan J. Heeger, Alan G. MacDiarmid, and Hideki Shirakawa for the discovery and subsequent development of this new class of materials. Thanks to intense research efforts, there are now a large variety of conducting polymers, with polyacetylene, polythiophene, polypyrrole and polyaniline being four of the most studied and promising types. They have showed that polyacetylene, which is the simplest polyconjugated system, can be made conductive by reaction with bromine or iodine vapors. Spectroscopic studies, that followed demonstrated without any ambiguity that this reaction is redox in nature and consists of the transformation of neutral polymer chains into polycarbocations with simultaneous insertion of the corresponding number of Br_3^- or I_3^- anions between the polymer chains in order to neutralize the positive charge generated on the polymer chain. This important discovery initiated an extensive and systematic research of various aspects of the chemistry and physics of conjugated polymers both in their neutral (undoped) and charged (doped) states. Undoped conjugated polymers are semiconductors with band gaps ranging from 1 to 4eV, therefore their room temperature conductivities are very low, typically of the order of 10^{-8} S/cm or lower. However, doping can leads to an increase in conductivity of polymer by many orders of magnitude (Skotheim, Elsenbaumer, and Renolds, 1998).

17.2.1 DOPING OF POLYMERS

The concept of doping is unique and distinguishes conducting polymers from all other types of polymers (Swager, 2002). During the doping process, an undoped conjugated polymer having small conductivity, typically in the range of 10^{-10}–10^{-5} S/cm, is converted to a doped conducting polymer, which is in "metallic" conducting regime (1–10^4 S/cm). The controlled addition of known, usually small (<10%) non-stiochiometric quantities of chemical species results in dramatic changes in electronic, electrical, magnetic, optical, and structural properties of the polymers. By controllably adjusting the doping level, the conductivity anywhere between that of the non-doped and fully doped form of the polymer can be easily obtained. The highest value reported to date has been obtained in iodine-doped polyacetylene (>10^5 S/cm) and the predicted theoretical limit is about 2×10^7, more than an order of magnitude higher than that of copper. Conductivity of other conjugated polymers reaches up to 10^3 S/cm. Trans-$(CH)_x$ form of polyacetylene (PA) and emeraldine base form of polyaniline (PANI) are shown in Figure 1, to illustrate the increases in the electrical conductivity of many orders of magnitude which can be obtained by doping (MacDiarmid, 2002).

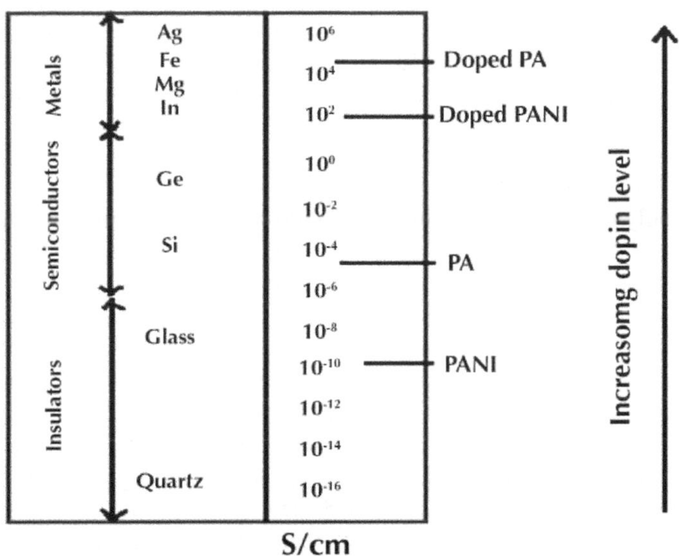

FIGURE 1 Conductivity of some metals, semiconductors, insulators and conjugated polymers.

In the doped state, the backbone of conducting polymer consists of a delocalized π-system. In the un-doped state, the polymer may have a conjugated backbone such as in trans-$(CH)_x$ which is retain in modified form after doping, or it may have a non-conjugated backbone, as in polyaniline (leucoemeraldine base form) which become truly conjugated only after p-doping, or a non-conjugated structure as in the emeraldine base form of polyaniline which becomes conjugated only after protonic acid doping. All conducting polymers and most of their derivatives shown in Figure 2, undergo either p- and/or n- redox doping by chemical and/or electrochemical processes during which the number of electrons associated with the polymer backbone changes. The molecules of the monomers of all these polymers have alternating double-single chemical bonds and hence they form polymers that are p-conjugated.

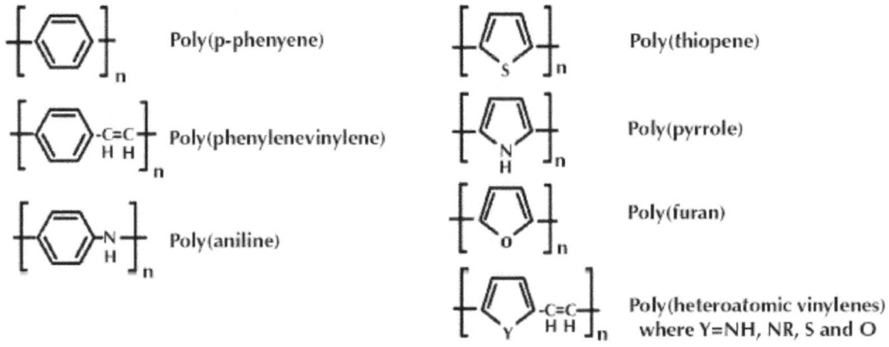

Poly(p-phenyene)

Poly(phenylenevinylene)

Poly(aniline)

Poly(thiopene)

Poly(pyrrole)

Poly(furan)

Poly(heteroatomic vinylenes) where Y=NH, NR, S and O

FIGURE 2 Conducting polymers undergo redox doping.

Charge injection onto conjugated, semiconducting macromolecular chains, doping leads to the wide variety of interesting and important phenomena. Reversible charge injection by doping can be accomplished in a number of ways for example chemical doping by charge transfer, electrochemical doping, doping by acid-base chemistry and photodoping.

17.2.2 CONDUCTION MECHANISM

Conducting polymers are unusual in that they do not conduct electrons via the same mechanisms used to describe classical semiconductors and hence their electronic properties cannot be explained well by standard band theory. The electronic conductivity of conductive polymers results from mobile charge carriers introduced into the conjugated π- system through doping. To explain the electronic phenomena in these organic conducting polymers, new concepts including solitons, polarons, and bipolarons have been proposed by solid-state physicists (Heeger, 2010; Bishop, Campbell, and Fesser, 1981; Bredas, Themans, Andre, Chance, and Silbey, 1984). The electronic structures of π-conjugated polymers with degenerate and nondegenerate ground states are different. In π-conjugated polymers with degenerate ground states, solitons are the important and dominant charge storage species. Polyacetylene, $(CH)_x$, is the only known polymer with a degenerate ground state due to its access to two possible configurations. The two structures differ from each other by the position of carbon–carbon single and double bonds. While polyacetylene can exist in two isomeric forms: cis and trans polyacetylene, the trans-acetylene form is thermodynamically more stable and the cis–trans isomerization is irreversible. A soliton can also be viewed as an excitation of the radical from one potential well to another well of the same energy. A neutral soliton occurs in pristine trans-polyacetylene when a chain contains an odd number of conjugated carbons, in which case there remains an unpaired π-electron, a radical, which corresponds to a soliton. In a long chain, the spin density in a neutral soliton (or charge density in a charged soliton) is not localized on one carbon but spread over several carbons which gives the soliton a width. Starting from one side of the soliton, the double bonds become gradually longer and the single bonds shorter, so that arriving at the other side, the alternation has completely reversed. This implies that the bond lengths do equalize in the middle of a soliton. The presence of a soliton leads to the appearance of a localized electronic level at mid-gap which is half occupied in the case of a neutral soliton and empty (doubly occupied) in the case of a positively (negatively) charged soliton. Similarly, in n-type doping, neutral chains are either chemically or electrochemically reduced to polycarbonium anions and simultaneously charge-compensating cations are inserted into the polymer matrix. In this case, negatively charged, spinless solitons are charge carriers.

Consider the structure of other conductive polymers such as poly(p-phenylene), poly(pyrrole), poly(thiophene), and poly(aniline). These polymers do not support soliton-like defects because the ground state energy of the quinoid form is substantially higher than the aromatic benzenoid structure. As a result, the charge defects on these polymers are different. As an example, consider the oxidation of polypyrrole (Figure 3).

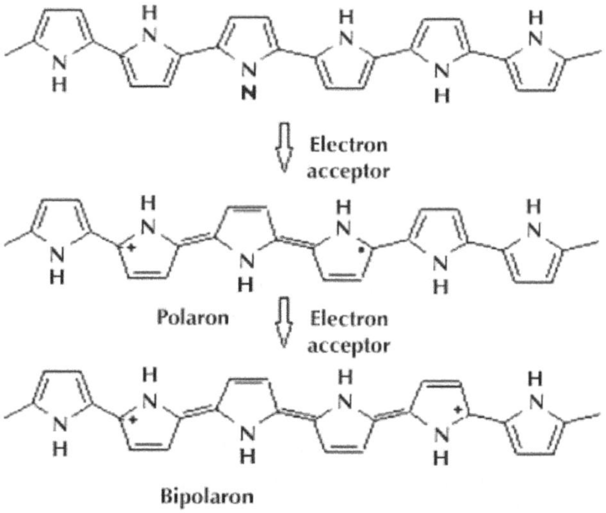

FIGURE 3 Polaron and bipolaron formation on π-conjugated backbone of polypyrrole.

Removal of an electron from a pyrrole unit leads to the formation of a polaron consisting of a tightly bound radical and cation. The binding arises from the increase in the energy (of the defect) with increasing radical-cation separation. The increase in energy is partly due to a loss of aromaticity. Calculations on poly(pyrrole) and poly(thiophene) indicate that two polarons in close proximity are unstable with respect to the formation of a bipolaron. The two free radicals combine leaving behind two cations separated by a quinoidal section of the polymer chain. However, the higher energy of the quinoid section between them binds them together resulting in correlated motion. The net effect is the formation of a doubly charged defect acting like a single entity and delocalized over several rings (3–5) that is a bipolaron. The formation of bipolarons implies a net free energy gain in forming a closed shell defect from two open shell structures. The quinoid form has a higher energy than the aromatic benzenoid form but its electron affinity is higher and the ionization potential lower. This leads to the formation of two localized states in the band gap. As the doping is increased, additional states are created in the gap and they finally evolve into two narrow bands. At low doping levels, the defects are polarons, which tend to combine at higher doping levels to form bipolarons.

Let us now consider some peculiarities of the polyaniline system. In the conventional conductive polymers, for example, polypyrrole, polythiophene, polyacetylene, and poly(p-phenylene), and so on, oxidative doping results in the removal of electrons from the bonding π- system. In polyaniline, in contrast, the initial removal of electrons is from the non-bonding nitrogen lone pairs. Unlike other conducting polymers, the quinoid form is not just simply an alternative resonance form. Its formation requires reduction and deprotonation so that it actually differs in chemical composition from the benzenoid form (Figure 4).

Benzenoid Quinoid

FIGURE 4 Benzenoid and quinoid form of polyaniline.

This peculiarity of the polyaniline structure makes doping by an acid-base reaction possible. In addition, the constituent parts of both the polaron and the bipolaron are very tightly bound owing to valence restrictions. The radical cations of the polaron are confined to a single aniline residue. The bipolaron is confined to, and identical with a (doubly protonated) quinone-diimine unit. This narrow confinement may destabilize bipolarons with respect to polarons owing to the coulomb repulsion between the cations. As indicated above, doping of polyaniline can be achieved via two routes. Doping by oxidation of the leucoemeraldine form, results in the formation of radical cations which may then convert to bipolarons. Alternatively, protonation of the emeraldine base form leads to the initial formation of bipolarons that may rearrange with neutral amine units to form radical cations as shown in Figure 5. The totally reduced leucoemeraldine state of polyaniline (PANI) can be oxidatively doped to the conductive emeraldine state either by chemical or electrochemical means. The progressive oxidative doping of leucoemeraldine base to emeraldine salt and finally to pernigraniline by electrochemical means, has been reported using cyclic voltammetry. The protonic doping involves the doping of a conjugated polymer to its conducting regime without any change in the number of electrons associated with the polymer.

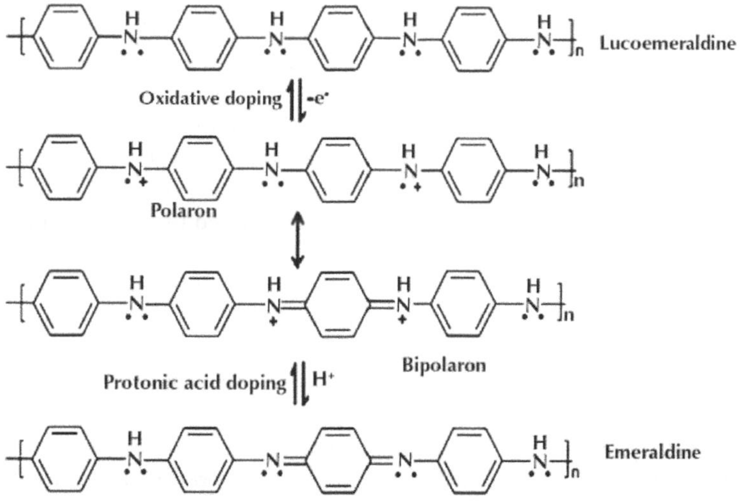

FIGURE 5 Oxidative and protonic acid doping in polyaniline.

The PANI was the first well-established example of such kind of doping. This was accomplished by treating the emeraldine base with aqueous protonic acids. For example, emeraldine base can be doped by HCL to yield the conducting emeraldine hydrochloride. The dication can form a resonance structure consisting of two separated polarons. The ability of PANI to exist in various forms via acid/base treatment and oxidation/reduction, either chemically or electrochemically, has made PANI the most tunable member of the conducting polymer. PANI finds wide variety of applications in different fields. Polyaniline with different forms find different uses like leucoemeraldine the completely reduced form find applications in electrochromic devices and in Li–PANI batteries, perningraniline is used for non-linear optics while emeraldine base consisting of 50% reduced, and 50% oxidized moieties is used in HCL sensors and for making thin films. Other applications of conducting PANI include electrostatic charge dissipation, electromagnetic interference (EMI) shielding, transparent packaging of electronic components, solar batteries, nonlinear optical display devices, "smart" fabrics and recording, and so on (Baughman and Shacklette, 1987).

17.2.3 CHARGE TRANSPORT

It is possible to differentiate between intra and intermolecular charge transport mechanisms in conducting polymers. Charge may be delocalized along the polymer backbone but without an efficient intermolecular tunneling or hopping mechanism, conductivity may still be low. Since conducting polymers are generally refractory materials, poor inter particle contacts may further limit conductivity in compressed pellets (Wegner, 1981). Unhomogeneous doping may also lead to a dependence on doping level, which does not reflect intrinsic behavior, but derives from the microscopic topology of the granular dispersion. The conductivity of a dispersion of conducting particles in an insulating matrix shows percolation effects (Horovitz, 1985). The composite shows a dramatic rise in conductivity above a threshold loading corresponding to the formation of an "infinite" network of particles in mutual contact. The complicating factor here is that differential thermal expansion coefficients may lead to pseudo temperature dependence. In conductive polymers, soliton, and polaron, or bipolaron states can overlap to form bands of states (Kivelson and Heeger, 1985). In principle, a metallic state may result when such bands are partially filled. Overlap of bipolaron bands with the valence band and a regular array of polarons are examples of such metallic states with a finite density of states at the Fermi level. Historically, the concept of a metallic state for the conductive form of polyaniline (PANI) was marked by revolutionary change. One of the initial suggestions by MacDiarmid (Chance and MacDiarmid, 1986) was that the metallic state in PANI derives from a delocalized bipolaron defect. On the other hand, Epstein and coworkers (Epstein, Ginder, Zhuo, Gibelow, Woo, Tanner, Richter, Huang, and MacDiarmid, 1987) favored a metallic state derived from a regular array of polarons. Since the observed temperature dependence for the conductivity is not metallic, they proposed a picture of metallic particles embedded in a dielectric medium. The conductivity of such a granular metal is determined by space charging limited tunnelling. In case of conductive polymers, it is observed that the D.C. electrical conductivity varies exponentially with absolute temperature. The temperature dependence of the conductivity has been described by an Arrhenius type equation (1):

$$\sigma = \sigma_0 e\left(\frac{-Eg}{2kT}\right) \tag{1}$$

where, σ = Electrical conductivity at temperature T,

Eg = Pre exponential conductivity,

σo = Activation energy,

k = Boltzmann constant = 1.38 x 10^{-34} J/K

T = Absolute temperature.

The logarithmic form of above equation is written as (2):

$$\log(\sigma) = \log(\sigma_0) - \frac{Eg}{2.303(2k)T} \tag{2}$$

The temperature dependence of electrical conductivity for conductive polymers was estimated in the range of applicability of Arrhenius type equation. The measured values were plotted as a function of reciprocal of temperature. The conductivity increases with temperature, however there are deviations at lower temp. Arrhenius behavior can be regarded as a good approximation to band theory related to a small temperature. The conduction mechanism in polymers could be explained by the well-known band structure. Since organic polymers are generally insulators, as a result there are no mobile charge carriers to support conduction. However, appropriate charge carriers may be generated in organic polymers by partial oxidation or partial reduction. Many investigators (Mott, 1980; Matare, 1984; Zeller, 1972) have suggested equations relevant to various modes for electrical conduction in composites.

In Greave's equation, conductivity is a function of temperature in terms of hopping conduction mechanism:

$$\sigma\sqrt{T} = e^{\left(\frac{-\beta}{\sqrt[4]{T}}\right)} \tag{3}$$

Equation (3) represents a variable range hopping conduction mechanism. In Matare's (Matare, 1984) equation (4), conductivity is a function of temperature in terms of electrical conduction including grain boundary barriers:

$$\sigma = A\sqrt{T}e^{\left(\frac{-Eg}{kT}\right)} \tag{4}$$

Zeller's equation (Zeller, 1972) for tunneling conduction is expressed by equation (5):

$$\sigma = \sigma_0 e^{\left(-\frac{A}{\sqrt{T}}\right)} \tag{5}$$

Zuo, Angelopoulos, MacDiarmid, and Epstein (1987) proposed PANI salt as a 3D - granular metal, however Lundberg suggested the electron state of PANI salt is localized and the conductivity is due to quasi – 1 D - variable range hopping (VRH). According to Wang, Scherr, MacDiarmid, and Epstein (1995), PANI salt represents class of quasi 1 D - disordered conductors consisting of bundles of well coupled chain in which electron states are 3D extended. 3D- metallic bundles correspond to crystalline region of the polymer and are separated from one another by the amorphous regions. For the inter bundles quasi 1 D – VRH, equation (6) can be applied for conductivity along the chain:

$$\sigma = \sigma_0 e^{\left(-\sqrt{\frac{T_0}{T}}\right)} \tag{6}$$

"T_o" – characteristic temperature, can be used to calculate the transport parameters such as charge localization length (α^{1}), most probable hopping distance (R), and charge hopping energy (w) using the following relations (7), (8) and (9):

$$T_0 = \frac{8\alpha}{N\left(E_f\right)Zk} \tag{7}$$

$$R = \frac{\sqrt{T_0}}{4\alpha\sqrt{T}} \tag{8}$$

$$W = Zk\frac{T_0}{16} \tag{9}$$

where, Z is the number of nearest neighboring chains, k is Boltzmann constant and N $()$ is the density of states of Fermi energy for both sign of spin.

17.2.4 NANOSTRUCTURE CONDUCTIVE POLYMERS

Nanotechnology has become an active field of research during the past decade because of its tremendous potential for a variety of applications. When the size of many established, well-studied materials is reduced to the nanoscale, radically improved or new surprising properties often emerge. Intrinsically conducting polymers have been studied extensively due to their intriguing electronic and redox properties and numerous potential applications in many fields since their discovery in 1977. To improve and extend their functions, the fabrication of nanostructure conducting polymers has attracted a great deal of attention because of the emergence of nanotechnology. Unlike traditional polymers, which are electrical insulators, conducting polymers (CPs) are semiconducting and can be doped into regions of metallic conductivity. Semiconducting and metallic polymers are the fourth generation of polymeric materials. Their electrical conductivities can be increased by many orders of magnitude which cover the whole insulator-semiconductor-metal range. Due to their special conduction mechanism, unique electrical properties, their reversible doping/dedoping process, their controllable chemical and electrochemical properties and their processability, a variety of conducting polymers especially their nanostructures, have recently received special attention in the areas of nanoscience and nanotechnology (Yun-Ze Long et al., 2011; Im, Kim, Lee, and Lee, 2010). By offering metal-like electrical and optical properties in addition to the inherent ease of processing and mechanical flexibility of polymers, innovative new devices and applications have been made possible by CPs. Moreover, the morphology of such materials can be tuned at the nanoscale.

Nanostructures conducting polymeric materials are of exceptional interest due to their potential applications in EMI shielding, sensors, actuators, transistors and displays. This nanostructuring of the materials, by designing their dimensions to be on the order of hundreds of nanometres or lower, often yields novel properties. Combining these two sources of innovation – conductive polymers and nanostructuring has been an area of intense study in recent years, yielding innovative applications. Conducting polymers can be synthesized in various sizes and morphologies, with a useful distinction to be drawn between bulk and nanostructure conducting polymer. The bulk, granular conducting polymer can be easily obtained through conventional processing methods, namely by dissolving the monomer in a strong acid and then polymerizing it through the addition of oxidant (Kang, Neoh, and Tan, 1998). For tubular conducting polymer the options are more varied, with a number of synthesis methods under active research and development (Skotheim and Reynolds (Eds.), 2006). Using an external template to form conducting polymer nanostructures is the most conceptually straight forward synthesis method of all known as template synthesis. A variety of conducting polymer nanostructures has been prepared by the template-free method (Reneker and Chun, 1996). Electrospinning is an effective approach to fabricate long polymer fibers with diameter from micrometers down to 100 nm or even a few nanometers by using strong electrostatic forces. This process evenly created a mat of uniform thickness without imparting a high degree of alignment to the deposited fibers (Kim, Yun, and Lee, 2010; Im, Jang, and Lee, 2009; Im, Park, and Lee, 2009). Electrospinning technique is useful for fabrication of polymeric nanofibers, but unfortunately most of the conducting polymers pose low solubility problems, making them difficult to electrospun in diameters smaller than 100 nm (Ding, Long, Shen, and Wan, 2010). This can be overcome by utilizing a polymer blend, however this process reduces the electrical conductivity of the resulting composite nanofiber. The most popular approaches, however, are based on self-assembly techniques, generally favored for their simplicity and high throughput, when nanostructures can be induced to self-assemble, there is no longer a need for external manipulation through templates, lithography or electrospinning. With the tremendous research efforts, there is now a large variety of CPs, with polyacetylene (PA), polythiophene (PTH), polypyrrole (PPY), and polyaniline (PANI) being four of the most studied and promising types. PANI exhibits high conductivity, excellent environmental stability, and is low-cost and straightforward to synthesize. It is also unique among CPs in that it has a reversible and relatively simple acid–base doping–dedoping pathway, useful for tuning its electrical and optical properties. Arguably the most promising method for synthesizing polyaniline nanostructures is self-assembly, which is very advantageous in its simplicity and volume. However, this self-assembly remains only partly understood, with a number of already established models such as "micelle theory" and "phenazine theory" at odds with more recent discoveries (nanosheet curling and nanoparticle agglomeration), leading to a fragmented understanding of this important topic. An expanded polyaniline nanostructure self-assembly model "Multi-Layer Theory" that goes beyond the scope of existing theories, thereby accommodating the more recent discoveries. The expanded synthesis framework is based on a multi-layered approach incorporating intrinsic morphologies. The three proposed intrinsic morphologies underpinning model are nanofibrils, nanosheets, and nanoparticles; the forces driving their subsequent self-assembly interactions are mainly p-p stacking, hydrogen bonding and charge–charge repulsion from protonation. These interactions between the three intrinsic morphologies give rise to observed growth, agglomeration, and curling behaviors that ultimately generate complex multi-layered nanostructures such as double-walled conducting polymer nanotubes. The challenges that motivate much of this research have to do with understanding and ultimately controlling the morphology and underlying structure of nanostructure conducting polymers.

A number of methods, such as hard and soft template as well as physical methodologies have been used to synthesize conducting polymer nanostructures. In the hard template method, a template membrane is usually required to guide the growth of the nanostructures within the pores or channels of the membrane as a hard-template, leading to completely controlled nanostructures in morphology and diameter, dominated by the pores or channels. The advantages of the hard template method are, it is a general tool to prepare material nanostructures including metal, semiconductor, and conducting polymers, which can be synthesized either chemically or electrochemically. The diameter of the nanostructure is controlled by the size of the pores or channels in the membrane, whereas the length and thickness of the nanostructure is usually adjusted by changing polymerization time. For this reason, the hard template method is the most commonly used and the most efficient approach for preparing well controlled and highly oriented nanostructures. On the other hand, disadvantages of the hard template method are obvious, which includes the complication of the preparation process and the post process can often destroy or disorder the formed nanostructures, the quantity of the nanostructures produced by this method is limited by the size of the template membrane, which thereby limits its applications in large scale nanostructure production. In general, the preparation of conducting polymer nanostructures by a hard template method is carried out by either chemical or electrochemical polymerization. Chemical hard template synthesis is accomplished by simply immersing a membrane in a solution of the desired monomer, dopant, and oxidant, and then allowing monomer polymerization within the pores served as the "nanoreactor". By controlling the polymerization time, different types of nanostructures can be produced. Short polymerizations lead to tubules with thin walls, while longer polymerizations can produce thick walled tubules or even fibers. For an electrochemical hard template synthesis, a metal film coated on one surface of the membrane is required in order to carry out electrochemical polymerization of the desired polymer within the pores of the membrane. Compared to chemical hard template method, the electrochemical hard template method is more complex and expensive, but it is more controllable through changing current density, applied potential, and polymerization time. On the other hand, large mass synthesis by the electrochemical hard template method is impossible because of the limiting size of the membrane used as the template. The porous membrane is the basic and most important part of the hard template method. A typical template synthesis process based on the porous membrane is illustrated in Figure 6, in which (a) porous membrane as the hard template to produce conducting polymer nanotubes and nanowires. Firstly, the growth of the conducting polymers is guided within the pores or channels of the membrane, and then the template is removed after the polymerization; (b) nanofibers as the hard template to produce conducting polymer nanotubes. Nucleation and growth of conducting polymers took place on nanofiber templates, and then the nanofibers were dissolved or depolymerized to obtain nanotubes; and (c) colloidal particles as the hard template to produce nanoporous membranes. The monomers of conducting polymers were polymerized in the voids between colloidal particles. When colloidal particles were removed, a three dimensional conducting polymer porous structure remained.

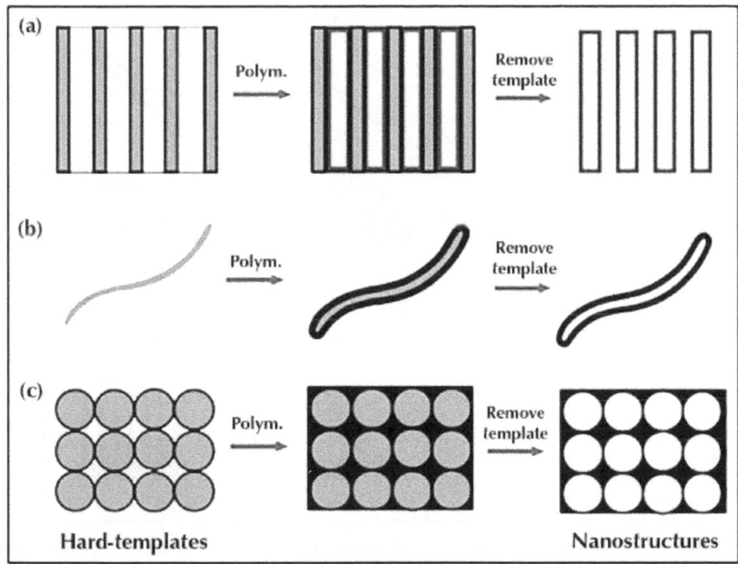

FIGURE 6 Schematic of the hard-template synthesis of different conducting polymer nanostructures.

Soft template method is also called as the template free or self-assembly method. It is another relatively simple, cheap, and powerful approach for fabricating conducting polymer nanostructures via a self-assembly process. Self-assembly is based on selective control of noncovalent interactions, such as hydrogen bonds, Van der Waals forces, p-p stacking interaction, metal coordination, and dispersive forces as the driving forces of self-assembly. To date, surfactants, colloidal particles, structure-directing molecules, oligomers, soap bubbles and colloids as soft-templates, as well as interfacial polymerization, have been employed to synthesize conducting polymer nanostructures.

Surfactant is a class of molecules that form thermodynamically stable aggregates of inherently nanoscale dimensions both in solution and at interfaces. Surfactant self-assembly in a solution has been investigated both theoretically and experimentally owing to its importance in synthesis of micro or nanoscale structures with controlled dimensions. The equilibrium size and shape of surfactant aggregates are controlled by the volume and length of the surfactant tail within the hydrophobic core of the aggregate and the effective area occupied by each surfactant head group at the surface of the aggregate. The favored aggregate morphology of a surfactant in a solution is spherical, cylindrical, or a flat bi-layer, depending on these parameters. The self-assembly ability of surfactants in a bulk solution therefore creates the possibility of surfactant micelles serving as soft-templates to form conducting polymer nanostructures (Figure 7).

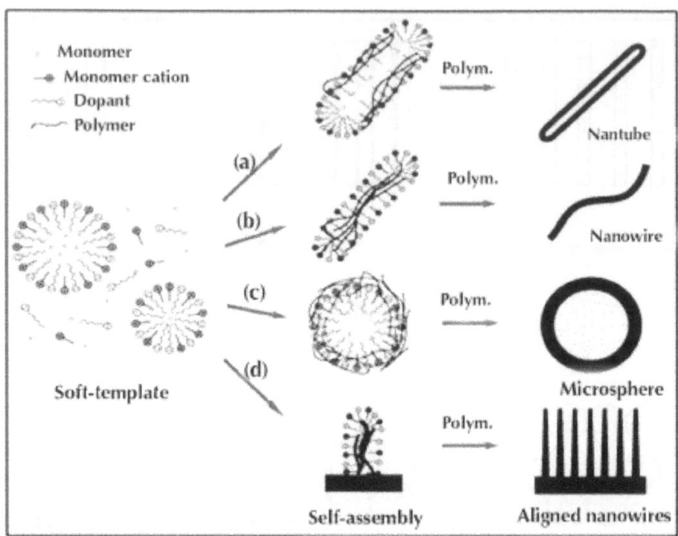

FIGURE 7 Schematic of the mechanism of the soft-template synthesis of different conducting polymer nanostructures.

Schematic of the mechanism of the soft-template synthesis of different conducting polymer nanostructures (a) micelles acted as soft-templates in the formation of nanotubes. Micelles were formed by the self-assembly of dopants, and the polymerization were carried out on the surface of the micelles; (b) nanowires formed by the protection of dopants. The polymerization were carried out inside the micelles; (c) monomer droplets acted as soft-templates in the formation of microsphere; and (d) polymerization on the substrate producing aligned nanowire arrays. Nanowires were protected by the dopants, and polymerization preferred to carry out on the tips of nanowires.

17.2.5 CONDUCTIVE POLYMER NANOCOMPOSITES

Conducting polymers provide tremendous scope for tuning of their electrical conductivity from semi conducting to metallic regime. The conductivity of the conducting polymer as stated earlier can be tuned by electrical manipulation of the polymer backbone by the nature of dopant, by the degree of doping, by blending with other polymers (co-polymer) and by making composites with inorganic materials. When conducting polymers are taken in the composite form, their properties are altered from those of basic materials. It has been shown that the conductivity of these heterogeneous systems depends on a number of factors such as the concentration of the conducting fillers, their shape, size, orientation, and interaction between filler molecules and host matrix. Since polymeric materials are generally thermally insulating, nanoscale fillers of superior thermal transport characteristics are incorporated into polymeric matrices to prepare nanocomposites that exhibit a good combination of processability and thermal conductivity (Wei-Li Song et al., 2012). The geometrical shape of the dispersant governs the ability of conductive network formation which results in large increase in conductivity. Also dispersant /matrix interactions and physical properties of the matrix influence the agglomeration of the dispersant phase which affects the properties of composites. Using natural reagents and polymers such as carbohydrates, lipids, and proteins, nature makes strong composites such as bones, shells, and wood. These are examples of nanocomposites, made by mixing two or more phases such as particles, layers or fibers, where at least one of the phases is in the 1-100nm size range. A nanocomposite is defined as a material with more than one solid phase, metal ceramic, or polymer, compositionally or structurally where at least one dimension falls in the nanometers range. Many composite materials are composed of just two phases; one is termed the matrix, which is continuous and surrounds the other phase, often called the dispersed phase. The properties of composites are a function of the properties of the constituent phases, their relative amounts, and the geometry of the dispersed phase. There have been great effort to fabricate nanocomposite to obtain unique physical properties since these properties become increasingly size dependent at low dimension.

The combination of the nanomaterial with polymer is very attractive not only to reinforce polymer but also to introduce new electronic properties based on the morphological modification or electronic interaction between the two components. The properties of a nanocomposite are greatly influenced by the size scale of its component phases and the degree of mixing between the two phases. Depending on the nature of the components used and the method of preparation, significant differences in composite properties may be obtained. Nanocomposites of conducting polymers have been prepared by various methods such as colloidal dispersions (Jarjeyes, Fries, and Bidan, 1995), electrochemical encapsulation (Huang and Matijevic, 1995), coating of inorganic polymers, in-situ polymerization with nanoparticles (Wu, Shin, and Yang, 2006) and have opened new avenues for material synthesis. The combination of the magnetic nanoparticles with conducting polymer leads to formation of ferromagnetic conducting polymer nanocomposites possessing unique combination of both electrical and magnetic properties. This property of the nanocomposites can be used an electromagnetic shielding material since the electromagnetic wave consist of an electric (E) and the magnetic field (H) right angle to each other. The ratio over E to H factor (impedance) has been subjugated in the shielding purpose.

The conducting ferromagnetic type of materials can effectively shield electromagnetic waves generated from an electric source, whereas electromagnetic waves from a magnetic source can be effectively shielded only by magnetic materials. The primary mechanism of EMI shielding is usually reflection. For reflection of the radiation by the shield, the shield must have mobile charge carriers (electrons or holes) which interact with the electromagnetic fields in the radiation. As a result, the shield tends to be electrically conducting, although a high conductivity is not required. For example, a volume resistivity in the order of 1Ω cm is typically sufficient. A secondary mechanism of EMI shielding is usually absorption. For significant absorption of the radiation, the shield should have electric and/or magnetic dipoles which interact with the electromagnetic fields in the radiation. Thus, having both conducting and magnetic components in a single system could be used as an EMI shielding material. The electrical permittivity and the magnetic permeability, depending upon the pulsation, are the major physical parameters that characterize the material as a microwave absorber. It has been shown recently that promising shielding performance could be achieved with composites comprising of an electrically non-conducting polymer matrix filled with conducting objects such as CNT-polymer composites, nickel ceramic composites and so on. However, high filler contents are required from 10–15% for fibers and 30% for spheres to reach a percolation level leading to microwave losses. These materials are generally compared to each other with respect to their imaginary part of permittivity or loss tangent. The design of large bandwidth structures requires multilayer structures in which the frequency behavior of these two parameters must be controlled. Despite the large number of conductive fillers, the radio electric properties of all these composites are generally same. Differences are due to the processing of the fillers in the matrix. They are often correlated with aggregation effects and inter-particle contacts processing of the fillers in the matrix. They are often correlated with aggregation effects and inter-particle contacts. Nevertheless, the use of dielectric materials obtained by conductive filler dispersion (carbon black, graphite fibers, metallic powders) is limited. As a matter of fact, material performances are dependent on the filler content as well as particle aggregation phenomena.

These composites require a high level of reproducibility and their behavior is linked to the control of electronic inter-particle transfer. The measured parameter (complex permittivity) depends on the texture of the percolation aggregates and consequently on the processing conditions. The percolation threshold depends on the particle shape (sphere, plates or fibers). This filler aggregation during the processing, observed with nanometric particles (carbon black) or micron sized fillers (metallic or carbon sphere), leads to a very difficult reproducibility of the material. Different approaches have been proposed to overcome these drawbacks. In this respect, conducting polymers represent a very attractive solution to the problem described above. These new materials present very promising properties because of the chemical nature of the macromolecular chains in which electronic conduction occurs at long range. These conducting polymers have some specific characteristics that make them far more interesting than traditional dielectric materials. Besides their reproducible properties, the chemistry of conducting polymers offers a great variety of methods of synthesis. The insertion of conductivity into various materials (insulating polymer matrix, reinforcing fabrics, honeycomb structure) is now possible which leads to complex structures (Dhawan, Singh, and Venkatachalam, 2002; Hakansson, Amiet, and Kaynak, 2006). They absorb radar waves and can match new environmental constraints (mechanical properties for example). The driving idea is based on the growing process at molecular scale of the conducting entity leading to a uniform macroscopic network in the material. Polyaniline, polypyrrole, or polythiophene possess this property. Work has to be done to reach the compromise between compatibility of the conducting polymers and its processing properties. In particular, dependence of their conductivity on frequency, many ideas have been attempted to adapt these phenomenon to microwave applications (Kaynak, Hakansson, and Amiet, 2009; Chandrasekhar and Naishadham, 1999; Hoang, Wojkiewicz, Miane, and Biscarro, 2007). The intrinsic conductivity of conjugated polymers in the field of microwave absorption (100MHz–20GHz) makes them viable materials.

17.3 PACKAGING APPLICATIONS

17.3.1 ELECTROMAGNETIC SHIELDING AND MICROWAVE ABSORPTION

Electromagnetic interference also called as radio frequency interference, is a serious issue caused by the rapid proliferation of electronics, wireless systems and development in navigation, space technology and so on. EMI not only affects the performance of the electric devices but may also be harmful to life forms, including humans. Therefore, some kind of shielding materials must be employed to prevent the electromagnetic noise or pollution. The term shielding is usually referred to a metallic enclosure or packaging by conductive materials that completely encloses an electronic product or the portion of that product. Therefore, it limits the amount of EMI (electromagnetic interference) radiation from the external environment that can penetrate the circuit and conversely, it influences how much EMI energy generated by the circuit can escape into the external environment. Electromagnetic shielding is the practice of reducing the electromagnetic field in a space by blocking the field with barriers made of conductive or magnetic materials. Shielding is typically applied to enclosures to isolate electrical devices from the "outside world", and to cables to isolate wires from the environment through which the cable runs. Electromagnetic shielding that blocks radio frequency electromagnetic radiation is also known as RF shielding. The shielding can reduce the coupling of radio waves, electromagnetic fields and electrostatic fields. A conductive enclosure used to block electrostatic fields is also known as a Faraday cage. The amount of reduction depends very much upon the material used, its thickness, the size of the shielded volume and the frequency of the fields of interest and the size, shape and orientation of apertures in a shield to an incident electromagnetic field.

There has been an intensive interest in the electromagnetic radiation shielding/absorbing field due to the tremendous growth of high-performance electrical apparatuses in commercial, military, and scientific fields. Electromagnetic radiation, which is a type of artificial electromagnetic environmental pollution, causes the performance degradation of electrical system/equipment as well as health threats. With increasing concerns regarding the electromagnetic interference (EMI), a variety of studies related to EMI shielding/absorbing materials have been carried out to improve the electromagnetic environment. Among various materials, typical metallic materials (for example, steel, copper, and aluminum) have been used as representative shielding materials with high EMI shielding efficiency (SE) that stems from their high conductivity and dielectric constant. However, these conventional shielding materials have serviceability disadvantages, such as their heavy weight, corrosion susceptibility, limited physical/mechanical flexibility, and poor processibility. Therefore, studies of novel materials, as an alternative to conventional shielding materials, are required to obtain superior serviceability as well as efficient EMI shielding/absorbing properties. Intrinsically conductive polymers (ICPs) have inspired much interest due to their lightweight, corrosion resistance, good processibility, and simple conductivity control. In particular, these materials have recently received significant attentions as one of the best candidates for EMI shielding/absorbing materials since they are associated with high conductivity, permittivity, and good environmental stability. Polyaniline (PANI), a typical ICP, is non-toxic and provides good thermal, chemical, environmental stability. Due to these potential abilities as EMI shielding/absorbing materials, a series of studies pertaining to the EMI shielding/absorbing behavior of PANI films or composites have been carried out by many researchers (Bhadra, Singha, and Khastgir, 2009). Makela, Pienimaa, Taka, Jussila, and Isotalo (1997), reported the EMI SE (shielding effectiveness) of thin PANI films with a thickness of 1–10mm in a frequency range of 0.1–1000 MHz. Wang and Jing (2007), also reported the conductivity and optical transmittance of thin films based on PANI nanofibers with 5–20 wt% loadings. However, a comprehensive study of PANI-coated transparent thin films has yet to be completed for a thorough understanding of the electromagnetic wave transport mechanism (for example, transmission, reflection, and absorption) and the potential use of these materials as EMI shielding/ absorbing materials. It is thus necessary to quantitatively and qualitatively investigate the EMI shielding/absorbing characteristic of the PANI-coated transparent thin films in order to clearly foresee their potential industrial developments. It is also intended to show the potential use of PANI-coated transparent thin films as an EMI shielding/absorbing material that satisfies both optical transmission and EMI SE requirements for transparent bodies such as windows or doorways of buildings.

Polymer composites containing conductive fillers were developed as an alternative EMI shielding material. These composites are light weight, cheap, resistant to corrosion and easily processable. Research in the past decade has established the ability of polymer composites made with electrically conducting polymers to be suitable as a shield against the electromagnetic interference. So far conducting composites were made by adding metallic fillers, C-black or metallic powders. However because of certain disadvantages like labor intensity, relatively high cost, and time consuming besides the galvanic corrosion phenomenon observed when dissimilar metals are joined, conducting polymer composites were being made which were found suitable for EMI shielding and for the dissipation of electrostatic charge. In recent time, extensive research, have been done for the development of new microwave shielding materials with high efficiency, lightweight and having high durability. This includes composites based on polymers, like hexagonal-ferrite/polymer, metal/polymer composites, and SWCNT-epoxy composites. To achieve higher values of shielding effectiveness much attention has been paid to polyaniline and their composites with ferrite particles, which possesses the moderate magnetization and conductivity.

17.3.3 ELECTROMAGNETIC SHIELDING THEORY

Shielding effectiveness (SE) can be defined as a measure of the reduction or attenuation in the electromagnetic field strength at a point in space caused by the insertion of a shield between the source and that point. According to the electromagnetic wave shielding mechanism of a material, the incident electromagnetic wave in the shielding material is split into four parts: a reflected wave, an absorbed wave, an internal reflected wave, and a transmitted wave. As shown in Figure 8, the shielding is a direct consequence of reflection, absorption, and multiple internal reflection losses at the existing interfaces, suffered by incident electromagnetic (EM) waves.

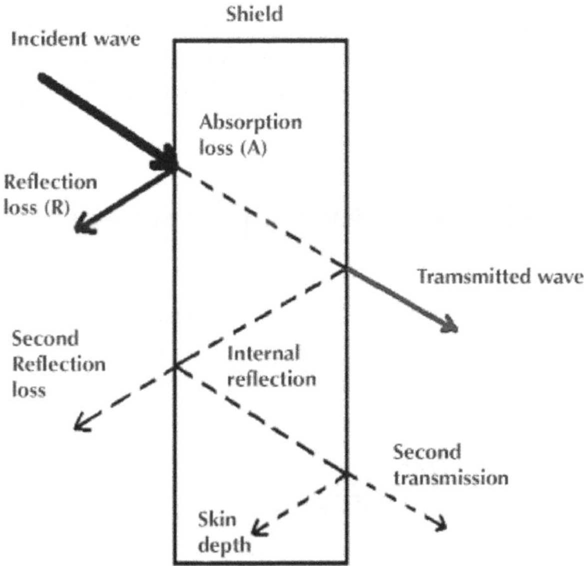

FIGURE 8 Interaction of electromagnetic waves with shield material.

The SE, which is expressed in dB, is described (Klemperer and Maharaj, 2009) as equation (10).

$$SE = SE_R + SE_A + SE_B \tag{10}$$

where, SE_R, SE_A, and SE_B denote the initial reflection losses from the surfaces of the material, the absorption and penetration loss within the material, and the internal reflection loss at the existing interface of the medium, respectively. SE_R and SE_A are caused by interrelations among the conductivity (s), permittivity (ε), and permeability of material (m), which are denoted as electromagnetic properties of the material. Due to the internal reflection loss caused by the constant contact with the wall surface by the material, SE_B has a very low value compared to other values. Thus, for practical purposes, only the SE_R and SE_A values must be calculated in most shielding environments. The total SE can be obtained by adding SE_R and SE_A. However, due to the difficulty in a direct measurement of the electromagnetic properties, SE is quantitatively expressed as S-parameters that represent the scattering of the wave (Lee, Song, Jang, Oh, Epstein, and Joo, 1999; Kwon and Lee, 2009). In the waveguide, when the incident wave at a point (port i) progresses towards another point (port j), the radiated wave experiences a shielding process (reflection, absorption, and transmission), and these wave scattering are evaluated by S_{ji}. Here, the scattering parameters S_{11} and S_{21} signify the amount of electromagnetic reflection and transmission coefficients, respectively. They are also expressed in dB. Absorption loss SE_A, is a function of the physical characteristics of the shield and is independent of the type of source field. Therefore, the absorption term SE_A is the same for all three waves. When an electromagnetic wave passes through a medium, its amplitude decreases exponentially. This decay or absorption loss occurs because currents induced in the medium produce ohmic losses and heating of the material. The distance required by the wave to be attenuated to 37% is defined as the skin depth. Therefore, the absorption term SE_A in decibel is given by the equation (11),

$$SE_A = 20\left(\frac{t}{\delta}\right)\log e = 8.69\left(\frac{t}{\delta}\right) = 131t\,(f\mu\sigma)^{1/2} \qquad (11)$$

where, t is the thickness of the shield in mm, f is frequency in MHz, μ is relative permeability (1 for copper), σ is conductivity relative to copper.

The skin depth δ can be expressed by equation (12),

$$\delta = \frac{1}{(\pi f \mu\sigma)^{1/2}} \qquad (12)$$

The absorption loss of one skin depth in a shield is approximately 9 dB. Skin effect is especially important at low frequencies, where the fields experienced are more likely to be predominantly magnetic with lower wave impedance than 377Ω. From the absorption loss point of view, a good material for a shield will have high conductivity and high permeability along with a sufficient thickness to achieve the required number of skin depths at the lowest frequency of concern. The reflection loss is related to the relative mismatch between the incident wave and the surface impedance of the shield. The computation of refection losses can be greatly simplified by considering shielding effectiveness for incident electric fields as a separate problem from that of electric, magnetic or plane waves.

The S-parameters can be measured in an attempt to quantitatively express the EMI SE of the material films. The S-parameters, expressed as a complex number: $S_{ji} = S^r_{ji} + jS^i_{ji}$, are used to denote the scattering of a wave (Baker-Jarvis, Vanzura, and Kissick, 1990). Here, the superscripts r and i signify the real and imaginary parts of the S-parameters, respectively. In addition, the S_{11} parameter and S_{21} parameter are related to the reflection and transmission of electromagnetic wave, respectively. By measuring the S-parameters, it is possible to determine the contribution of the reflection and the absorption of the total shielding effectiveness. The reflection and transmittance ratios of the films can be obtained according to the equation (13) for an accurate evaluation of the SE values,

$$SE = 10\log\left(\frac{P_1}{P_2}\right) = 20\log\left(\frac{V_1}{V_2}\right) \qquad (13)$$

where, *P1* (or *V1*) and *P2* (or *V2*) are the received powers (or respective voltage levels) of the load and reference specimens, respectively. The materials with larger values of *SE* are good for EMI shielding.

17.2.4 CONDUCTIVE POLYMERS AND NANOCOMPOSITES FOR EMI SHIELDING

Various types of materials have been used in EMI shielding, including metals, carbon materials, conducting polymers, and nanocomposites. Depending upon the shielding efficiency (SE) at different frequency ranges, electromagnetic shielding materials are used as packaging for the encapsulation of different microelectronic devices, computer housings, switches, connector gaskets, and so on. The required conductivity levels are approximately 10^{-3}–10^{-7} S/m for electrostatic dissipating and greater than 10^{-2} S/m for electromagnetic shielding applications. High conductivity and high dielectric constant of the materials contribute to high SE. Typical metals such as copper or aluminum have been used as conductive filler which have high conductivity and dielectric constant (Kanatzidis, Wu, Marcy, and Kannewurf, 1989). However, this has got certain disadvantages such as heavy weight, physical rigidity, easy corrosion, and poor processability in corners and tips. Recently, the EMI shielding and microwave absorption properties of conducting polymer nanocomposites have attracted increased attention owing to their good electrical conductivity, high dielectric constant, ease of processability, low density, high surface area to volume ratio and unique shielding mechanism of absorption(Paul and Pillai, 2000). Among the conducting polymers, polyaniline (PANI) has been studied most extensively since it has got unique doping mechanism, excellent physico-chemical properties, good stability and its raw material can be obtained easily (Dhawan, Singh, and Rodrigues, 2003). For the attainment of high shielding efficiency with good mechanical strength low filler loading is desirable. According to the electromagnetic wave percolation theory, if the dimension of the conductive filler is in a nanometer regime and retains a high aspect ratio, with low loading of the filler, easily forms a conductive network. At a certain threshold value of the filler, the particles or fibers are sufficiently close-packed to form unbroken conducting pathway through the composite and the conductivity of the material increases sharply. This minimum concentration of the conductive filler for the formation of conductive network is termed as the percolation threshold concentration. Several strategies are reported for the preparation of nanostructure PANIs. One way to improve the bulk properties of PANI is to confine it in an inorganic layered material. The interest in these materials is due to their synergistic effect arising from the intimate mixing between inorganic and organic components at a molecular level.

The electromagnetic interference (EMI) attenuation offered by a shield may depend on three mechanisms. The first is usually the reflection of the wave from the shield. The second is the absorption of the wave as it passes through the shield. The third is due to the multiple reflections of the waves at various surfaces or interfaces within the shield. The multiple reflections, however, requires the presence of large surface areas (for example, a porous or foam material) or interface areas (for example, a composite material containing fillers that have large surface area) in the shield. Carbon nanotube based polymer nanocomposites are reported to be excellent shielding materials for electromagnetic interference, especially at high frequency (Lee, Woo, Park, Hahm, Wu, and Kim, 2002). Filling a matrix of engineering plastic with an electrically conducting material or a matrix of conducting polymer with a filler in nanoscale combines the availability of a housing made of shielding material with the advantages of traditional compounding of this composite. Traditionally, metal or carbon black particles have been used as electrically conducting filler materials. A high level of these fillers can be detrimental for the processability, density, and surface quality of the material, the costs and mechanical properties of the molded product, and may cause wear to the processing equipment. Furthermore, materials which contain carbon black are limited in their color. Therefore, the aim at developing novel filler materials such as intrinsically conducting polyaniline polymers and conducting carbon nanotubes, with a filler content that is as low as possible. In this way, conductivity will be provided to the material while the original plastic processing properties will remain the same.

When the concentration of electrically conducting particles in a composite exceeds a certain level, the so-called percolation limit, the particles come into contact with each other and form a continuous path in the material for electrons to travel. In this way, the composite material has become electrically conducting. The conductivity of the filler material will be the upper limit for the electrical conductivity of the entire composite. The percolation limit depends on the shape of the conducting particles. For traditional spherical shaped fillers at a random distribution, approximately 10–20% has to be added before the composite will be electrically conducting. Higher the aspect ratio (length-to-width ratio) of the particles, lower the concentration for percolation to take place. Carbon nanotubes with a diameter of a few nanometers and a length of micrometers (that is, a high aspect ratio) can form a conducting network at much lower volume fractions and potentially lower costs than cheaper, traditional fillers as carbon fiber and carbon black. Carbon nanotubes, especially those of a multi-wall composition, which can best be described as multiple layers of graphite rolled in them, are known to conduct electricity. For intrinsically electrically conducting plastics, conjugated polymers form the basis.

The values of the shielding effectiveness as measured using 101GHz transmitter set up (Koul, Chandra, and Dhawan, 2000) showed that higher loading levels of PANI in the composite matrix gives a higher shielding effectiveness of the order of 60 dB whereas lower loadings of PANI in the composite shows a shielding effectiveness of 11.35 dB. A loading level of 5% PANI gives a shielding effectiveness of the order of 6.23 dB which cannot be effectively used for the controlling of interference in the mm range but can be used for the dissipation of static charge. Low loading levels of polyaniline in the composite can be used for the dissipation of electrostatic charge whereas high loading levels can be used for the shielding of electromagnetic interference at 101GHz.

Carbon nanotubes (CNTs) have been a focus of considerable research since their first observation by Iijima. Their remarkable properties include their interesting mechanical, electrical, thermal, and conductive behavior, and they offer tremendous opportunities for the development of fundamentally new material systems for application in nano scale devices and materials particularly. Similarly, composites based on polymers and CNTs have the potential to make an impact on various applications ranging from general low-cost circuits and displays to power devices, microelectromechanical systems, super capacitors, solar cell sensors, and displays. Despite their appealing properties, the low solubility of CNTs in most organic solvents and their poor compatibility within a polymer matrix make the uniform dispersion of CNTs in the polymer matrix very difficult, limiting their applications. Recent study shows that modifications with the conducting polymers can enhance the sensing properties and dispersion of CNTs. Polyaniline (PANI) has great potential for use in commercial applications because it is cheap to produce, environmentally stable, exhibits enhanced conductivity, and changes its color depending on the redox states.

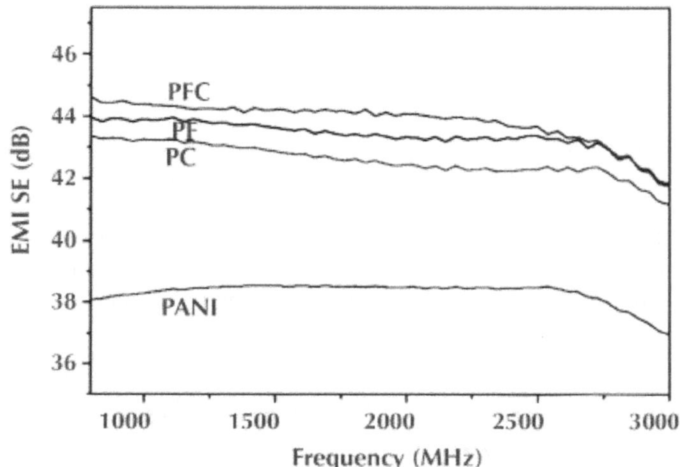

FIGURE 9 EMI shielding efficiency of PANI, PC, PF, PFC samples.

In this study, highly conducting PANI and multi-walled carbon nanotube (MW-CNT) nanocomposites were prepared by in-situ polymerization for electromagnetic interference shielding. The advantages of using both polyaniline and carbon nanotubes as conducting fillers in nanocomposites were also investigated for EMI shielding. The idea here was that the carbon provides long, continuous electrical conducting pathways, and the small contact surfaces between the nanotubes would be increased by using the conducting polymer as "glue" at the contact points between the nanotubes. This would provide good conductivity at low filler fractions. Hybrid materials of polyaniline complex and carbon nanotubes, having surface resistances lower than 50Ω, were obtained by in-situ synthesis of the complex. To improve the ferromagnetic properties of the nanocomposite, ferromagnetic materials like γ-Fe_2O_3 magnetic particles were used together. In case of PANI sample, the average EMI SE was around 38 dB (Figure 9). EMI SE increased by forming composites with additives up to 43 and 44 dB due to the effects of CNT or Fe_2O_3 additives, respectively. The highest EMI SE was observed in PFC sample showing an average EMI SE about 45 dB. Therefore, it was concluded that the use of Fe_2O_3 and CNT additives together led the synergistic effect for improving EMI SE effectively.

The Im et al., (2010) prepared polyaniline-based fibers fabricated with multi-walled carbon nanotubes and polyethylene oxide by the electrospinning method employed as an electromagnetic interference shielding material. They reported EMI SE of these materials in the frequency range of 800–4000MHz. To improve the electromagnetic interference shielding efficiency, the dispersion and adhesion of the carbon nanotubes in the polyaniline electrospun fibers were enhanced through surface modification by the direct fluorination method. This fluorination treatment enhanced the electron donor–acceptor reaction between the fibers and nanotubes, causing adhesive bonding for high electromagnetic interference shielding efficiency. The electrical conductivity improved with the addition of the carbon nanotubes and the fluorination treatment of the carbon nanotubes, reaching up to 4.8×10^3 S/m. EMI SE of PANI/PEO fibers (sample 1) was 8.5 dB, and did not significantly change with variations in the frequency. 100 wt% of pristine and fluorinated MWCNTs embedded PANI/PEO fibers (sample 2) showed an improved EMI SE based on the effects of adding the conductive MWCNTs. The effects of the fluorination treatment were also observed. The EMI SE of 100 wt% of pristine MWCNTs embedded PANI/PEO fibers abruptly decreased over 3000 MHz, while the EMI SE of 100 wt% of fluorinated MWCNTs embedded PANI/PEO fibers (sample 3) was higher across the whole frequency range and did not decrease as dramatically at high frequencies. This may be attributed to the enhanced adhesion between the PANI polymer and the MWCNTs provided by fluorination. The highest EMI SE was observed in 120 wt% of fluorinated MWCNTs embedded PANI/PEO fibers (sample 4), which is based on the effects of the largest amount of added MWCNTs and the fluorination treatment (Figure 10),

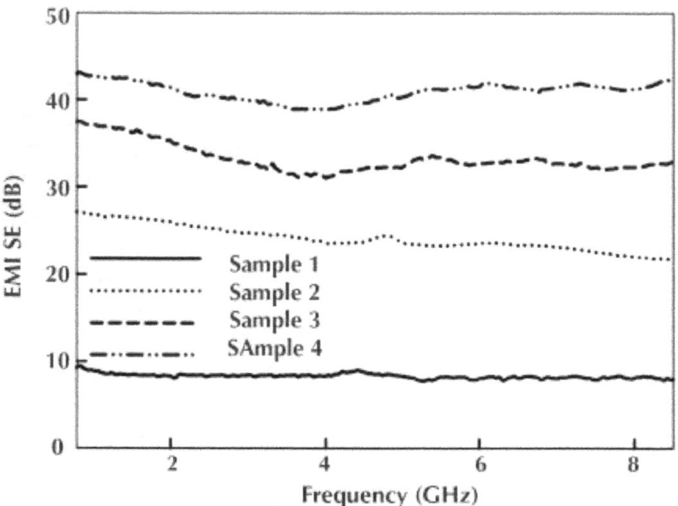

FIGURE 10 EMI shielding efficiency of the samples.

The Dhawan et al., (2009) synthesized conducting ferrimagnetic polyaniline nanocomposites embedded with g-Fe_2O_3 and titanium dioxide nanoparticles via a microemulsion polymerization. The microwave absorption properties of nanocomposite in 12.4–18GHz (Ku-band) frequency range showed shielding effectiveness due to absorption (SEA) value of −45 dB, which is much higher than polyaniline composite with iron oxide and polyaniline–TiO_2 composites. The higher EMI shielding is mainly arising due to combined effect of g-Fe_2O_3 and TiO_2 that leads to more dielectric and magnetic losses which consequently contributed to higher values of shielding effectiveness.

Ferrites are usually non-conductive ferrimagnetic compounds derived from iron oxides such as hematite (Fe_2O_3) or magnetite (Fe_3O_4) as well as oxides of other metals. In terms of their magnetic properties, the different ferrites are often classified as "soft" or "hard", which refers to their low or high magnetic coercivity. Soft ferrites are ferrimagnetic materials with spinel structure and general formula $MeFe_2O_4$ (Me = divalent metal ion, for example, Zn, Ni, Co, Cu, and so on). The spinel structure has a close-packed structure of oxygen anions. Metallic cations, magnetic, and nonmagnetic, reside on the interstices of the close-packed oxygen lattice. In the spinel structure these cations form tetrahedra and octahedra sub-lattices that are in themselves arranged in a close packed arrangement. The small amount of iron deficiency in the composition of ferrites can increase the electrical resistivity further by reducing the amount of Fe^{2+} ions and inhibiting electron hopping between Fe^{2+} and Fe^{3+} ions at the octahedral sites of the spinel unit cell. The NiZn composites ferrites magnetic properties are determined by their chemical composition, crystallinity, porosity, grain size, and so on. The NiZn composites ferrites are widely used as soft magnetic materials for high frequency applications due to their high electrical resistivity and low hysteresis losses. Ferrites can absorb variably electromagnetic radiation in microwave bands cast in various forms, for example, sheets, paints, films, ceramic tiles, powders, and loads in matrix composites or mixed with conducting material. Looking to these special characteristics of ferrites, our research group have synthesized nanocomposites of PANI with nickel ferrite ($NiFe_2O_4$), zinc ferrite ($ZnFe_2O_4$) and nickel zinc ferrite ($NiZnFe_2O_4$) for EMI shielding. We have synthesized nanoferrites by refluxing method (Nandapure, Kondawar, Sawadh, and Nandapure, 2012) and polyaniline nanoferrite composites by in-situ polymerization method. Figure 11 shows the EMI shielding efficiency of these nanocomposites,

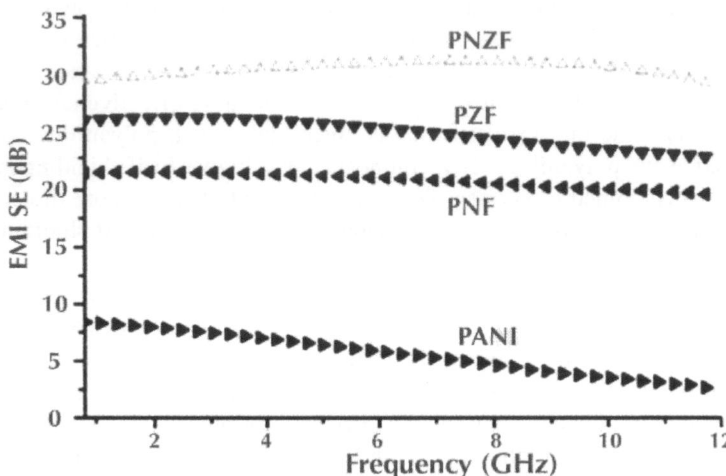

FIGURE 11 EMI shielding efficiency of PANI and PANI-ferrite nanocompositesr.

The conductivity of PANI-ferrite nanocomposites was found to be higher enough suitable for EMI shielding as an example of the synergistic effect of the two components. PANI-ferrite nanocomposites can be applied for the shielding purposes in the frequency range 1–12 GHz for their total shielding effectiveness in the range of 2.5–32 dB. The results showed that EMI ES of PANI-$NiZnFe_2O_4$ nanocomposite (PNZF) was found to be higher than those of pure PANI, PANI-$NiFe_2O_4$ nanocomposite (PNF), PANI-$ZnFe_2O_4$ nanocomposite (PZF). The high value of EMI ES of PANI-$NiZnFe_2O_4$ nanocomposite (PNZF) may be due to high conductivity and high magnetization as that of superparamagnetic material compared to ferromagnetic PANI-$NiFe_2O_4$ nanocomposite (PNF) and paramagnetic PANI-$ZnFe_2O_4$ nanocomposite (PZF).

Although only a few reports on the application of conductive polymer nanocomposites for microwave absorption and EMI shielding are available, it is believed that conductive polymer nanocomposites could be important candidates in this field.

17.4 CONCLUDING REMARKS

Efforts are being made to reduce the electromagnetic radiation effect using EMI shielding materials as an intelligent packaging for electronic gadgets. With increasing concerns regarding the electromagnetic interference (EMI), a variety of studies related to EMI shielding/absorbing materials have been carried out to improve the electromagnetic environment. The enhancement in the microwave shielding and absorption properties of the conducting polymer nanocomposites has been achieved by the incorporation of nanofiller in the polymer matrix. The conducting paths of conducting polymer due to nanoparticles embedded into polymer matrix increase the absorption of the electromagnetic wave to a large extent. The dependence of EMI SE on magnetic permeability and conductivity demonstrates that better absorption value can be obtained for materials with higher conductivity and magnetization. The incorporation of magnetic and dielectric fillers in the polymer matrix leads to better absorbing material which makes them futuristic radar absorbing material. The microwave absorption property of the composites strongly depends on the intrinsic properties of nanofillers in the polymer matrix. The higher EMI shielding may be obtained by combined effect of different nanofillers in polymer matrix that leads to more dielectric and magnetic losses which consequently contributed to higher values of shielding effectiveness.

KEYWORDS

- **Conducting Polymers**
- **Conduction Mechanism**
- **Electromagnetic Interference**
- **Polyacetylene, (CH)x**
- **Food Packaging**

REFERENCES

1. Al-Saleh, M. H. and Sundararaj, U., *Carbon*, **47**, 1738 (2009).
2. Baker-Jarvis, J., Vanzura, E. J., and Kissick, W. A., *IEEE Transactions on Microwave Theory and Technique*, **38**, 1096 (1990).
3. Baughman, R. H. and Shacklette, L. W., *Synth. Met.*, **17**, 173 (1987).
4. Bhadra, S., Singha, N. K., and Khastgir, D., *Curr. Appl. Phys.*, **9**, 396 (2009).
5. Bishop, A. R., Campbell, D. K., and Fesser, K., *Mol. Cryst. Liq. Cryst.*, **77**, 253 (1981).
6. Bredas, J. L., Themans, B., Andre, J. M., Chance, R. R., and Silbey, R., *Synth. Met.*, **9**, 265 (1984).
7. Chance, J. C. and MacDiarmid, A. G., *Synth. Met.*, **13**, 193 (1986).
8. Chandrasekhar, P. and Naishadham, K., *Synth. Met.*, **105**, 115 (1999).
9. Dhawan, S. K., Kuldeep, Singh, and Bakhshi, A. K., *Anil Ohlan, Synthetic Metals*, **159**, 2259 (2009).
10. Dhawan, S. K., Singh, N., and Rodrigues, D., *Sci. Technol. Adv. Mater*, **4**, 105 (2003).
11. Dhawan, S. K., Singh, N., and Venkatachalam, S., *Synth. Met.*, **129**, 261 (2002).
12. Ding, H. J., Long, Y. J., Shen, J. Y., and Wan, M. X., *J. Phys. Chem. B*, **114–115** (2010).
13. Epstein, A. J., Ginder, J. M., Zhuo, F., Gibelow, R. W., Woo, H. S., Tanner, D. B., Richter, A. F., Huang, W. S., and MacDiarmid, A. G., *Synth. Met.*, **18**, 303 (1987).
14. Hakansson, E., Amiet, A., and Kaynak, A., *Synth. Met.*, **156**, 917 (2006).
15. Heeger, A. J., *Chem. Soc. Rev.*, **39**, 2354 (2010).
16. Hoang, N. H., Wojkiewicz, J. L., Miane, J. L., and Biscarro, R. S., *Polym. Adv. Technol.*, **18**, 257 (2007).
17. Holze, R., *J. Appl. Electrochem.*, **39**, 953 (2009).
18. Horovitz, B., *Phys. Rev. Lett*, **55**, 1429 (1985).
19. Huang, C. L. and Matijevic, E., *J. Mater. Res.*, **10**, 1327 (1995).
20. Im, J. S., Jang, J. S., and Lee, Y. S., *J. Ind. Eng. Chem.*, **15**, 914 (2009).
21. Im, J. S., Kim, J. G., Lee, S. H., and Lee, Y. S., *Colloids and Surfaces A: Physicochem. Eng. Aspects*, **364**, 151 (2010).
22. Im, J. S., Park, S. J., and Lee, Y. S., *Int. J. Hydrogen Energy*, **34**, 1423 (2009).
23. Jadhav, R. S., Hundiwale, D. G., and Mahulikar, P. P., *J. Coat. Technol. Res.*, **7** (4), 449 (2010).
24. Jang, J., Emissive Mater. Nanomater, **199**, 189 (2006).
25. Jarjeyes, O., Fries, P. H. and Bidan, G., *Synth. Met.*, **69**, 43 (1995).
26. Kanatzidis, M. G., Wu, C. G., Marcy, H. O., and Kannewurf, C. R., *J. Am. Chem. Soc.*, **4**, 139 (1989).

27. Kang, E. T., Neoh, K. G., and Tan, K. L., *Prog. Polym. Sci.*, **23**, 277 (1998).
28. Kaynak, A., Hakansson, E., and Amiet, A., *Synth. Met.*, **159**, 1373 (2009).
29. Kim, S. J., Yun, S. M., and Lee, Y. S., *J. Ind. Eng. Chem.*, **16**, 273 (2010).
30. Kivelson, S. and Heeger, A. J., *Phys. Rev. Lett*, **55**, 308 (1985).
31. Klemperer, C. J. V. and Maharaj, D., *Compos. Struct*, **91**, 467 (2009).
32. Koul, S., Chandra, R., and Dhawan, S. K., *Polymer*, **41**, 9305 (2000).
33. Kwon, S. H. and Lee, H. K., *Computers, Materials & Continua*, **12**, 197 (2009).
34. Laslaua, C., Zujovica, Z., and Sejdica, J. T., *Progress in Polymer Science*, **35**, 1403 (2010).
35. Lee, B. O., Woo, W. J., Park, H. S., Hahm, H. S., Wu, J. P., and Kim, M. S., *J. Mater. Sci.*, **37**, 1839 (2002).
36. Lee, C. Y., Song, H. G., Jang, K. S., Oh, E. J., Epstein, A. J., and Joo, J., *Synthetic Metals*, **102**, 1346 (1999).
37. MacDiarmid, A. G., *Synthetic Metals*, **125**, 11 (2002).
38. Makela, T., Pienimaa, S., Taka, T., Jussila, S., and Isotalo, H., *Synth. Met.*, **85**, 1335 (1997).
39. Matare, M. F., *J. Appl. Phys.*, **56**, 2605 (1984).
40. Mott, N. F., *J. Non. Cryst. Solids*, **1**, 1 (1980).
41. Nandapure, A. I., Kondawar, S. B., Sawadh, P. S., and Nandapure, B. I., Physica B, **407**, 1104 (2012).
42. Paul, R. K. and Pillai, C. K. S., *Synth Met.*, **27**, 114 (2000).
43. Reneker, D. H. and Chun, I., *Nanotechnology*, **7**, 216 (1996).
44. Saini, P., Choudhary, V., Singh, B. P., Mathur, R. B., and Dhawan, S. K., *Mater. Chem. Phys.*, **113**, 919 (2009).
45. Skotheim, T. A. and Reynolds, J. R., (Eds.) *Conjugated polymers: theory, synthesis, properties, and characterization, 3rd (ed.)*, Cleveland: CRC Press, (2006).
46. Skotheim, T. A., Elsenbaumer, R. L., and Renolds, J. R., *Handbook of Conducting Polymers, 2nd (Ed.)*, Marcel Dekker, New York, (1998).
47. Stejskal, J., *Polym. Int.*, **58**, 872 (2009).
48. Swager, T. M., *Chem. Res. Toxicol.*, **15**, 125 (2002).
49. Trojanowicz, M., *Microchim. Acta*, **143**, 75 (2003).
50. Wang, L. L., Tay, B. K., See, K. Y., Sun, Z., Tan, L. K., and Lua, D., *Carbon*, **47**, 1905 (2009).
51. Wang, Y. and Jing, X., *Mater. Sci. Enging: B Adv. Funct. Solid-State Materials*, **138**, 95 (2007).
52. Wang, Y. and Jing, X., *Polymers for Advanced Technologies*, **16**, 344 (2005).
53. Wang, Z. H., Scherr, E. M., MacDiarmid, A. G., and Epstein, A. J., *Phys. Rev.*, **62**, 976 (1995).
54. Wegner, G., *Chem. Int. (Ed.)*, **20**, 361 (1981).
55. Wei-Li, Song, et al., *Polymer*, **53**, 3910 (2012).
56. Wu, K. H., Shin, M., and Yang, C. C., *J. Polym. Sci. Part A – Polym. Chem.*, **44**, 2657 (2006).
57. Yun-Ze, Long, et al., *Progress in Polymer Science*, **36**, 1415 (2011).
58. Zeller, H. R., *Phys. Rev. Lett*, **28**, 1452 (1972).
59. Zuo, F., Angelopoulos, M., MacDiarmid, A. G., and Epstein, A. J., *Phys. Rev.*, **26** (B3), 475 (1987).

INDEX

C

D

E

F

G

H

I

L

M

R

S

T

V

W

Z

Milton Keynes UK
Ingram Content Group UK Ltd.
UKHW031138141024
449569UK00024B/1249